W9-CFQ-361

Real Analysis

Real Analysis

Gabriel Klambauer

Department of Mathematics University of Ottawa

American Elsevier Publishing Company, Inc.
New York London Amsterdam

AMERICAN ELSEVIER PUBLISHING COMPANY, INC.
52 Vanderbilt Avenue, New York, N.Y. 10017

ELSEVIER PUBLISHING COMPANY
335 Jan Van Galenstraat, P.O. Box 211
Amsterdam, The Netherlands

International Standard Book Number 044-00133-6
Library of Congress Card Number 72-93078

Library of Congress Cataloging in Publication Data

Klambauer, Gabriel.
Real analysis.
1. Mathematical analysis. 2. Integrals, Generalized. 3. Measure theory. I. Title.
QA300.K54 515 72-93078
ISBn 0-444-00133-6

Manufactured in the United States of America

To
CAT,
John, Francis, and Peter

Contents

GABRIEL KLAMBAUER is Associate Professor of Mathematics at the University of Ottawa. From 1959 to 1965 he served successively as Research Assistant at Wayne State University, Lecturer at the University of Windsor and Teaching Fellow at McMaster University. Prior to assuming his present post in 1971, he was Assistant Professor of Mathematics at the University of Ottawa. Dr. Klambauer received his B. Sc. from the University of Windsor in 1956, his M. A. from Wayne State University in 1958 and his Ph. D. from McMaster University in 1966. He is a member of the American Mathematical Society and the Canadian Mathematical Congress.

Preface

This book treats basic matters in contemporary real analysis and quite properly focuses on integration theory. Its scope and content are primarily designed to meet the needs of beginning graduate students in mathematics and to assist those who wish to review the subject at hand before taking comprehensive examinations. The requirements for reading this book have been deliberately kept at a modest level; this feature serves the purpose of making it readily accessible to advanced students in their final year of undergraduate mathematics and to prove useful to graduate students in applied fields with a serious interest in modern analysis.

Concerning the prerequisites the following particulars deserve mention. We assume that the student reader is familiar with the real number system and Riemann integration and has some acquaintance with the metric space setting for ideas of continuity and convergence of real functions; the first six or seven chapters in the second edition of "Principles of Mathematical Analysis" by W. Rudin (McGraw-Hill Book Company, New York, 1964) provide excellent background in this respect. A very good reference for Cantor's construction of the real numbers from the rational numbers is volume I of "Modern Algebra" by B. L. van der Waerden (Frederick Ungar Publishing Co., New York, 1953); the matter referred to is in Chapter IX. We also assume some knowledge of modern algebra as encountered in the usual undergraduate courses.

The style of presentation is concise and the approach is to proceed from concrete situations to the more abstract theory. Students prefer to learn quickly and efficiently; abstract blasts turning into a litany of specialized results choke the enthusiam of the learner.

Within the stated objectives the book is reasonably comprehensive. The first four chapters deal with the Lebesgue theory of measure and integration of real functions and in a way constitute a critical study of differential and integral calculus. Chapters 5 and 6 treat abstract measure and integration theory. Chapter 7 is on topological and metric spaces with an emphasis on those topics which are most relevant to analysis. In a narrower sense, Chapter 7 provides background for the remaining chapters. Chapters 8 and 9 present Stone's formulation of Daniell integration; Chapter 8 is short and culminates in the Riesz representation theorem;

Chapter 9, on the other hand, gives a rather detailed account. Chapter 10 is on normed linear spaces; it rounds off our presentation and provides a signpost to the next area of study, namely, the field of functional analysis.

The exercise sections at the ends of the chapters are of course very much an integral part of the book; to entice the reader we have often sketched proofs to the exercises. Among the exercises there is a very formidable one indeed (see Problem 40, Chapter 3); it relates to the famous Riemann hypothesis (the linear hull of the set of functions f_α with $0 < \alpha \leqq 1$ is dense in $L^2(0, 1)$ if and only if the Riemann hypothesis is true!).

We feel that Chapter 9 lends itself well to a seminar-type treatment; we found that students welcome such involvement and stimulation. Chapter 9 may also serve as reading assignment for the more advance students taking the course. In recommending Chapter 9 for such special treatment, we feel that we owe the student reader a sketch of an introduction and we do this now.

The Stone-Daniell theory of integration commences with a postulated "elementary integral" defined for a class of "elementary functions," and proceeds by defining in terms of the elementary integral, a "norm," or upper integral, for all (extended) real-valued functions. A pseudometric (it satisfies all conditions of a metric except the $\rho(x,y) = 0$ need not imply $x = y$) is introduced in the class of all functions whose norm is finite; upon proper identification, the resulting function space is seen to be complete. Defining integrable functions and integral, it is seen that these concepts have many of their usual properties, in particular, as they relate to monotone and bounded convergence and the integrability of certain (Baire) functions of integrable functions. A noteworthy aspect of this theory is the definition of measurability. A function f is called measurable if and only if $\operatorname{med}(f, g, h)$ is integrable whenever g and h are integrable, where $\operatorname{med}(a, b, c)$ is the intermediate of the three numbers a, b, and c. The class of measurable functions is seen to have most of its customary closure properties. In general, however, it need not be true that the constant function 1 be measurable. In the special case in which 1 is measurable, the theory is essentially identical with the classical measure-theoretic approach to integration. The role of the theory of measure in Stone's theory is pursued far enough to obtain an analogue of Egorov's theorem. The representation of positive linear functionals on the space of continuous functions vanishing in a neighborhood of infinity in a locally compact Hausdorff space and the role of L^p spaces in Stone's theory are discussed. Adjoining to the postulates an assumption which in measure-theoretic language amounts to σ-finiteness, an analogue of the Lebesgue decomposition theorem and Stone's version of the Radon-Nikodym theorem are considered. From the latter theorem the change of variable formula is derived. We also consider iterated integrals and describe Stone's version of Fubini's theorem. For completeness of presentation, Chapter 9 begins with some pertinent facts from the theory of uniform approximation of continuous functions. In short, Chapter 9 is pretty well a complete course on integration.

Finally, I am pleased to express my sense of gratitude to Dr. Antoine D'Iorio, Dean, Faculty of Science and Engineering, University of Ottawa. I also wish to acknowledge the friendship and help of my colleagues Drs. E. L. Cohen, S. Jansen, V. Linis, and D. B. J. Tomiuk and of my students Louis Culumovic, Keith May, and Basil Spanoyiannopoulos.

Gabriel Klambauer

Preliminaries

We commence with a brief review of certain basic concepts and facts which are part of the formal prerequisites for the study of this book and which were explained in the Preface.

Let x and y be any objects. Then the *ordered pair* (x,y) is defined as the set $\{\{x\},\{x,y\}\}$. It is easy to check the fundamental property of ordered pairs:

$$(x,y) = (u,v) \text{ if and only if } x = u \text{ and } y = v.$$

More generally we may define in a similar way an ordered n-tuple $(x_1,\ldots,x_n)$ with the property $(x_1,\ldots,x_n) = (y_1,\ldots,y_n)$ if and only if $x_1 = y_1,\ldots, x_n = y_n$.

A *relation* is defined as a set of ordered pairs.

Let X and Y be given sets. Then

$$X \times Y = \{(x,y) : x \in X \text{ and } y \in Y\}$$

is called the *Cartesian product of X and Y*.

Let X be a given set. A relation $\sim$ is called an *equivalence relation* on X if it is

(i) *reflexive*, i.e., $x \sim x$ for all $x \in X$,
(ii) *symmetric*, i.e., $x \sim y$ implies $y \sim x$,
(iii) *transitive*, i.e., $x \sim y$ and $y \sim z$ imply $x \sim z$.[1]

The *domain* of a relation is the set of all first coordinates of its members; the *range* is the set of all second coordinates. If $\sim$ is an equivalence relation on X, then we define $E_x = \{y \in X : y \sim x\}$ and call E_x the *equivalence class containing the element x*.

A *function* f is defined to be a relation, such that if $(x,y) \in f$ and $(x,z) \in f$ then $y = z$. If f is a function and $(x,y) \in f$, then we write $y = f(x)$; we say that y is the value of f at x, or that y is the image of x under f.

The notation $f\colon X \to Y$ is interpreted as "f is a function from the set X into the set Y"; X is the domain of f and the range of f is a subset of Y, not necessarily the whole of Y.

Let $f\colon X \to Y$. Then f is called *one-one* if $f(x_1) = f(x_2)$ implies $x_1 = x_2$, for every $x_1, x_2 \in X$. If the range of f is the whole of Y, then f is called *onto*. A function which is both one-one and onto is called a *one-one correspondence*.

[1] As usual $x \sim y$ means $(x,y) \in \sim$.

The terms "injective," "surjective," and "bijective" are sometimes used instead of "one-one," "onto," and "one-one correspondence," respectively.

Let $f: X \to Y$ be a one-one correspondence. Since f is onto, if $y \in Y$ then there exists $x \in X$ such that $y = f(x)$. This x is unique, since f is one-one. Hence there is an inverse function $g: Y \to X$ such that $g(f(x)) = x$, for all $x \in X$, and $f(g(y)) = y$ for all $y \in Y$. It is usual to write $g = f^{-1}$.

Let $f: X \to Y$, and let $A \subset X$, $B \subset Y$. Then the *image of the set A under the function f* is defined to be $f(A) = \{f(x) : x \in A\}$. The *inverse image of B* is defined to be $f^{-1}(B) = \{x : x \in X \text{ and } f(x) \in B\}$. Note that $f^{-1}(B)$ is a subset of X; it is not necessary that f should be a one-one correspondence in order that we may write $f^{-1}(B)$.

We now list the basic properties of image and inverse image:

Let $f: X \to Y$, and suppose $\{A_t\}$ is a class of subsets of X and $\{B_t\}$ is a class of subsets of Y. Then

(i) $A_t \subset A_s$ implies $f(A_t) \subset f(A_s)$;
(ii) $B_t \subset B_s$ implies $f^{-1}(B_t) \subset f^{-1}(B_s)$;
(iii) $f(\cup A_t) = \cup f(A_t)$;
(iv) $f(\cap A_t) \subset \cap f(A_t)$;
(v) $f^{-1}(\cup B_t) = \cup f^{-1}(B_t)$;
(vi) $f^{-1}(\cap B_t) = \cap f^{-1}(B_t)$;
(vii) $A_t \subset f^{-1}(f(A_t))$;
(viii) $f(f^{-1}(B_t)) \subset B_t$.

Let $f: X \to Y$ and $g: Y \to Z$, where f, g are any functions and X, Y, Z are any sets. Then we define the composite function $g \circ f : X \to Z$ by $(g \circ f)(x) = g(f(x))$ for each $x \in X$.

Let X be a set and let ρ be a function from $X \times X$ into the set of all real numbers R^1 such that for all $x, y, z \in X$ we have:

(i) $\rho(x,y) \geqq 0$;
(ii) $\rho(x,y) = 0$ if and only if $x = y$;
(iii) $\rho(x,y) = \rho(y,x)$;
(iv) $\rho(x,z) \leqq \rho(x,y) + \rho(y,z)$ [the triangle inequality].

Then ρ is called a *metric* (or *distance-function*) *for* X; $\rho(x,y)$ is called the *distance from x to y*, and the pair (X,ρ) is called a *metric space*.

A *pseudometric* satisfies all conditions of a metric except that $\rho(x,y) = 0$ need not imply $x = y$. If ρ is a pseudometric for X, then the pair (X,ρ) is called a *pseudometric space*.

There is a standard method of turning a pseudometric space into a metric space; we describe this method now.

Let (X,ρ) be a pseudometric space and define $x \sim y$ to mean that $\rho(x,y) = 0$. Then $\sim$ is an equivalence relation on X and if E_x is the equivalence class containing x and $E = \{E_x : x \in X\}$, then

$$d(E_x, E_y) = \rho(x,y)$$

is well defined on $E \times E$ and (E,d) is a metric space.

While the foregoing procedure gives a metric space, beginning with a pseudometric space, it is to be observed that this metric space has somewhat more complicated elements, viz. equivalence classes, than the original space.

A sequence $(x_n)_{n=1}^{\infty}$ in a metric space (X,ρ) is called *convergent* (to x) if there exists an $x \in X$ such that $\rho(x_n, x) \to 0$ as $n \to \infty$. We then write

$$x = \lim_{n\to\infty} x_n$$

or $x_n \to x$ as $n \to \infty$ and call x the *limit of the sequence* $(x_n)_{n=1}^{\infty}$.

A sequence $(x_n)_{n=1}^{\infty}$, where $x_n \in X$ for every n, is called a *Cauchy sequence* in a metric space (X,ρ) if for every $\varepsilon > 0$ there exists a $k = k(\varepsilon)$ such that

$$\rho(x_n, x_m) < \varepsilon \text{ for all } n,\ m > k.$$

A convergent sequence in a metric space has a unique limit. Every convergent sequence in a metric space is a Cauchy sequence, but not conversely, in general. If a Cauchy sequence in a metric space has a convergent subsequence, then the whole sequence is convergent.

A metric space (X,ρ) is called *complete* if every Cauchy sequence converges (to a point of X).

Let $x \in X$, where (X,ρ) is a metric space. Then, for $r > 0$, $S(x,r) = \{y \in X : \rho(x,y) < r\}$ is a *neighborhood* (or *open sphere*) of center x and radius r.

Let (X,ρ) be a metric space and A, B nonempty subsets of X. Define

$$\rho(x,A) = \inf\{\rho(x,a) : a \in A\},$$

$$\rho(A,B) = \inf\{\rho(a,b) : a \in A, b \in B\},$$

$$\rho(A) = \sup\{\rho(a,a') : a, a' \in A\}.$$

We call $\rho(x,A)$ the *distance between the point* $x \in X$ *and the set* A, $\rho(A,B)$ the *distance between the sets* A *and* B and $\rho(A)$ the *diameter of the set* A.

A set A in a metric space is called *bounded* if it has a finite diameter; otherwise it is *unbounded.*

Let (X,ρ) be a metric space. Then $G \subset X$ is called *open* if for every point $x \in G$ there exists an open sphere $S(x,r)$ with $r > 0$ such that $S(x,r) \subset G$.

Observe that $S(x,r)$, where $x \in X$ and $r > 0$, is open. In a metric space (X,ρ) the empty set $\varnothing$ and X are open; moreover, the union of any collection of open sets is open and the intersection of a finite number of open sets is open.

Let A be a set in the metric space (X,ρ) and x be a point of X, not necessarily belonging to A. Then x is called a *limit* (or *accumulation*) *point of* A if every open sphere with center x and radius $r > 0$ contains a point of A different from x.

A set in (X,ρ) is called *closed* if its complement is open. A set in (X,ρ) is closed if and only if it contains all its limit points.

The *interior* of A in (X,ρ) is the union of all open sets contained in A and the *closure* of A is the intersection of all closed sets containing A. We denote by $\mathring{A}$ the interior of A and by $\bar{A}$ the closure of A.

A subset A of a metric space (X,ρ) is called *dense* in X if $\bar{A} = X$ and a metric space is called *separable* if it contains a countable dense subset.

Let (X,d) and (Y,ρ) be metric spaces and $f\colon X \to Y$. Suppose that $b \in Y$. We say that *f has limit b at x_0* $\in X$ if for every $\varepsilon > 0$ there exists a $\delta = \delta(\varepsilon) > 0$ such that $0 < d(x,x_0) < \delta$ implies $\rho(f(x),b) < \varepsilon$. We say that f is *continuous at x_0* if for every $\varepsilon > 0$ there exists a $\delta = \delta(\varepsilon, x_0) > 0$ such that $d(x,x_0) < \delta$ implies $\rho(f(x),f(x_0)) < \varepsilon$. It can be seen that $f\colon X \to Y$ is continuous at every point of X if and only if the inverse image of every open set in Y is an open set in X.

Let (X,d) and (Y,ρ) be metric spaces. Then $f\colon X \to Y$ is called *uniformly continuous on* X if for every $\varepsilon > 0$ there exists a $\delta = \delta(\varepsilon) > 0$, depending only on ε, such that $d(x,x') < \delta$ implies $\rho(f(x),f(x')) < \varepsilon$, where x, x' are in X.

Let $\mathcal{S}$ be a class of sets A. Then the class $\mathcal{S}$ is called a *cover* for a set X if $X \subset \bigcup \{A : A \in \mathcal{S}\}$. Any subclass of $\mathcal{S}$ which also covers X is called a *subcover* of $\mathcal{S}$. A cover in which each member is an open set is said to be an *open cover*.

Let (X,ρ) be a metric space. Then a subset A of X is said to be *compact* if each open cover of A admits a finite subcover.

The continuous image of a compact set A in a metric space is compact. A continuous real-valued function on a compact set is bounded and attains its bounds.

In a metric space a compact set is closed and bounded, but not conversely in general. If the metric space (X,ρ) is compact then it is complete, but not conversely in general.

A *linear space* over K (where K denotes either the field of real numbers R^1 or the field of complex numbers C^1) is a nonempty set X with a function $+$ on $X \times X$ into X, and a function $\cdot$ on $K \times X$ into X such that for all α, β in K and all elements x, y, z in X we have:

(i) $x + y = y + x$,
(ii) $(x + y) + z = x + (y + z)$,
(iii) there exists $\theta \in X$ such that $x + \theta = x$,
(iv) there exists $-x \in X$ such that $x + (-x) = \theta$,
(v) $1 \cdot x = x$
(vi) $\alpha \cdot (x + y) = \alpha \cdot x + \alpha \cdot y$,
(vii) $(\alpha + \beta) \cdot x = \alpha \cdot x + \beta \cdot x$,
(viii) $\alpha \cdot (\beta \cdot x) = (\alpha\beta) \cdot x$.

In the following we shall simply denote $\alpha \cdot x$ by αx. If $K = R^1$ we speak of a *real linear space* and if $K = C^1$ we speak of a *complex linear space*.

A finite subset $\{x_1, \ldots, x_n\}$ of a linear space X is called a *linearly independent set* if a relation of the form $\alpha_1 x_1 + \alpha_2 x_2 + \ldots + \alpha_n x_n = \theta$ implies $\alpha_1 = \alpha_2 = \ldots = \alpha_n = 0$. An expression of the form $\alpha_1 x_1 + \alpha_2 x_2 + \ldots + \alpha_n x_n$ is called a *linear combination of the elements* $x_1, x_2, \ldots, x_n$.

EXERCISES

1. Let $x > 0$ and define

$$x_1 = x, \quad x_n = x^{x_{n-1}}$$

for $n \geqq 2$. Show that the sequence $(x_n)_{n=1}^{\infty}$ converges if and only if $(1/e)^e \leqq x \leqq (e)^{1/e}$.

2. Let $2n$ points be distributed in space. Show that one may draw at least n^2 line segments connecting these points without obtaining a triangle. (Only the given points are to be considered as vertices of a triangle.)

3. Let $a_1, a_2, a_3, \ldots$ be a sequence of positive integers, all greater than 1. Show that any real number α is uniquely expressible in the form

$$\alpha = c_0 + \sum_{j=1}^{\infty} \frac{c_j}{a_1 a_2 \cdots a_j},$$

with integers c_j satisfying the inequalities $0 \leqq c_j \leqq a_j - 1$ for all $j \geqq 1$, and $c_j < a_j - 1$ for infinitely many j. Is α an irrational number if each prime number divides infinitely many of the a_j and an infinite number of the c_j are positive?

4. For any two real numbers a and b, $|a + b| \leqq |a| + |b|$. (Solution: Clearly, $-|a| \leqq a \leqq |a|$ and $-|b| \leqq b \leqq |b|$ and so $-(|a| + |b|) \leqq a + b \leqq |a| + |b|$. But the two inequalities $-B \leqq A \leqq B$ and $|A| \leqq B$ are equivalent.)

Chapter 1

Lebesgue Measure on the Real Line R^1

Let Λ be a class of subsets of the real line R^1. We say that Λ is a *sequential covering class* if for every $E \subset R^1$ there is a sequence $(E_n)_{n=1}^{\infty}$ of sets from Λ such that

$$E \subset \bigcup_{n=1}^{\infty} E_n .$$

If

$$E \subset \bigcup_{n=1}^{\infty} E_n,$$

we shall say that $(E_n)_{n=1}^{\infty}$ *covers* E. In order that finite unions may appear as special cases of infinite unions, we shall assume that the empty set $\varnothing$ is in Λ.

The covering class itself need not be countable. In fact one of the most important examples of a sequential covering class is the family of all open intervals on the real line R^1. Evidently, this is an uncountable class of sets, but every set on the real line is contained in some countable union of open intervals. We shall denote by the letter Γ the family of all countable collections of open intervals on the real line R^1.

The length $\lambda(I)$ of an interval I with endpoints a and b will be defined, as usual, by

$$\lambda(I) = b - a.$$

We shall extend this notion of length to more complicated subsets of R^1 by means of the family Γ of all countable collections of open intervals.

For an arbitrary member γ of Γ, the sum

$$\lambda^*(\gamma) = \sum_{I \in \gamma} \lambda(I)$$

is either ∞ or a nonnegative real number. Moreover, $\lambda^*(\gamma)$ depends only on γ and not on the order used in the process of summation.

Let E be an arbitrary subset of R^1. Consider the subfamily

$$\mathfrak{C}(E) = \{\gamma \in \Gamma : \gamma \text{ covers } E\}$$

of Γ. $\mathfrak{C}(E)$ is nonempty. We obtain a well-defined number

$$\mu^*(E) = \inf\{\lambda^*(\gamma) : \gamma \in \mathfrak{C}(E)\}$$

of the extended reals R_e^1, which we shall call the *Lebesgue outer measure* of E. Clearly, $\mu^*(E)$ is either ∞ or a nonnegative real number and the assignment $E \to \mu^*(E)$ defines a set function from the power set of the real line R^1 into the set of nonnegative extended real numbers.

The following three propositions are immediate consequences of the definition of Lebesgue outer measure.

PROPOSITION 1. *For the empty set* $\varnothing$, $\mu^*(\varnothing) = 0$.

PROPOSITION 2. $\mu^*(E) = 0$ *for every singleton subset* $E = \{x\}$ *of* R^1.

PROPOSITION 3. *For any two subsets A and B of* R^1, $A \subset B$ *implies* $\mu^*(A) \leqslant \mu^*(B)$.

Let τ_r be the "translation" function $\tau_r(x) = x + r$ for every $x \in R^1$ and r a fixed real. We show the invariance of μ^* under translations of the real line R^1.

PROPOSITION 4. *The function* μ^* *is translation invariant, i.e., for every* $r \in R^1$ *and* $E \subset R^1$, *we have*

$$\mu^*(\tau_r(E)) = \mu^*(E).$$

Proof. If I is an interval with endpoints a and b, then $\tau_r(I)$ is an interval with endpoints $a + r$ and $b + r$. Hence $\lambda(\tau_r(I)) = \lambda(I)$.

For every $\gamma \in \Gamma$, let

$$\tau_r(\gamma) = \{\tau_r(I) : I \in \gamma\} \in \Gamma.$$

If γ covers E, then $\tau_r(\gamma)$ covers $\tau_r(E)$. Since

$$\lambda^*(\tau_r(\gamma)) = \sum_{I \in \gamma} \lambda(\tau_r(I)) = \sum_{I \in \gamma} \lambda(I) = \lambda^*(\gamma)$$

it follows that

$$\mu^*(\tau_r(E)) \leqslant \mu^*(E).$$

Since the translation τ_{-r} sends $\tau_r(E)$ back onto E, we also have

$$\mu^*(E) \leqslant \mu^*(\tau_r(E))$$

and the proof is finished.

We shall next establish that the Lebesgue outer measure of an interval is the length of the interval; in other words, the function μ^* is an extension of the function λ.

PROPOSITION 5. *For every interval I, we have $\mu^*(I) = \lambda(I)$.*

Proof. We first assume that I is a closed bounded interval, say $[a, b]$. Since for every $\varepsilon > 0$, the open interval $(a - \varepsilon, b + \varepsilon)$ covers I, we have that

$$\mu^*(I) \leqslant \lambda(a - \varepsilon, b + \varepsilon) = b - a + 2\varepsilon.$$

As this is true for every $\varepsilon > 0$, it follows that

$$\mu^*(I) \leqslant b - a = \lambda(I).$$

Next we wish to show that

$$\mu^*(I) \geqslant b - a.$$

This is equivalent to proving that

$$\lambda^*(\gamma) \geqslant b - a$$

for every $\gamma \in \mathfrak{C}(I)$.

By the Heine-Borel Theorem, every $\gamma \in \mathfrak{C}(I)$ contains a finite subcollection which covers I. Therefore, it is sufficient to establish that

$$\lambda^*(\gamma) \geqslant b - a$$

for every finite $\gamma \in \mathfrak{C}(I)$. We shall do this now.

Let γ be any finite collection of open intervals which covers I. Since $a \in I$, there exists an open interval (a_1, b_1) in γ with $a_1 < a < b_1$. If $b_1 \leqslant b$, then $b_1 \in I$. Since b_1 is not covered by the open interval (a_1, b_1), there is an open interval (a_2, b_2) in γ with $a_2 < b_1 < b_2$. Continuing in this manner, we shall obtain open intervals

$$(a_1, b_1), (a_2, b_2), \ldots, (a_n, b_n)$$

in γ satisfying $a_j < b_{j-1} < b_j$ for every $j = 2, 3, \ldots, n$.

Since γ is finite, the process must terminate with an (a_n, b_n) satisfying $b < b_n$. Thus we have

$$\lambda^*(\gamma) \geqslant \sum_{j=1}^{n} \lambda(a_j, b_j) = b_n - (a_n - b_{n-1}) - \ldots - (a_2 - b_1) - a_1$$
$$> b_n - a_1 > b - a.$$

This completes the proof of the proposition for the special case $I = [a, b]$ with $b - a < \infty$.

Next, let I be a bounded interval. Given any $\varepsilon > 0$, there exists a closed interval $J \subset I$, such that

$$\lambda(J) > \lambda(I) - \varepsilon.$$

Hence

$$\lambda(I) - \varepsilon < \lambda(J) = \mu^*(J) \leqslant \mu^*(I) \leqslant \mu^*(\bar{I}) = \lambda(\bar{I}) = \lambda(I),$$

where $\bar{I}$ denotes the closure of I. Thus for each $\varepsilon > 0$,

$$\lambda(I) - \varepsilon < \mu^*(I) \leqslant \lambda(I)$$

and so

$$\mu^*(I) = \lambda(I).$$

Finally, let I be an unbounded interval. Then, for every real number r, I contains a bounded interval J of length $\lambda(J) \geqslant r$. By Proposition 3, this implies that

$$\mu^*(I) \geqslant \mu^*(J) = \lambda(J) \geqslant r.$$

Since this holds for every $r \in R^1$, we must have that

$$\mu^*(I) = \infty = \lambda(I)$$

and the proof is finished.

PROPOSITION 6. *The set function μ^* is countably subadditive, i.e., for every countable family Δ of subsets of R^1, we have*

$$\mu^*\Big(\bigcup_{E \in \Delta} E\Big) \leqslant \sum_{E \in \Delta} \mu^*(E).$$

Proof. We take

$$\Delta = \{E_n : n \in N\},$$

where N denotes the set of natural numbers, and we let $\varepsilon > 0$ be arbitrary. Then, for each $n \in N$, there exists a $\gamma_n \in \mathfrak{C}(E_n)$ with

$$\lambda^*(\gamma_n) \leqslant \mu^*(E_n) + \frac{\varepsilon}{2^n}$$

by the definition of $\mu^*(E_n)$. Letting

$$\gamma = \bigcup_{n \in N} \gamma_n,$$

we have

$$\lambda^*(\gamma) \leqslant \sum_{n \in N} \lambda^*(\gamma_n) \leqslant \sum_{n \in N} \mu^*(E_n) + \varepsilon.$$

Since γ_n covers E_n for each $n \in N$, γ covers the union

$$E = \bigcup_{n \in N} E_n.$$

Hence

$$\mu^*(E) \leqslant \lambda^*(\gamma) \leqslant \sum_{n \in N} \mu^*(E_n) + \varepsilon.$$

But this holds for every $\varepsilon > 0$ and so

$$\mu^*(E) \leqslant \sum_{n \in N} \mu^*(E_n),$$

completing the proof.

PROPOSITION 7. *If E is a countable subset of R^1, then $\mu^*(E) = 0$.*

Proof. The proposition follows from Propositions 2 and 6.

PROPOSITION 8. *Every interval is uncountable.*

Proof. The proposition is a corollary of Propositions 5 and 7.

The function μ^* is countably subadditive by Proposition 6. It would of course be desirable if μ^* were also *countably additive*, that is, if

$$\mu^*\Big(\bigcup_{E \in \Delta} E\Big) = \sum_{E \in \Delta} \mu^*(E)$$

for every countable family Δ of disjoint subsets of R^1. If we insist on countable additivity, we have to restrict the domain of definition of the function μ^* to some subset $\mathfrak{M}$ of the power set $\mathcal{P}(R^1)$. The members of $\mathfrak{M}$ are called the *Lebesgue measurable subsets* of R^1, or *measurable* subsets of R^1, for short.

Following Carathéodory, we define the measurable subsets of R^1 as follows. A subset E of R^1 is said to be *Lebesgue measurable* if, for every subset X of R^1, we have

$$\mu^*(X) = \mu^*(X \cap E) + \mu^*(X - E).$$

Since $X = (X \cap E) \cup (X - E)$ it follows from Proposition 6 that

$$\mu^*(X) \leqslant \mu^*(X \cap E) + \mu^*(X - E).$$

Hence we have the following proposition.

PROPOSITION 9. *A subset E of R^1 is measurable if and only if, for every subset X of R^1, we have*

$$\mu^*(X) \geqslant \mu^*(X \cap E) + \mu^*(X - E).$$

On account of the set-theoretic relations

$$X \cap (R^1 - E) = X - E, \quad X - (R^1 - E) = X \cap E,$$

we have the following proposition.

PROPOSITION 10. *If a subset* E *of* R^1 *is measurable, then so is its complement* $E^c = R^1 - E$.

PROPOSITION 11. *If two subsets* A *and* B *of* R^1 *are measurable, so is* $A \cap B$.

Proof. Let X denote any subset of R^1. Since B is measurable, we have

$$\mu^*(X \cap A) = \mu^*(X \cap A \cap B) + \mu^*[(X \cap A) - B].$$

Next, since $X - (A \cap B) = (X - A) \cup [(X \cap A) - B]$, it follows from Proposition 6 that

$$\mu^*[X - (A \cap B)] \leqslant \mu^*(X - A) + \mu^*[(X \cap A) - B].$$

Consequently, we obtain

$$\begin{aligned} \mu^*[X \cap (A \cap B)] &+ \mu^*[X - (A \cap B)] \\ &\leqslant \mu^*(X \cap A \cap B) + \mu^*(X - A) + \mu^*[(X \cap A) - B] \\ &= \mu^*(X \cap A) + \mu^*(X - A) \\ &= \mu^*(X) \end{aligned}$$

because A is measurable. According to Proposition 9, this proves that $A \cap B$ is measurable. The proof is finished.

Using De Morgan's formulas, we get the following as an immediate consequence of Propositions 10 and 11.

PROPOSITION 12. *The class* $\mathfrak{M}$ *of all measurable subsets of the real line* R^1 *is a Boolean algebra of sets, i.e., the class is closed under complementation, finite union, and finite intersection.*

PROPOSITION 13. *If* $E_1, E_2, \ldots, E_n$ *are pairwise disjoint measurable subsets of* R^1, *then*

$$\mu^*(X \cap (\bigcup_{j=1}^{n} E_j)) = \sum_{j=1}^{n} \mu^*(X \cap E_j)$$

holds for every subset X *of* R^1.

Proof. The proposition is trivial for $n = 1$. Let $n > 1$ and suppose that the proposition is true for less than n measurable subsets of R^1. Let X be an arbitrary subset of R^1. Since E_n is measurable, we have

$$\mu^*(Y) = \mu^*(Y \cap E_n) + \mu^*(Y - E_n)$$

for every subset Y of R^1. In particular, we may take

$$Y = X \cap (\bigcup_{j=1}^{n} E_j).$$

Since $E_1, E_2, \ldots, E_n$ are disjoint, we have

$$Y \cap E_n = X \cap E_n, \quad Y - E_n = X \cap (\bigcup_{j=1}^{n-1} E_j).$$

Hence we get

$$\mu^*(Y) = \mu^*(X \cap E_n) + \mu^*(X \cap \bigcup_{j=1}^{n-1} E_j).$$

By our inductive assumption,

$$\mu^*(X \cap \bigcup_{j=1}^{n-1} E_j) = \sum_{j=1}^{n-1} \mu^*(X \cap E_j).$$

Hence

$$\mu^*(Y) = \mu^*(X \cap E_n) + \sum_{j=1}^{n-1} \mu^*(X \cap E_j) = \sum_{j=1}^{n} \mu^*(X \cap E_j)$$

and the proof is finished.

PROPOSITION 14. *For every sequence* $\{E_j : j \in N\}$ *of members of a Boolean algebra* Δ *of subsets of a set* S *of the real line* R^1, *there exists a sequence* $\{D_j : j \in N\}$ *of disjoint members of* Δ *satisfying*

$$D_j \subset E_j \qquad (j \in N)$$

$$\bigcup_{j \in N} D_j = \bigcup_{j \in N} E_j.$$

Proof. Let $D_1 = E_1$ and for $n \geqq 2$ let

$$D_n = E_n - (E_1 \cup \ldots \cup E_{n-1}) = E_n \cap [S - (E_1 \cup \ldots \cup E_{n-1})].$$

Since Δ is an algebra of subsets of S, D_n is a member of Δ for every $n \in N$. By the construction, we obviously have

$$D_j \subset E_j \qquad (j \in N)$$

$$D_j \cap D_k = \varnothing \qquad (j \neq k).$$

The first implies

$$\bigcup_{j \in N} D_j \subset \bigcup_{j \in N} E_j.$$

It remains to prove that

$$\bigcup_{j\in N} E_j \subset \bigcup_{j\in N} D_j.$$

For this purpose, let x be any member of

$$\bigcup_{j\in N} E_j.$$

Let n denote the least natural number satisfying $x \in E_n$. Then we have

$$x \in E_n - (E_1 \cup \ldots \cup E_{n-1}) = D_n \subset \bigcup_{j\in N} D_j.$$

This completes the proof.

PROPOSITION 15. *The set* $\mathfrak{M}$ *of all measurable subsets of the real line* R^1 *is a* σ*-algebra of subsets of* R^1, *i.e.,* $\mathfrak{M}$ *is a Boolean algebra of sets closed under countable unions.*

Proof. By Proposition 12, $\mathfrak{M}$ is an algebra of sets. It remains to prove that the union E of any sequence $\Omega = \{E_j : j \in N\}$ of measurable subsets E_j of R^1 is measurable. Because of Proposition 14, we may assume that the members of Ω are disjoint.

For each $n \in N$, let

$$H_n = \bigcup_{j=1}^{n} E_j \subset E.$$

Since $\mathfrak{M}$ is an algebra of sets, H_n is measurable. Hence we have

$$\begin{aligned}\mu^*(X) &= \mu^*(X \cap H_n) + \mu^*(X - H_n)\\ &\geqslant \mu^*(X \cap H_n) + \mu^*(X - E)\end{aligned}$$

for every subset X of R^1. By Proposition 13 we have

$$\mu^*(X \cap H_n) = \sum_{j=1}^{n} \mu^*(X \cap E_j).$$

Hence

$$\mu^*(X) \geqslant \sum_{j=1}^{n} \mu^*(X \cap E_j) + \mu^*(X - E).$$

Since this holds for every $n \in N$, we must have

$$\mu^*(X) \geqslant \sum_{j=1}^{\infty} \mu^*(X \cap E_j) + \mu^*(X - E).$$

By Proposition 6 we also have

$$\sum_{j=1}^{\infty} \mu^*(X \cap E_j) \geqslant \mu^*(X \cap E).$$

Consequently, we obtain

$$\mu^*(X) \geqslant \mu^*(X \cap E) + \mu^*(X - E).$$

By Proposition 9, this means that E is measurable and the proof is complete.

PROPOSITION 16. *The set function* $\mu^* : \mathcal{P}(R^1) \to R_e^1$ *is countably additive on the subset* $\mathfrak{M}$ *of* $\mathcal{P}(R^1)$, *that is, for every countable family* Δ *of disjoint measurable subsets of* R^1, *we have*

$$\mu^*(\bigcup_{E \in \Delta} E) = \sum_{E \in \Delta} \mu^*(E).$$

Proof. In case Δ is finite, say, $\Delta = \{E_1, \ldots, E_n\}$, then Proposition 13 with $X = R^1$ gives

$$\mu^*(\bigcup_{j=1}^{n} E_j) = \sum_{j=1}^{n} \mu^*(E_j).$$

Next, let $\Delta = \{E_j : j \in N\}$ be a sequence of disjoint measurable subsets of R^1. Since

$$\bigcup_{j=1}^{\infty} E_j \supset \bigcup_{j=1}^{n} E_j,$$

we get by Proposition 3 and the finite case that

$$\mu^*(\bigcup_{j=1}^{\infty} E_j) \geqslant \mu^*(\bigcup_{j=1}^{n} E_j) = \sum_{j=1}^{n} \mu^*(E_j).$$

Since this holds for every $n \in N$, we must have

$$\mu^*(\bigcup_{j=1}^{\infty} E_j) \geqslant \sum_{j=1}^{\infty} \mu^*(E_j).$$

By Proposition 6, we also have

$$\mu^*(\bigcup_{j=1}^{\infty} E_j) \leqslant \sum_{j=1}^{\infty} \mu^*(E_j).$$

Consequently, we obtain

$$\mu^*(\bigcup_{j=1}^{\infty} E_j) = \sum_{j=1}^{\infty} \mu^*(E_j).$$

This finishes the proof.

We define a set function μ

$$\mu : \mathfrak{M} \to R_e^1$$

by taking the restriction $\mu = \mu^*|\mathfrak{M}$ to the subset $\mathfrak{M}$ of the domain of definition $\mathcal{P}(R^1)$ of μ^*. This function μ is called the *Lebesgue measure function* for the real line R^1. For every $E \in \mathfrak{M}$, the extended real number

$$\mu(E) = \mu^*(E)$$

is called the *Lebesgue measure* (or *L-measure*, for short) of the measurable subset E of R^1.

PROPOSITION 17. *Every subset* E *of* R^1 *with* $\mu^*(E) = 0$ *is measurable*; *clearly*, $\mu(E) = 0$.

Proof. Let X be any subset of R^1. Since $X \cap E \subset E$, it follows from Proposition 3 that

$$0 \leqslant \mu^*(X \cap E) \leqslant \mu^*(E) = 0$$

and hence $\mu^*(X \cap E) = 0$. On the other hand, since $X \supset X - E$, we have

$$\mu^*(X) \geqslant \mu^*(X - E) = \mu^*(X \cap E) + \mu^*(X - E).$$

By Proposition 9, this shows that E is measurable. The proof is complete.

Remark. Every countable subset E of R^1 is measurable with Lebesgue measure $\mu(E) = 0$. If $A \subset B$ and B has Lebesgue measure zero, then A is Lebesgue measurable and has Lebesgue measure zero; this property is the so-called "*completeness*" *property* of L-measure.

Noting that the power set of R^1 is a σ-algebra we will next consider all the σ-algebras of subsets of R^1 containing every open subset of R^1. The intersection $\mathcal{B}$ of all these σ-algebras is also a σ-algebra; it is the smallest σ-algebra of subsets of R^1 containing all open sets of R^1. The members of $\mathcal{B}$ are called the *Borel subsets* of the real line R^1.

PROPOSITION 18. $\mathcal{B} \subset \mathfrak{M}$, *that is, every Borel subset of* R^1 *is measurable.*

Proof. By the definition of $\mathcal{B}$, we have to show that every open subset of R^1 is measurable. But it suffices to verify that every open interval (a, b) is measurable since every open set of R^1 can be represented as the union of a countable collection of disjoint open intervals uniquely, except for the order of summation. Since

$$(a, b) = (a, \infty) \cap (-\infty, b),$$

it remains to show that (a, ∞) and $(-\infty, b)$ are measurable.

To prove that (a, ∞) is measurable, let X be any subset of R^1 and let

$$A = X \cap (a, \infty), \quad B = X - (a, \infty) = X \cap (-\infty, a].$$

According to Proposition 9, we only need to show that

$$\mu^*(X) \geqslant \mu^*(A) + \mu^*(B).$$

In case $\mu^*(X) = \infty$, there is nothing to prove. Assume $\mu^*(X) < \infty$ and let $\varepsilon > 0$ by any real number. By the definition of $\mu^*(X)$, there exists a countable collection γ of open intervals covering X and with

$$\lambda^*(\gamma) = \sum_{I \in \gamma} \lambda(I) \leqslant \mu^*(X) + \varepsilon.$$

For every open interval $I \in \gamma$, each of the sets $I_A = I \cap (a, \infty)$ and $I_B = I \cap (-\infty, a]$ is either empty or an interval. Moreover, we have

$$\lambda(I) = \lambda(I_A) + \lambda(I_B) = \mu^*(I_A) + \mu^*(I_B).$$

Since A is covered by the collection $\{I_A : I \in \gamma\}$, we have

$$\mu^*(A) \leqslant \mu^*(\bigcup_{I \in \gamma} I_A) \leqslant \sum_{I \in \gamma} \mu^*(I_A).$$

Similarly, we have

$$\mu^*(B) \leqslant \mu^*(\bigcup_{I \in \gamma} I_B) \leqslant \sum_{I \in \gamma} \mu^*(I_B).$$

Consequently, we obtain

$$\begin{aligned}\mu^*(A) + \mu^*(B) &\leqslant \sum_{I \in \gamma} (\mu^*(I_A) + \mu^*(I_B)) \\ &= \sum_{I \in \gamma} \lambda(I) \leqslant \mu^*(X) + \varepsilon.\end{aligned}$$

Since this holds for every $\varepsilon > 0$, we must have

$$\mu^*(A) + \mu^*(B) \leqslant \mu^*(X).$$

This completes the proof.

From the definition of $\mathscr{B}$ we easily obtain the following.

PROPOSITION 19. *Each of the following subsets of the real line* R^1 *is measurable*:

(*a*) *every open set*;
(*b*) *every closed set*;
(*c*) *every set which is the union of any countable collection of closed sets (such sets are termed* F_σ *sets)*;
(*d*) *every set which is the intersection of any countable collection of open sets (such sets are termed* G_δ *sets)*.

In the next two propositions we shall establish that μ is conditionally continuous from above and below.

PROPOSITION 20. *Let* E_n *be an infinite decreasing sequence of measurable sets on* R^1, *i.e.,* $E_{n+1} \subset E_n$ *for* $n \in N$, *and let* $\mu(E_1)$ *be finite [of course,* $\mu(E_n) < \infty$ *for at least one n will suffice as well]. Then*

$$\mu(\bigcap_{k=1}^{\infty} E_k) = \lim_{n\to\infty} \mu(E_n).$$

Remark. If for example $E_n = \{x \in R^1 : x \geqslant n\}$ for $n = 1, 2, \ldots,$ then $E_1 \supset E_2 \supset E_3 \supset \ldots$ and $\mu(E_n) = \infty$ and

$$\lim_{n\to\infty} \mu(E_n) = \infty.$$

However,

$$\bigcap_{n=1}^{\infty} E_n = \varnothing$$

and thus

$$\mu(\bigcap_{n=1}^{\infty} E_n) = 0.$$

This shows that the condition $\mu(E_n) < \infty$ for some n is necessary in the proposition.

Proof (of the proposition). Let

$$E = \bigcap_{k=1}^{\infty} E_k$$

and let $F_k = E_k - E_{k+1}$. Then

$$E_1 - E = \bigcup_{k=1}^{\infty} F_k,$$

and the sets F_k are pairwise disjoint. Hence

$$\mu(E_1 - E) = \sum_{k=1}^{\infty} \mu(F_k) = \sum_{k=1}^{\infty} \mu(E_k - E_{k+1}).$$

But

$$\mu(E_1) = \mu(E) + \mu(E_1 - E)$$

and

$$\mu(E_k) = \mu(E_{k+1}) + \mu(E_k - E_{k+1})$$

as $E \subset E_1$ and $E_{k+1} \subset E_k$. Since $\mu(E_k) \leqslant \mu(E_1) < \infty$, we have

$$\mu(E_1 - E) = \mu(E_1) - \mu(E)$$

and

$$\mu(E_k - E_{k+1}) = \mu(E_k) - \mu(E_{k+1}).$$

Thus

$$\begin{aligned}\mu(E_1) - \mu(E) &= \sum_{k=1}^{\infty} (\mu(E_k) - \mu(E_{k+1})) \\ &= \lim_{n\to\infty} \sum_{k=1}^{n} (\mu(E_k) - \mu(E_{k+1})) \\ &= \lim_{n\to\infty} (\mu(E_1) - \mu(E_n)) \\ &= \mu(E_1) - \lim_{n\to\infty} \mu(E_n).\end{aligned}$$

As $\mu(E_1) < \infty$, we have

$$\mu(E) = \lim_{n\to\infty} \mu(E_n).$$

The proof is complete.

PROPOSITION 21. *A monotone increasing sequence of measurable sets in R^1 satisfies the relation*

$$\mu(\bigcup_{n=1}^{\infty} A_n) = \lim_{n\to\infty} \mu(A_n).$$

Proof. If $\mu(A_{n_0}) = \infty$ for some integer n_0, then

$$\mu(\bigcup_{n=1}^{\infty} A_n) \geqq \mu(A_{n_0}) = \infty,$$

$\mu(A_n) = \infty$ for each $n \geqslant n_0$ and the equality holds.

Hence, let us consider the case when $\mu(A_n) < \infty$ for $n = 1, 2, \ldots$. Since

$$\bigcup_{n=1}^{\infty} A_n = A_1 \cup (A_2 \cap A_1^c) \cup (A_3 \cap A_2^c) \cup \ldots,$$

where A_j^c denotes the complement of A_j in R^1, is the union of disjoint measurable sets, we have

$$\begin{aligned}\mu\Big(\bigcup_{n=1}^{\infty} A_n\Big) &= \mu(A_1) + \mu(A_2 \cap A_1^c) + \mu(A_3 \cap A_2^c) + \ldots \\ &= \mu(A_1) + \lim_{n\to\infty} \sum_{k=1}^{n} (\mu(A_{k+1}) - \mu(A_k)) \\ &= \lim_{n\to\infty} \mu(A_{n+1}).\end{aligned}$$

This completes the proof.

Concluding Remarks

In connection with Proposition 18 we wish to point out that not every Lebesgue measurable set is a Borel set and that therefore the inclusion $\mathcal{B} \subset \mathfrak{M}$ is proper. In fact, M. Souslin has constructed what he called analytic sets, some of which are Lebesgue measurable but are not Borel sets. Here we are only concerned with the existence of Lebesgue measurable sets that are not Borel sets. To this end consider Cantor's ternary set P. The set P has the same cardinality as the set of all real numbers, namely c. The Lebesgue measure of the set P is zero, however, and thus each member of the power set of P is measurable and in fact has measure zero. The cardinality of this power set is 2^c and is thus greater than c. On the other hand, the cardinality of the class of all Borel sets in R^1 is c. To see this, observe that there is a continuum of, at most, countable systems of intervals with rational endpoints. Since each Borel set can be obtained by means of a finite or countable number of operations performed over these intervals, it follows that there is a continuum of Borel sets.

Finally we construct a subset of R^1 that is nonmeasurable in the sense of Lebesgue's theory. The example in question is due to Zermelo and its construction makes use of the axiom of choice.

AXIOM OF CHOICE: *Let $\{H_\alpha\}$ be any system of pairwise disjoint, nonempty sets. Then there exists a subset L of the set*

$$\bigcup_{\alpha} H_\alpha,$$

which has exactly one point in common with each of the sets H_α (thus L consists of "representative" elements chosen from the sets H_α).

Zermelo's example of a nonmeasurable set: Consider the half-open interval $[0, 1)$. Two points of $[0, 1)$ will be called equivalent if the distance between them is a rational number. This equivalence relation defines a partition of the points of $[0, 1)$ into classes. We choose from each class a representative point; these representatives form a set E by the axiom of choice. We shall show that E is not measurable.

Let us arrange the rational points of the interval $[0, 1)$ in a sequence $r_1, r_2, \ldots$. Call E_n the set formed as follows: Shift E to the left by the distance r_n, and then shift back to the right, by the distance 1, that part which has come to the left of $x = 0$. Clearly $E_n = \{x : x \in [0, 1), x + r_n - [x + r_n] \in E\}$ for $n = 1, 2, \ldots$, where $[t]$ denotes the integral part of t.

The sets E_n are pairwise disjoint. For if E_n and E_m $(n \neq m)$ have a common point x, then we have

$$y_n = x + r_n - [x + r_n] \in E,$$

$$y_m = x + r_m - [x + r_m] \in E,$$

hence $y_n - y_m = r_n - r_m + \text{integer} = \text{rational number}$; since E contains precisely one point from each equivalence class, this implies $y_n - y_m = 0$, $r_n = r_m + \text{integer}$, which is impossible, because r_n, r_m both lie in $[0, 1)$ and are different. Thus the sets E_n are disjoint indeed.

Now, if E were measurable, E_n would also be measurable, with the same measure. This follows from the fact that measurability and measure are invariant under translation.

Thus, if E were measurable, all the sets E_n would also be measurable, with $\mu(E_n) = \mu(E)$, and since

$$\bigcup_{n=1}^{\infty} E_n = [0, 1),$$

we would have, by the countable additivity of measure,

$$\sum_{n=1}^{\infty} \mu(E_n) = 1.$$

This is absurd: Indeed, an infinite series of equal terms cannot have the sum one. This shows that E is not measurable.

Remarks. Note that the foregoing proof can easily be modified to show that every set of positive measure contains a nonmeasurable subset.

A geometric description of Cantor's ternary set P is as follows: Divide $[0,1]$ into three equal parts and remove the middle-third open interval $(1/3, 2/3)$. At the second stage, remove the middle third of the two remaining intervals $[0, 1/3]$, $[2/3, 1]$. Proceeding analogously, remove at the n-th stage the union M_n of the middle thirds of the 2^{n-1} intervals present. The set P is defined to be

$$P = [0, 1] - \bigcup_{n=1}^{\infty} M_n.$$

Considering the triadic expansion

$$\sum_{i=1}^{\infty} n_i/3^i \qquad (n_i = 0 \text{ or } 1 \text{ or } 2)$$

of the real numbers in $[0, 1]$, we see that P consists of those which do not require the use of "1," i.e., $n_i = 0$ or 2. But the set of all sequences of 0's and 2's is of cardinality c.

EXERCISES

1. (a) If A is a measurable set with $\mu(A) < \infty$ and if B is any set such that $B \supset A$, show that

$$\mu^*(B - A) = \mu^*(B) - \mu(A).$$

(b) If A has $\mu^*(A)$ finite and there is a measurable subset $\underset{\vee}{A} \subset A$ with

$$\mu(\underset{\vee}{A}) = \mu^*(A),$$

then A must be measurable. (A is sometimes called an *equimeasurable kernel* of A.)

(c) Let A be any set in R^1. Show that for each $\varepsilon > 0$ there is an open set G such that

$$A \subset G \text{ and } \mu(G) \leqq \mu^*(A) + \varepsilon.$$

Also show that there is a measurable set $\hat{A}$ (actually a G_δ set) such that

$$A \subset \hat{A} \text{ and } \mu(\hat{A}) = \mu^*(A).$$

($\hat{A}$ is sometimes called an *equimeasurable superset* of A.)

(d) Let M be a measurable set with $\mu(M) < \infty$. Then a set $A \subset M$ is measurable if and only if

$$\mu(M) = \mu^*(A) + \mu^*(M - A).$$

2. Define h on $[0, \infty]$ by putting

$$h(t) = \begin{cases} 1 - e^{-t} & \text{if } 0 \leqslant t < \infty \\ 1 & \text{if } t = \infty. \end{cases}$$

Let $\mathfrak{M}$ be the class of all measurable sets in R^1. For $A, B \in \mathfrak{M}$ define

$$d(A, B) = h(\mu(A \triangle B)),$$

where $A \triangle B = (A - B) \cup (B - A)$.

(a) Identifying A and B for which $\mu(A \triangle B) = 0$, show that $(\mathfrak{M}, d)$ is a complete metric space.[1]

(b) Verify that the mappings from $\mathfrak{M} \times \mathfrak{M}$ to $\mathfrak{M}$

$$(A, B) \to A \cup B$$

$$(A, B) \to A \triangle B$$

$$(A, B) \to A \cap B$$

and the mapping from $\mathfrak{M}$ to $\mathfrak{M}$

$$A \to R^1 - A$$

are all continuous.

3. For an arbitrary subset E of R^1, prove that the following five statements are equivalent:
   (i) E is measurable;
   (ii) given $\varepsilon > 0$, there exists an open subset G of R^1 with $E \subset G$ and $\mu^*(G - E) < \varepsilon$;
   (iii) given $\varepsilon > 0$, there exists a closed subset F of R^1 with $F \subset E$ and $\mu^*(E - F) < \varepsilon$;
   (iv) there is a G_δ subset W (sometimes called an equimeasurable superset of E) of R^1 with $E \subset W$ and $\mu^*(W - E) = 0$;
   (v) there is an F_σ subset K (sometimes called an equimeasurable kernel of E) of R^1 with $K \subset E$ and $\mu^*(E - K) = 0$.

   Moreover, if $\mu^*(E)$ is finite, the above statements are equivalent to the following
   (vi) given $\varepsilon > 0$, there is a finite union V of open intervals such that

$$\mu^*(V \triangle E) < \varepsilon,$$

   where $V \triangle E = (V - E) \cup (E - V)$.

4. For an arbitrary subset of R^1, one defines the *Lebesgue inner measure* $\mu_*(E)$ of E by $\mu_*(E) = \sup\{\mu(X) : X \in \mathfrak{M} \text{ and } X \subset E\}$ where $\mathfrak{M}$ denotes the class of all measurable subsets of R^1. It is clear that

$$\mu_*(E) \leqslant \mu^*(E).$$

   Prove:
   (a) If E is measurable, then $\mu_*(E) = \mu^*(E)$.
   (b) If $\mu_*(E) = \mu^*(E) < \infty$, then E is measurable.
   (c) If E is contained in a bounded closed interval $I = [a, b]$, then

[1] Topics in the theory of topological and metric spaces will be discussed in Chapter 7.

$$\mu_*(E) = b - a - \mu^*(I - E).$$

5. Let μ^* denote Lebesgue outer measure on the real line. Show that for any two sets A and B on the real line the following inequality holds:

$$\mu^*(A \cup B) + \mu^*(A \cap B) \leqq \mu^*(A) + \mu^*(B).$$

Verify also that $\mu^*(B) = 0$ implies $\mu^*(A \cup B) = \mu^*(A)$.

6. Assume that A and B have positive distance, i.e., assume that

$$d(A, B) = \inf\{|x - y| : x \in A, y \in B\}$$

is strictly positive. Verify that

$$\mu^*(A \cup B) = \mu^*(A) + \mu^*(B).$$

Is it possible that

$$\mu^*(A \cup B) < \mu^*(A) + \mu^*(B)$$

holds with A and B mutually disjoint? Explain.

7. (a) Let G and H be open sets with $H \subset G$ and $\mu^*(H) < \infty$; prove that

$$\mu^*(G - H) = \mu^*(G) - \mu^*(H).$$

(b) If $(S_k)_{k=1}^{\infty}$ is a monotone increasing sequence, then

$$\mu^*(\bigcup_k S_k) = \lim_k \mu^*(S_k).$$

(c) For an open set G let $S \subset G$ and $T \cap G = \varnothing$, then

$$\mu^*(S \cup T) = \mu^*(S) + \mu^*(T).$$

8. (a) Let A be a bounded closed set of R^1 with $\mu(A) > 0$. Show that there is a $\delta > 0$ such that every number z satisfying $|z| < \delta$ can be represented as the difference $x - y$ of two points x and y in the set A.
[*Hints*: Let V_n be the open bounded set of points v such that $|v - a| < 1/n$ for some $a \in A$. The sequence $(V_n)_{n=1}^{\infty}$ is decreasing and

$$\bigcap_{n=1}^{\infty} V_n = A$$

because A is closed (note that $R^1 - A$ is open). Since

$$\lim_{n\to\infty} \mu(V_n) = \mu(A) > 0,$$

we can find an n such that $\frac{2}{3}\mu(V_n) < \mu(A)$. Let $\delta = 1/n$. If $|z| < \delta$, then the set $A_z = \{a - z : a \in A\}$ is contained in V_n. Since $\mu(A_z) = \mu(A)$, we get

$$\mu(V_n - (A \cap A_z)) = \mu((V_n - A) \cup (V_n - A_z))$$
$$\leqq \mu(V_n - A) + \mu(V_n - A_z) = 2\mu(V_n) - 2\mu(A)$$
$$< \tfrac{2}{3}\mu(V_n).$$

i.e., $A \cap A_z$ has positive measure and is thus not empty. Hence there is $x \in A$ such that $y = x - z \in A$.]

(b) Given a number $0 < b \leqq 1$, are there points $c_1 \in P$ and $c_2 \in P$, where P denotes Cantor's ternary set, such that $c_2 - c_1 = b$? Yes! Show also: Every subinterval $[a, b]$ of $[0, 1]$ contains an interval (a', b') free of points of P such that its length satisfies $b' - a' \geqslant (1/5)(b - a)$.

9. Let $T(x) = ax + b$ with $a \neq 0$. Show: $\mu^*[T(E)] = |a|\mu^*(E)$ for each E in R^1 and $T(E)$ is measurable if and only if E is measurable.

10. Let $(A_n)_{n=1}^{\infty}$ be a sequence of sets in R^1. We define the limit superior and the limit inferior of $(A_n)_{n=1}^{\infty}$ as follows:

$$\overline{\lim_{n\to\infty}}\, A_n = \bigcap_{k=1}^{\infty} \bigcup_{n=k}^{\infty} A_n$$
$$\underline{\lim_{n\to\infty}}\, A_n = \bigcup_{k=1}^{\infty} \bigcap_{n=k}^{\infty} A_n .$$

(a) Show that if A_n is measurable for each n, then

$$\mu(\underline{\lim_{n\to\infty}}\, A_n) \leqslant \underline{\lim_{n\to\infty}}\, \mu(A_n).$$

(b) If, moreover, $\mu(A_n \cup A_{n+1} \cup \ldots) < \infty$ for at least one n, then

$$\mu(\overline{\lim_{n\to\infty}}\, A_n) \geqslant \overline{\lim_{n\to\infty}}\, \mu(A_n).$$

(c) If (A_n) is a convergent sequence (i.e., if lim sup and lim inf coincide) of measurable sets with each $A_n \subset B$, where $\mu^*(B) < \infty$, then

$$\lim_{n\to\infty} \mu(A_n)$$

exists and

$$\mu(\lim_{n\to\infty} A_n) = \lim_{n\to\infty} \mu(A_n).$$

11. Consider the following enumeration of proper fractions

$$\frac{0}{1}, \frac{1}{1}, \frac{0}{2}, \frac{1}{2}, \frac{2}{2}, \frac{0}{3}, \frac{1}{3}, \frac{2}{3}, \ldots .$$

Here we have ordered the proper fractions into groups according to increasing denominators and, within each group, according to increasing numerators and we have also listed fractions in their unreduced form. It is clear that in this enumeration the fraction p/q has index

$$k = k(p,q) = p + \frac{1}{2}q(q+1).$$

For each k, we now cover the k-th term in this enumeration of rational numbers between 0 and 1 by an interval of length 2^{-k} in such a way that the k-th term is at the center of its covering interval. The length of the interval cover is

$$\sum_{k=1}^{\infty} \frac{1}{2^k} = 1.$$

Show that the point $\sqrt{2}/2$ does not belong to any covering interval however.

[*Hint*: Suppose $\xi = \sqrt{2}/2$ belongs to one of the covering intervals. Then there would be integers p and q ($q \geqslant p \geqslant 0, q \geqslant 1$) such that

$$\left|\frac{p}{q} - \xi\right| < \frac{1}{2}\frac{1}{2^{p+(1/2)q(q+1)}} \leqslant \frac{1}{2^{(1/2)q(q+1)+1}}.$$

But for every p and q with $q \geqslant 1$, we have, since $\sqrt{2}$ is an irrational number, $|2p^2 - q^2| \geqslant 1$ and therefore

$$\left|\frac{p}{q} - \frac{\sqrt{2}}{2}\right| = \frac{|(p/q)^2 - (1/2)|}{(p/q) + (\sqrt{2}/2)} = \frac{|2p^2 - q^2|}{2pq + q^2\sqrt{2}} > \frac{1}{4q^2}$$
$$\geqslant \frac{1}{2^{(1/2)q(q+1)+1}}.\Bigg]$$

12. Show that every open set of the real line R^1 is the union of a countable family of disjoint open intervals. This representation is in fact unique up to the order in which the terms appear in the union.
[*Hints*: Let U be an arbitrary open subset of R^1. Let $x \in U$, $a = \inf\{y \in R^1 : (y,x) \subset U\}$, $b = \sup\{z \in R^1 : (x,z) \subset U\}$, with $a, b \in R^1_e$. Then the open interval $I_x = (a,b)$ contains the real number x and $I_x \subset U$. Next, verify that $a \notin U$ and $b \notin U$. Now consider the family of open intervals $\Omega = \{I_x : x \in U\}$. Since $x \in I_x \subset U$ for every $x \in U$, we have

$$U = \bigcup_{x \in U} I_x.$$

Show that any two open intervals in the family Ω must be disjoint. If we identify coincident open intervals in Ω, we obtain a collection Λ of disjoint open intervals whose union is the given open set U. To see that Λ is a countable family of open intervals, show that every collection of disjoint nonempty open sets of R^1 is countable. Consider an arbitrary collection $\Lambda = \{U_\beta : \beta \in M\}$ of disjoint nonempty open sets in R^1 and show that the set M of indices is countable. Let $\beta \in M$. Since U_β is nonempty, there is a real number $x \in U_\beta$. Since U_β is open, there is a $\delta > 0$ such that U_β contains the open interval $(x - \delta, x + \delta)$. Let Q be the set of all rationals and $j(\beta) \in Q$ be such that $x - \delta < j(\beta) < x + \delta$. Hence $j(\beta) \in Q \cap U_\beta$. Doing this for every $\beta \in M$ we see that the assignment $\beta \to j(\beta)$ defines a function $j : M \to Q$. Since the U_β's are disjoint, j is an injective map. Since Q is countable, there is an injective map $i : Q \to N$, where N is the set of natural numbers. Define $k = i \circ j : M \to N$. Clearly, k is injective and hence M is countable.

Finally, if there were two distinct representations of U in terms of countable unions of disjoint open intervals, then there would exist a point $t \in U$ which would belong to the component interval I in the first representation and a component interval $J \neq I$ in the second representation. But then one of these component intervals, e.g., J, would extend beyond the other; it would follow from this that one of the endpoints of I belongs to J, which is impossible, inasmuch as the endpoints of I do not belong to U.]

13. Show that every family Ω of open sets of the real line R^1 contains a countable subfamily Δ with

$$\bigcup_{U \in \Delta} U = \bigcup_{U \in \Omega} U.$$

[*Hints*: Let

$$W = \bigcup_{U \in \Omega} U$$

and $x \in W$. Then there is an open set $U \in \Omega$ with $x \in U$. Since U is open, there exists a real number $\delta > 0$ such that the open interval $(x - \delta, x + \delta) \subset U$. Since the rational numbers are dense in the set of real numbers, there are rationals a and b satisfying $x - \delta < a < x < b < x + \delta$. Thus we obtain an open interval $I_x = (a, b)$ with rational endpoints and satisfying $x \in I_x \subset U \subset W$. Hence we get a family $\Lambda = \{I_x : x \in W\}$ of open intervals with

$$W = \bigcup_{x \in W} I_x.$$

If we identify coincident open intervals in the family Λ, we obtain a family Γ of distinct open intervals with rational endpoints and

$$W = \bigcup_{J \in \Gamma} J.$$

This family Γ is clearly countable. According to the construction of the family Λ, every open interval $J \in \Gamma$ is contained in an open set $U_J \in \Omega$. Thus we obtain a countable subfamily $\Delta = \{U_J : J \in \Gamma\}$ of Ω with

$$W = \bigcup_{J \in \Gamma} J \subset \bigcup_{J \in \Gamma} U_J \subset W.$$

This implies

$$\bigcup_{U \in \Delta} U = \bigcup_{U \in \Omega} U.$$

Note that we have also shown: Every open set of the real line is the union of a collection of open intervals with rational endpoints.]

14. Every open cover $\mathfrak{C}$ of a bounded closed subset E of the real line R^1 contains a finite subcover (Heine-Borel).
[*Hints*: Consider first the special case where E is a bounded closed interval $[a, b]$. Let H be the set of all real numbers x such that $a < x \leqq b$ and that the closed interval $[a, x]$ can be covered by a finite subcollection of $\mathfrak{C}$. Since $\mathfrak{C}$ is an open cover of $[a, b]$, H is nonempty. Since b is an upper bound of H, H has a supremum $c \leqq b$. Since $\mathfrak{C}$ is an open cover of $[a, b]$ and c is in $[a, b]$, there is an open set $U \in \mathfrak{C}$ with $c \in U$. Since U is open, there is a positive real number δ such that U contains the open interval $(c - \delta, c + \delta)$. Since $c - \delta$ is not an upper bound of H, there exists a point x in H with $c - \delta < x$. By the definition of H, there is a finite subcollection of $\mathfrak{C}$, say $\{U_1, \ldots, U_n\}$, which covers $[a, x]$. It follows that the finite subcollection $\{U, U_1, \ldots, U_n\}$ covers the closed interval $[a, c + \delta]$. Since c is an upper bound of H, this implies $c = b$ and hence $b \in H$. This completes the proof for the special case.
Now we consider the general case where E is any bounded closed subset of R^1 and $\mathfrak{C}$ is any open cover of E. Since E is bounded, there is a bounded closed interval $[a, b]$ which contains E. Now the collection

$$\mathfrak{D} = \mathfrak{C} \cup (R^1 - E)$$

of open sets covers R^1 and hence also $[a, b]$. By what we have proved already, $\mathfrak{D}$ contains a finite subcollection $\mathcal{E}$ which is a cover of $[a, b]$ and therefore of E. If $\mathcal{E}$ does not contain the open set $R^1 - E$, then $\mathcal{E}$ is a subcollection $\mathfrak{C}$; otherwise

$$\mathcal{G} = \mathcal{E} - (R^1 - E)$$

is a finite subcollection of $\mathscr{E}$ and is a cover of E. This finishes the proof.

Observe: The sequence of open intervals $[(1/2^n), (3/2^n)]$ with $n = 1, 2, \ldots$ covers $(0, 1]$ and the sequence of closed intervals

$$\left[\frac{2^n - 1}{2^n}, \frac{2^{n+1} - 1}{2^{n+1}}\right]$$

with $n = 1, 2, \ldots$ and the closed interval $[1, 2]$ cover the interval $[0, 2]$; can we pick finite subcovers?]

15. In this exercise we review certain facts about metric spaces which will prove useful later.

(i) Let (X, d) be a metric space. If x, x', y and y' are any four points of the space, verify that the following inequality holds:

$$|d(x,y) - d(x',y')| \leqq d(x,x') + d(y,y').$$

Observe that this means: The mapping d of $X \times X$ into R^1 is uniformly continuous.

DEFINITION. *Let (X, d) be a metric space and let A and B be two nonempty subsets of X. The number*

$$d(A, B) = \inf_{x \in A, y \in B} d(x, y)$$

is called the distance between the sets A and B. In particular we denote by $d(x, A)$ the distance between the set $\{x\}$ and the set A.

(ii) The statements $d(x, A) = 0$ and $x \in \bar{A}$ are equivalent.

(iii) If A is a fixed subset, then the function $d(x, A)$ is uniformly continuous on X because $|d(x, A) - d(y, A)| \leqq d(x, y)$.

(iv) Let A be a compact subset of (X, d) and $B \subset X$. Show that there is a point $p \in A$ such that $d(p, B) = d(A, B)$. If the subset B is compact as well, what conclusions can be drawn? Suppose that A and B are two nonempty closed subsets of the real line and suppose that A is bounded; are there points $p \in A$ and $q \in B$ such that $d(p, q) = |p - q| = d(A, B)$? Yes!

(v) Let A and B be two disjoint closed subsets of (X, d). Then there exist disjoint open sets G and H such that $A \subset G$ and $B \subset H$. To show this, try the subsets

$$\{x \in X : d(x, A) - d(x, B) \lessgtr 0\}$$

for G and H, respectively. (The foregoing shows that a metric space is a normal topological space; in the proof of Proposition 8 in Chapter 4 we shall make use of the fact that R^1 is normal.)

16. Let E be the set of points in $(0, 1)$ such that x is in E if and only if the decimal expansion of x does not contain the digit 7. Show that E has Lebesgue measure 0.
17. Let E be a subset of the real line having the property: there is a positive q less than 1 such that, for every interval (a, b), the set $E \cap (a, b)$ can be covered by countably many intervals whose total length is at most $q(b - a)$. Then E is a set of measure zero.
18. Show that the following two definitions are equivalent:
    (a) A set of points E on the real line has measure zero if it can be covered by a finite or infinite sequence of intervals whose total length (i.e., the sum of individual lengths) is arbitrarily small, i.e., smaller than any preassigned $\varepsilon > 0$.
    (b) A set of points F on the real line has measure zero if it can be covered by a sequence of intervals, of finite total length, in such a way that every point of the set lies in the interior of an infinite number of these intervals.
19. Let A be a compact subset of R^1; let $\mathfrak{G} = \{G_\alpha\}$ be an arbitrary open covering thereof. Show that associated with A and $\mathfrak{G}$ is a number $\lambda = \lambda(A; \mathfrak{G})$ with the following property: For every pair of points x_1 and x_2 of A for which $|x_1 - x_2| < \lambda$ there is a member G_α of $\mathfrak{G}$ containing them both. The number λ is called a *Lebesgue number* of the covering. Clearly, λ is not unique.
20. Consider the open interval $(0, 1)$ and the family $\mathfrak{G}$ of intervals

$$G_n = \left(\frac{1}{2^{n+1}}, \frac{1}{2^{n-1}}\right), \quad n = 1, 2, \ldots.$$

    Show that no Lebesgue number λ exists for this covering.
21. Construct an open covering for $[1, \infty)$ having no Lebesgue number.
22. Construct an open covering for $[0, 1]$ having the Lebesgue number $\lambda = (10)^{-1}$.
23. Is it true that a set A in R^1 is compact if and only if it has a Lebesgue number?
24. In the exercise section of Chapter 7 we are going to see that the set D of discontinuities of a function f: $R^1 \to R^1$ is of type F_σ. Since the set of all irrationals is not of type F_σ, there is no function from R^1 into R^1 which is continuous at each rational but discontinuous at each irrational point. In terms of the foregoing results the following examples are of great interest.
    (a) If the number $x \in (0, 1)$ is rational and is expressed in reduced form as the fraction p/q with $q > 0$, then set $f(x) = 1/q$; if the number $x \in (0, 1)$ is irrational, then let $f(x) = 0$. One can easily see that f is continuous at all irrational points of $(0, 1)$ and is

discontinuous at all rational points of $(0, 1)$. Putting $f(0) = 1$, one can readily extend the function f to a function g mapping R^1 into itself such that g will be continuous at all irrational points and discontinuous at all rational points of R^1.

(b) Let (x) denote the positive or negative excess of x above the integer nearest to it, and if x be halfway between two successive integers, let $(x) = 0$. Let a function g be defined for the interval $(0, 1)$ as the limit of

$$\frac{(x)}{1} + \frac{(2x)}{4} + \frac{(3x)}{9} + \ldots + \frac{(nx)}{n^2}$$

as $n \to \infty$. Observe that the right- and left-hand limits of g at $x = p/(2n)$, with p and $2n$ relatively prime, satisfy

$$g(x+0) = g(x) - \frac{1}{2n^2}\left(1 + \frac{1}{9} + \frac{1}{25} + \ldots\right) = g(x) - \frac{\pi^2}{16n^2},$$

$$g(x-0) = g(x) + \frac{1}{2n^2}\left(1 + \frac{1}{9} + \frac{1}{25} + \ldots\right) = g(x) + \frac{\pi^2}{16n^2},$$

and that everywhere else $g(x + 0) = g(x)$, $g(x - 0) = g(x)$. Note that the set K, of points at which the jump is $\geqq k$, is finite for each positive value of k; the number of points of K is the number of irreducible proper fractions having even denominators $2n$, such that $(\pi^2)/(8n^2) \geqq k$. The set of all points of discontinuity of g is everywhere dense in the interval $(0, 1)$. (The foregoing two functions are due to B. Riemann.)

(c) The function which is equal to x when x is irrational and to $\{(1 + p^2)/(1 + q^2)\}^{1/2}$ when x is a rational fraction p/q is discontinuous for all negative and for positive rational values of x, but continuous for positive irrational values.

(d) Let the numbers of the interval $(0, 1)$ be expressed as finite or infinite decimals $x = 0.a_1 a_2 \ldots a_n \ldots$, and let

$$f(x) = \left(\frac{a_1}{10}\right)^2 + \left(\frac{a_2}{100}\right)^2 + \ldots.$$

The function f is monotone, and is discontinuous for every value of x represented by a finite decimal. The set K of points at which the jump is $\geqq k$ (with $k > 0$) is finite. The function f defined by $f(x) = 0.\,0a_1\,0a_2\,0a_3 \ldots$ has similar properties.

(e) State the familiar ε, δ-definition for the continuity of a real-valued function of a real variable at a point. Find the correct form of its negation.

(f) Give an example of a function on R^1 continuous only at one point.

25. Construct a Cantor set in $[0, 1]$ having measure $1/2$ (i.e., a set whose complement in $[0, 1]$ consists of intervals having total length $1/2$). Now show that for each $\varepsilon > 0$ there is a Cantor set in $[0, 1]$ having measure $1 - \varepsilon$.

Note to the Student. Many interesting questions pertaining to the study of measure and integration theory are discussed in *Counterexamples in Analysis* by B. R. Gelbaum and J. M. H. Olmsted, Holden-Day, San Francisco, Cal., 1964. Another interesting book is *Measure and Category* by J. C. Oxtoby, Springer-Verlag, New York, Heidelberg, Berlin, 1971 where, on pp. 43-44, you will find help for doing Problem 2 above. Amongst journals, *The American Mathematical Monthly*, a publication of the *Mathematical Association of America*, is famous for its elementary and advanced problem sections; a great number of these problems, as well as classroom notes, deal with the subject of real analysis.

Chapter 2

Lebesgue Measurable Functions on the Real Line

PROPOSITION 1. *Let E denote a Lebesgue measurable subset of the real line R^1. For an arbitrary function f mapping E into the extended real line R_e^1 the following four conditions are equivalent:*

(*i*) *For each real number r the set $\{x \in E : f(x) > r\}$ is measurable.*
(*ii*) *For each real number r the set $\{x \in E : f(x) \geqslant r\}$ is measurable.*
(*iii*) *For each real number r the set $\{x \in E : f(x) < r\}$ is measurable.*
(*iv*) *For each real number r the set $\{x \in E : f(x) \leqslant r\}$ is measurable.*

Proof. Since the class $\mathfrak{M}$ of Lebesgue measurable subsets of the real line R^1 is a σ-algebra of sets, the implications

$$(i) \to (ii) \to (iii) \to (iv) \to (i)$$

are consequences of the following relations, respectively:

$$f^{-1}[r, \infty] = \bigcap_{n=1}^{\infty} f^{-1}\left(r - \frac{1}{n}, \infty\right],$$

$$f^{-1}[-\infty, r) = E - f^{-1}[r, \infty],$$

$$f^{-1}[-\infty, r] = \bigcap_{n=1}^{\infty} f^{-1}\left[-\infty, r + \frac{1}{n}\right),$$

$$f^{-1}(r, \infty] = E - f^{-1}[-\infty, r].$$

This completes the proof.

A function f: $E \to R_e^1$ for a Lebesgue measurable set E is said to be a *Lebesgue measurable function*, or *measurable function* for short, if and only if it satisfies one of the four conditions of Proposition 1.

An immediate consequence of the foregoing definition is the following proposition.

PROPOSITION 2. *Let f be a measurable function. Then, for every interval I, the inverse image $f^{-1}(I)$ is a measurable set. Also, if r is an extended real number, then $\{x : f(x) = r\}$ is a measurable set.*

Let $P(x)$ denote a statement concerning the point $x \in E$ of a measurable set in R^1. Then we say that $P(x)$ holds *almost everywhere* if and only if there is a subset D of E with $\mu^*(D) = 0$ such that $P(x)$ holds for every point x of $E - D$. As we know, D is then measurable and is of measure 0. Hence $P(x)$ holds almost everywhere (abbreviated: a.e.) if and only if the set of points where it fails to hold is a set of measure zero.

For example, a function f on a measurable subset E of R^1 is said to be *continuous almost everywhere* in E if and only if there is a subset D of E of measure 0 such that f is continuous at every point of $E - D$.

In the same way, two measurable functions $f, g : E \to R_e^1$ are said to be *equal almost everywhere* if and only if there exists a subset D of E of measure 0 such that $f(x) = g(x)$ for every $x \in E - D$.

The phrase "almost everywhere" means that we regard as negligible sets of measure zero.

PROPOSITION 3. *If a function f: $E \to R_e^1$ on a measurable set E is continuous almost everywhere in E, then f is measurable. In particular, if f is continuous on E, then it is measurable.*

Proof. Since f: $E \to R_e^1$ is continuous almost everywhere in E, there is a subset D of E with $\mu^*(D) = 0$ such that f is continuous at every point of the set $C = E - D$. To verify that f is measurable, let r denote any given real number. It is enough to prove that the inverse image

$$B = f^{-1}(r, \infty] = \{x \in E : f(x) > r\}$$

of the interval $(r, \infty]$ is measurable.

Indeed, let x be an arbitrary point in $B \cap C$. Then $f(x) > r$ and f is continuous at x. Hence there is an open interval U_x containing x such that $f(y) > r$ holds for every point y of $E \cap U_x$. Let

$$U = \bigcup_{x \in B \cap C} U_x .$$

Since $x \in E \cap U_x \subset B$ holds for every $x \in B \cap C$, we have

$$B \cap C \subset E \cap U \subset B.$$

This implies $B = (E \cap U) \cup (B \cap D)$. Since U is open in R^1, it is measurable and hence $E \cap U$ is measurable. But

$$\mu^*(B \cap D) \leqslant \mu^*(D) = 0,$$

$B \cap D$ is also measurable (by Proposition 17, Chapter 1). Hence B is measurable and the proof is finished.

PROPOSITION 4. *Let f, g: $E \to R_e^1$, where E is measurable, and suppose that $f = g$ almost everywhere in E. If f is measurable, then so is g.*

Proof. Let $D = \{x \in E : f(x) \neq g(x)\}$. Since f and g are equal almost everywhere in E, we have that $\mu^*(D) = 0$. Let r be an arbitrary real number and put

$$A = D \cap g^{-1}[-\infty, r], \quad B = D \cap g^{-1}(r, \infty].$$

Since $\mu^*(A) = 0 = \mu^*(B)$, the sets A and B are measurable. As f is measurable, the set $f^{-1}(r, \infty]$ is measurable and hence $g^{-1}(r, \infty] = \{f^{-1}(r, \infty] - A\} \cup B$ is measurable. Since r is any real number, this shows that g is measurable.

PROPOSITION 5. *If f: $E \to R_e^1$, with E a measurable set, is a measurable function and if D is a measurable subset of E, then the restriction $g = f|D$ is a measurable function as well.*

Proof. Let r be an arbitrary real number. Since $g = f|D$, we have

$$g^{-1}(r, \infty] = D \cap f^{-1}(r, \infty].$$

But the set $f^{-1}(r, \infty]$ is measurable and the set D is measurable by hypothesis. By Proposition 11 of Chapter 1, $g^{-1}(r, \infty]$ must be measurable. Since r is an arbitrary real number, g: $D \to R_e^1$ must be a measurable function also.

PROPOSITION 6. *For any two measurable functions f and g defined on a measurable set E, the functions*

$$f \vee g, f \wedge g : E \to R_e^1$$

defined by

$$(f \vee g)(x) = \max\{f(x), g(x)\},$$
$$(f \wedge g)(x) = \min\{f(x), g(x)\},$$

for every $x \in E$, are measurable.

Proof. Let $u = f \vee g$ and $v = f \wedge g$. To show that u and v are measurable, let r be any real number. We have that

$$u^{-1}(r, \infty] = \{f^{-1}(r, \infty]\} \cup \{g^{-1}(r, \infty]\}$$
$$v^{-1}(r, \infty] = \{f^{-1}(r, \infty]\} \cap \{g^{-1}(r, \infty]\}.$$

As f and g are measurable, the sets $f^{-1}(r, \infty]$ and $g^{-1}(r, \infty]$ are measurable. But the union and the intersection of two measurable sets is again measurable. This proves that u and v are measurable.

PROPOSITION 7. *For any two measurable functions f and g on a measurable set E, the following sets are measurable:*

$$A(f,g) = \{x \in E : f(x) < g(x)\}$$
$$B(f,g) = \{x \in E : f(x) \leqslant g(x)\}$$
$$C(f,g) = \{x \in E : f(x) = g(x)\}.$$

Proof. If $f(x) < g(x)$, then there is a rational r in the set of all rationals Q such that

$$f(x) < r < g(x).$$

Hence we have

$$A(f,g) = \bigcup_{r \in Q} \{f^{-1}[-\infty, r) \cap g^{-1}(r, \infty]\}.$$

But Q is countable and the set of L-measurable sets in R^1 is a σ-algebra. Thus $A(f,g)$ is measurable.

To see that $B(f,g)$ and $C(f,g)$ are measurable, we merely note that

$$B(f,g) = E - A(g,f)$$
$$C(f,g) = B(f,g) \cap B(g,f).$$

The proof is finished.

PROPOSITION 8. *For any real number c and any two measurable functions f,g: $E \to R^1$ on a measurable set E, the functions*

$$f + c,\ cf,\ f + g,\ fg : E \to R^1$$

are measurable. (Here we restrict the range of the functions f and g to R^1 because $\infty + (-\infty)$ and $(-\infty) + \infty$ are not defined in R_e^1).

Proof. Let $h = f + c$ and let r denote an arbitrary real number. Then

$$h^{-1}(r, \infty] = \{x \in E : h(x) = f(x) + c > r\}$$
$$= \{x \in E : f(x) > r - c\} = f^{-1}(r - c, \infty].$$

Since $f^{-1}(r - c, \infty]$ is measurable, we have that $h = f + c$ is measurable.

Consider $w = cf$. In case $c = 0$, w is the constant function 0 and hence is measurable because it is a continuous function. In case $c > 0$ we have

$$w^{-1}(r, \infty] = \{x \in E : w(x) = cf(x) > r\}$$
$$= \left\{x \in E : f(x) > \frac{r}{c}\right\} = f^{-1}\left(\frac{r}{c}, \infty\right].$$

In case $c < 0$, we have

$$w^{-1}(r, \infty] = \{x \in E : w(x) = cf(x) > r\}$$
$$= \left\{x \in E : f(x) < \frac{r}{c}\right\} = f^{-1}\left[-\infty, \frac{r}{c}\right).$$

Since $f^{-1}(r/c, \infty]$ and $f^{-1}[-\infty, r/c)$ are measurable, it follows that $w = cf$ is measurable.

Next, consider $v = f + g$. Then we have

$$v^{-1}(r, \infty] = \{x \in E : v(x) = f(x) + g(x) > r\}$$
$$= \{x \in E : -f(x) < g(x) - r\}$$
$$= A(-f, g - r).$$

Since $A(-f, g - r)$ is measurable by Proposition 7, this shows that $v = f + g$ is measurable.

Finally, we consider the function $p = fg$. First we note that the function $q = f^2$ is measurable. Indeed, we have

$$q^{-1}(r, \infty] = \{x \in E : q(x) = [f(x)]^2 > r\}$$
$$= \begin{cases} f^{-1}(\sqrt{r}, \infty] \cup f^{-1}[-\infty, -\sqrt{r}) & \text{if } r \geqslant 0 \\ E & \text{if } r < 0. \end{cases}$$

Similarly, g^2 and $(f + g)^2$ are measurable. Consequently,

$$p = fg = \frac{1}{2}[(f + g)^2 - f^2 - g^2]$$

is also measurable and the proof is finished.

N.B. If f is measurable and nonvanishing, then $1/f$ is also measurable. Indeed, we have

$$\text{for } r > 0 : \left\{x : \frac{1}{f(x)} > r\right\} = \left\{x : 0 < f(x) < \frac{1}{r}\right\},$$

$$\text{for } r < 0 : \left\{x : \frac{1}{f(x)} > r\right\}$$
$$= \left\{x : f(x) > 0\right\} \cup \left\{x : f(x) < \frac{1}{r}\right\},$$

$$\text{for } r = 0 : \left\{x : \frac{1}{f(x)} > r\right\} = \{x : f(x) > 0\}.$$

By a *simple function* on a measurable subset of E of R^1 we mean a measurable function f: $E \to R^1$ such that its image $f(E)$ is a finite subset of R^1.

By definition every simple function is bounded.

Examples of simple functions are:

(a) *characteristic function* χ_D of a measurable set D:

$$\chi_D : E \to R^1$$

where E and D are measurable, $D \subset E$ and

$$\chi_D(x) = \begin{cases} 1 \text{ if } x \in D \\ 0 \text{ if } x \in E - D. \end{cases}$$

(b) *step functions*, i.e., finite real linear combinations of characteristic functions of an interval partition.

Let f: $E \to R^1$ be an arbitrary simple function and let $c_1, c_2, \dots, c_n$ denote the nonzero real numbers in its image $f(E)$. For each $j = 1, 2, \dots, n$, let

$$D_j = f^{-1}(c_j) \subset E.$$

Then we have

$$f = \sum_{j=1}^{n} c_j \chi_{D_j}.$$

If $f(E)$ contains no nonzero real numbers, then $f = 0$ and is the characteristic function $\chi_\varnothing$ of the empty set $\varnothing$ of E. Hence we obtain Proposition 9.

PROPOSITION 9. *Every simple function on E is a linear combination of characteristic functions of measurable subsets of E, and conversely.*

PROPOSITION 10. *For any real number c and any two finite simple functions* $f, g : E \to R^1$, *the four functions*

$$f + c,\ cf,\ f + g \text{ and } fg$$

are simple.

We now establish an important approximation theorem.

PROPOSITION 11. *For every measurable function* f: $E \to R^1_e$ *there exists a sequence*

$$\{f_n : E \to R^1 : n \in N\}$$

of simple functions f_n which converges pointwise to f. In case f is nonnegative, the sequence $(f_n)_{n=1}^{\infty}$ of approximating simple functions can be so selected that

$$0 \leqslant f_n \leqslant f_{n+1}$$

holds for every $n \in N$. In case f is a bounded function the approximating sequence of simple functions can be so chosen that the convergence is uniform.

Proof. Assume that $f \geqslant 0$ and define

$$C_n = \{x \in E : f(x) \geqslant n\}$$

$$D_{n,k} = \left\{x \in E : \frac{k-1}{2^n} \leqslant f(x) < \frac{k}{2^n}\right\}$$

for every $n \in N$ and every $k = 1, 2, \ldots, n2^n$. Let

$$f_n = n\chi_{C_n} + \sum_{k=1}^{n2^n} \frac{k-1}{2^n} \chi_{D_{n,k}}$$

for every $n \in N$. It is clear that all of the sets C_n and $D_{n,k}$ are measurable, and so each f_n is a simple function. It is also easy to see that

$$0 \leqslant f_1 \leqslant f_2 \leqslant \ldots \leqslant f, \quad |f_n| \leqslant n,$$

and

$$|f(x) - f_n(x)| < \frac{1}{2^n}$$

for all $x \in E - C_n$. It follows that

$$\lim_{n \to \infty} f_n(x) = f(x)$$

for every $x \in E$. Moreover, if there exists a $K \in R^1$ such that $|f| \leqslant K$, then

$$\sup_{x \in E} |f(x) - f_n(x)| \leqslant \frac{1}{2^n}$$

for all $n \geqslant K$; therefore $f_n \to f$ uniformly on E if f is bounded.

In the general case, we apply the preceding construction to the nonnegative functions

$$f^+ = f \vee 0 \text{ and } f^- = -(f \wedge 0),$$

and obtain sequences $(f_n^+)_{n=1}^{\infty}$ and $(f_n^-)_{n=1}^{\infty}$, respectively. Since

$$f = f^+ - f^-,$$

we get a sequence

$$f_n = f_n^+ - f_n^- : E \to R^1, \, n \in N$$

of simple functions with the required properties. This finishes the proof.

Remark. In connection with the foregoing proposition we note the following fact as well: If $N(f) = \{x \in R^1 : f(x) \neq 0\}$ and $N(f)$ is the union of a countable class of measurable sets of finite measure, then $(f_n)_{n=1}^{\infty}$ may be constructed so as to have the additional property that for each n, $N(f_n)$ has finite measure. Conversely, if for each simple function f_n, $N(f_n)$ has finite measure, and if for each x, $f_n(x) \to f(x)$, then $N(f)$ is the union of a countable class of measurable sets of finite measure.

Indeed. To prove the first statement, let

$$\{x \in R^1 : f(x) \neq 0\} = \bigcup_{k=1}^{\infty} E_k$$

and let

$$A_n = \bigcup_{k=1}^{n} E_k$$

for each n. Then the sequence $(A_n)_{n=1}^{\infty}$ is increasing and so is the sequence of characteristic functions $(\chi_{A_n})_{n=1}^{\infty}$. If $(f_n)_{n=1}^{\infty}$ is the sequence constructed in the proof of the foregoing proposition, then the sequence $(\sigma_n)_{n=1}^{\infty}$ defined by

$$\sigma_n = f_n \chi_{A_n}$$

has all the required properties.

To prove the second statement, we note that if $f_n(x) = 0$ for every n, then $f(x) = 0$; so

$$\{x \in R^1 : f(x) \neq 0\} \subset \bigcup_{n=1}^{\infty} \{x \in R^1 : f_n(x) \neq 0\}.$$

Setting

$$E_n = \{x \in R^1 : f(x) \neq 0\} \cap \{x \in R^1 : f_n(x) \neq 0\},$$

we have

$$\{x \in R^1 : f(x) \neq 0\} = \bigcup_{n=1}^{\infty} E_n;$$

the sets E_n are measurable, and each has finite measure.

PROPOSITION 12. *Let $(f_n)_{n=1}^{\infty}$ be a sequence of measurable functions with $f_n : E \to R_e^1$ for each n. Then the functions $f_1 \vee \ldots \vee f_n, f_1 \wedge \ldots \wedge f_n$,*

$$\sup_n f_n, \inf_n f_n, \overline{\lim_n} f_n = \inf_{n \in N} \sup_{k \geqq n} f_k, \underline{\lim_n} f_n = \sup_{n \in N} \inf_{k \geqq n} f_k$$

are all measurable.

Proof. Let $h(x) = (f_1 \vee \ldots \vee f_n)(x)$. Then

$$\{x \in E : h(x) > r\} = \bigcup_{j=1}^{n} \{x \in E : f_j(x) > r\}.$$

Hence, the measurability of the f_j's implies that of h. Similarly, if we set

$$g(x) = \sup_{n \in N} f_n(x),$$

the identity

$$\{x \in E : g(x) > r\} = \bigcup_{n=1}^{\infty} \{x \in E : f_n(x) > r\}$$

shows that g is measurable also. A similar argument establishes the corresponding statements for the infima; we could of course also use the identity

$$\inf_n f_n(x) = -\sup_n(-f_n(x)).$$

The rest of the proposition follows now from what we have already proved.

If $(f_n)_{n=1}^{\infty}$ is a sequence of measurable functions, converging pointwise to the function f, then

$$\overline{\lim_{n \to \infty}} f_n = \underline{\lim_{n \to \infty}} f_n = f.$$

Hence the following proposition.

PROPOSITION 13. *If $(f_n)_{n=1}^{\infty}$ is a sequence of measurable functions converging pointwise to the function f, then f is also measurable.*

PROPOSITION 14. *Let E be a measurable set of finite measure and f be a measurable function on E which is finite almost everywhere. Then, given any* $\varepsilon > 0$, *there exists a measurable function g which is bounded such that*

$$\mu\{x \in E : f(x) \neq g(x)\} < \varepsilon.$$

Proof. We put

$$A_k = \{x \in E : |f(x)| > k\}$$

and

$$Q = \{x \in E : |f(x)| = \infty\}.$$

By assumption $\mu(Q) = 0$. Moreover,

$$A_1 \supset A_2 \supset A_3 \supset \ldots \text{ and } Q = \bigcap_{k=1}^{\infty} A_k$$

so that for $k \to \infty$ we have

$$\mu(A_k) \to \mu(Q) = 0.$$

Hence there is a k_0 such that $\mu(A_{k_0}) < \varepsilon$. We define on E the function g as follows

$$g(x) = \begin{cases} f(x) & \text{if } x \in E - A_{k_0} \\ 0 & \text{if } x \in A_{k_0}. \end{cases}$$

This function g is measurable and bounded (actually $|g| \leqslant k_0$). Moreover

$$\{x \in E : f(x) \neq g(x)\} = A_{k_0}. \qquad \text{(Q.E.D.)}$$

Given the measurable functions $f_n : E \to R_e^1 (n = 1, 2, \ldots)$ and f: $E \to R_e^1$ we say that $(f_n)_{n=1}^{\infty}$ *converges almost everywhere* to f on E if there exists a set $D \subset E$ with $\mu(D) = 0$ such that $f_n \to f$ pointwise on $E - D$; we write $f_n \to f$ a.e. on E.

PROPOSITION 15. *Let E be a measurable subset of R^1 of finite measure and let f and $(f_n)_{n=1}^{\infty}$ be measurable functions defined and finite almost everywhere on E. Suppose that $f_n \to f$ a.e. in E. Then for each pair of positive real numbers δ and ε, there exists a measurable set $A \subset E$ and an integer n_0 such that $\mu(E - A) < \varepsilon$ and $|f(x) - f_n(x)| < \delta$ for all $x \in A$ and $n \geqslant n_0$.*

Proof. Let $B = \{x \in E : f(x)$ is finite, $f_n(x)$ is finite for all $n \in N$, and $f_n(x) \to f(x)\}$. By hypothesis, $\mu(E - B) = 0$. For each $m \in N$, let $B_m = \{x \in E : |f(x) - f_n(x)| < \delta$ for all $n \geqslant m\}$. We have $B_1 \subset B_2 \subset B_3 \subset \ldots$ and

$$\bigcup_{m=1}^{\infty} B_m = B.$$

Therefore $E - B_1 \supset E - B_2 \supset E - B_3 \supset \ldots$ and

$$\bigcap_{m=1}^{\infty} (E - B_m) = E - B.$$

Since $\mu(E - B_1) \leqslant \mu(E) < \infty$, it follows that $\mu(E - B_m) \to \mu(E - B) = 0$ as $m \to \infty$. Thus choose $n_0 \in N$ such that $\mu(E - B_{n_0}) < \varepsilon$ and set $A = B_{n_0}$.

The following result is due to Lebesgue.

PROPOSITION 16. *Let E, f and $(f_n)_{n=1}^{\infty}$ be as in Proposition 15. For each $\delta > 0$ and each $n \in N$, let*

$$C_n = C_n(\delta) = \{x \in E : |f(x) - f_n(x)| \geqslant \delta\}.$$

Then $\mu(C_n) \to 0$ as $n \to \infty$.

Proof. Choose arbitrary positive numbers δ and ε. By Proposition 15, there exists a measurable set A and $n_0 \in N$ such that $\mu(E - A) < \varepsilon$ and $|f(x) - f_n(x)| < \delta$ for all $x \in A$ and for all $n \geqslant n_0$. Thus $n \geqslant n_0$ implies $C_n \subset E - A$. Therefore $n \geqslant n_0$ implies $\mu(C_n) \leqslant \mu(E - A) < \varepsilon$. Since ε is arbitrary, it follows that $\mu(C_n) \to 0$ as $n \to \infty$.

We now establish a result of Egorov.

PROPOSITION 17. *Let E, f and $(f_n)_{n=1}^{\infty}$ be as in Proposition 15. Then for each $\varepsilon > 0$ there exists a measurable set F such that $\mu(E - F) < \varepsilon$ and $f_n \to f$ uniformly on F.*

Proof. Choose a positive number ε. By Proposition 15, for each $m \in N$ there exists a measurable set A_m and an integer $n_m \in N$ such that $\mu(E - A_m) < \varepsilon/2^m$ and $|f(x) - f_n(x)| < 1/m$ for all $x \in A_m$ and all $n \geqslant n_m$. Define F by

$$F = \bigcap_{m=1}^{\infty} A_m.$$

Then

$$E - F = \bigcup_{m=1}^{\infty} (E - A_m),$$

and so

$$\mu(E - F) \leqslant \sum_{m=1}^{\infty} \mu(E - A_m) < \sum_{m=1}^{\infty} \frac{\varepsilon}{2^m} = \varepsilon.$$

Also $n \geqslant n_m$ implies that

$$\sup_{x \in F} |f(x) - f_n(x)| \leqslant \sup_{x \in A_m} |f(x) - f_n(x)| \leqslant \frac{1}{m}$$

for every $m \in N$. Thus $f_n \to f$ uniformly on F.

Remark. Propositions 16 and 17 depend very much on the assumption that $\mu(E) < \infty$. Suppose, for example, that $E = R^1$ and $f_n = \chi_{[n,n+1]}$ and $f = 0$. Then $f_n(x) \to f(x) = 0$ for all $x \in R^1$. But $\mu\{x \in R^1 : |f(x) - f_n(x)| \geqslant 1\} = \mu([n, n+1]) = 1 \nrightarrow 0$, and so $f_n \nrightarrow f$ in measure. Also, if F is measurable and $\mu(R^1 - F) < 1$, then for each $n \in N$ there exists $x_n \in F \cap [n, n+1]$, and so $|f(x_n) - f_n(x_n)| = 1$, that is, $f_n \nrightarrow f$ uniformly on F.

The foregoing Proposition 17 motivates the following definition:

Given functions $f_n : E \to R_e^1$ $(n = 1, 2, \ldots)$ and f: $E \to R_e^1$ each of which is measurable and finite a.e. on E, we say that f_n *converges almost uniformly* to f on E if, for each $\varepsilon > 0$, there is a measurable set F such that $\mu(E - F) < \varepsilon$ and $f_n \to f$ uniformly on F; we write $f_n \to f$ a. uniformly.

An example of another kind of convergence is *uniform convergence almost everywhere*. The sequence $(f_n)_{n=1}^{\infty}$ converges to f on E uniformly a.e. if there is a measurable set F such that $\mu(E - F) = 0$ and $f_n \to f$ uniformly on F; we write $f_n \to f$ uniformly a.e. on E.

For illustration consider the following example. Let $E = [0, 1]$, $f_n(x) = x^n$. This example shows that it is possible for a sequence to converge almost uniformly on E while it does not converge uniformly almost everywhere on E. It is clear, however, that convergence uniformly almost everywhere implies almost uniform convergence.

Concluding Remarks

On the relation between continuity and measurability for real-valued functions on R^1 there is a very interesting theorem due to N.N. Lusin, which we will now discuss. We begin with two lemmas.

LEMMA 1. *If f is a continuous function on a closed set F of R^1, then there exists a continuous extension $\hat{f}$ of f to R^1. Moreover, if $|f| \leqslant M$ on F, then $|\hat{f}| \leqslant M$ on R^1.*

Indeed. There exist disjoint open intervals (a_n, b_n) such that $R^1 - F$, being an open set, can be represented uniquely up to the order of summation by the sum-set of these intervals. Define $\hat{f}$ on each of these intervals (a_n, b_n) linearly by

$$\hat{f}(x) = f(a_n) + \frac{f(b_n) - f(a_n)}{b_n - a_n}(x - a_n).$$

LEMMA 2. *For any $\varepsilon > 0$ and any simple function s, there exists a continuous function f and a closed set F such that $\mu(R^1 - F) < \varepsilon$ and f agrees with s on F.*

Indeed. Let

$$s = \sum_{k=1}^{n} c_k \chi_{E_k},$$

where χ_{E_k} is the characteristic function of a measurable set E_k and the numbers $c_1, c_2, \ldots, c_n$ are the distinct values assumed by s. Let F_k, $k = 1, 2, \ldots, n$, be closed sets such that

$$E_k \supset F_k \text{ and } \mu(E_k - F_k) < \frac{\varepsilon}{n}.$$

Then the set

$$F = \bigcup_{k=1}^{n} F_k$$

is closed and

$$\mu(R^1 - F) = \sum_{k=1}^{n} \mu(E_k - F_k) < \varepsilon.$$

Construct the continuous function f by letting it agree with s on F (it is then continuous on F) and extending it to R^1 by applying Lemma 1.

LUSIN'S THEOREM. *Let f be an almost everywhere finite-valued measurable function on R^1. Then, for every $\varepsilon > 0$, there exists a closed set F such that f is continuous on F and $\mu(R^1 - F) < \varepsilon$.*

Proof. We first consider the case where f is defined only on a bounded measurable set E. Let $(f_n)_{n=1}^{\infty}$ be a sequence of simple functions converging pointwise to f. By Lemma 2, there exist closed sets F_n such that $\mu(E - F_n) < \varepsilon/2^{n+1}$, and f_n is continuous on F_n. Let

$$F_0 = \bigcap_{n=1}^{\infty} F_n.$$

Then all f_n are continuous on F_0 and

$$\mu(E - F_0) = \mu(E - \bigcap_{n=1}^{\infty} F_n) = \mu(\bigcup_{n=1}^{\infty} (E - F_n)) \leqslant \sum_{n=1}^{\infty} \frac{\varepsilon}{2^{n+1}}$$
$$= \frac{\varepsilon}{2}.$$

By Proposition 17, which applies since $\mu(E) < \infty$, we can find a set F such that $F_0 \supset F$, $\mu(F_0 - F) < \varepsilon/2$ and $f_n \to f$ uniformly on F. Then f, as the uniform limit of continuous functions on F, is also continuous on F. Moreover

$$\mu(E - F) = \mu(E - F_0) + \mu(F_0 - F) < \frac{\varepsilon}{2} + \frac{\varepsilon}{2} = \varepsilon.$$

We can extend this result from E to R^1 by writing

$$R^1 = \bigcup_n E_n,$$

where the E_n's are disjoint bounded measurable sets. Each E_n contains a set G_n, closed in R^1, such that $\mu(E_n - G_n) < \varepsilon/2^n$ and f is continuous on each G_n. If we take

$$E_{2n} = [n, n+1)$$

$$E_{2n+1} = [-n, -n+1),$$

then the set

$$G_0 = \bigcup_n G_n$$

is closed and f is continuous on G_0. In addition,

$$\mu(R^1 - G_0) = \sum_{n=1}^{\infty} \mu(E_n - G_n) < \sum_{n=1}^{\infty} \frac{\varepsilon}{2^n} = \varepsilon.$$

The proof is finished.

Remark. In Exercise 27 at the end of Chapter 3 we shall come across the following theorem of N. N. Lusin: If f: $[a, b] \to R_e^1$ is measurable and finite a.e. and $b - a < \infty$, then for any $\delta > 0$ there exists a continuous function h on $[a, b]$ such that $\mu\{x \in [a, b] : f(x) \neq h(x)\} < \delta$. Moreover, if K is a bound for f, then K is also a bound for h.

EXERCISES

1. Show the converse of Proposition 17, namely, if $(f_n)_{n=1}^{\infty}$ and f are measurable and finite a.e. on E and $f_n \to f$ almost uniformly on E then $f_n \to f$ a.e. in E.
2. Let f be a differentiable function on $[a, b]$. Show that f' is measurable on $[a, b]$.
3. Let f be an extended real-valued function on a measurable set E. Prove that f is measurable if and only if
   (i) $f^{-1}(\{\infty\})$ and $f^{-1}(\{-\infty\})$ are both measurable and
   (ii) $f^{-1}(B)$ is measurable for every Borel set B.
4. Let $f, g : R^1 \to R_e^1$ with f measurable and g continuous. Verify that the composition $g \circ f$ is a measurable function. Can the roles of continuous function and measurable function be reversed in the above composition?

5. Let f be a measurable function and let h be any extended real-valued function defined on R_e^1 such that

$$h^{-1}([r, \infty]) \cap R^1$$

is a Borel set for all real numbers r. Prove that $h \circ f$ is measurable.

6. Let f: $R^1 \to R^1$ such that $f(x + t) = f(x) + f(t)$. Prove that
   (a) f is continuous everywhere if it is continuous at a single point; if f is continuous, then $f(x) = f(1)x$.
   (b) f is continuous if it is measurable. [*Hint*: Use Lusin's Theorem.]
   [The article "Additive Functions" by A. Wilansky in *Lectures on Calculus*, K. O. May, ed., Holden-Day, San Francisco, Cal., 1967, makes for interesting reading in connection with this exercise.]

7. A function f: $R^1 \to R^1$ is said to be a Baire function if for every open set $G, f^{-1}(G)$ is a Borel set. In case f assumes infinite values, we merely add the requirement that $f^{-1}(\{\infty\})$ and $f^{-1}(\{-\infty\})$ be Borel sets. The set of all Baire functions on R^1 we call the class of Baire functions on R^1.

   The class of Baire functions on R^1 is frequently defined as the smallest class which contains the continuous functions and is closed under the operation of taking pointwise limits.

   Prove that the two definitions are equivalent.

8. A measurable function f: $R^1 \to R_e^1$ is said to be essentially bounded if $\mu\{x : |f(x)| > r\} = 0$ for some real number r. In this case we define the *essential supremum* of f by $\text{ess sup}|f| = \inf\{r : \mu\{x : |f(x)| > r\} = 0\}$. Note that, if $\text{ess sup}|f| = c$, then

$$E = \{x : |f(x)| > c\} = \bigcup_{k=1}^{\infty} \left\{x : |f(x)| > c + \frac{1}{k}\right\}$$

so that $\mu(E) = 0$ and $|f(x)| \leqslant c$ outside E. Thus $|f(x)| \leqslant c$ a.e., and if we define

$$f^*(x) = \begin{cases} f(x) & \text{if } |f(x)| \leqslant c, \\ 0 & \text{if } |f(x)| > c, \end{cases}$$

then $|f^*(x)| \leqslant c$ for all x and $f^* = f$ a.e. Furthermore, $\{x : |f^*(x)| > c - \varepsilon\}$ has positive measure for all $\varepsilon > 0$, so that it is nonempty and we must have $\sup|f^*| = c$. It is clear that, if $f = g$ a.e., then ess sup f = ess sup g, so that we can interpret ess sup as a functional on the subset L^∞ of the essentially bounded functions of the class of all measurable functions in which we have identified any two functions which coincide a.e. If we define

$$(af + bg)(x) = \begin{cases} af(x) + bg(x) \text{ when } f(x), g(x) \in R^1, \\ 0 \text{ otherwise,} \end{cases}$$

it is clear that L^∞ is a linear subspace over the reals of the class of all measurable functions.

(a) Show that

$$d_\infty(f,g) = \text{ess sup}|f - g|$$

defines a metric in L^∞ for

(i) $d_\infty(f,g) = d_\infty(g,f)$;

(ii) $d_\infty(f,g) = 0$ if and only if $f = g$ a.e.;

(iii) $d_\infty(f,g) \leqslant d_\infty(f,h) + d_\infty(h,g)$.

(b) Show that, if $(f_n)_{n=1}^\infty$ and f are in L^∞, then $d_\infty(f_n,f) \to 0$ as $n \to \infty$ if and only if $f_n \to f$ uniformly a.e.

(c) Is (L^∞, d_∞) a complete metric space? Is it separable?

[This problem has covered matters that will prove useful in the next chapter.]

9. Define what is meant by saying that a sequence $(f_n)_{n=1}^\infty$ of a.e. finite-valued measurable functions is a Cauchy sequence in the sense of almost uniform convergence and show that this implies the existence of a measurable limit function f such that $f_n \to f$ a. uniformly.

10. Consider the half-open interval $[0, 1)$ and let T be the transformation

$$T(x) = 2x (\text{mod } 1),$$

that is,

$$T(x) = \begin{cases} 2x & \text{if } 0 \leqq x < \frac{1}{2}, \\ 2x - 1 & \text{if } \frac{1}{2} \leqq x < 1. \end{cases}$$

Verify that the inverse image under T of every measurable subset of $[0, 1)$ is again measurable and that in fact the inverse image of every such measurable set has the same measure as the original set—in other words, T is a measure-preserving transformation.

11. Let f be a measurable, a.e. finite, real-valued function on the interval $[a, b]$. Show that there exists a monotone decreasing function g on $[a, b]$ such that for all real numbers r

$$\mu(\{t \in [a, b] : f(t) > r\}) = \mu(\{t \in [a, b] : g(t) > r\}).$$

12. Let f be as in the foregoing problem. Show the existence and uniqueness of a real number h such that

$$\mu(\{t \in [a,b] : f(t) \geqq h\}) \geqq \frac{b-a}{2},$$

$$\mu(\{t \in [a,b] : f(t) \geqq H\}) < \frac{b-a}{2} \qquad \text{for } H > h.$$

13. Let E be a measurable set and f be a measurable function on E.
    (a) Show that the sets $\{x \in E : f(x) = r\}$, with r any real number, are measurable by observing that

$$\{x \in E : f(x) = r\} = \bigcap_n \left\{x \in E : r - \tfrac{1}{n} \leqq f(x) < r + \tfrac{1}{n}\right\}.$$

    (b) Show that the sets $\{x \in E : f(x) = r\}$ cannot be used for a characterization of the measurability of f. [*Hint*: Consider a nonmeasurable set $A \subset [0, 1]$ and put

$$f(x) = \begin{cases} x \text{ on } A, \\ -x \text{ on } [0,1] - A. \end{cases}$$

For each r, the set $\{x \in [0,1] : f(x) = r\}$ consists of at most one point and hence is measurable. However, the set

$$\{x \in [0,1] : 0 \leqq f(x) \leqq 1\} = A$$

is nonmeasurable and thus f is nonmeasurable.]

14. Construct a continuous increasing function f: $[0, 1] \to [0, 1]$ which maps a set of measure $1/2$ onto a set of measure 0.
15. (a) Show that for any irrational α, there exist infinitely many rationals p/q such that $|\alpha - (p/q)| < 1/q^2$.
    [*Hints*: Let n be any positive integer. Consider the $n + 1$ real numbers

$$(*) \qquad 0, \alpha - [\alpha], 2\alpha - [2\alpha], \ldots, \quad n\alpha - [n\alpha]$$

(here $[r]$ denotes the greatest integer contained in r) and their distribution in the intervals $[(j/n), (j+1)/n)$ with $j = 0, 1, 2, \ldots, n - 1$. These n intervals cover the interval $[0, 1)$ and hence contain the $n + 1$ numbers (*). It is clear that two of the numbers (*) lie in the same interval; this is the "pigeonhole principle": if there are $n + 1$ objects in n boxes, there must be at least one box containing more than one object. Call these numbers $n_1\alpha - [n_1\alpha]$ and $n_2\alpha - [n_2\alpha]$ with $0 \leqq n_1 < n_2 \leqq n$. Since the intervals are of length $1/n$ and are not closed at both ends, we must have that

$$|n_2\alpha - [n_2\alpha] - n_1\alpha + [n_1\alpha]| < \frac{1}{n}.$$

Write q for the positive integer $n_2 - n_1$ and p for $[n_2\alpha] - [n_1\alpha]$; we have that $|q\alpha - p| < 1/n$ with $q \leqq n$. Hence, given any positive integer n, there exist integers q and p with $n \geqq q > 0$ such that

$$(**) \qquad |q\alpha - p| < \frac{1}{n} \text{ or } \left|\alpha - \frac{p}{q}\right| < \frac{1}{nq}.$$

The latter relation implies the inequality stated in the problem, because $n \geqq q$ implies that

$$\frac{1}{nq} \leqq \frac{1}{q^2}.$$

To show that there are infinitely many pairs (p, q), suppose, on the contrary, that there are only finitely many, say $(p_1, q_1), \ldots, (p_m, q_m)$. We prove that this supposition is false by finding another pair (p, q) satisfying (**). We define ε as the minimum of

$$\left|\alpha - \frac{p_1}{q_1}\right|, \ldots, \left|\alpha - \frac{p_m}{q_m}\right|.$$

Since α is irrational, ε is positive. Choose n so that $1/n < \varepsilon$, and then by the first part of the proof which led to (**) we can find a rational p/q so that

$$\left|\alpha - \frac{p}{q}\right| < \frac{1}{nq} \leqq \frac{1}{n} < \varepsilon.$$

By the definition of ε it follows that p/q is different from p_i/q_i for $i = 1, 2, \ldots, m$.]

(b) Let α be any irrational number. Then the set of all numbers of the form $\alpha x + y$, with x and y integers, is dense in R^1. (We have to show that given any real number r and any positive $\varepsilon > 0$, there are integers x and y such that $|\alpha x + y - r| < \varepsilon$ holds. From part (*a*) we see that the inequality $|\alpha x + y| < \varepsilon < 1$ can be satisfied by nonzero integers x and y. Hence let

$$\alpha x_1 + y_1 = \sigma, \quad |\sigma| < \varepsilon.$$

Let g be the largest integer contained in r/σ; then $g \leqq r/\sigma < g + 1$. Thus, putting $x = gx_1$ and $y = gy_1$, we get

$$|\alpha x + y - r| = |\alpha g x_1 + g y_1 - r| = |g\sigma - r| < |\sigma| < \varepsilon.)$$

16. Let D be a dense subset of R^1. Show that the following conditions on an extended real-valued function are equivalent:
    (i) f is measurable;
    (ii) $f^{-1}((r, \infty]) \in \mathfrak{M}$ for all $r \in D$;
    (iii) $f^{-1}([r, \infty]) \in \mathfrak{M}$ for all $r \in D$;
    (iv) $f^{-1}([-\infty, r)) \in \mathfrak{M}$ for all $r \in D$;
    (v) $f^{-1}([-\infty, r]) \in \mathfrak{M}$ for all $r \in D$.
17. Let f be a measurable function and g be a Baire function. Show that $g(f)$ is measurable. (Observe that this is a corollary of Problem 3; for a definition of Baire functions see Problem 7. It may be of interest to mention here yet another definition of the class of Baire functions on R^1. Let B_0 be the set of all continuous real-valued functions on R^1. If α is an ordinal number such that $0 < \alpha < \Omega$, with Ω the smallest nondenumerable ordinal number, define B_α to be the family of all functions f such that f is the pointwise limit of some sequence

$$(f_n) \subset \cup \{B_\beta : \beta \text{ is an ordinal number, } \beta < \alpha\}.$$

The functions in B_α are known as the *Baire functions of type* α. The class of Baire functions on R^1 is the set $\bigcup_{\alpha<\Omega} B_\alpha$.)

Chapter 3

The Lebesgue Integral on the Real Line R^1 and the Lebesgue Function Spaces

We first introduce the Lebesgue integral over a subset E of R^1 with $\mu(E) < \infty$.

Let $S(E)$ denote the set of all simple functions $s : E \to R^1$, where E has finite measure. For every $s \in S(E)$, its image $s(E)$ is a finite set of real numbers by definition. Let $c_1, \ldots, c_n$ denote the nonzero real numbers in $s(E)$. Then we have

$$s = \sum_{k=1}^{n} c_k \chi_{D_k}$$

where

$$D_k = s^{-1}(c_k) \qquad (k = 1, 2, \ldots, n).$$

This special representation for s as a linear combination of characteristic functions of measurable subsets of E we shall call the *canonical representation*; it is characterized by the fact that the subsets D_k are disjoint and the real numbers c_k are nonzero and distinct.

We define a function

$$I_E : S(E) \to R^1$$

by taking

$$I_E(s) = \sum_{k=1}^{n} c_k \mu(D_k)$$

for any $s \in S(E)$ with canonical representation and, for completeness, we also set

$$I_E(0) = 0.$$

PROPOSITION 1. *If $E_1, \ldots, E_m$ are disjoint measurable subsets of E with $\mu(E) < \infty$, then every linear combination*

$$s = \sum_{k=1}^{m} a_k \chi_{E_k}$$

with real coefficients $a_1, \ldots, a_m$ *is a simple function and*

$$I_E(s) = \sum_{k=1}^{m} a_k \mu(E_k).$$

Proof. It is clear that s is a simple function. Let $c_1, \ldots, c_n$ denote the nonzero real numbers in $s(E)$. For each $j = 1, 2, \ldots, n$, we let

$$D_j = \bigcup_{a_k = c_j} E_k = \{x \in E : s(x) = c_j\} = s^{-1}(c_j)$$

and therefore we have the canonical representation

$$s = \sum_{j=1}^{n} c_j \chi_{D_j}.$$

Thus, we obtain

$$I_E(s) = \sum_{j=1}^{n} c_j \mu(D_j) = \sum_{j=1}^{n} c_j [\sum_{a_k = c_j} \mu(E_k)]$$

(by the additivity of μ). Hence

$$I_E(s) = \sum_{k=1}^{m} a_k \mu(E_k).$$

This completes the proof.

PROPOSITION 2. *The function* $I_E : S(E) \to R^1$ *with* $\mu(E) < \infty$ *has the following properties*:

(*a*) *For every real number* c *and every simple function* $s : E \to R^1$, *we have*

$$I_E(cs) = cI_E(s)$$

(*b*) *For any two simple functions* $s, h : E \to R^1$ *we have*

$$I_E(s + h) = I_E(s) + I_E(h).$$

(*c*) *For any two simple functions* $s, h : E \to R^1$ *satisfying* $s \leqslant h$ *almost everywhere in* E, *we have*

$$I_E(s) \leqslant I_E(h).$$

Proof. Part (*a*) is trivial. To prove part (*b*), let $c_1, \ldots, c_m$ and $d_1, \ldots, d_n$, denote the nonzero numbers in $s(E)$ and $h(E)$, respectively. For each $k = 1, \ldots, m$ and each $j = 1, \ldots, n$, let

$$E_{kj} = s^{-1}(c_k) \cap h^{-1}(d_j),$$
$$e_{kj} = c_k + d_j.$$

Then $s + h = \sum_{k,j} e_{kj}\chi_{E_{kj}}$ and by Proposition 1 we have

$$\begin{aligned} I_E(s + h) &= \sum_{k,j} e_{kj}\mu(E_{kj}) = \sum_{k=1}^{m} c_k[\sum_{j=1}^{n} \mu(E_{kj})] + \sum_{j=1}^{n} d_j[\sum_{k=1}^{m} \mu(E_{kj})] \\ &= \sum_{k=1}^{m} c_k\mu[s^{-1}(c_k)] + \sum_{j=1}^{n} d_j\mu[h^{-1}(d_j)] \\ &= I_E(s) + I_E(h). \end{aligned}$$

To prove part (*c*) we observe that $I_E(\psi) \geqslant 0$ holds for every simple function $\psi : E \to R^1$ satisfying $\psi \geqslant 0$ almost everywhere in E. Hence we get

$$I_E(h) - I_E(s) = I_E(h - s) \geqslant 0$$

and the proof is finished.

Consider an arbitrarily given bounded function f: $E \to R^1$ on a subset E of R^1 with $\mu(E) < \infty$ and let $\alpha = \inf\{f(E)\}$ and $\beta = \sup\{f(E)\}$. Then the constant functions α and β on E are simple functions satisfying $\alpha \leqslant f \leqslant \beta$.

Consider the subset $L(f) = \{s \in S(E) : s \leqslant f\}$ of the set $S(E)$ of all simple functions on E. For any $s \in L(f)$ we have $s \leqslant f \leqslant \beta$ and by part (*c*) of Proposition 2, we have $I_E(s) \leqslant I_E(\beta) = \beta\mu(E)$. This implies that the nonempty subset $\{I_E(s) : s \in L(f)\}$ of R^1 has an upper bound $\beta\mu(E)$ in R^1 and hence

$$\sup\{I_E(s) : s \in L(f)\}$$

exists; we shall call it the *lower Lebesgue integral* of f over E and write $\underline{\int}_E f$. Similarly, by considering the subset

$$U(f) = \{h \in S(E) : h \geqslant f\}$$

of $S(E)$ we define the *upper Lebesgue integral* of f over E as the real number

$$\inf\{I_E(h) : h \in U(f)\}$$

and we write $\overline{\int}_E f$.

Since $\alpha \in L(f)$, $\beta \in U(f)$, and $s \leqslant f \leqslant h$ holds for every $s \in L(f)$ and every $h \in U(f)$, we obtain

$$\alpha\mu(E) \leqslant \underline{\int}_E f \leqslant \overline{\int}_E f \leqslant \beta\mu(E).$$

N.B. We shall show that an arbitrary bounded function $f\colon E \to R^1$, with $\mu(E) < \infty$ is measurable if and only if

$$\underline{\int}_E f = \overline{\int}_E f.$$

PROPOSITION 3. *For every simple function $f\colon E \to R^1$ with $\mu(E) < \infty$ we have*

$$\underline{\int}_E f = I_E(f) = \overline{\int}_E f.$$

Proof. Since f is a simple function we have that $f \in L(f)$ and $f \in U(f)$. Hence we get that

$$I_E(f) \leqslant \underline{\int}_E f \leqslant \overline{\int}_E f \leqq I_E(f).$$

But this amounts to what we set out to show.

PROPOSITION 4. *An arbitrary bounded function $f\colon E \to R^1$ with $\mu(E) < \infty$ is measurable if and only if*

$$\underline{\int}_E f = \overline{\int}_E f.$$

Proof. Necessity. Assume that $f\colon E \to R^1$ is a bounded measurable function. Pick real numbers α and β such that $\alpha \leqslant f(x) < \beta$ holds for every $x \in E$. Let n denote any natural number and let

$$\delta = \frac{\beta - \alpha}{n}.$$

Since f is measurable it follows that the subset

$$E_k = \{x \in E : \alpha + (k-1)\delta \leqslant f(x) < \alpha + k\delta\}$$

of E is measurable for every $k = 1, \ldots, n$. Since these subsets are disjoint and their union is E it follows that

$$\mu(E) = \sum_{k=1}^{n} \mu(E_k).$$

We define two simple functions $\phi_n, \psi_n : E \to R^1$ by setting

$$\phi_n = \sum_{k=1}^{n} [\alpha + (k-1)\delta]\chi_{E_k}$$

and

$$\psi_n = \sum_{k=1}^{n} [\alpha + k\delta]\chi_{E_k}.$$

Since $\phi_n(x) \leqslant f(x) < \psi_n(x)$ holds for every $x \in E$, we have $\phi_n \in L(f)$ and $\psi_n \in U(f)$. This implies

$$\underline{\int}_E f \geqslant I_E(\phi_n) = \sum_{k=1}^{n} [\alpha + (k-1)\delta]\mu(E_k)$$

$$\overline{\int}_E f \leqslant I_E(\psi_n) = \sum_{k=1}^{n} [\alpha + k\delta]\mu(E_k).$$

Hence

$$0 \leqslant \overline{\int}_E f - \underline{\int}_E f \leqslant I_E(\psi_n) - I_E(\phi_n) = \delta \sum_{k=1}^{n} \mu(E_k)$$

$$= \frac{\beta - \alpha}{n} \mu(E).$$

Since this holds for every $n \in N$ and $\mu(E) < \infty$ we must have that

$$\underline{\int}_E f = \overline{\int}_E f.$$

This proves the necessity.

Sufficiency. Assume that the upper and lower integral of f are equal. For every $n \in N$, there exists a $\phi_n \in L(f)$ and a $\psi_n \in U(f)$ such that

$$I_E(\phi_n) > \underline{\int}_E f - \frac{1}{n}, \quad I_E(\psi_n) < \overline{\int}_E f + \frac{1}{n}.$$

Hence we obtain that

$$0 \leqslant I_E(\psi_n - \phi_n) = I_E(\psi_n) - I_E(\phi_n) \leqq \frac{2}{n}.$$

We now define two functions $\phi, \psi \colon E \to R^1$ as follows:

$$\phi(x) = \sup_{n \in N}\{\phi_n(x)\}, \quad \psi(x) = \inf_{n \in N}\{\psi_n(x)\}.$$

We know that ϕ and ψ are measurable because ϕ_n and ψ_n are simple and hence measurable. Since $\phi(x) \leqslant f(x) \leqslant \psi(x)$ holds for every $x \in E$, it will be enough to show that ϕ and ψ are equal almost everywhere in E because this will then imply that f is measurable. Let m be an arbitrary natural number and consider the subset

$$D_m = \left\{x \in E : \psi(x) - \phi(x) > \frac{1}{m}\right\}$$

of E. Since $\phi_n \leqslant \phi \leqslant \psi \leqslant \psi_n$, this set D_m is contained in the subset

$$D_{mn} = \left\{x \in E : \psi_n(x) - \phi_n(x) > \frac{1}{m}\right\}$$

of E for every $n \in N$. Since

$$\frac{1}{m}\chi_{D_{mn}}(x) \leqslant \psi_n(x) - \phi_n(x)$$

holds for every $x \in E$, it follows from part (*c*) of Proposition 2 that

$$\frac{1}{m}\mu(D_{mn}) = I_E\left[\frac{1}{m}\chi_{D_{mn}}\right] \leqslant I_E(\psi_n - \phi_n) \leqslant \frac{2}{n}.$$

This implies that

$$\mu(D_m) \leqslant \mu(D_{mn}) \leqslant \frac{2m}{n}.$$

Since this holds for every $n \in N$, we must have

$$\mu(D_m) = 0.$$

Since this holds for every $m \in N$, it follows that the subset

$$D = \{x \in E : \psi(x) - \phi(x) > 0\} = \bigcup_{m=1}^{\infty} D_m$$

of E is of measure 0. Hence $\phi = \psi$ a.e. and the proof is finished.

For an arbitrary bounded measurable function f: $E \to R^1$ with $\mu(E) < \infty$, the common value of $\underline{\int}_E f$ and $\overline{\int}_E f$ is called the *Lebesgue integral*, or *integral* for short, of f over E and is denoted by $\int_E f$. If we denote by $BM(E)$ the set of all bounded measurable functions on E, then the mapping

$$\int_E : BM(E) \to R^1$$

is an extension of the function I_E, that is

$$\int_E \Big| S(E) = I_E,$$

where $S(E)$ is the set of all simple functions on E.

PROPOSITION 5. *If a bounded function* f: $[a,b] \to R^1$ *with* $b - a < \infty$ *is Riemann integrable on* $[a,b]$, *then* f *is measurable and*

$$\int_{[a,b]} f = \int_a^b f = (R) \int_a^b f(x)\,dx.$$

Proof. Let $S^{\#} = S^{\#}([a,b])$ denote the set of all step functions $\sigma\colon [a,b] \to R^1$. The lower Riemann integral and the upper Riemann integral of an arbitrary bounded function $f\colon [a,b] \to R^1$ can be defined as follows:

$$d = (R) \underline{\int_a^b} f(x)\,dx = \sup\{I_E(\sigma) : \sigma \in L^{\#}(f)\},$$

$$D = (R) \overline{\int_a^b} f(x)\,dx = \inf\{I_E(\sigma) : \sigma \in U^{\#}(f)\},$$

where $E = [a,b]$ and

$$L^{\#}(f) = \{\sigma \in S^{\#} : \sigma \leqslant f\}$$

$$U^{\#}(f) = \{\sigma \in S^{\#} : \sigma \geqslant f\}.$$

The function f is said to be Riemann integrable over $[a,b]$ if and only if $d = D$ and their common value is called the Riemann integral

$$(R) \int_a^b f(x)\,dx$$

of f over $[a,b]$. Since every step function is a simple function, we have

$$d = (R) \underline{\int_a^b} f(x)\,dx \leqslant \underline{\int_a^b} f \leqslant \overline{\int_a^b} f \leqslant (R) \overline{\int_a^b} f(x)\,dx = D.$$

This completes the proof.

N.B. Not every bounded measurable function on the interval $[a,b]$ is Riemann integrable, of course. For example, let $f\colon [a,b] \to R^1$ be the function

$$f(x) = \begin{cases} 1 & \text{if } x \text{ is rational} \\ 0 & \text{if } x \text{ is irrational}. \end{cases}$$

Then f is certainly bounded and measurable (because $f = 0$ a.e.) on $[a,b]$ and

$$\int_{[a,b]} f = 0.$$

However, $(R) \underline{\int}_a^b f(x)\,dx = 0$ and $(R) \overline{\int}_a^b f(x)\,dx = b - a$. We shall show later in this chapter that a bounded function f: $[a, b] \to R^1$ is Riemann integrable if and only if it is continuous almost everywhere in $[a, b]$.

PROPOSITION 6. *The function* $\int_E : BM(E) \to R^1$ *with* $\mu(E) < \infty$ *has the following properties*:

(*a*) *For every* $f \in BM(E)$ *we have the "mean value property"*

$$\inf\{f(E)\}\mu(E) \leqslant \int_E f \leqslant \sup\{f(E)\}\mu(E).$$

(*b*) *For every real number* c *and every* $f \in BM(E)$, *we have*

$$\int_E (cf) = c \int_E f.$$

(*c*) *For arbitrary* $f, g \in BM(E)$, *we have*

$$\int_E (f + g) = \int_E f + \int_E g.$$

(*d*) *For every* $f, g \in BM(E)$ *satisfying* $f \leqslant g$ *almost everywhere, we have*

$$\int_E f \leqslant \int_E g.$$

Moreover, we note that (c) *implies*: *If* $f \in BM(E)$, *then*, $f^+, f^- \in BM(E)$ *and* $\int_E f = \int_E f^+ - \int_E f^-$ *and* $\int_E |f| = \int_E f^+ + \int_E f^-$. *We observe also that* (*d*) *implies*: *If* $f, g \in BM(E)$ *and* $f = g$ *a.e. in* E, *then* $\int_E f = \int_E g$ *and if* $f \in BM(E)$, *then* $|\int_E f| \leqslant \int_E |f|$.

Proof. Properties (a) and (*b*) are obvious from the definition of the Lebesgue integral.

To verify property (*c*), take any $\varepsilon > 0$. There are simple functions $\phi \leqslant f$, $\psi \geqslant f$, $\gamma \leqslant g$, and $\eta \geqslant g$ satisfying

$$I_E(\phi) > \int_E f - \varepsilon, \quad I_E(\psi) < \int_E f + \varepsilon,$$
$$I_E(\gamma) > \int_E g - \varepsilon, \quad I_E(\eta) < \int_E g + \varepsilon.$$

Since $\phi + \gamma \leqslant f + g \leqslant \psi + \eta$, we have

$$\int_E (f + g) \geqslant I_E(\phi + \gamma) = I_E(\phi) + I_E(\gamma) > \int_E f + \int_E g - 2\varepsilon$$
$$\int_E (f + g) \leqslant I_E(\psi + \eta) = I_E(\psi) + I_E(\eta) < \int_E f + \int_E g + 2\varepsilon.$$

But these estimates hold for every $\varepsilon > 0$ and therefore we must have that $\int_E (f+g) = \int_E f + \int_E g$.

To prove (*d*) it is enough to establish that $\int_E (g-f) \geqslant 0$ holds. But for every simple function $\psi \geqslant g - f$ we have $\psi \geqslant 0$ almost everywhere in E. Hence $I_E(\psi) \geqslant 0$ by Proposition 2, part (*c*). Thus

$$\int_E (g-f) = \inf\{I_E(\psi) : \psi \in U(g-f)\} \geqslant 0.$$

N.B. For an arbitrary function $f\colon E \to R^1$, we let $f^+ = f \vee 0$, $f^- = -(f \wedge 0)$. Then we have $f = f^+ - f^-$, $|f| = f^+ + f^-$.

Remark. If E_1 and E_2 are disjoint measurable subsets of E and $E = E_1 \cup E_2$ and if $f \in BM(E)$, then $\int_E f = \int_{E_1} f + \int_{E_2} f$ because $\chi_{E_1 \cup E_2} = \chi_E = \chi_{E_1} + \chi_{E_2}$ and $\int_E$ is a linear function.

We note the following interesting application: If $\phi \geqslant 0$ on E, $\mu(E) < \infty$ and ϕ is integrable on E, then, for any positive real number r we have Chebyshev's Inequality:

$$\mu\{x \in E : \phi(x) \geqslant r\} < \frac{1}{r} \int_E \phi.$$

To see this, we let

$$E_0 = \{x \in E : \phi(x) \geqslant r\},$$

then

$$\int_E \phi = \int_{E_0} \phi + \int_{E-E_0} \phi \geqslant \int_{E_0} \phi \geqslant r\mu(E_0).$$

From the above inequality we have at once: If $\mu(E) < \infty$ and $\int_E |f|$ exists and is zero, then $f = 0$ a.e. on E.

Indeed, we have that $\mu\{x \in E : |f(x)| \geqslant 1/n\} \leqslant n \int_E |f| = 0$ for all $n \in N$. Hence

$$\mu\{x \in E : f(x) \neq 0\} \leqslant \sum_{n=1}^{\infty} \mu\{x \in E : |f(x)| \geqslant 1/n\} = 0.$$

We now establish the so-called "Bounded Convergence Theorem."

PROPOSITION 7. *Let f_n be a sequence of measurable functions defined on a set E of finite measure, and suppose that there is a real number M such that $|f_n(x)| \leqslant M$ for all n and all x. If*

$$f(x) = \lim_{n\to\infty} f_n(x)$$

for each x in E, then

$$\int_E f = \lim_{n\to\infty} \int_E f_n .$$

Proof. Recalling Proposition 15 of Chapter 2, we may, for the purpose of our present proof, assert that given $\varepsilon > 0$ there is an integer n_0 and a measurable set A with $\mu(E - A) < \varepsilon/4M$ such that for all $x \in A$ and $n \geqslant n_0$ we have $|f_n(x) - f(x)| < [\varepsilon/2\mu(E)]$. Thus

$$\left|\int_E f_n - \int_E f\right| = \left|\int_E (f_n - f)\right| \leqslant \int_E |f_n - f|$$
$$= \int_{E-A} |f_n - f| + \int_A |f_n - f| < \frac{\varepsilon}{2} + \frac{\varepsilon}{2} = \varepsilon.$$

Hence

$$\int_E f_n \to \int_E f$$

and the proof is complete.

We now consider the Lebesgue integral for nonnegative functions over any measurable set E.

Let f be a nonnegative measurable function defined on a measurable set E where $\mu(E)$ is not necessarily finite, we define

$$\int_E f = \sup_{h \leqslant f} \int_E h$$

where h is a bounded measurable function such that $\mu\{x : h(x) \neq 0\} < \infty$.

PROPOSITION 8. *If f and g are nonnegative measurable functions, then*

(*a*) $\int_E cf = c \int_E f,\ c > 0.$
(*b*) $\int_E (f + g) = \int_E f + \int_E g$
(*c*) *If $f \leqslant g$ a.e., then* $\int_E f \leqslant \int_E g$.

Proof. We shall only prove part (*b*). If $h(x) \leqslant f(x)$ and $k(x) \leqslant g(x)$, we have $h(x) + k(x) \leqslant f(x) + g(x)$ and so

$$\int_E h + \int_E k \leqslant \int_E (f + g).$$

Taking suprema we have

$$\int_E f + \int_E g \leqslant \int_E (f + g).$$

On the other hand, let v be a bounded measurable function which vanishes outside a set of finite measure and which is not greater than $f + g$. Then we define the functions h and k by setting

$$h(x) = \min\{f(x), v(x)\}$$

and

$$k(x) = v(x) - h(x).$$

We have $0 \leqslant h(x) \leqslant f(x)$ and $0 \leqslant k(x) \leqslant g(x)$ for $x \in E$ while h and k are bounded by the bound for v and vanish where v vanishes. Hence $\int_E v \leqslant \int_E h + \int_E k \leqslant \int_E f + \int_E g$, and so $\int_E f + \int_E g \geqslant \int_E (f + g)$. This completes the proof.

Next we establish "Fatou's Lemma."

PROPOSITION 9. *If f_n is a sequence of nonnegative measurable functions and $f_n(x) \to f(x)$ almost everywhere on a set E, then*

$$\int_E f \leqslant \varliminf_{n\to\infty} \int_E f_n .$$

Proof. Without loss of generality we may assume that the convergence is everywhere, since integrals over sets of measure zero can be seen to be zero. Let h be a bounded measurable function which is not greater than f and which vanishes outside a set E_0 of finite measure. Define a function h_n by setting

$$h_n(x) = \min\{h(x), f_n(x)\}.$$

Then h_n is bounded by the bound for h and vanishes outside E_0. Since $h_n(x) \to h(x)$ for each $x \in E_0$ we have by Proposition 7 that

$$\int_E h = \int_{E_0} h = \lim_{n\to\infty} \int_{E_0} h_n \leqslant \varliminf_{n\to\infty} \int_E f_n .$$

Taking the supremum over h we get

$$\int_E f \leqslant \varliminf_{n\to\infty} \int_E f_n$$

and the proof is finished.

We now consider the so-called "Monotone Convergence Theorem."

PROPOSITION 10. *Let f_n be an increasing sequence of nonnegative measurable functions, and let*

$$f = \lim_{n\to\infty} f_n$$

on E. Then

$$\int_E f = \lim_{n\to\infty} \int_E f_n .$$

Proof. By Proposition 9 we have that

$$\int_E f \leqslant \varliminf_{n\to\infty} \int_E f_n .$$

But for each n we have $f_n \leqslant f$ and so $\int_E f_n \leqslant \int_E f$. This however implies that

$$\varlimsup_{n\to\infty} \int_E f_n \leqslant \int_E f.$$

Hence

$$\int_E f = \lim_{n\to\infty} \int_E f_n$$

and the proof is complete.

Remark. Proposition 10 has the following direct consequences:

If u_k is a sequence of nonnegative measurable functions on E and

$$f = \sum_{k=1}^{\infty} u_k,$$

then

$$\int_E f = \sum_{k=1}^{\infty} \int_E u_k .$$

If f is a nonnegative function and E_n is a disjoint sequence of measurable sets with

$$E = \bigcup_{n=1}^{\infty} E_n,$$

then

$$\int_E f = \sum_{n=1}^{\infty} \int_{E_n} f.$$

(Indeed, let $u_k = f \cdot \chi_{E_k}$. Then $f \cdot \chi_E = \sum u_k$ and everything is clear by the first part of the Remark.)

BASIC DEFINITION. *A nonnegative measurable function f is called integrable over the measurable set E if $\int_E f$ exists and is finite.*

PROPOSITION 11. *Let f and g be two nonnegative measurable functions. If f is integrable over E and $g(x) < f(x)$ on E, then g is also integrable on E and*

$$\int_E (f-g) = \int_E f - \int_E g.$$

Proof. By Proposition 8

$$\int_E f = \int_E (f-g) + \int_E g.$$

Since the left-hand side is finite, the terms on the right-hand side must also be finite (they are nonnegative) and so g is integrable.

We show now that the Lebesgue integral is "absolutely continuous."

PROPOSITION 12. *Let f be a nonnegative function which is integrable over a set E. Then given $\varepsilon > 0$ there is a $\delta > 0$ such that for every set $A \subset E$ with $\mu(A) < \delta$ we have $\int_A f < \varepsilon$.*

Proof. Suppose not. Then for some $\varepsilon > 0$ we can find sets A of arbitrarily small measure such that $\int_A f \geqslant \varepsilon$, and in particular there must be a measurable set A_n for each n such that $\int_{A_n} f \geqslant \varepsilon$ and $\mu(A_n) < 1/2^n$. Let $g_n = f \cdot \chi_{A_n}$. Then $g_n \to 0$ except on the set

$$\bigcap_{n=1}^{\infty} \bigcup_{k=n}^{\infty} A_k.$$

Since

$$\mu[\bigcup_{k=n}^{\infty} A_k] < 1/2^{n-1},$$

we have $g_n \to 0$ a.e. Let $f_n = f - g_n$. Then f_n is a sequence of nonnegative functions and $f_n \to f$ a.e. By Proposition 9 we have

$$\int_E f \leqslant \varliminf_{n\to\infty} \int_E f_n \leqslant \int_E f - \varlimsup_{n\to\infty} \int_E g_n \leqslant \int_E f - \varepsilon.$$

But this can hold only in case $\int_E f = \infty$ and so f is not integrable. Q.E.D.

Now we consider the general Lebesgue integral and say that a measurable function $f = E \to R_e^1$ is *Lebesgue integrable*, or *integrable*, for short, over E if both f^+ and f^- are integrable over E. In this case we define

$$\int_E f = \int_E f^+ - \int_E f^-.$$

It is clear that if f is integrable, so is $|f|$, because $|f| = f^+ + f^-$. However, the converse is false. For example, let $E = [0, 1]$ and E_0 be some nonmeasurable subset of E. Put

$$f(x) = \begin{cases} 1 & \text{if } x \in E_0, \\ -1 & \text{if } x \in E - E_0. \end{cases}$$

Then f even fails to be measurable.

A direct consequence of this definition is the next proposition.

PROPOSITION 13. *Let f and g be integrable over E.*

(*a*) *The function $(af + bg)$ with $a, b \in R^1$, is integrable over E and*

$$\int_E (af + bg) = a \int_E f + b \int_E g.$$

(*b*) *If $f \leqslant g$ a.e. on E, then $\int_E f \leqslant \int_E g$.*

(*c*) *If A and B are disjoint measurable subsets of E, then*

$$\int_{A \cup B} f = \int_A f + \int_B f.$$

Remark. If f is integrable over E, then f must be finite a.e.

Indeed, let $A = f^{-1}(\{\infty\})$ and $B = f^{-1}(\{-\infty\})$. To see that $\mu(A) = 0$, let n be any natural number. Since $n\chi_A \leqslant f^+$, we have

$$\int_E f^+ \geqslant I_E(n\chi_A) = n\mu(A).$$

Since f^+ is integrable and this holds for all $n \in N$, we have $\mu(A) = 0$. Similarly, we can show that $\mu(B) = 0$. It is also clear that if $\mu(E) = 0$, then $\int_E f = 0$ for any f.

PROPOSITION 14. *Let $f \geqslant 0$ and integrable over E. If $\int_E f = 0$, then $f = 0$ a.e.*

Proof. If $\{x \in f(x) > 0\}$ has positive measure, then there is an integer n such that, if

$$A = \left\{x \in E : f(x) > \frac{1}{n}\right\},$$

then $\mu(A) > 0$ by the continuity of measure. But

$$\frac{1}{n}\chi_A \leqslant f \circ \chi_A \leqslant f$$

so that

$$\int_E f \geqslant \frac{1}{n} \int_E \chi_A = \frac{1}{n}\mu(A) > 0.$$

Hence, if $f \geqslant 0$ and $\int_E f = 0$, we must have

$$\mu\{x \in E : f(x) > 0\} = 0. \qquad \text{Q.E.D.}$$

We now establish the so-called "Dominated Convergence Theorem."

PROPOSITION 15. *Let g be integrable over E and let $(f_n)_{n=1}^{\infty}$ be a sequence of measurable functions such that $|f_n| \leqslant g$ on E for all $n \in N$ and suppose $(f_n)_{n=1}^{\infty}$ converges to some function f: $E \to R_e^1$ almost everywhere on E. Then f is integrable and*

$$\int_E f = \lim_{n\to\infty} \int_E f_n.$$

Proof. The function $g - f_n$ is nonnegative and so by Proposition 9

$$\int_E (g - f) \leqslant \varliminf_{n\to\infty} \int_E (g - f_n).$$

Since $|f| \leqslant g$, f is integrable and we have

$$\int_E g - \int_E f \leqslant \int_E g - \varlimsup_{n\to\infty} \int_E f_n$$

so that

$$\int_E f \geqslant \varlimsup_{n\to\infty} \int_E f_n.$$

Similarly, considering $g + f_n$ we get

$$\int_E f \leqslant \varliminf_{n\to\infty} \int_E f_n$$

and the theorem follows.

N.B. The foregoing proposition easily yields the fact that the Lebesgue integral is countably additive, i.e., if $(E_n)_{n=1}^{\infty}$ is a sequence of disjoint measurable sets such that

$$E = \bigcup_{n=1}^{\infty} E_n,$$

then

$$\int_E f = \sum_{n=1}^{\infty} \int_{E_n} f$$

holds for every measurable function $f: E \to R_e^1$ which is Lebesgue integrable.

Let us also note in passing that the presence of the dominating function g in Proposition 15 is essential. For example, if $E = R^1$ and

$$f_n(x) = \begin{cases} n & \text{if } x \in (0, 1/n], \\ 0 & \text{elsewhere,} \end{cases}$$

then $\int_E f_n = n(1/n) = 1$ for all $n \in N$ while $f_n(x) \to 0$ as $n \to \infty$ for all $x \in R^1$. Thus

$$\lim_{n\to\infty} \int_E f_n = 1 \neq 0 = \int_E \lim_{n\to\infty} f_n .$$

An interesting application of Proposition 15 is the following:

Let f be a real-valued function on $[a, b]$ such that the derivative f' exists everywhere on $[a, b]$. If f' is bounded, then f' is integrable and

$$f(x) - f(a) = \int_a^x f'(t)\, dt \text{ for } a \leqslant x \leqslant b.$$

Proof. Since f' is finite on $[a, b]$ f must be continuous. We extend the domain of definition of f to $[a, b + 1]$ by letting

$$f(x) = f(b) + (x - b) f'(b)$$

for $b < x \leqslant b + 1$.

It is clear that the extended function is continuous on $[a, b + 1]$ and has finite derivative on $[a, b + 1]$.

For $x \in [a, b]$ and $n \in N$ we set

$$\phi_n(x) = n\left[f\left(x + \frac{1}{n}\right) - f(x)\right].$$

For every $x \in [a, b]$ we have

$$\lim_{n\to\infty} \phi_n(x) = f'(x).$$

Since each function ϕ_n is continuous and therefore measurable, f' is measurable. Since f' is also bounded, f' is integrable.

By the mean-value theorem of ordinary calculus

$$\phi_n(x) = n\left[f\left(x+\tfrac{1}{n}\right) - f(x)\right] = f'\left(x+\tfrac{\theta}{n}\right) \qquad (0<\theta<1),$$

and therefore all functions ϕ_n are bounded by one number. Thus

$$\int_a^b f'(x)\,dx = \lim_{n\to\infty} \int_a^b \phi_n(x)\,dx.$$

But

$$\int_a^b \phi_n(x)\,dx = n\int_a^b f\left(x+\tfrac{1}{n}\right)dx - n\int_a^b f(x)\,dx$$

$$= n\int_b^{b+1/n} f(x)\,dx - n\int_a^{a+1/n} f(x)\,dx.$$

Applying the mean-value theorem once again, we get

$$\int_a^b \phi_n(x)\,dx = f\left(b+\tfrac{\theta_n'}{n}\right) - f\left(a-\tfrac{\theta_n''}{n}\right)$$

$$(0<\theta_n'<1;\quad 0<\theta_n''<1),$$

and, since f is continuous, we obtain

$$\lim_{n\to\infty} \int_a^b \theta_n(x)\,dx = f(b) - f(a).$$

This means that

$$f(b) - f(a) = \int_a^b f'(x)\,dx.$$

Replacing b by an arbitrary x in $[a, b]$, we obtain what we set out to do and the proof is finished.

Let $f\colon E \to R_e^1$ and $f_n : E \to R_e^1$ $(n = 1, 2, \ldots)$ be measurable functions. We say that $(f_n)_{n=1}^{\infty}$ *converges in measure* to f on E if, for each $\varepsilon > 0$, there exists an integer n_0 such that for all $n \geqslant n_0$ we have

$$\lim_{n\to\infty} \mu(\{x \in E : |f(x) - f_n(x)| \geqslant \varepsilon\}) = 0.$$

Note that the foregoing definition only makes sense for measurable functions which are finite a.e. and that the limit in measure is unique in the sense that if $f_n \to f$ in measure and $f_n \to g$ in measure, then $f = g$ a.e.

It is a straightforward exercise to verify that almost uniform convergence implies convergence in measure, i.e., if f and $(f_n)_{n=1}^{\infty}$ are measurable functions which are finite almost everywhere, then $f_n \to f$ in measure, if $f_n \to f$ almost uniformly.

PROPOSITION 16. (*Theorem of F. Riesz*) *Let $(f_n)_{n=1}^{\infty}$ be a sequence of measurable functions which converges in measure to f on E. Then there exists a subsequence $(f_{n_k})_{k=1}^{\infty}$ which converges to f almost everywhere.*

Proof. Choose $n_1 \in N$ such that

$$\mu(\{x \in E : |f(x) - f_{n_1}(x)| \geqslant 1\}) < \frac{1}{2}.$$

Suppose that $n_1, n_2, \ldots, n_k$ have been chosen. Then pick n_{k+1} such that $n_{k+1} > n_k$ and

$$\mu\left(\left\{x \in E : |f(x) - f_{n_{k+1}}(x)| \geqslant \frac{1}{k+1}\right\}\right) < \frac{1}{2^{k+1}}.$$

Let

$$A_j = \bigcup_{k=j}^{\infty} \left\{x \in E : |f(x) - f_{n_k}(x)| \geqslant \frac{1}{k}\right\}$$

for each $j \in N$. Clearly we have $A_1 \supset A_2 \supset A_3 \supset \ldots$. Next let

$$B = \bigcap_{j=1}^{\infty} A_j.$$

Since

$$\mu(A_1) < \sum_{k=1}^{\infty} (1/2^k) < \infty,$$

it follows that

$$\mu(B) = \lim_{j\to\infty} \mu(A_j) \leqslant \lim_{j\to\infty} \sum_{k=j}^{\infty} (1/2^k) = 0,$$

that is, $\mu(B) = 0$. Next, let

$$x \in E - B = \bigcup_{j=1}^{\infty} (E - A_j).$$

Then there is a j_x such that

$$x \in E - A_{j_x} = \bigcap_{k=j_x}^{\infty} \left\{t \in E : |f(t) - f_{n_k}(t)| < \frac{1}{k}\right\}.$$

Given $\varepsilon > 0$, choose k_0 such that $k_0 \geqslant j_x$ and $1/k_0 \leqslant \varepsilon$. Then $k \geqslant k_0$ implies that $|f(x) - f_{n_k}(x)| < 1/k \leqslant \varepsilon$ and this proves that

$$f_{n_k}(x) \to f(x)$$

for all $x \in E - B$. The proof is finished.

Remark. There are sequences of functions that converge in measure and do not converge almost everywhere.

For example, let $E = [0, 1]$ and let $f_n = \chi_{[(j/2^k),(j+1)/2^k]}$, where $n = 2^k + j$, $0 \leqslant j < 2^k$. Evidently,

$$\mu(\{x \in [0, 1] : |f_n(x)| \geqslant \varepsilon\}) \leqslant \frac{1}{2^k} \to 0$$

as $n \to \infty$ for every $\varepsilon > 0$. Thus $f_n \to 0$ in measure. On the other hand, if $x \in [0, 1]$, the sequence $(f_n(x))_{n=1}^{\infty}$ contains an infinite number of zeros and an infinite number of ones. Thus the sequence of functions in question converges nowhere on $[0, 1]$.

We now consider certain spaces of functions, called *L^p-spaces*, the construction of which is based on the theory of integration.

Let p be a positive real number, and let f be a real-valued measurable function defined a.e. on a measurable set E of R^1. Suppose that $|f|^p$ is integrable on E. We define the symbol $\|f\|_p$ by

$$\|f\|_p = (\int_E |f|^p)^{1/p}.$$

A measurable function defined a.e. on a measurable set E of R^1 is said to be of class $L^p(E)$ if and only if $|f|^p$ is integrable over E. (Note that f is integrable over E if and only if $|f|$ is, i.e., if and only if $f \in L^1(E)$, because f is measurable.)

Let X be a linear space over R^1. Suppose that there is a function $x \to \|x\|$ with domain of definition X and range contained in R^1 such that

(i) $\|0\| = 0$ and $\|x\| > 0$ if x is not the zero vector 0 of the space X;
(ii) $\|\alpha x\| = |\alpha| \cdot \|x\|$ for all $x \in X$ and $\alpha \in R^1$;
(iii) $\|x + y\| \leqslant \|x\| + \|y\|$ for all $x, y \in X$.

The pair $(X, \| \ \|)$ is called a *real normed linear space* and $\| \ \|$ is called a *norm.*

We shall see that for $1 \leqslant p < \infty$, the function $f \to \|f\|_p$ on $L^p(E)$ satisfies all the axioms for a norm set down in the foregoing definition, except for the requirement: $\|f\|_p > 0$ if $f \neq 0$. (Note that if $A \subset E$ and $\mu(A) = 0$, then $\chi_A \neq 0$ but $\|\chi_A\|_p = 0$.) We shall refer to $\|f\|_p$ as the *L^p-norm* of f and agree to identify functions that are equal almost everywhere.

For any positive real number p such that $p \neq 1$, we define $q = p/(p-1)$ (thus $1/p + 1/q = 1$) and we say that p and q are *conjugate indices.*

PROPOSITION 17. *If $p > 0$ and a and b are any nonnegative real numbers, then*

$$ab \leqslant \frac{a^p}{p} + \frac{b^q}{q}.$$

Equality only occurs if $a^p = b^q$.

Proof. By putting $p = 1/\alpha > 0$ and $q = 1/\beta > 0$, $a = x^\alpha$, $b = y^\beta$, we see that the proposition is a direct consequence of the following algebraic inequality: If $x > 0$, $y > 0$, $\alpha > 0$, $\beta > 0$, $\alpha + \beta = 1$, then

$$x^\alpha y^\beta \leqslant \alpha x + \beta y,$$

where the inequality is strict unless $x = y$. To verify the inequality we observe that

$$\alpha \log x + \beta \log y \leqslant \log(\alpha x + \beta y)$$

because the logarithmic function is strictly convex.

Another method to establish the foregoing algebraic inequality is to consider the function

$$\phi(t) = t^\gamma - \gamma t$$

for $t > 0$ and $0 < \gamma < 1$. One sees from elementary calculus that $\phi(t) \leqslant \phi(1)$.

We now establish Hölder's Inequality for $p > 1$.

PROPOSITION 18. *Let $f \in L^p(E)$ and $g \in L^q(E)$, where $p > 1$. Then $fg \in L^1(E)$, and we have*

(*i*)
$$\left|\int_E fg\right| \leqslant \int_E |fg|,$$

and

(*ii*)
$$\int_E |fg| \leqslant \|f\|_p \|g\|_q;$$

and so also

(*iii*)
$$\left|\int_E fg\right| \leqslant \|f\|_p \|g\|_q.$$

Proof. We begin with the proof of (ii). If f or g is zero a.e., then (ii) is obvious. Otherwise, using Proposition 17, we have that

$$\frac{|f(t)|}{\|f\|_p}\frac{|g(t)|}{\|g\|_q} \leqslant \frac{1}{p}\frac{|f(t)|^p}{\|f\|_p^p} + \frac{1}{q}\frac{|g(t)|^q}{\|g\|_q^q}$$

for all $t \in E$ such that $f(t)$ and $g(t)$ are defined, i.e., for almost all t in E. Thus we have

$$\frac{1}{\|f\|_p\|g\|_q}\int_E |fg| \leqslant \frac{1}{p\|f\|_p^p}\int_E |f|^p + \frac{1}{q\|g\|_q^q}\int_E |g|^q = \frac{1}{p} + \frac{1}{q} = 1$$

and this proves (ii).

Inequalities (i) and (iii) are immediate consequences of inequality (ii).

Remark. In Proposition 18 we get equality in (ii) if and only if

$$\frac{|f(t)|}{\|f\|_p}\frac{|g(t)|}{\|g\|_q} = \frac{1}{p}\frac{|f(t)|^p}{\|f\|_p^p} + \frac{1}{q}\frac{|g(t)|^q}{\|g\|_q^q}$$

for almost all $t \in E$. By Proposition 17 this happens if and only if

$$\frac{|f|^p}{\|f\|_p^p} = \frac{|g|^q}{\|g\|_q^q}$$

almost everywhere.

We now prove Minkowski's Inequality.

PROPOSITION 19. *For $1 \leqslant p < \infty$ and $f, g \in L^p(E)$, we have*

$$\|f + g\|_p \leqslant \|f\|_p + \|g\|_p.$$

Proof. The inequality is trivial for $p = 1$. We suppose that $p > 1$. Then

$$\begin{aligned}|f + g|^p &\leqslant (|f| + |g|)^p \leqslant [2\max\{|f|, |g|\}]^p = 2^p\max\{|f|^p, |g|^p\}\\ &\leqslant 2^p(|f|^p + |g|^p).\end{aligned}$$

This shows that $|f + g|^p$ is in $L^1(E)$, i.e., $f + g \in L^p(E)$. By Proposition 18 we get

$$\begin{aligned}\|f + g\|_p^p &= \int_E |f + g|^p \leqslant \int_E |f + g|^{p-1}|f| + \int_E |f + g|^{p-1}|g|\\ &\leqslant \Big(\int_E |f|^p\Big)^{1/p}\Big(\int_E |f + g|^{(p-1)q}\Big)^{1/q}\\ &\quad + \Big(\int_E |g|^p\Big)^{1/p}\Big(\int_E |f + g|^{(p-1)q}\Big)^{1/q}\\ &= (\|f\|_p + \|g\|_p)\|f + g\|_p^{p/q}.\end{aligned}$$

This shows that

$$\|f+g\|_p^{p-p/q} \leqslant \|f\|_p + \|g\|_p.$$

But $p - p/q = 1$ and so the inequality also holds when $p > 1$ and the proof is complete.

Since $\|\alpha f\|_p = |\alpha| \cdot \|f\|_p$ is obvious, we can summarize the foregoing in the next proposition.

PROPOSITION 20. *For $1 \leqslant p < \infty$, $L^p(E)$ is a normed linear space over R^1, where we agree that $f = g$ means $f(x) = g(x)$ for almost all x in E.*

PROPOSITION 21. *For $1 \leqslant p < \infty$, $L^p(E)$ with the metric*

$$d(f, g) = \|f - g\|_p$$

is a complete metric space.

Proof. Let $(f_n)_{n=1}^{\infty}$ be a Cauchy sequence in $L^p(E)$, i.e., given any $\varepsilon > 0$, there is a natural number n_0 such that for all $n > n_0$ and all $m > n_0$ we have $\|f_n - f_m\|_p < \varepsilon$. The sequence of numbers $(f_n(x))_{n=1}^{\infty}$ may converge at no point x of E (as the example following Proposition 16 illustrates) but we can find a subsequence of $(f_n)_{n=1}^{\infty}$ that does converge a.e. in E. Indeed, we choose $(f_{n_j})_{j=1}^{\infty}$ to be any subsequence of $(f_n)_{n=1}^{\infty}$ for which $n_1 < n_2 < \ldots < n_j < \ldots$ and

$$\sum_{j=1}^{\infty} \|f_{n_{j+1}} - f_{n_j}\|_p = \alpha < \infty.$$

This can be done; for example, we can select an increasing sequence of n_j's such that $\|f_m - f_{n_j}\|_p < 1/2^j$ for all $m \geqslant n_j$. Next, we define

$$g_j = |f_{n_1}| + |f_{n_2} - f_{n_1}| + \ldots + |f_{n_{j+1}} - f_{n_j}|$$

for $k = 1, 2, \ldots$. It is clear that

$$\|g_j^p\|_1 = \|g_j\|_p^p = (\| |f_{n_1}| + |f_{n_2} - f_{n_1}| + \ldots + |f_{n_{j+1}} - f_{n_j}| \|_p)^p$$

$$\leqslant (\|f_{n_1}\|_p + \sum_{k=1}^{j} \|f_{n_{k+1}} - f_{n_k}\|_p)^p \leqslant (\|f_{n_1}\|_p + \alpha)^p < \infty.$$

Let

$$g = \lim_{j \to \infty} g_j.$$

By the Monotone Convergence Theorem and the above estimate

$$\int_E g^p = \int_E \lim_{j\to\infty} g_j^p = \lim_{j\to\infty} \int_E g_j^p < \infty.$$

This shows that g is in $L^p(E)$, i.e.,

$$\int_E [|f_{n_1}| + \sum_{k=1}^{\infty} |f_{n_{k+1}} - f_{n_k}|]^p < \infty.$$

The nonnegative integrand in the foregoing integral must be finite a.e. in E and therefore the series

$$\sum_{k=1}^{\infty} |f_{n_{k+1}} - f_{n_k}|$$

converges a.e. in E. Evidently the series

$$f_{n_1}(x) + \sum_{k=1}^{\infty} (f_{n_{k+1}}(x) - f_{n_k}(x))$$

also converges a.e. in E. The j-th partial sum of this series is $f_{n_{j+1}}(x)$, and so the sequence $(f_{n_j}(x))_{j=1}^{\infty}$ converges a.e. in E to a real number $f(x)$. Define $f(x)$ to be 0 for all other x in E. It is clear that f is a measurable real-valued function on all of E.

We now show that f is the limit in $L^p(E)$ of the sequence $(f_n)_{n=1}^{\infty}$ and this will then prove that $L^p(E)$ is complete in the metric d induced by the L^p-norm $\| \ \|_p$. Given $\varepsilon > 0$, let i be so large that

$$\|f_s - f_t\|_p < \varepsilon \text{ for } s, t \geqslant n_i.$$

Then, for $j \geqslant i$ and $m > n_i$, we have

$$\|f_m - f_{n_j}\|_p < \varepsilon.$$

By the generalized Fatou's Lemma (see Exercise 12 at the end of this chapter) we get

$$\int_E |f - f_m|^p = \int_E \varliminf_{j\to\infty} |f_{n_j} - f_m|^p \leqslant \varliminf_{j\to\infty} \int_E |f_{n_j} - f_m|^p \leqslant \varepsilon^p.$$

Hence, for each $m > n_i$, the function $f - f_m$ belongs to $L^p(E)$ and so $f = (f - f_m) + f_m$ belongs to $L^p(E)$. Moreover,

$$\lim_{n\to\infty} \|f - f_n\|_p = 0.$$

The proof is finished.

PROPOSITION 22. *If* $\mu(E) < \infty$ *and* $0 < p < q < \infty$, *then* $L^q(E) \subset L^p(E)$ *and*

$$\|f\|_p \leqslant \|f\|_q(\mu(E))^{(1/p)-(1/q)}$$

holds for $f \in L^q(E)$.

Proof. Let $r = q/p > 1$ and $s = r/(r-1)$. For any $f \in L^q(E)$ we have

$$\int_E |f|^p \leqslant (\int_E |f|^{pr})^{1/r}(\int_E 1^s)^{1/s}$$
$$= (\int_E |f|^q)^{p/q}(\mu(E))^{(q-p)/q}.$$

It follows that $f \in L^q(E)$, and that

$$\|f\|_p \leqslant \|f\|_q(\mu(E))^{(q-p)/pq} = \|f\|_q(\mu(E))^{(1/p)-(1/q)}.$$

This finishes the proof.

In Exercise 8 of Chapter 2 we introduced the space $L^\infty(E)$ of all essentially bounded measurable functions on a measurable set E with the metric d_∞. Letting $d_\infty(f,g) = \|f - g\|_\infty$, the norm $\|f\|_\infty$ will be the essential supremum of $|f|$.

PROPOSITION 23. *Suppose that* $f \in L^1(E)$ *and* g *is essentially bounded. Then* $fg \in L^1(E)$ *and*

$$|\int_E fg| \leqslant \|g\|_\infty \|f\|_1 .$$

Proof. Let a be a nonnegative real number and

$$A = \{x \in E : |g(x)| > a\}.$$

Define also the sets

$$B = \{x \in E : |f(x)| > 0\}$$

and

$$B_n = \{x \in E : |f(x)| \geqslant 1/n\}$$

for $n \in N$. Obviously

$$B = \bigcup_{n=1}^{\infty} B_n$$

and $\mu(B_n) < \infty$ for all $n \in N$. By the Monotone Convergence Theorem, we have

$$\int_E |fg| = \int_{E-B} |fg| + \lim_{n\to\infty} \int_{B_n \cap A} |fg| + \int_{B\cap(E-A)} |fg|$$

$$\leqslant 0 + 0 + a \int_{B\cap(E-A)} |f| \leqslant a \int_E |f| = a\|f\|_1.$$

Taking the infimum over all such a, we obtain what we have set out to do.

Remarks. Toward the end of Chapter 5 we shall come back to the subject of L^p-spaces and look at some deeper aspects of their structure.

In the following addendum to this chapter we wish to compare the Riemann and the Lebesgue integral and briefly discuss two other matters of interest.

Addendum

We start with some preliminaries.

Let f: $R^1 \to R^1$ be a bounded function. If J is any bounded open interval in R^1, we define

$$\omega(f; J) = \sup_{x\in J} f(x) - \inf_{x\in J} f(x).$$

We call $\omega(f; J)$ the oscillation of f over J.

It is clear that $\omega(f; J) \geqslant 0$ for any interval J. It is also easy to see that the oscillation is a monotone mapping in the sense that if J_1 and J_2 are bounded open intervals and $J_1 \subset J_2$, then $\omega(f; J_1) \leqslant \omega(f; J_2)$.

If $a \in R^1$ we define

$$\omega(f; a) = \inf \omega(f; J),$$

where the infimum is taken over all bounded open intervals J containing a.

One can easily see that if f is continuous at $a \in R^1$, then $\omega(f; a) = 0$, and if f is not continuous at a, then $\omega(f; a) > 0$.

Suppose that $r > 0$ and let E_r be the set of all $a \in R^1$ such that $\omega(f; a) \geqslant 1/r$. We want to verify that E_r is a closed set.

Let x be any accumulation point of E_r. We have to show that $x \in E_r$, that is, we must show that $\omega(f; x) \geqslant 1/r$. To do this, it is enough to verify that if J is a bounded open interval containing x, then $\omega(f; J) \geqslant 1/r$, since $\omega(f; x)$ is the infimum of such $\omega(f; J)$. Since the open interval J contains the accumulation point x of E_r, J must contain a point y of E_r. But then $\omega(f; J) \geqslant \omega(f; y) \geqslant 1/r$ and we have what we set out to do.

Remark. The set D of points in R^1 at which f is not continuous is of type F_σ.

Indeed, if $x \in D$, then $\omega(f; x) > 0$. For some natural number n we must have $\omega(f; x) \geqslant 1/n$. This shows that

$$D \subset \bigcup_{n=1}^{\infty} E_{1/n}$$

when $E_{1/n}$ is defined as above. Conversely, if

$$x \in \bigcup_{n=1}^{\infty} E_{1/n},$$

then $\omega(f; x) > 0$ and so $x \in D$. Hence

$$D = \bigcup_{n=1}^{\infty} E_{1/n}.$$

But each $E_{1/n}$ is closed and therefore D is a countable union of closed sets.

After these preparations we now take up the following characterization of the Riemann integral:

THEOREM 1. *Let f be a bounded function on the closed bounded interval $[a, b]$. Then f is Riemann integrable if and only if f is continuous at almost every point in $[a, b]$.*

Proof. Suppose first that f is Riemann integrable on $[a, b]$. We wish to verify that the set E of points in $[a, b]$ at which f is not continuous is of measure zero. But $x \in E$ if and only if $\omega(f; x) > 0$. Hence

$$E = \bigcup_{m=1}^{\infty} E_m,$$

where E_m is the set of all $x \in [a, b]$ such that $\omega(f; x) \geqslant 1/m$. To see that E is of measure zero, it is sufficient to prove that each E_m is of measure zero.

We take a fixed m. Since f is Riemann integrable, given $\varepsilon > 0$ there is a subdivision τ of $[a, b]$ such that the difference of the upper and the lower Riemann sums

$$U(f; \tau) - L(f; \tau)$$

is less than $\varepsilon/2m$. Thus, if $I_1, \ldots, I_n$ are the closed component intervals of τ, we have

$$\sum_{k=1}^{n} \omega(f; I_k) \cdot |I_k| = U(f; \sigma) - L(f; \tau)$$

and hence

$$\sum_{k=1}^{n} \omega(f; I_k) \cdot |I_k| < \frac{\varepsilon}{2m}$$

where $|I_k|$ denotes the length of the interval I_k. Now $E_m = E_m^* \cup E_m^{**}$, where E_m^* is the set of points of E_m which are points of the subdivision τ, and $E_m^{**} = E - E_m^*$. Evidently, $E_m^* \subset J_1 \cup \ldots \cup J_p$, where the J_j are open subintervals such that $|J_1| + \ldots + |J_p| < \varepsilon/2$, since there is only a finite number of points of subdivision.

But if $x \in E_m^{**}$, then x is an interior point of some I_k. Hence

$$\omega(f; I_k) \geqslant \omega(f; x) \geqslant \frac{1}{m}.$$

If we denote by $I_{k_1}, \ldots, I_{k_r}$ those component intervals of τ which contain a point of E_m^{**} (in their interior), we have

$$\frac{1}{m}(|I_{k_1}| + \ldots + |I_{k_r}|) \leqslant \omega(f; I_{k_1}) \cdot |I_{k_1}| + \ldots + \omega(f; I_{k_r}) \cdot |I_{k_r}|.$$

Since

$$\sum_{k=1}^{n} \omega(f; I_k) \cdot |I_k| < \frac{\varepsilon}{2m},$$

we have

$$\frac{1}{m}(|I_{k_1}| + \ldots + |I_{k_r}|) < \frac{\varepsilon}{2m}$$

or

$$|I_{k_1}| + \ldots + |I_{k_r}| < \frac{\varepsilon}{2}.$$

Since E_m^{**} is covered by the interior of $I_{k_1}, \ldots, I_{k_r}$, and since E_m^* is covered by $J_1, \ldots, J_p$, it follows that $E_m = E_m^* \cup E_m^{**}$ is of measure zero.

To prove the converse we first note that if $\omega(f; x) < a$ for each x in a closed bounded interval J, then there is a subdivision π of J such that

$$U(f; \pi) - L(f; \pi) < a|J|.$$

Indeed. For each $x \in J$ there is an open subinterval I_x containing x such that $\omega(f; \bar{I}_x) < a$. Since J is compact, a finite number of these I_x will cover J. Let π be the set of endpoints of these I_x. If $I_1, \ldots, I_n$ are the component intervals of π, we have $\omega(f; I_k) < a$ $(k = 1, 2, \ldots, n)$ and our assertion follows at once.

Now we suppose that f is continuous at almost every point of $[a, b]$. We wish to deduce that f is Riemann integrable on $[a, b]$. Given $\varepsilon > 0$ we choose a natural number m such that $(b - a)/m < \varepsilon/2$. If E_m is defined as in the first part of the proof, then, by hypothesis, E_m is of measure zero.

Hence

$$E_m \subset \bigcup_{n=1}^{\infty} I_n,$$

where each I_n is an open subinterval of $[a,b]$ and

$$\sum_{n=1}^{\infty} |I_n| < \frac{\varepsilon}{2\omega(f;[a,b])}.$$

Note that we may assume that $\omega(f;[a,b]) > 0$. But E_m is closed in R^1 and hence is a closed subset of $[a,b]$ and is therefore compact. Consequently, a finite number of I_n's, say $I_{n_1}, \ldots, I_{n_k}$, cover E_m. Now

$$[a,b] - (I_{n_1} \cup \ldots \cup I_{n_k})$$

is a union of closed intervals $J_1, \ldots, J_p$. That is,

$$[a,b] = I_{n_1} \cup \ldots \cup I_{n_k} \cup J_1 \cup \ldots \cup J_p.$$

Since no interval J_j $(j = 1, 2, \ldots, p)$ contains a point of E_m, there is a subdivision π_j of J_j such that

$$U(f;\pi_j) - L(f;\pi_j) < \frac{1}{m}|J_j|$$

by what we have proved above. Now we define the subdivision π of $[a,b]$ as $\pi = \pi_1 \cup \ldots \cup \pi_p$. Then the component intervals of π are the component intervals of $\pi_1, \ldots, \pi_p$ together with $\bar{I}_{n_1}, \ldots, \bar{I}_{n_k}$. Hence

$$\begin{aligned} U(f;\pi) - L(f;\pi) &= \sum_{j=1}^{p} \{U(f;\pi_j) - L(f;\pi_j)\} \\ &\quad + \sum_{i=1}^{k} \omega(f;\bar{I}_{n_i}) \cdot |I_{n_i}| \\ &< \frac{1}{m}\sum_{j=1}^{p} |J_j| + \sum_{i=1}^{k} \omega(f;\bar{I}_{n_i})|I_{n_i}| \\ &\leqslant \frac{b-a}{m} + \omega(f;[a,b]) \sum_{i=1}^{k} |I_{n_i}| \\ &< \frac{\varepsilon}{2} + \omega(f;[a,b]) \cdot \frac{\varepsilon}{2\omega(f;[a,b])} = \varepsilon \end{aligned}$$

and the proof is complete.

The foregoing theorem shows that Riemann integrable functions have to be continuous almost everywhere. We have also seen an example of a Lebesgue integrable function that is continuous nowhere. We now want to show that Lebesgue integrable functions can be approximated in a certain sense by very smooth functions nevertheless.

THEOREM 2. *Given a Lebesgue integrable function* $f: R^1 \to R^1_e$ *and any* $\varepsilon > 0$. *Then there is a finite interval* (a, b) *and a bounded function* $g: R^1 \to R^1$ *such that* g *vanishes outside* (a, b), *is infinitely often differentiable and*

$$\int_{(-\infty,\infty)} |f - g| < \varepsilon.$$

Proof. The proof can be carried out conveniently in the following four steps.

(i) First, we find a bounded closed interval $[a, b]$ and a bounded measurable function f_1 which vanishes outside $[a, b]$ and is such that

$$\int_{(-\infty,\infty)} |f - f_1| < \frac{\varepsilon}{4}.$$

This can easily be done by considering the sequence of functions

$$g_n(x) = \begin{cases} f(x) & \text{if } x \in [-n, n] \quad \text{and } |f(x)| \leqslant n, \\ n & \text{if } x \in [-n, n] \quad \text{and } f(x) > n, \\ -n & \text{if } x \in [-n, n] \quad \text{and } f(x) < -n, \\ 0 & \text{if } x \notin [-n, n]. \end{cases}$$

Then $g_n(x) \to f(x)$ for all x and $|g_n| \leqslant |f|$. Thus

$$\int_{(-\infty,\infty)} |f - g_n| \to 0 \text{ as } n \to \infty$$

and we can pick a sufficiently large n_0 and put $f_1 = g_{n_0}$.

(ii) Next, we approximate f_1 by a simple function f_2 which vanishes outside $[a, b]$ and satisfies

$$\int_{(-\infty,\infty)} |f_1 - f_2| < \frac{\varepsilon}{4}.$$

This is possible by the definition of the Lebesgue integral and Proposition 11 of Chapter 2.

(iii) A simple function is a finite linear combination of characteristic functions of disjoint measurable sets. If each of these characteristic functions in turn can be approximated by a finite linear combination of characteristic functions of disjoint intervals, then it will be clear that f_2 can be approximated by f_3, a step function of the form

$$f_3 = \sum_{j=1}^{n} c_j \chi_{J_j},$$

where each J_j is a finite interval and

$$\int_{(-\infty,\infty)} |f_2 - f_3| < \frac{\varepsilon}{4}.$$

But this is possible by Exercise 3 (part (v)) at the end of Chapter 1.

(iv) In order to obtain the required infinitely often differentiable function g for which

$$\int_{(-\infty,\infty)} |f_3 - g| < \frac{\varepsilon}{4},$$

it will be sufficient to find a function for one of the components χ_{J_j} of f_3. Given $\delta > 0$, let $J = (a, b)$, and put

$$h(x) = \begin{cases} 0 & \text{if } x \leqslant a - \delta/2, \\ \exp\{4\delta^{-2} - 4(\delta^2 - 4(x-a)^2)^{-1}\} & \text{if } a - \delta/2 < x \leqslant a, \\ 1 & \text{if } a < x < b, \\ \exp\{4\delta^{-2} - 4(\delta^2 - 4(b-x)^2)^{-1}\} & \text{if } b \leqslant x < b + \delta/2, \\ 0 & \text{if } x \geqslant b + \delta/2. \end{cases}$$

It is easy to check that h is infinitely often differentiable and

$$\int |\chi_J - h| < \delta$$

since $|\chi_J - h| \leqslant 1$ and $\{x : \chi_J(x) \neq h(x)\}$ is contained in the two intervals $[a - (\delta/2), a]$ and $[b, b + (\delta/2)]$. The proof is finished.

Finally, we want to establish that the Fourier coefficients

$$a_n = \frac{1}{\pi} \int_{(-\pi,\pi)} f(t) \cos nt \, dt$$

$$b_n = \frac{1}{\pi} \int_{(-\pi,\pi)} f(t) \sin nt \, dt$$

of an arbitrary Lebesgue integrable function tend to zero as $n \to \infty$. To this end we prove the following more general theorem.

THEOREM 3. *Let $[a, b]$ be a finite interval and suppose that $(h_n)_{n=1}^{\infty}$ is a sequence of Lebesgue measurable functions on $[a, b]$ with uniform bound K and such that*

$$\lim_{n\to\infty} \int_{(a,c)} h_n(t) \, dt = 0$$

holds for each c ($a \leqslant c \leqslant b$). Then, for any Lebesgue integrable function f on $[a, b]$, we have that

$$\lim_{n\to\infty} \int_{(a,b)} f(t) h_n(t) \, dt = 0.$$

Proof. If $[\alpha, \beta] \subset [a, b]$, then clearly

$$\lim_{n\to\infty} \int_{(\alpha,\beta)} h_n(t) \, dt = 0.$$

Suppose that f is an arbitrary continuous function. Given $\varepsilon > 0$, we partition (a, b) in such a way that the oscillation of f on each subinterval $[x_{k-1}, x_k]$ is smaller than ε. Then

$$\int_{(a,b)} f(t) h_n(t) \, dt = \sum_{k=0}^{m-1} \int_{(x_k, x_{k+1})} (f(t) - f(x_k)) h_n(t) \, dt + \sum_{k=0}^{m-1} f(x_k) \int_{(x_k, x_{k+1})} h_n(t) \, dt$$

for $a = x_0 < x_1 < \ldots < x_m = b$. But

$$\left| \int_{(x_k, x_{k+1})} (f(t) - f(x_k)) h_n(t) \, dt \right| \leqslant K(x_{k+1} - x_k)$$

where $|h_n(t)| \leqslant K$ for all n and $t \in [a, b]$. Hence

$$\left| \sum_{k=0}^{m-1} \int_{(x_k, x_{k+1})} (f(t) - f(x_k)) h_n(t) \, dt \right| \leqslant K\varepsilon(b - a).$$

But

$$\sum_{k=0}^{m-1} f(x_k) \int_{(x_k,x_{k+1})} h_n(t)\,dt \to 0$$

as $n \to \infty$ because

$$\int_{(\alpha,\beta)} h_n(t)\,dt \to 0$$

as $n \to \infty$, as was noted at the beginning of the proof. Thus there exists an integer n_0 such that, for $n \geqslant n_0$, we have

$$\left|\sum_{k=0}^{m-1} f(x_k) \int_{(x_k,x_{k+1})} h_n(t)\,dt\right| < \varepsilon$$

and hence, for $n \geqslant n_0$, we have

$$\left|\int_{(a,b)} f(t)h_n(t)\,dt\right| < \varepsilon(K(b-a)+1)$$

and this proves the theorem when f is continuous.

Next, we let f be a bounded measurable function with $|f(x)| \leqslant M$ for all $x \in [a,b]$. For every $\varepsilon > 0$ there exists by Exercise 27 at the end of this chapter a continuous function g such that $\mu\{x \in [a,b] : f(x) \neq g(x)\} < \varepsilon$. In addition, we see that the function g is bounded by the same constant M, where M is a bound of the function f.

We have that

$$\int_{(a,b)} f(t)h_n(t)\,dt = \int_{(a,b)} (f(t)-g(t))h_n(t)\,dt + \int_{(a,b)} g(t)h_n(t)\,dt.$$

But

$$\left|\int_{(a,b)} (f(t)-g(t))h_n(t)\,dt\right| = \left|\int_E (f(t)-g(t))h_n(t)\,dt\right|$$

where

$$E = \{x \in [a,b] : f(x) \neq g(x)\}.$$

Therefore

$$\left|\int_{(a,b)} (f(t)-g(t))h_n(t)\,dt\right| < 2KM\varepsilon.$$

But, as we know already from the first part of our proof,

$$\int_{(a,b)} g(t)h_n(t)\,dt \to 0$$

as $n \to \infty$. Therefore, there exists an integer n_0 such that, if $n \geqslant n_0$,

$$\left|\int_{(a,b)} g(t)h_n(t)\,dt\right| < \varepsilon$$

and so, for all $n \geqslant n_0$, we have

$$\left|\int_{(a,b)} f(t)h_n(t)\,dt\right| < (2KM + 1)\varepsilon$$

and this establishes the theorem in the special case when f is bounded and measurable.

Finally, let f be an arbitrary integrable function. Since f is then finite almost everywhere on $[a, b]$, we can find a bounded function g such that, for each $\delta > 0$,

$$\mu\{x \in [a, b] : f(x) \neq g(x)\} < \delta$$

by Proposition 14 of Chapter 2. We may assume that g is zero on the set $\{x \in [a, b] : f(x) \neq g(x)\}$. Doing this, we observe that

$$\int_{(a,b)} f(t)h_n(t)\,dt = \int_{(a,b)} (f(t) - g(t))h_n(t)\,dt + \int_{(a,b)} g(t)h_n(t)\,dt$$

can be estimated by using the uniform continuity of the Lebesgue integral, namely, for each $\varepsilon > 0$ there is a $\delta > 0$, such that for any measurable set $A \subset [a, b]$ of measure $\mu(A) < \delta$ we have $\int_A |f| < \varepsilon$. We get

$$\left|\int_{(a,b)} (f(t) - g(t))h_n(t)\,dt\right| = \left|\int_E f(t)h_n(t)\,dt\right| \leqslant K\varepsilon,$$

where

$$E = \{x \in [a, b] : f(x) \neq g(x)\}.$$

But $\left|\int_{(a,b)} g(t)h_n(t)\,dt\right| < \varepsilon$ for sufficiently large n by what we have proved already. Hence, there exists an integer n_0, such that for $n \geqslant n_0$, we have

$$\left|\int_{(a,b)} f(t)h_n(t)\,dt\right| < (K + 1)\varepsilon$$

and the proof is complete.

EXERCISES

1. Show by suitable examples that one may have strict inequality in Fatou's Lemma and that the Monotone Convergence Theorem need not hold for decreasing sequences of functions. [*Hint*: Consider $f_n(x) = nx^{n-1}$ in $(0, 1)$ and $g_n = \chi_{(n,\infty)}$ in R^1.]
2. Using the Monotone Convergence Theorem, derive Fatou's Lemma.
3. For $x > 0$, let

$$f(x) = \frac{d}{dx}\left(x^2 \sin \frac{\pi}{x^2}\right) = 2x \sin \frac{\pi}{x^2} - \frac{2\pi}{x} \cos \frac{\pi}{x^2}.$$

Since the Riemann integral of f over $(\varepsilon, 1)$ is $-\varepsilon^2 \sin(\pi/\varepsilon^2)$ we have that the Cauchy-Riemann integral of f over $(0, 1)$ vanishes. Show that f is not Lebesgue integrable over $(0, 1)$ by verifying that

$$\int_{(0,1)} \frac{1}{x}\left|\cos \frac{\pi}{x^2}\right| dx = \infty.$$

[*Hint*: Let $a_n = (n + 1/3)^{-(1/2)}$ and $b_n = (n - 1/3)^{-(1/2)}$; then $a_n \leqslant x \leqslant b_n$ implies $n\pi - (1/3)\pi \leqslant (\pi/x^2) \leqslant n\pi + (1/3)\pi$, and consequently $|\cos(\pi/x^2) \geqslant 1/2$.]

4. Let f: $R^1 \to R^1_e$ be a nonnegative integrable function and define

$$F(x) = \int_{(-\infty,x)} f(t)\, dt.$$

Show that F is continuous using the Monotone Convergence Theorem.

5. Let f: $[a, b] \to R^1_e$ be integrable on a bounded closed interval $[a, b]$ such that

$$\int_{[a,x]} f = 0$$

for every $x \in [a, b]$. Show that $f = 0$ a.e. on $[a, b]$.

6. Let f and g be integrable on E. Show that if

$$\int_A f = \int_A g$$

for every measurable $A \subset E$, then $f = g$ a.e. in E.

7. Let f be measurable on E. Show that
   (a) f is integrable on E if $|f| \leqslant g$ a.e. on E for some integrable function g on E.
   (b) f is integrable on E if $g \leqslant f \leqslant h$ a.e. on E for some integrable functions g and h on E.

8. If $f = g$ a.e. on E and if g is integrable on E, verify that f is integrable on E and

$$\int_E f = \int_E g.$$

9. For $\mu(E) < \infty$, let f be measurable on E and f^2 be integrable over E. Verify that f is then integrable over E. Give a suitable example to show that the condition $\mu(E) < \infty$ is necessary.
10. If E is a measurable subset of a bounded interval $[a, b]$ and $\mu(E) > 0$, show that the difference set $D(E)$, i.e., the set of all numbers of the form $x - y$ for $x, y \in E$, contains an open interval. (Compare this with Problem 5 of Chapter 1.)
[*Hints*: Let

$$F(x) = \int_a^b \chi_E(t)\chi_E(x + t)\, dt$$

for $-\infty < x < \infty$. Then F is continuous at 0 and $F(0) > 0$. There is a $\delta > 0$ such that $F(x) > 0$ for $-\delta < x < \delta$. If $x \in (-\delta, \delta)$, there is a t_0, depending on x, such that $\chi_E(t_0)\chi_E(x + t_0) > 0$. This in turn implies that $t_0 \in E$ and $x + t_0 \in E$. Since $x = (x + t_0) - t_0$, we get that $x \in D(E)$, showing that $(-\delta, \delta) \subset D(E)$.]
11. Do Propositions 9, 10, and 15 remain valid if we replace convergence a.e. by convergence in measure?
12. Generalize Fatou's Lemma as follows: Let $(f_n)_{n=1}^{\infty}$ be a sequence of nonnegative measurable functions. Then we have that

$$\int \varliminf_{n\to\infty} f_n \leqslant \varliminf_{n\to\infty} \int f_n.$$

[*Hint*: Let $g_k = \inf\{f_k, f_{k+1}, \ldots\}$. Then g_k is measurable, $g_k \leqslant g_{k+1}$ and $g_k \leqslant f_k$.]
13. Let f be an integrable function over R^1 and g be a bounded measurable function on R^1 with $|g| \leqslant M$.
    (a) Show that gf is integrable over R^1 and

$$\left|\int_{R^1} gf\right| \leqslant M \int_{R^1} |f|.$$

    (b) Show that

$$\int_{R^1} |g(x)(f(x + h) - f(x))|\, dx \to 0 \text{ as } h \to 0.$$

[*Hint*: Approximate f by a function that is uniformly continuous and zero outside a bounded interval.]

14. Let $(f_n)_{n=1}^{\infty}$ be a sequence of measurable functions on E and such that

$$\sum_{n=1}^{\infty} \int_E |f_n| < \infty.$$

Show that

$$\sum_{n=1}^{\infty} f_n$$

is integrable over E and

$$\int_E \sum_{n=1}^{\infty} f_n = \sum_{n=1}^{\infty} \int_E f_n .$$

[*Hint*: Use the Dominated Convergence Theorem.]

15. Let $f_n(x) = nx^{n-1} - (n+1)x^n$ on $(0, 1)$. Show that

$$\int_{(0,1)} \sum_{n=1}^{\infty} f_n \neq \sum_{n=1}^{\infty} \int_{(0,1)} f_n$$

and verify directly that

$$\sum_{n=1}^{\infty} \int_{(0,1)} |f_n| = \infty.$$

16. Let E be a measurable set of finite measure. Let M denote the set of all measurable functions on E. For $f, g \in M$ define

$$d(f, g) = \int_E \frac{|f - g|}{1 + |f - g|}.$$

(a) Show that d is a metric on M and that $d(f, g) = 0$ if and only if $f = g$ a.e. on E.

(b) Identifying functions equal a.e., verify that (M, d) is a complete metric space.

(c) Prove that $d(f_n, f) \to 0$ if and only if $f_n \to f$ in measure on E. Observe: If $r > 0$ and $B_r = \{t \in E : |f(t)| \geqslant r\}$, then $(r/(1 + r))\mu(B_r) \leqslant d(f, 0) \leqslant \mu(B_r) + (r/(1 + r))\mu(E - B_r)$.

17. Let f: $R^1 \to R^1$ be an additive measurable function. Let $x \neq 0$ and put

$$\phi(t) = f(t) - \frac{f(x)}{x} t \text{ and } \psi(t) = \frac{1}{1 + |\phi(t)|}.$$

Clearly, ϕ and ψ have period x and

$$\int_{(0,x)} \frac{dt}{1+|\phi(t)|} = \int_{(0,x)} \frac{dt}{1+2|\phi(t)|}.$$

Show that $\phi(t) = 0$ almost everywhere by virtue of the fact that

$$\int_{(0,x)} \frac{|\phi(t)|\,dt}{(1+2|\phi(t)|)(1+|\phi(t)|)} = 0.$$

[N.B. This means that $f(t) = (f(x)/x)t$ for almost all t. In particular, for $x = 1$, we see that $f(t) = f(1)t$ for almost all t. Hence there exists for each $x \neq 0$ a $t_0 \neq 0$ such that $f(t_0) = (f(x)/x)t_0$ and $f(t_0) = f(1)t_0$, from which we get that $f(x) = f(1)x$. Clearly this equality also holds for $x = 0$.]

18. Let $f_n \to f$ a.e. in E and $\|f_n\|_1 \to \|f\|_1$. Show that
    (i) $\int_A |f_n| \to \int_A |f|$ for all measurable subsets A of E;
    (ii) $\|f - f_n\|_1 \to 0$.
    [*Hints*: To prove (i) let A any measurable subset of E. By Fatou's Lemma

$$\begin{aligned}\lim_{n\to\infty} \int_A |f_n| &\geqslant \int_A |f| = \int_E |f| - \int_{E-A} |f| \\ &\geqslant \int_E |f| - \lim_{n\to\infty} \int_{E-A} |f_n| = \overline{\lim_{n\to\infty}}\left(\int_E |f_n| - \int_{E-A} |f_n|\right) \\ &= \overline{\lim_{n\to\infty}} \int_A |f_n|.\end{aligned}$$

Hence

$$\lim_{n\to\infty} \int_A |f_n|$$

exists and (i) is true.

To prove (ii), let $\varepsilon > 0$. Pick $A \subset E$ such that $\mu(A) < \infty$ and $\int_{E-A} |f| < \varepsilon/5$. By Proposition 12 obtain a $\delta > 0$ such that if $D \subset E$ and $\mu(D) < \delta$, then $\int_D |f| < \varepsilon/5$. Now use Egorov's Theorem to find a measurable set B such that $B \subset A$ and $\mu(A \cap B^c) < \delta$ and $f_n \to f$ uniformly on B. Pick n_0 such that for all $n \geqslant n_0$

$$\mu(B) \cdot \sup_{x\in B} |f(x) - f_n(x)| < \frac{\varepsilon}{5}.$$

Applying part (i), we get

$$\overline{\lim_{n\to\infty}} \int_E |f - f_n|$$

$$= \overline{\lim_{n\to\infty}}\left[\int_{A^c} |f-f_n| + \int_{A\cap B^c} |f-f_n| + \int_B |f-f_n|\right]$$

$$\leqslant \overline{\lim_{n\to\infty}}\left[\int_{A^c} |f| + \int_{A^c} |f_n| + \int_{A\cap B^c} |f| + \int_{A\cap B^c} |f_n| + \frac{\varepsilon}{5}\right]$$

$$= 2\int_{A^c} |f| + 2\int_{A\cap B^c} |f| + \frac{\varepsilon}{5} < \varepsilon.]$$

19. Let $1 \leqslant p < \infty$ and suppose that $(f_n)_{n=1}^{\infty}$ is a sequence in $L^p(E)$. Let f be a measurable function on E such that f is finite a.e. in E and $f_n \to f$ a.e. in E. Show that $f \in L^p(E)$ and $\|f - f_n\|_p \to 0$ if and only if
(i) for each $\varepsilon > 0$ there is a set $A_\varepsilon \subset E$ such that $\mu(A_\varepsilon) < \infty$ and

$$\int_{E-A_\varepsilon} |f_n|^p < \varepsilon \text{ for all } n \in N$$

and

(ii)
$$\lim_{\mu(D)\to 0} \int_D |f_n|^p = 0$$

uniformly in n, i.e., for each $\varepsilon > 0$ there is a $\delta > 0$ such that $D \subset E$ and $\mu(D) < \delta$ imply

$$\int_D |f_n|^p < \varepsilon \text{ for all } n \in N.$$

[*Hints*: To show the necessity of (i), let $\varepsilon > 0$ be given. Choose n_0 such that $\|f - f_n\|_p < \varepsilon$ for all $n \geqslant n_0$ and pick measurable subsets B_ε and C_ε of E of finite measure such that

$$\int_{E-B_\varepsilon} |f|^p < \varepsilon \text{ and } \int_{E-C_\varepsilon} |f_n|^p < \varepsilon$$

for $n = 1, 2, \ldots, n_0$. Then set $A_\varepsilon = B_\varepsilon \cup C_\varepsilon$. The necessity of (ii) is proved in the same way by using the absolute continuity of the integral (see Proposition 12). Next, suppose that (i) and (ii) hold. Use (i), Fatou's Lemma and Minkowski's Inequality to reduce the problem to the case that $\mu(E) < \infty$. For $\varepsilon > 0$, take δ as in (ii). Using Egorov's Theorem, find $B \subset E$ such that $\mu(B) < \delta$ and $f_n \to f$ uniformly on $E - B$. Use Fatou's Lemma to show that $\int_B |f|^p < \varepsilon$. Then use Minkowski's Inequality to verify that $\int_E |f - f_n|^p < 3^p\varepsilon$ for sufficiently large n. Thus infer that $f = (f - f_n) + f_n \in L^p(E)$ and $\|f - f_n\|_p \to 0$.]

20. Let p, $(f_n)_{n=1}^{\infty}$, and f be as in the foregoing exercise. Suppose that $f_n \to f$ a.e. in E. For each $(n, k) \in N \times N$, let

$$B_{n,k} = \{x \in E : |f_n(x)|^p \geqslant k\}.$$

(a) Suppose that for each $\varepsilon > 0$ there is a set $A_\varepsilon \subset E$ such that $\mu(A_\varepsilon) < \infty$ and $\int_{E-A_\varepsilon} |f_n|^p < \varepsilon$ and for all $n \in N$.
Prove that the following conditions are equivalent:
(i) $f \in L^p(E)$ and $\|f - f_n\|_p \to 0$;
(ii) if $(E_k)_{k=1}^{\infty}$ are measurable subsets of E and $E_1 \supset E_2 \supset \dots$ and

$$\bigcap_{k=1}^{\infty} E_k = \varnothing,$$

then

$$\lim_{k\to\infty} \int_{E_k} |f_n|^p = 0$$

uniformly in n;

(iii)
$$\lim_{k\to\infty} \int_{B_{n,k}} |f_n|^p = 0$$

uniformly in n; moreover, uniformly in n,

(iv)
$$\lim_{\mu(D)\to 0} \int_D |f_n|^p = 0.$$

(b) Prove that condition (ii) of part (a) implies conditions (i) and (ii) of Exercise 19.
[*Hints* to prove part (a): Conditions (i) and (iv) are equivalent by Exercise 19. To show that (i) implies (ii), take $\varepsilon > 0$ and $n_0 \in N$ such that $\|f_n - f\|_p < \varepsilon$ for all $n \geqslant n_0$. Then for $n \geqslant n_0$, we get

$$\begin{aligned}\Big(\int_{E_k} |f_n|^p\Big)^{1/p} &\leqslant \Big(\int_{E_k} |f|^p\Big)^{1/p} + \Big(\int_{E_k} |f_n - f|^p\Big)^{1/p}\\ &< \Big(\int_{E_k} |f|^p\Big)^{1/p} + \varepsilon;\end{aligned}$$

we then apply the Dominated Convergence Theorem to $(|f|^p \chi_{E_k})_{k=1}^{\infty}$ to show that

$$\left(\int_{E_k} |f|^p\right)^{1/p} < 2\varepsilon$$

for $k \geqslant k_0$ and all $n \geqslant n_0$. If $n = 1, 2, \ldots, n_0$, then

$$\int_{E_k} |f_n|^p \leqslant \int_{E_k} \max\{|f_1|^p, \ldots, |f_{n_0}|^p\}$$

and dominated convergence implies (ii).
Suppose now that (ii) holds. Let

$$E_k = \bigcup_{n=k}^{\infty} B_{n,k}.$$

Then $E_1 \supset E_2 \supset \ldots$ and

$$\overline{\lim_{n\to\infty}} |f_n(x)| = \infty$$

on

$$\bigcap_{k=1}^{\infty} E_k.$$

Hence

$$\mu\left(\bigcap_{k=1}^{\infty} E_k\right) = 0.$$

Let

$$F_k = E_k \cap \left(\bigcap_{k=1}^{\infty} E_k\right)^c,$$

where the superscript c denotes complementation. Using (ii) pick a k_0 such that if $k \geqslant k_0$, then for all n,

$$\int_{F_k} |f_n|^p < \varepsilon.$$

If $n \geqslant k_0$ we have $B_{n,k_0} \subset E_{k_0}$ and so for $n \geqslant k_0$ and $k \geqslant k_0$, we get

$$\int_{B_{n,k}} |f_n|^p \leqslant \int_{B_{n,k_0}} |f_n|^p \leqslant \int_{E_{k_0}} |f_n|^p = \int_{F_{k_0}} |f_n|^p < \varepsilon.$$

For $n = 1, 2, \ldots, k_0 - 1$, we have

$$|f_n|^p \leqslant \max\{|f_1|^p, \ldots, |f_{k_0-1}|^p\} = g$$

and so

$$\int_{B_{n,k}} |f_n|^p \leqslant \int_{B_k} g,$$

where

$$B_k = \bigcup_{n=1}^{k_0-1} B_{n,k} = \{x \in E : g(x) \geqslant k\}.$$

Thus dominated convergence applies and so (iii) holds if (ii) holds. Finally, suppose that (iii) holds and we want to show that (iv) then follows. Choose k_0 so large that if $k \geqslant k_0$ and $n \in N$, we have

$$\int_{B_{n,k}} |f_n|^p < \varepsilon^p.$$

If D is a measurable subset of E and $\mu(D) < (1/k_0)\varepsilon^p$, then

$$(\int_D |f_n|^p)^{1/p} \leqslant (\int_{D \cap B_{n,k}} |f_n|^p)^{1/p} + (\int_{D \cap B_{n,k}^c} |f_n|^p)^{1/p} < \varepsilon + \varepsilon.]$$

21. Let $\mu(E) < \infty$ and f be any bounded measurable function on E. Verify that

$$\lim_{p \to \infty} \|f\|_p = \inf\{a \in R^1 : a > 0, \mu(\{x \in E : |f(x)| > a\}) = 0\}.$$

22. Let $1 \leqslant p < \infty$ and consider any $L^p(E)$. Show that for every $f \in L^p(E)$ and every $\varepsilon > 0$
    (a) there exists a simple function s in $L^p(E)$ such that $|s| \leqslant |f|$ and $\|s - f\|_p < \varepsilon$;
    (b) there is a step function, i.e., a function σ of the form

$$\sum_{k=1}^{n} \alpha_k \chi_{I_k},$$

where $\alpha_1, \ldots, \alpha_n$ are real numbers and each I_k is a bounded interval, such that $\|\sigma - f\|_p < \varepsilon$;
    (c) there is a continuous function h vanishing outside a closed bounded set of R^1 such that

$$\|h - f\|_p < \varepsilon.$$

23. Let $1 \leqslant p < \infty$ and let $f \in L^p(E)$. Let f_h denote

$$f_h(t) = f(t + h).$$

Show that

$$\lim_{h\downarrow 0}\|f_h - f\|_p = \lim_{h\uparrow 0}\|f_h - f\|_p = 0.$$

24. We have considered various types of convergence: [unif.], [a.un.], [a.e.], [meas.], and [L^p-norm]. In the following diagrams an arrow from one type of convergence to the other indicates implication, i.e., $[A] \to [B]$ means that if a sequence converges in the sense of $[A]$ convergence, then it converges in the sense of $[B]$ convergence. Check these diagrams to see that for each arrow there is a proposition or supply a proof. Moreover, if an arrow is missing, try to construct a counterexample. [N.B. A good reference on convergence theorems is *Introduction to Measure and Integration* by M. E. Munroe (Addison-Wesley, Reading, Mass.) or the second edition, *Measure and Integration*; nearly an entire chapter is devoted to the subject at hand.]
25. Let f: $[a, b] \to R_e^1$ be measurable and finite a.e. and $b - a < \infty$. Show that if ε and δ are given positive numbers, there exists a bounded continuous function g: $[a, b] \to R^1$ such that

$$\mu(\{x \in [a, b] : |g(x) - f(x)| \geqq \delta\}) < \varepsilon.$$

[*Hints*: Suppose first that f is bounded. If $|f(x)| < M$ for each x, pick a positive integer n such that $M/n < \delta$. Let $G_i = (((i-1)/n)M, (i/n)M)$ $i = -n+1, -n+2, \ldots, n$, and let $E_i = f^{-1}(G_i)$. Pick a closed set $A_i \subset E_i$ so that $\mu(A_i) > \mu(E_i) - \varepsilon/2n$. Let

$$A = \bigcup_i A_i$$

and define a function h: $A \to R^1$ by setting $h(x) = (i/n)M$ if $x \in A_i$. Is h continuous on A? What can be said about $\mu([a, b] - A)$? Show that there is a continuous function g such that $g(x) = h(x)$ if $x \in A$. Show that g meets the requirement of the problem. Now deal with the case when f is not bounded. Begin by showing that there is a bounded function w: $[a, b] \to R^1$ such that $\mu(\{x \in [a, b] : f(x) \neq w(x)\}) < \frac{1}{2}\varepsilon$ and complete the proof.]
26. Let f and $[a, b]$ be as in Problem 25. Show that there is a sequence of bounded continuous functions $(f_n)_{n=1}^{\infty}$ on $[a, b]$ such that $f_n \to f$ a.e. in $[a, b]$. [*Hint*: Use Problem 25 and Proposition 16.]
27. Let f and $[a, b]$ be as in Problem 25. Show that for any $\delta > 0$ there is a continuous function h on $[a, b]$ such that $\mu\{x \in [a, b] : f(x) \neq h(x)\} < \delta$. In particular, if K is a bound for f, then K is also a bound for h [N. N. Lusin].
[*Hint*: Use Problem 26 and Egorov's Theorem (Chapter 2, Proposition 17).]

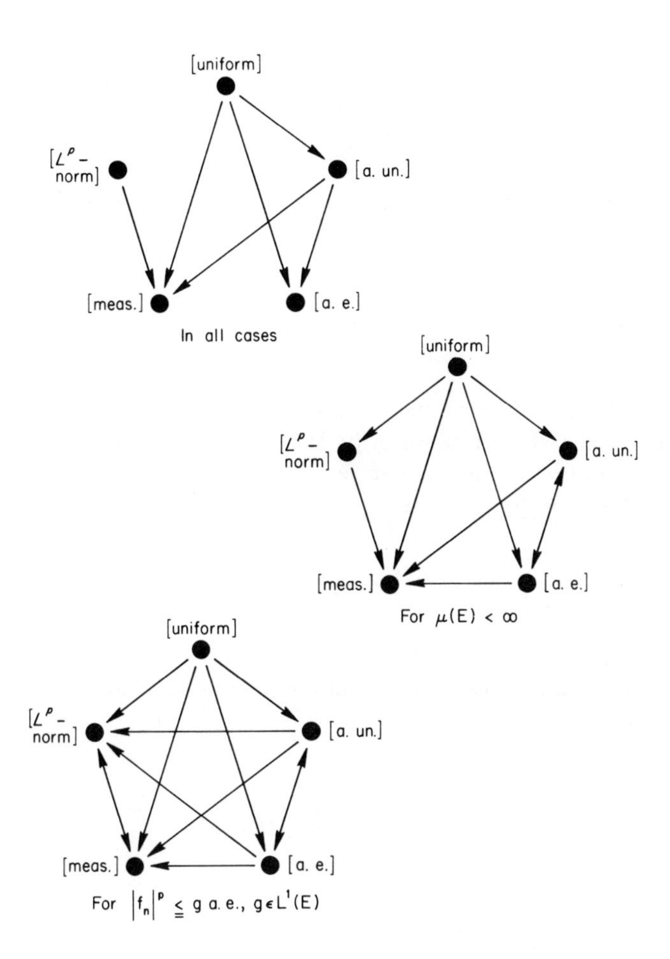

Problem 24: Convergence Diagrams.

28. Let f be continuous on the open set D of R^1. Assume that the integrals of f^+ and f^- over D both diverge to ∞. Show that given any number r there is a sequence of compact sets $K_1 \subset K_2 \subset \ldots$ such that $D = K_1 \cup K_2 \cup \ldots$ and

$$\lim_{n\to\infty} \int_{K_n} f = r.$$

29. Let f and g be nonnegative measurable functions on E with $\mu(E) < \infty$. If $E_y = \{x \in E : g(x) \geqq y\}$, show that

$$\int_E f(x)g(x)\,dx = \int_0^\infty h(y)\,dy, \text{ where } h(y) = \int_{E_y} f(x)\,dx.$$

30. Let $E_1, E_2, \ldots, E_n$ be n measurable sets which are situated on the unit interval $[0, 1]$. If each point of $[0, 1]$ belongs to at least q of these sets, show that at least one of these sets must have measure $\geqq q/n$.
[*Hint*: For $x \in [0, 1]$ define

$$f(x) = \sum_{k=1}^{n} \chi_{E_k}(x)$$

and integrate f over $[0, 1]$.]

31. Let f be an integrable function on the interval $[a, b]$ and suppose that f is zero outside $[a, b]$. If

$$g(x) = \frac{1}{2h} \int_{x-h}^{x+h} f(t)\,dt,$$

show that

$$\int_a^b |g(x)|\,dx \leqq \int_a^b |f(x)|\,dx \text{ and } \lim_{h\to 0} \int_a^b |g(x) - f(x)|\,dx = 0.$$

32. Let $[a, b]$ be a bounded interval and f a positive integrable function on it. Let $0 < q \leqq b - a$ and S be all measurable subsets $E \subset [a, b]$ for which $\mu(E) \geqq q$. Show:

$$\inf_{E\in S}\{\int_E f(x)\,dx\} > 0.$$

33. Using the Dominated Convergence Theorem, prove that

$$\lim_{n\to\infty} \int_0^n \left(1 - \frac{x}{n}\right)^n x^s\, dx = \int_0^\infty e^{-x} x^s\, dx,\ s < -1.$$

34. Let f_n be measurable and uniformly bounded on $[0, 1]$. Let

$$F_n(x) = \int_0^x f_n(t)\, dt.$$

Show that there exists a subsequence F_{n_i} that converges uniformly on $[0, 1]$.

35. Show that if f is integrable and nonnegative on $(0, 1)$, and if

$$\int_0^1 f^n(x)\, dx$$

exists for $n = 1, 2, \ldots$ with a numerical value independent of n, then f is the characteristic function of a measurable set.

36. Show that $(e^{-x} \sin x)/x$ is Lebesgue integrable over $(0, \infty)$.

37. Prove or disprove (by counterexample) the following statements:
    (a) For every Lebesgue integrable function f on the unit interval, there is a Riemann integrable function g such that $f(x) = g(x)$ for almost every x.
    (b) With f as above, and $\varepsilon > 0$, there is a Riemann integrable function g such that f and g differ on a set of measure $< \varepsilon$.

38. Consider $L^p(-\infty, \infty)$, $1 \leqq p \leqq \infty$. Show that there are no inclusion relations among these spaces.

39. Let E have finite positive measure on the real line, and let $E + E$ denote the algebraic sum of E with itself, i.e., the set of points $x + y$ where x, y are in E. Show that $E + E$ contains an interval.

40. For any α with $0 < \alpha \leqq 1$, let f_α denote the function

$$f_\alpha(x) = \left[\frac{\alpha}{x}\right] - \alpha\left[\frac{1}{x}\right], \qquad 0 < x < 1;$$

here $[t]$ denotes the largest integer contained in t. We may view these functions as elements of the space $L^2(0,1)$ of (Lebesgue) square-integrable functions on the interval (0,1). The problem is to decide whether or not the linear hull (i.e., the set of all finite linear combinations with real coefficients) of the set of functions f_α with $0 < \alpha \leqq 1$ is dense in the space $L^2(0, 1)$. [Some information concerning the foregoing problem and its significance can be found in: G. Klambauer, "A Note on Nyman's Function System," *Proc. Amer. Math. Soc.* **13** (1962), 312 - 314.]

41. (i) Let a and b denote positive numbers with $b < a$. Form the consecutive arithmetic and geometric means:

$$a_1 = \frac{a+b}{2} \qquad b_1 = \sqrt{ab}$$

$$a_2 = \frac{a_1+b_1}{2} \qquad b_2 = \sqrt{a_1 b_1}$$

$$a_3 = \frac{a_2+b_2}{2} \qquad b_3 = \sqrt{a_2 b_2}$$

etc. Show that the sequences of consecutive arithmetic and geometric means are monotone and converge to the same limit, say, $\lambda(a, b)$.

(ii) Let a and b be as in part (i). Put

$$G = \int_0^{\pi/2} \frac{dt}{\sqrt{a^2\cos^2 t + b^2 \sin^2 t}}.$$

Show that

$$G = \int_0^{\pi/2} \frac{dt}{\sqrt{a_n^2\cos^2 t + b_n^2 \sin^2 t}}$$

for $n = 1, 2, 3, \ldots$, where

$$a_n = \frac{a_{n-1}+b_{n-1}}{2}, \qquad b_n = \sqrt{a_{n-1} b_{n-1}}\,.$$

Observing that

$$\frac{\pi}{2a_n} < G < \frac{\pi}{2b_n},$$

verify the equation

$$G = \frac{\pi}{2\lambda(a,b)}.$$

[*Hint*: Try the substitution

$$\sin t = \frac{2a\sin s}{(a+b)+(a-b)\sin^2 s}.]$$

42. (i) If $f \in L^p \cap L^q$, where $0 < p < q < \infty$, and if $p < r < q$, then $f \in L^r$. Also, the function θ defined by

$$\theta(r) = \log(\|f\|_r^r)$$

on $[p, q]$ is convex, i.e., $0 < \alpha < 1$ implies

$$\theta(\alpha p + (1 - \alpha)q) \leqq \alpha\theta(p) + (1 - \alpha)\theta(q).$$

[*Hints*: Let $r = \alpha p + (1 - \alpha)q$, $0 < \alpha < 1$. Using Hölder's Inequality with $1/\alpha$ we have

$$\int |f|^r = \int |f|^{\alpha p+(1-\alpha)q} \leqq \left(\int |f|^{\alpha p(1/\alpha)}\right)^{\alpha}\left(\int |f|^{(1-\alpha)q(1/(1-\alpha))}\right)^{1-\alpha}$$
$$= \left(\int |f|^p\right)^{\alpha}\left(\int |f|^q\right)^{1-\alpha}$$

(note that $|f|^{\alpha p} \in L^{1/\alpha}$ and $|f|^{(1-\alpha)q} \in L^{1/(1-\alpha)}$). Hence we have that $f \in L^{\alpha p+(1-\alpha)q}$ and

$$\|f\|_{p+(1-\alpha)q}^{p+(1-\alpha)q} \leqq (\|f\|_p^p)^{\alpha}(\|f\|_q^q)^{1-\alpha}.$$

Taking logarithms on both sides of this inequality, we see that θ is convex on $[p, q]$.]

(ii) It can happen that L^p is a proper subset of L^1 if $p > 1$. Consider $f(x) = (x(\log(2/x))^2)^{-1}$ on $[0, 1]$.

(iii) It can happen that a function belongs to L^q but not to

$$\bigcup_{p \neq q} L^p.$$

Consider $f(x) = (x(1 + |\log x|)^2)^{-1}$ with using Lebesgue integrals over $[0, \infty)$.

(iv) If $\mu(E) < \infty$, $L^\infty(E) \subset L^p(E)$ for each $p > 0$. If $f \in L^\infty(E)$, then

$$\lim_{p\to\infty} \|f\|_p = \text{ess sup}|f| \quad (= \|f\|_\infty).$$

[*Hints*: If $\|f\|_\infty = 0$, then $\|f\|_p = 0$ for every $p > 0$. Otherwise, let a be any positive number less than $\|f\|_\infty$ and let $E_a = \{x : |f(x)| > a\}$. Then $\|f\|_p \geqq a(\mu(E_a))^{1/p}$. Moreover, $0 < \mu(E_a) < \infty$, because a $< \|f\|_\infty$ and $f \in L^p$. Therefore

$$\liminf_{p\to\infty} \|f\|_p \geqq a$$

and so

$$\liminf_{p\to\infty} \|f\|_p \geqq \|f\|_\infty.$$

On the other hand, $|f|^p \leqq |f|^q(\|f\|_\infty)^{p-q}$ and therefore $\|f\|_p \leqq (\|f\|_q)^{q/p}(\|f\|_\infty)^{1-q/p}$, giving the desired inequality

$$\limsup_{p\to\infty} \|f\|_p \leqq \|f\|_\infty.]$$

(v) If $f \in L^\infty \cap L^p$, for some $p > 0$, then $f \in L^q$ if $q > p$. Moreover,

$$\lim_{q\to\infty} \|f\|_q = \text{ess sup}|f| \quad (= \|f\|_\infty).$$

43. If $m \leqq f(x) \leqq M$ for all $x \in E$ and if f is measurable and g is integrable, then there is a number a such that $m \leqq a \leqq M$ and $\int_E f|g| = a \int_E |g|$.

44. Suppose that f is a bounded, measurable function on a set E of finite measure and let $A < f(x) < B$ for all $x \in E$. Consider a partition $A = y_0 < y_1 < \ldots < y_n = B$ of the interval $[A, B]$ and let $e_k = \{x \in E : y_k \leqq f(x) < y_{k+1}\}$. Define lower and upper sums s and S, respectively, as follows:

$$s = \sum_{k=0}^{n-1} y_k \, \mu(e_k), \qquad S = \sum_{k=0}^{n-1} y_{k+1} \, \mu(e_k).$$

It is clear that $0 \leqq S - s \leqq \lambda\mu(E)$, where

$$\lambda = \max_k (y_{k+1} - y_k).$$

Show: Let s_0 and S_0 correspond to some partition of $[A, B]$; if we add a new point $\bar{y}$ to the partition and form the corresponding sums s and S, then $s_0 \leqq s$ and $S \leqq S_0$. Every lower sum s is smaller than every upper sum S. If V is the infimum of all upper sums S, and U is the supremum of all lower sums s, then $0 \leqq V - U \leqq \lambda\mu(E)$. For $\lambda \to 0$ the lower sums s and the upper sums S converge to the same value; indeed, this common value is equal to the Lebesgue integral of f over E. Observe also that in the foregoing analysis the choice of the numbers A and B is inessential; picking another pair of numbers A^* and B^* such that $A^* < f(x) < B^*$ does not change the value of the integral.

45. (i) Let $(f_n)_{x=1}^\infty$ be a decreasing sequence of nonnegative measurable functions on E with $f_1 \in L^1(E)$. Show that

$$\int_E \lim_{n\to\infty} f_n = \lim_{n\to\infty} \int_E f_n < \infty.$$

(ii) Let x be any real number, $x > 1$; show that

$$\lim_{n\to\infty} n(x^{1/n} - 1) = \log x$$

by considering the function $f_n(t) = t^{(1/n)-1}$ with $1 < t < x$.

46. Let p and q be finite positive numbers. Show that

$$\int_0^1 \frac{x^{p-1}}{1 + x^q}\, dx = \frac{1}{p} - \frac{1}{p+q} + \frac{1}{p+2q} - \frac{1}{p+3q} + \cdots.$$

[*Hint*: If $0 < x < 1$, then

$$x^{p-1}(1 + x^q)^{-1} = x^{p-1}(1 - x^q + x^{2q} - x^{3q} + \ldots) = \sum_{n=0}^{\infty} f_n(x),$$

where $f_n(x) = (1 - x^q)x^{p-1+2nq} > 0$. Now use the Remark following Proposition 10.]

47. (Hölder's Inequality for $0 < p < 1$): Let $0 < p < 1$ and suppose that f and g are nonnegative functions in $L^p(E)$ and $L^q(E)$, respectively. Show that

$$\int_E fg \geqq \left(\int_E f^p\right)^{1/p}\left(\int_E g^q\right)^{1/q}$$

unless $\int_E g^q = 0$ (note that $q < 0$).

48. For $0 < p < 1$, let f and g be nonnegative functions in $L^p(E)$. Verify that

$$\|f + g\|_p \geqq \|f\|_p + \|g\|_p.$$

49. Prove that in any set of ten different two digit numbers one can select two disjoint subsets such that the sum of numbers in each of the subsets is the same.

Chapter 4

Differentiation and Absolute Continuity

Let $a \in R^1$ and $\delta > 0$. If g is a real-valued function on the open interval $(a, a + \delta)$, we define the *lower right limit* and the *upper right limit of g at a*, respectively, as follows:

$$\varliminf_{h \downarrow a} g(h) = \sup\{\inf\{g(h) : a < h < t\} : a < t \leqslant a + \delta\}$$

and

$$\varlimsup_{h \downarrow a} g(h) = \inf\{\sup\{g(h) : a < h < t\} : a < t \leqslant a + \delta\}.$$

Similarly, if g is a real-valued function on $(a - \delta, a)$, we define the *lower left limit* and the *upper left limit of g at a* as

$$\varliminf_{h \uparrow a} g(h) = \sup\{\inf\{g(h) : t < h < a\} : a - \delta \leqslant t < a\}$$

and

$$\varlimsup_{h \uparrow a} g(h) = \inf\{\sup\{g(h) : t < h < a\} : a - \delta \leqslant t < a\},$$

respectively.

Let $a \in R^1$ and $\delta > 0$. If f is a real-valued function defined on the half-open interval $[a, a + \delta)$, we put

$$D_+ f(a) = \varliminf_{h \downarrow 0} \frac{f(a + h) - f(a)}{h}$$

and

$$D^+ f(a) = \varlimsup_{h \downarrow 0} \frac{f(a + h) - f(a)}{h}.$$

If f is a real-valued function defined on $(a - \delta, a]$, we put

$$D_- f(a) = \varliminf_{h \uparrow 0} \frac{f(a + h) - f(a)}{h}$$

and

$$D^- f(a) = \overline{\lim_{h \uparrow 0}} \frac{f(a+h) - f(a)}{h}.$$

These four extended real numbers are called the *Dini derivatives of f at a*; $D_+ f(a)$ is the *lower right derivative*, $D^+ f(a)$ is the *upper right derivative*, $D_- f(a)$ *is the lower left derivative* and $D^- f(a)$ is the *upper left derivative.*

The inequalities

$$D_+ f(a) \leqslant D^+ f(a)$$

and

$$D_- f(a) \leqslant D^- f(a)$$

always hold.

If $D^+ f(a) = D_+ f(a)$ then f is said to have a *right derivative at a* and we write $f'_+(a)$ for the common value $D^+ f(a) = D_+ f(a)$. The left derivative of f at a is defined analogously, and is written $f'_-(a)$. If $f'_+(a)$ and $f'_-(a)$ exist and are equal, then f is said to be *differentiable at a*, and we write $f'(a)$ for the common value $f'_+(a) = f'_-(a)$ and we call the number $f'(a)$ the *derivative of f at a.*

N.B. In the definition of $f'(a)$ given above we have not excluded ∞ or $-\infty$ as a possible value for it.

PROPOSITION 1. *Let (a, b) be any open interval of R^1 and let f be an arbitrary real-valued function on (a, b). Then there are only countably many points $x \in (a, b)$ such that $f'_-(x)$ and $f'_+(x)$ both exist (they may be infinite) and are not equal.*

Proof. Let

$$A = \{x \in (a, b) : f'_-(x) \text{ exists}, f'_+(x) \text{ exists}, f'_+(x) < f'_-(x)\}$$

and let

$$B = \{x \in (a, b) : f'_-(x) \text{ exists}, f'_+(x) \text{ exists}, f'_+(x) > f'_-(x)\}.$$

For each $x \in A$, choose a rational number r_x such that $f'_+(x) < r_x < f'_-(x)$. Next, select rational numbers s_x and t_x such that $a < s_x < t_x < b$,

$$\frac{f(y) - f(x)}{y - x} > r_x \text{ if } s_x < y < x,$$

and

$$\frac{f(y) - f(x)}{y - x} < r_x \text{ if } x < y < t_x.$$

Combining these we have

$$(*) \qquad f(y) - f(x) < r_x(y - x)$$

whenever $y \neq x$ and $s_x < y < t_x$. Thus, we obtain a function ϕ from A into the countable set Q^3, where Q denotes the set of all rational numbers, defined by $\phi(x) = (r_x, s_x, t_x)$. We will show that A is countable by showing that ϕ is one-to-one. Assume that there are distinct x and y in A such that $\phi(x) = \phi(y)$. Then $(s_y, t_y) = (s_x, t_x)$ and x and y are both in this interval. It follows from (*) that

$$f(y) - f(x) < r_x(y - x)$$

and

$$f(x) - f(y) < r_y(x - y).$$

Since $r_x = r_y$, adding the foregoing two inequalities gives $0 < 0$. This is a contradiction, so that ϕ is one-to-one and A is countable.

A similar argument shows that B is a countable set and the proof is finished.

Let $E \subset R^1$. A collection Ω of closed intervals of R^1, each having positive length, is said to be a *Vitali cover of* E if for each $x \in E$ and each $\varepsilon > 0$ there exists an interval $I \in \Omega$ such that $x \in I$ and $\mu(I) < \varepsilon$ (i.e., each point of E is in arbitrarily short intervals of Ω).

We now establish the so-called "Covering Theorem of Vitali."

PROPOSITION 2. *Let E be any subset of R^1 and let Ω be any nonempty Vitali cover of E. Then there is a mutually disjoint countable family $\{I_n\} \subset \Omega$ such that*

$$\mu^*(E \cap (\bigcup_{n=1}^{\infty} I_n)^c) = 0,$$

where the superscript c denotes complementation relative to R^1. In addition, if $\mu^*(E) < \infty$ then for each $\varepsilon > 0$ there is a mutually disjoint finite family $\{I_1, I_2, \ldots, I_p\} \subset \Omega$ such that

$$\mu^*(E \cap (\bigcup_{n=1}^{p} I_n)^c) < \varepsilon.$$

Proof. We first consider the case where $\mu^*(E) < \infty$. We choose an open set V containing E and such that $\mu^*(V) < \infty$. Let $\Omega_0 = \{I \in \Omega : I \subset V\}$. Evidently Ω_0 is a Vitali cover of E. Let $I_1 \in \Omega_0$. If $E \subset I_1$, the construction is complete. Otherwise we continue by induction as follows. Suppose that $I_1, I_2, \ldots, I_n$ have been chosen and are pairwise disjoint. If

$$E \subset \bigcup_{k=1}^{n} I_k,$$

the construction is finished. Otherwise, we write

$$A_n = \bigcup_{k=1}^{n} I_k, \quad U_n = V \cap A_n^c.$$

Clearly A_n is closed, U_n is open and $U_n \cap E$ is nonempty. Let

$$\delta_n = \sup\{\mu(I) : I \in \Omega_0, I \subset U_n\}.$$

Choose $I_{n+1} \in \Omega_0$ such that $I_{n+1} \subset U_n$ and $\mu(I_{n+1}) > \frac{1}{2}\delta_n$. If our process does not terminate after a finite number of steps (in which case there is nothing left to prove), then it yields an infinite sequence $(I_n)_{n=1}^{\infty}$ of mutually disjoint members of Ω_0. Let

$$A = \bigcup_{n=1}^{\infty} I_n.$$

We must verify that $\mu^*(E \cap A^c) = 0$. For each $n \in N$, let J_n be the closed interval having the same midpoint as I_n and such that

$$\mu(J_n) = 5\mu(I_n).$$

We have

$$(\#) \qquad \mu\Big(\bigcup_{n=1}^{\infty} J_n\Big) \leqslant \sum_{n=1}^{\infty} \mu(J_n) = 5 \sum_{n=1}^{\infty} \mu(I_n) = 5\mu(A) \leqslant 5\mu^*(V) < \infty.$$

Therefore,

$$\lim_{p\to\infty} \mu\Big(\bigcup_{n=p}^{\infty} J_n\Big) = 0.$$

Thus, to show that $\mu^*(E \cap A^c) = 0$, it is enough to show that

$$E \cap A^c \subset \bigcup_{n=p}^{\infty} J_n$$

for every $p \in N$. Fixing $p \in N$ and letting $x \in E \cap A^c$, we have that $x \in E \cap A_p^c \subset U_p$, and so there is an $I \in \Omega_0$ such that $x \in I \subset U_p$. It is clear that $\delta_n < 2\mu(I_{n+1})$ and (#) shows that $\mu(I_n) \to 0$ as $n \to \infty$. Hence there is an integer n such that $\delta_n < \mu(I)$.

Therefore, by the definition of δ_n, there is an integer n such that

$$I \not\subset U_n;$$

let q be the smallest such integer. It is clear that $p < q$. We conclude that

$$I \cap A_q \neq \varnothing \text{ and } I \cap A_{q-1} = \varnothing .$$

It follows that

$$I \cap I_q \neq \varnothing$$

and, since $I \subset U_{q-1}$, we have

$$\mu(I) \leqslant \delta_{q-1} < 2\mu(I_q).$$

But $\mu(J_q) = 5\mu(I_q)$ and so the above shows that

$$I \subset J_q \subset \bigcup_{n=p}^{\infty} J_n,$$

implying that

$$x \in \bigcup_{n=p}^{\infty} J_n .$$

Hence we have that

$$E \cap A^c \subset \bigcup_{n=p}^{\infty} J_n$$

and so $\mu^*(E \cap A^c) = 0$.

Now let $\varepsilon > 0$ be given and we choose an integer p so large that

$$\sum_{n=p+1}^{\infty} \mu(I_n) < \varepsilon.$$

Then

$$E \cap A_p^c \subset (E \cap A^c) \cup \Big(\bigcup_{n=p+1}^{\infty} I_n\Big),$$

and so

$$\mu^*(E \cap A_p^c) \leqslant 0 + \mu\Big(\bigcup_{n=p+1}^{\infty} I_n\Big) < \varepsilon.$$

The proof is finished if $\mu^*(E) < \infty$.

We now consider the case when $\mu^*(E) = \infty$. For each $n \in N$, we let

$$E_n = E \cap (n, n+1)$$

and

$$\Omega_n = \{I \in \Omega : I \subset (n, n+1)\}.$$

Evidently Ω_n is a Vitali cover of E_n. Applying what we have established already, we find a countable pairwise disjoint family $\Lambda_n \subset \Omega_n$ such that $\mu^*(E_n \cap (\cup \Lambda_n)^c) = 0$ for each $n \in N$. Let

$$\Lambda = \bigcup_{n=-\infty}^{\infty} \Lambda_n .$$

Then Λ is a countable pairwise disjoint subfamily of Ω and

$$E \cap (\cup \Lambda)^c \subset N \cup [\bigcup_{n=-\infty}^{\infty} (E_n \cap (\cup \Lambda_n)^c)].$$

We see that

$$\mu^*(E \cap (\cup \Lambda)^c) \leqslant \mu(N) + \sum_{n=-\infty}^{\infty} 0 = 0.$$

The proof is finished. (The foregoing proof is due to S. Banach; see *Fund. Math.* **6** (1924), 170-188.)

PROPOSITION 3. *Let* $[a, b]$ *be a closed interval in* R^1 *and let* f *be a real-valued monotone function on* $[a, b]$. *Then* f *has a finite derivative almost everywhere on* $[a, b]$.

Proof. We assume that f is nondecreasing. This implies no loss of generality; if f were nonincreasing we would simply consider $-f$.

Let

$$E = \{x : a \leqslant x < b,\ D_- f(x) < D^+ f(x)\}.$$

We will first verify that $\mu(E) = 0$. For every pair of positive rational numbers w and v such that $w < v$, let

$$E_{w,v} = \{x \in E : D_- f(x) < w < v < D^+ f(x)\}.$$

It is clear that

$$E = \cup \{E_{w,v} : w, v \in Q, 0 < w < v\},$$

where Q denotes the set of all rational numbers. Since E is a countable union, it is enough to show that $\mu(E_{w,v}) = 0$ for all $0 < w < v$ in Q. Let us assume the contrary, namely, that there exist positive rational numbers w and v, $w < v$, such that $\mu(E_{w,v}) = \alpha > 0$. Let ε be such that

$$0 < \varepsilon < \frac{\alpha(v - w)}{w - 2v}.$$

We choose an open set $G \supset E_{w,v}$ such that $\mu(G) < \alpha + \varepsilon$. For each $x \in E_{w,v}$, there exist arbitrarily small positive numbers h such that $[x, x + b] \subset G \cap [a, b]$ and

$$f(x + h) - f(x) < wh. \tag{1}$$

The family Ω of all such closed intervals is a Vitali cover of $E_{w,v}$, and therefore by Proposition 2 there exists a finite, pairwise disjoint subfamily $\{[x_i, x_i + h_i]\}_{i=1}^m$ of Ω such that

$$\mu(E_{w,v} \cap (\bigcup_{i=1}^{m} [x_i, x_i + h_i])^c) < \varepsilon.$$

Let

$$V = \bigcup_{i=1}^{m} (x_i, x_i + h_i).$$

Then we have

$$\mu(E_{w,v} \cap V^c) < \varepsilon. \tag{2}$$

The inclusion $V \subset G$ implies that

$$\sum_{i=1}^{m} h_i = \mu(V) \leqslant \mu(G) < \alpha + \varepsilon,$$

and therefore inequality (1) yields that

$$\sum_{i=1}^{m} (f(x_i + h_i) - f(x_i)) < w \sum_{i=1}^{n} h_i < w(\alpha + \varepsilon). \tag{3}$$

Again, for all $y \in E_{w,v} \cap V$, there exist arbitrarily small positive numbers k such that $[y, y + k] \subset V$ and

$$f(y + k) - f(y) > vk. \tag{4}$$

The collection of all such closed intervals is a Vitali cover of $E_{w,v} \cap V$, and so there is a finite, pairwise disjoint family $\{[y_j, y_j + k_j]\}_{j=1}^n$ of such intervals with the property that

$$\mu(E_{w,v} \cap V \cap (\bigcup_{j=1}^{n} [y_j, y_j + k_j])^c) < \varepsilon.$$

This inequality together with the inequality (2) implies that

$$\alpha = \mu(E_{w,v}) \leqslant \mu(E_{w,v} \cap V^c) + \mu(E_{w,v} \cap V) < \varepsilon + (\varepsilon + \sum_{j=1}^{n} k_j). \tag{5}$$

Now, using (4) and (5), we get

$$v(\alpha - 2\varepsilon) < v \sum_{j=1}^{n} k_j < \sum_{j=1}^{n} (f(y_j + k_j) - f(y_j)). \tag{6}$$

Since

$$\bigcup_{j=1}^{n} [y_j, y_j + k_j] \subset \bigcup_{i=1}^{m} [x_i, x_i + h_i]$$

and f is nondecreasing, we also have

$$\sum_{j=1}^{n} (f(y_j + k_j) - f(y_j)) \leqslant \sum_{i=1}^{m} (f(x_i + h_i) - f(x_i)). \tag{7}$$

Combining (6), (7), and (3), we obtain

$$v(\alpha - 2\varepsilon) < w(\alpha + \varepsilon),$$

which contradicts our choice of ε. Thus $\mu(E) = 0$ and so $f'_{+}(x)$ exists a.e. on $[a, b]$. Similarly $f'_{-}(x)$ exists a.e. on $[a, b]$. Now we use Proposition 1 and see that $f'(x)$ exists a.e. on $[a, b]$.

It remains to verify that the set B of points x in (a, b) for which $f'(x) = \infty$ has measure zero. Let β be an arbitrary positive number. For each $x \in B$, there exist arbitrary small positive numbers h such that $[x, x + h] \subset (a, b)$ and

$$f(x + h) - f(x) > \beta h. \tag{8}$$

By Proposition 2 there exists a countable pairwise disjoint family $\{[x_n, x_n + h_n]\}$ of these intervals such that

$$\mu(B \cap (\bigcup_n [x_n, x_n + h_n])^c) = 0.$$

From this fact and (8) we get that

$$\beta\mu(B) \leqslant \beta \sum_n h_n < \sum_n (f(x_n + h_n) - f(x_n)) \leqslant f(b) - f(a).$$

Thus

$$\beta\mu(B) \leqslant f(b) - f(a) \text{ for all } \beta \in R^1,$$

which shows that $\mu(B) = 0$ and the proof is finished.

To see that the foregoing result is in some sense the best possible one, we establish Proposition 4.

PROPOSITION 4. *Let A be any subset of $[a, b]$ of measure zero. Then we can always construct a continuous nondecreasing function ψ such that for all $x \in A$ we have $\psi'(x) = \infty$.*

Proof. For each $n \in N$ there is a bounded open set G_n for which

$$G_n \supset A,\ \mu(G_n) < \frac{1}{2^n}\cdot$$

We put

$$\phi_n(x) = \mu(G_n \cap [a, x]).$$

The function ϕ_n is nondecreasing, nonnegative, and continuous and satisfies the inequality

$$\phi_n(x) < \frac{1}{2^n}\cdot$$

Thus

$$\psi(x) = \sum_{n=1}^{\infty} \phi_n(x)$$

is also nonnegative, nondecreasing, and continuous.

If $x_0 \in A$, then for sufficiently small $|h|$ the interval $[x_0, x_0 + h]$ is contained in G_n (for a fixed n). For this h we have (for simplicity we choose $h > 0$):

$$\begin{aligned}\phi_n(x_0 + h) &= \mu\{(G_n \cap [a, x_0]) \cup (G_n \cap [x_0, x_0 + h])\}\\ &= \phi_n(x_0) + h.\end{aligned}$$

It follows that

$$\frac{\phi_n(x_0 + h) - \phi_n(x_0)}{h} = 1.$$

Thus, for any natural number K and any sufficiently small $|h|$ we have that

$$\frac{\psi(x_0 + h) - \psi(x_0)}{h} \geqslant \sum_{n=1}^{K} \frac{\phi_n(x_0 + h) - \phi_n(x_0)}{h} = K$$

and therefore $\psi'(x_0) = \infty$. The proof is finished.

PROPOSITION 5. *Let f be a real-valued nondecreasing function on $[a, b]$. Then f' is measurable on $[a, b]$ and*

$$\int_a^b f'(x)\,dx \leqslant f(b) - f(a).$$

Proof. We extend the domain of definition of f by putting

$$f(x) = f(b) \text{ for } x > b.$$

Let

$$f_n(x) = n\left\{f\left(x + \frac{1}{n}\right) - f(x)\right\}$$

for $n = 1, 2, 3, \ldots$ and $a \leqslant x \leqslant b$. Then $(f_n)_{n=1}^{\infty}$ is a sequence of nonnegative measurable functions and

$$\lim_{n\to\infty} f_n(x) = f'(x)$$

for almost all $x \in (a, b)$ by Proposition 2. Thus f' is measurable. Since f' is nonnegative and (a, b) is bounded, f' is intergrable over (a, b). By Fatou's Lemma

$$\int_a^b f' \leqslant \varliminf_{n\to\infty} \int_a^b f_n = \varliminf_{n\to\infty}\left\{n \int_a^b \left[f\left(x + \frac{1}{n}\right) - f(x)\right] dx\right\}.$$

But

$$\int_a^b f\left(x + \frac{1}{n}\right) dx = \int_{a+1/n}^{b+1/n} f(x)\, dx.$$

This can easily be shown to hold for Lebesgue integrals. Note, however, that f is monotone and we can in fact regard these integrals as Riemann integrals because monotone functions have at most a countable number of discontinuities.

Continuing with the proof, we see that

$$\begin{aligned}\int_a^b \left[f\left(x + \frac{1}{n}\right) - f(x)\right] dx &= \int_b^{b+1/n} f(x)\, dx - \int_a^{a+1/n} f(x)\, dx \\ &= \frac{1}{n} f(b) - \int_a^{a+1/n} f(x)\, dx \\ &\leqq \frac{1}{n}(f(b) - f(a)).\end{aligned}$$

Therefore, we have the estimate

$$\int_a^b f' \leqq f(b) - f(a).$$

The proof is finished.

To see that strict inequality may in fact occur in the foregoing proposition we consider the following:

Example. Let P denote Cantor's ternary set. We arrange the complementary intervals into groups as follows: The first group contains the interval $(1/3, 2/3)$, the second the two intervals $(1/9, 2/9)$ and $(7/9, 8/9)$, the third group the four intervals $(1/27, 2/27)$, $(7/27, 8/27)$, $(19/27, 20/27)$, and $(25/27, 26/27)$, etc. The n-th group contains then 2^{n-1} intervals.

We define the function ω as follows:

$$\omega(x) = \frac{1}{2} \text{ for } x \in \left(\frac{1}{3}, \frac{2}{3}\right),$$

$$\omega(x) = \frac{1}{4} \text{ for } x \in \left(\frac{1}{9}, \frac{2}{9}\right), \quad \omega(x) = \frac{3}{4} \text{ for } x \in \left(\frac{7}{9}, \frac{8}{9}\right).$$

In the four intervals of the third group we set the function ω consecutively equal to $1/8$, $3/8$, $5/8$, and $7/8$. In the 2^{n-1} intervals of the n-th group we set ω consecutively equal to

$$\frac{1}{2^n}, \frac{3}{2^n}, \frac{5}{2^n}, \ldots, \frac{2^n - 1}{2^n}.$$

The function ω is in this way defined on the open set $[0, 1] - P$ and is seen to be constant on each component interval and is also seen to be nondecreasing on $[0, 1] - P$.

We extend the domain of definition of ω to all of $[0,1]$ by putting

$$\omega(0) = 0, \ \omega(1) = 1$$

and

$$\omega(x_0) = \sup\{\omega(x) : x \in [0, 1] - P, x < x_0\}.$$

We see that ω is now defined on all of $[0,1]$ and is nondecreasing throughout $[0,1]$.

Indeed, ω is continous on $[0,1]$. This follows from the fact that the set of values of the function ω on the set $[0, 1] - P$ is everywhere dense in $[0, 1]$; for if f is a nondecreasing function and has a point of discontinuity at x_0, then at least one of the intervals $(f(x_0 - 0), f(x_0))$ and $(f(x_0), f(x_0 + 0))$ must be free of values of f.

Thus, ω is a continuous and nondecreasing function on $[0,1]$. But $\omega'(x) = 0$ for $x \in [0, 1] - P$ and $\mu(P) = 0$. Hence

$$\omega' = 0 \text{ a.e. on } [0, 1].$$

We conclude that

$$\int_0^1 \omega'(x)\,dx = 0 < 1 = \omega(1) - \omega(0).$$

We now consider the following consequence of the a.e. differentiability of monotone functions.

PROPOSITION 6. *Let $(f_n)_{n=1}^{\infty}$ be a sequence of nondecreasing (or nonincreasing) real-valued functions on an interval $[a, b]$ such that*

$$\sum_{n=1}^{\infty} f_n(x) = s(x)$$

exists and is finite in $[a, b]$. Then

$$s'(x) = \sum_{n=1}^{\infty} f'_n(x)$$

a.e. in (a, b).

Proof. We shall assume that all f_n are nondecreasing. By considering the functions $f_n - f_n(a)$, we may further suppose that $f_n \geqslant 0$. Thus

$$s = \sum_{n=1}^{\infty} f_n$$

is nonnegative and nondecreasing. The derivative $s'(x)$ exists and is finite for almost all $x \in (a, b)$ by Proposition 3.

We consider the partial sums $s_n = f_1 + f_2 + \ldots + f_n$ and the remainders $r_n = s - s_n$. Each f_n has a finite derivative a.e. and hence there is a set $A \subset (a, b)$ such that

$$\mu(A^c \cap (a, b)) = 0,$$

$$s'_n(x) = f'_1(x) + f'_2(x) + \ldots + f'_n(x) < \infty$$

for all $x \in A$ and all n, and $s'(x)$ exists and is finite for $x \in A$. For any $x \in (a, b)$ and every $h > 0$ such that $x + h \in (a, b)$, it follows from the equality

$$\frac{s(x+h) - s(x)}{h} = \frac{s_n(x+h) - s_n(x)}{h} + \frac{r_n(x+h) - r_n(x)}{h}$$

that

$$\frac{s_n(x+h) - s_n(x)}{h} \leqslant \frac{s(x+h) - s(x)}{h}.$$

Hence $s_n'(x) \leqslant s'(x)$ for all $x \in A$. Evidently $s_n'(x) \leqslant s_{n+1}'(x)$, and so we have

$$s_n'(x) \leqslant s_{n+1}'(x) \leqslant s'(x)$$

for $x \in A$ and $n = 1, 2, \ldots$. Hence

$$\lim_{n\to\infty} s_n'(x) = \sum_{j=1}^{\infty} f_j'(x)$$

exists a.e. and it remains to show that

$$\lim_{n\to\infty} s_n'(x) = s'(x) \text{ a.e.}$$

Since the sequence $(s_n'(x))_{n=1}^{\infty}$ is nondecreasing for each $x \in A$, it is enough to show that $(s_n')_{n=1}^{\infty}$ has a subsequence which converges a.e. to s'. To this end, let $n_1, n_2, \ldots, n_k, \ldots$ be an increasing sequence of integers such that

$$\sum_{k=1}^{\infty} [s(b) - s_{n_k}(b)] < \infty.$$

For each n_k and for every $x \in (a, b)$, we have

$$0 \leqslant s(x) - s_{n_k}(x) \leqslant s(b) - s_{n_k}(b).$$

The terms $s(x) - s_{n_k}(x)$ are bounded by the terms $s(b) - s_{n_k}(b)$ of a convergent series of nonnegative terms and so

$$\sum_{k=1}^{\infty} [s(x) - s_{n_k}(x)]$$

converges. The terms of this series are monotone functions that have finite derivatives a.e. Therefore, the reasoning used above to show that

$$\sum_{j=1}^{\infty} f_j'(x)$$

converges a.e. also proves that

$$\sum_{k=1}^{\infty} [s'(x) - s_{n_k}'(x)]$$

converges a.e. It naturally follows that

$$\lim_{k\to\infty} s_{n_k}'(x) = s'(x)$$

a.e. and the proof is finished.

Let f be a real-valued function defined on $[a, b]$. We set

$$V_a^b f = \sup\{\sum_{k=1}^{n} |f(x_k) - f(x_{k-1})| : a = x_0 < x_1 < \ldots < x_n = b\}.$$

The extended real number $V_a^b f$ is called the *total variation of f* over $[a, b]$. If $V_a^b f < \infty$, then f is said to be of *finite variation over* $[a,b]$.

PROPOSITION 7. *If f is a real-valued function of finite variation over* $[a,b]$, *then it has a finite derivative a.e. on* $[a,b]$.

Proof. To see this we note that a function f of finite variation is the difference of two nondecreasing functions and therefore we can use Proposition 3.

Indeed. We write

$$f(x) = V_a^x f - (V_a^x f - f(x)),$$

where we define $V_a^a f = 0$. It is clear that the function $x \to V_a^x f$ is nondecreasing. The function $x \to V_a^x f - f(x)$ is also nondecreasing, for, if $x_0 > x$, then

$$V_a^{x_0} f - f(x_0) - (V_a^x f - f(x)) = V_x^{x_0} f - (f(x_0) - f(x)) \geqslant 0.$$

The proof is finished.

PROPOSITION 8. *If f is integrable on* $[a, b]$, *then the function F defined by*

$$F(x) = \int_a^x f(t)\, dt$$

is a continuous function of finite variation on $[a,b]$. *In fact,*

$$V_a^b F = \int_a^b |f(t)|\, dt.$$

Proof. For $x' > x$, the equality

$$|F(x') - F(x)| = |\int_x^{x'} f(t)\, dt|$$

holds and it is clear from Proposition 12 of Chapter 3, that F is continuous on $[a, b]$. If $a = x_0 < x_1 < \ldots < x_n = b$, then

$$\sum_{k=1}^{n} |F(x_k) - F(x_{k-1})| = \sum_{k=1}^{n} |\int_{x_{k-1}}^{x_k} f(t)\, dt| \leqslant \sum_{k=1}^{n} \int_{x_{k-1}}^{x_k} |f(t)|\, dt$$
$$= \int_a^b |f(t)|\, dt.$$

Hence

$$V_a^b F \leqslant \int_a^b |f(t)|\, dt$$

holds, and so F has finite variation over $[a, b]$.

To show that

$$V_a^b F = \int_a^b |f(t)|\, dt$$

we set for the purpose of this proof $(a, b) = E$ and

$$P = \{x \in E : f(x) \geqslant 0\}, \quad N = \{x \in E : f(x) < 0\}.$$

Then

$$\int_a^b |f(t)|\, dt = \int_P f(t)\, dt - \int_N f(t)\, dt.$$

Let $\varepsilon > 0$. By Proposition 12 of Chapter 3 there exists a $\delta > 0$ such that for an arbitrary measurable set $A \subset [a, b]$ with $\mu(A) < \delta$ we have

$$\int_A |f(t)|\, dt < \varepsilon.$$

Let P_1 and N_1 be closed sets that are contained in P and N, respectively, such that

$$\mu(P - P_1) < \delta, \quad \mu(N - N_1) < \delta.$$

Then

$$\int_a^b |f(t)|\, dt < \int_{P_1} f(t)\, dt - \int_{N_1} f(t)\, dt + 2\varepsilon.$$

By Exercise 15, part (v), of Chapter 1 we can select open sets P_2 and N_2 such that

$$P_2 \supset P_1,\ N_2 \supset N_1, \text{ and } P_2 \cap N_2 = \varnothing$$

and we may also assume that both P_2 and N_2 are contained in (a,b). We now pick two bounded open sets P_3 and N_3 such that $P_3 \supset P_1$ and $N_3 \supset N_1$ and $\mu(P_3 - P_1) < \delta$ and $\mu(N_3 - N_1) < \delta$. We put

$$P_0 = P_3 \cap P_2, \quad N_0 = N_3 \cap N_2.$$

The sets P_0 and N_0 have the following property: They are open, disjoint, and are contained in (a,b); they contain P_1 and N_1, respectively, and moreover,

$$\mu(P_0 - P_1) < \delta$$
$$\mu(N_0 - N_1) < \delta.$$

Hence

$$\int_a^b |f(t)|\,dt < \int_{P_0} f(t)\,dt - \int_{N_0} f(t)\,dt + 4\varepsilon.$$

Since P_0 is open, it can be represented uniquely up to the order of summation by the union of a countable family of disjoint open intervals. The union of a sufficiently large finite number of these component intervals yields a set P_{00} such that

$$\mu(P_0 - P_{00}) < \delta.$$

We therefore get that

$$\int_{P_0} f(t)\,dt - \int_{P_{00}} f(t)\,dt < \varepsilon.$$

Let

$$P_{00} = \bigcup_{k=1}^{n} (\alpha_k, \beta_k).$$

Then

$$\int_{P_{00}} f(t)\,dt = \sum_{k=1}^{n} \int_{\alpha_k}^{\beta_k} f(t)\,dt = \sum_{k=1}^{n} [F(\beta_k) - F(\alpha_k)]$$

and therefore

$$\int_{P_0} f(t)\,dt < \sum_{k=1}^{n} [F(\beta_k) - F(\alpha_k)] + \varepsilon.$$

In the same manner we can find a finite set of component intervals

$$\{(\sigma_j, \tau_j) : 1 \leqslant j \leqslant m\}$$

for the set N_0 such that

$$\int_{N_0} f(t) > \sum_{j=1}^{m} [F(\tau_j) - F(\sigma_j)] - \varepsilon.$$

Combining the foregoing results, we obtain

$$\int_a^b |f(t)|\,dt < \sum_{k=1}^{n} [F(\beta_k) - F(\alpha_k)] - \sum_{j=1}^{m} [F(\tau_j) - F(\sigma_j)] + 6\varepsilon.$$

Therefore

$$\int_a^b |f(t)|\,dt < \sum_{k=1}^{n} |F(\beta_k) - F(\alpha_k)| + \sum_{j=1}^{m} |F(\tau_j) - F(\sigma_j)| + 6\varepsilon.$$

The intervals (α_k, β_k) are pairwise disjoint; moreover, they do not intersect the pairwise disjoint interval (σ_j, τ_j). We have

$$\sum_{k=1}^{n} |F(\beta_k) - F(\alpha_k)| + \sum_{j=1}^{m} |F(\tau_j) - F(\sigma_j)| \leqslant V_a^b F.$$

Thus

$$\int_a^b |f(t)|\,dt < V_a^b F + 6\varepsilon$$

and so

$$\int_a^b |f(t)|\,dt = V_a^b F,$$

since ε is arbitrary. This completes the proof.

PROPOSITION 9. *If f is integrable on $[a,b]$ and*

$$\int_a^x f(t)\,dt = 0$$

for all $x \in [a,b]$, then $f = 0$ a.e. in $[a,b]$.

Proof. If the proposition is false then at least one of the sets

$$\{t : f(t) < 0\}, \qquad \{t : f(t) > 0\}$$

has positive measure. If $\mu\{t : f(t) > 0\} > 0$ then we can find a $\delta > 0$ for which $\mu(E) > 0$, where $E = \{t : f(x) > \delta\}$. We now choose a closed set $F \subset E$ with $\mu(F) > 0$, and consider the open set $G = (a,b) - F$. Then

$$0 = \int_a^b f(t)\,dt = \int_F f(t)\,dt + \int_G f(t)\,dt.$$

Now G is the disjoint union of a countable collection $\{(a_n, b_n)\}_{n=1}^{\infty}$ of open intervals and

$$\int_{a_n}^{b_n} f(t)\,dt = 0$$

for each n. But

$$\int_G f(t)\,dt = \sum_{n=1}^{\infty} \int_{a_n}^{b_n} f(t)\,dt$$

so that $\int_F f(t)\,dt = 0$ and this contradicts

$$\int_F f(t)\,dt \geqslant \delta\mu(F) > 0. \qquad \text{Q.E.D.}$$

PROPOSITION 10. *If f is bounded and measurable on $[a,b]$ and $F : [a, b] \to R^1$ satisfy*

$$F(x) = F(a) + \int_a^x f(t)\,dt,$$

then $F' = f$ a.e. in $[a, b]$.

Proof. By Proposition 8, F is of finite variation over $[a, b]$ and so F' exists a.e. in $[a, b]$. Let $|f| \leqslant K$. Then setting

$$f_n(x) = \frac{F(x+h) - F(x)}{h}$$

with $h = 1/n$, we have

$$f_n(x) = \frac{1}{h} \int_x^{x+h} f(t)\,dt,$$

and so $|f_n| \leqslant K$. Since $f_n(x) \to F'(x)$ a.e., the Bounded Convergence Theorem implies that

$$\begin{aligned}
\int_a^c F'(x)\,dx &= \lim_{n\to\infty} \int_a^c f_n(x)\,dx = \lim_{h\to 0} \frac{1}{h} \int_a^c [F(x+h) - F(x)]\,dx \\
&= \lim_{h\to 0}\left[\frac{1}{h} \int_c^{c+h} F(x)\,dx - \frac{1}{h} \int_a^{a+h} F(x)\,dx\right] \\
&= F(c) - F(a) = \int_a^c f(x)\,dx,
\end{aligned}$$

since F is continuous. Hence

$$\int_a^c [F'(x) - f(x)]\, dx = 0$$

for all $c \in [a, b]$ and so $F'(x) = f(x)$ a.e. by Proposition 9. Q.E.D.

PROPOSITION 11. *Let f be integrable on $[a,b]$ and suppose $F : [a, b] \to R^1$ satisfies*

$$F(x) = F(a) + \int_a^x f(t)\, dt.$$

Then $F' = f$ a.e. in $[a, b]$.

Proof. From the definition of the integral it is sufficient to prove the proposition when $f \geqslant 0$. Let

$$f_n(x) = \min\{n, f(x)\}.$$

Then $f - f_n \geqslant 0$ and so

$$G_n(x) = \int_a^x [f(t) - f_n(t)]\, dt$$

is an increasing function of x that must have a finite derivative almost everywhere and this derivative will be nonnegative. By Proposition 10,

$$\frac{d}{dx} \int_a^x f_n(t)\, dt = f_n(x) \text{ a.e.,}$$

and so

$$F'(x) = \frac{d}{dx} G_n(x) + \frac{d}{dx} \int_a^x f_n(t)\, dt \geqslant f_n(x) \text{ a.e.}$$

Since n is arbitrary,

$$F'(x) \geqslant f(x) \text{ a.e.}$$

Hence

$$\int_a^b F'(x)\, dx \geqslant \int_a^b f(x)\, dx = F(b) - F(a)$$

and by Proposition 5 we have

$$\int_a^b F'(x)\,dx = F(b) - F(a) = \int_a^b f(x)\,dx$$

and

$$\int_a^b [F'(x) - f(x)]\,dx = 0.$$

Since $F'(x) - f(x) \geqslant 0$ a.e., this implies that $F'(x) - f(x) = 0$, and so $F'(x) = f(x)$ a.e. The proof is finished.

If f is a real-valued function on $[a, b]$, then f is said to be *absolutely continuous* on $[a, b]$ if, given $\varepsilon > 0$, there is a $\delta > 0$ such that

$$\sum_{k=1}^{n} |f(b_k) - f(a_k)| < \varepsilon$$

for every finite, pairwise disjoint, family $\{(a_k, b_k)\}_{k=1}^{n}$ of open subintervals of $[a, b]$ for which

$$\sum_{k=1}^{n} (b_k - a_k) < \delta.$$

We observe that an absolutely continuous function is continuous and that every indefinite integral

$$F(x) = \int_a^x f(t)\,dt$$

of an integrable function f on $[a, b]$ is absolutely continuous by Proposition 12 of Chapter 3.

PROPOSITION 12. *If f is absolutely continuous on $[a,b]$, then it is of finite variation.*

Proof. Since f is absolutely continuous on $[a, b]$, there is a $\delta > 0$ such that for every finite, pairwise disjoint, family $\{(a_k, b_k)\}_{k=1}^{n}$ of open subintervals of $[a, b]$ of total length

$$\sum_{k=1}^{n} (b_k - a_k) < \delta$$

we have

$$\sum_{k=1}^{n} |f(b_k) - f(a_k)| < 1.$$

Partition $[a, b]$ by the points

$$a = c_0 < c_1 < \ldots < c_m = b$$

such that

$$c_{k+1} - c_k < \delta \qquad (k = 0, 1, \ldots, m - 1).$$

Then, for every subdivision of $[c_k, c_{k+1}]$ the sum of the absolute values of the increments of f over the intervals of subdivision is smaller than 1; hence

$$V_{c_k}^{c_{k+1}} f \leqslant 1$$

and therefore

$$V_a^b f \leqslant m.$$

This proves the proposition.

Remark. It is easy to see from the foregoing proposition that not every continuous function can be absolutely continuous. For example, let

$$f(x) = \begin{cases} x \cos \pi/2x & \text{for } 0 < x \leqslant 1 \\ 0 & \text{for } x = 0. \end{cases}$$

Choosing the partition

$$0 < \frac{1}{2n} < \frac{1}{2n-1} < \cdots < \frac{1}{3} < \frac{1}{2} < 1,$$

we see that

$$V_0^1 f = \infty.$$

From Proposition 12 we immediately obtain the next proposition.

PROPOSITION 13. *If f is absolutely continuous on $[a,b]$, then f has a finite derivative a.e. in $[a,b]$.*

PROPOSITION 14. *If f is absolutely continuous on $[a,b]$ and $f'(x) = 0$ a.e. in (a, b), then f is a constant.*

Proof. We will show that $f(c) = f(a)$ for all $c \in (a, b]$. Thus let $c \in (a, b]$ and $\varepsilon > 0$ be arbitrary. Select a number $\delta > 0$ corresponding to a given ε for which the condition in the definition of absolute continuity is satisfied. Let $E = \{x \in (a,c) : f'(x) = 0\}$. Clearly $\mu(E) = c - a$. For each $x \in E$ there exists arbitrarily small $h > 0$ such that $[x, x + h] \subset (a, c)$ and

$$|f(x + h) - f(x)| < \frac{\varepsilon h}{c - a}. \tag{1}$$

The family of all such intervals $[x, x + h]$ is a Vitali cover of E, and so, by Proposition 2 there exists a finite pairwise disjoint family $\{[x_k, x_k + h_k]\}_{k=1}^n$ of these intervals such that

$$\mu(E \cap (\bigcup_{k=1}^{n} [x_k, x_k + h_k])^c) < \delta.$$

Then

$$\mu\{(a, c)\} = \mu(E) < \delta + \sum_{k=1}^{n} h_k. \tag{2}$$

We can assume that $x_1 < x_2 < \ldots < x_n$. Then it follows from (2) that the sum of the lengths of the open intervals

$$(a, x_1), \quad (x_1 + h_1, x_2), \ldots, (x_n + h_n, c)$$

complementary to

$$\bigcup_{k=1}^{n} [x_k, x_k + h_k]$$

is less than δ, and so, in view of our choice of δ, we have

$$\begin{aligned} &|f(a) - f(x_1)| + \sum_{k=1}^{n-1} |f(x_k + h_k) - f(x_{k+1})| \\ &\quad + |f(x_n + h_n) - f(c)| < \varepsilon. \end{aligned} \tag{3}$$

Combining inequalities (1) and (3) we get

$$\begin{aligned} |f(a) - f(c)| &\leqslant |f(a) - f(x_1)| + \sum_{k=1}^{n-1} |f(x_k + h_k) - f(x_{k-1})| \\ &\quad + |f(x_n + h_n) - f(c)| + \sum_{k=1}^{n} |f(x_k + h_k) - f(x_k)| \\ &< \varepsilon + \sum_{k=1}^{n} \frac{\varepsilon h_k}{c - a} \leqslant 2\varepsilon. \end{aligned}$$

Since ε is arbitrary, it follows that $f(c) = f(a)$. *Q.E.D.*

PROPOSITION 15. *Let f be absolutely continuous on $[a, b]$. Then f' is integrable over $[a, b]$ and*

$$f(x) = f(a) + \int_a^x f'(t)\,dt$$

for every $x \in [a, b]$.

Proof. From Propositions 12, 13, and 5 it follows that f' is integrable. Let

$$g(x) = \int_a^x f'(t)\,dt.$$

Then g is absolutely continuous and $g'(x) = f'(x)$ a.e. by Proposition 10. Thus the function $h = f - g$ is absolutely continuous and $h'(x) = f'(x) - g'(x)$ a.e. By Proposition 14, h is a constant. Therefore

$$f(x) = h(x) + g(x) = h(a) + \int_a^x f'(t)\,dt = f(a) + \int_a^x f'(t)\,dt$$

for all $x \in [a, b]$ and the proof is finished.

Summarizing the foregoing results we get the following.

PROPOSITION 16. *A function f on $[a, b]$ has the form*

$$f(x) = f(a) + \int_a^x g(t)\,dt$$

for some integrable function g on $[a, b]$ if and only if f is absolutely continuous on $[a, b]$. In this case we have $g(x) = f'(x)$ a.e. on (a, b).

The formula for *integration by parts* holds for absolutely continuous functions and Lebesgue integrals.

PROPOSITION 17. *Let f and g be integrable functions over $[a, b]$, let*

$$F(x) = \alpha + \int_a^x f(t)\,dt,$$

and let

$$G(x) = \beta + \int_a^x g(t)\,dt.$$

Then

$$\int_a^b G(t)f(t)\,dt + \int_a^b g(t)F(t)\,dt = F(b)G(b) - F(a)G(a).$$

Proof. Let A and B denote the largest values of $|F(x)|$ and $|G(x)|$ on $[a, b]$. Then

$$|F(v)G(v) - F(w)G(w)| \leqslant A|G(v) - G(w)| + B|F(v) - F(w)|$$

shows that FG is also absolutely continuous on $[a, b]$. Hence FG is differentiable a.e. and

$$(FG)' = FG' + F'G$$

as can easily be seen. By Proposition 10 we have that $G' = g$ a.e. and $F' = f$ a.e. and the formula for integration by parts follows from Proposition 15.
Q.E.D.

Remark. The foregoing proposition in conjunction with Proposition 16 yields: If f and g are absolutely continuous functions on $[a, b]$, then

$$\int_a^b f(t)g'(t)\,dt + \int_a^b f'(t)g(t)\,dt = f(b)g(b) - f(a)g(a).$$

Given a set $A \subset R^1$ and let $x \in R^1$. Consider the ratio

$$\frac{\mu^*(I \cap A)}{\mu^*(I)}$$

for all intervals I containing x; here $\mu^*(E)$ denotes the Lebesgue outer measure of a subset E on the real line. If this ratio converges to a limit as $\mu^*(I) \to 0$, then this limit is called the *density of A at x* and we denote it by $\tau(x, A)$. The point x is called a *point of density of A if* $\tau(x, A) = 1$ and a *point of dispersion of A* if $\tau(x, A) = 0$.

The following result is a direct consequence of Proposition 16.

Proposition 18. *If $A \subset R^1$ and A is measurable, then*

$$\tau(x, A) = \begin{cases} 1 \text{ for almost all } x \in A, \\ 0 \text{ for almost all } x \in R^1 - A. \end{cases}$$

Proof. Suppose $a < x < b$. Then the characteristic function of A, i.e., χ_A, is integrable over $[a, b]$. Hence

$$F(x) = \int_a^x \chi_A(t)\,dt$$

is differentiable almost everywhere and

$$F'(x) = \begin{cases} 1 \text{ for almost all } x \text{ in } [a, b] \cap A, \\ 0 \text{ for almost all } x \text{ in } [a, b] \cap (R^1 - A). \end{cases}$$

But if x is such that $F'(x) = 1$, there is for each $\varepsilon > 0$ a $\delta > 0$ such that

$$1 \geqslant \frac{\mu([x, x + h] \cap A)}{h} > 1 - \varepsilon \text{ for } 0 < h < \delta,$$

$$1 \geqslant \frac{\mu([x - k, x] \cap A)}{k} > 1 - \varepsilon \text{ for } 0 < k < \delta;$$

hence

$$1 \geqslant \frac{\mu([x-k, x+h] \cap A)}{h+k} > 1 - \varepsilon \text{ for } 0 < h, k < 1,$$

which is exactly the condition for $\tau(x, A) = 1$.

A similar proof shows that, at points x where $F'(x) = 0$, we have $\tau(x, A) = 0$. Another argument could of course be based on the equation:

$$1 = \frac{\mu([x-k, x+h] \cap A)}{h+k} + \frac{\mu([x-k, x+h] \cap (R^1 - A))}{h+k}$$

for all h and k. Q.E.D.

Concluding Remarks.
The Proof of Proposition 3 was based on Vitali's Covering Theorem. There are various other proofs of Proposition 3. One of these is the ingenious proof due to F. Riesz, which is sketched in Problem 26.

EXERCISES

1. Show directly that the function ω on $[0, 1]$ constructed in the example following Proposition 5 is not absolutely continuous on $[0, 1]$.
2. Let g be a strictly increasing, absolutely continuous function on $[a, b]$ with $g(a) = c$, $g(b) = d$.
   (a) Show that for any open set $G \subset [c, d]$ $\mu(G) = \int_E g'(x)\,dx$, where $E = g^{-1}(G)$.
   (b) Let $H = \{x : g'(x) \neq 0\}$. If A is a subset of $[c, d]$ with $\mu(A) = 0$, then $g^{-1}(A) \cap H$ has measure zero.
   (c) If B is a measurable subset of $[c, d]$, then $F = g^{-1}(B) \cap H$ is measurable and

$$\mu(B) = \int_F g' = \int_a^b \chi_B(g(x))g'(x)\,dx.$$

   (d) If f is a nonnegative measurable function on $[c, d]$, then $(f \circ g)g'$ is measurable on $[a, b]$ and

$$\int_c^d f(t)\,dt = \int_a^b f(g(x))g'(x)\,dx.$$

   (e) If f is an integrable function on $[c, d]$, then

$$\int_c^d f(t)\,dt = \int_a^b f(g(x))g'(x)\,dx.$$

(This is the change-of-variable formula for Lebesgue integrals.)

3. A real-valued function ϕ defined on (a, b) is called *convex* if for each $x, y \in (a, b)$ and each λ, $0 \leqslant \lambda \leqslant 1$, we have

$$\phi[\lambda x + (1 - \lambda)y] \leqslant \lambda\phi(x) + (1 - \lambda)\phi(y).$$

(i) Let $x_1 \leqslant x_3 \leqslant x_2$. Show that

$$\frac{\phi(x_3) - \phi(x_1)}{x_3 - x_1} \leqslant \frac{\phi(x_2) - \phi(x_1)}{x_2 - x_1} \leqslant \frac{\phi(x_2) - \phi(x_3)}{x_2 - x_3}$$

[*Hint*: Let

$$x_3 = \frac{x_2 - x_3}{x_2 - x_1}x_1 + \frac{x_3 - x_1}{x_2 - x_1}x_2.\Big]$$

(ii) Let $t_j \leqslant t_{j+1}$ for $j = 1, 2, \ldots, 6$. Show that

$$\frac{\phi(t_{j+1}) - \phi(t_j)}{t_{j+1} - t_j} \leqslant \frac{\phi(t_{j+2}) - \phi(t_{j+1})}{t_{j+2} - t_{j+1}}.$$

(iii) Show that ϕ is absolutely continuous on each closed subinterval $[c, d] \subset (a, b)$.

(iv) Verify that the right and left derivatives $\phi'_+(x)$ and $\phi'_-(x)$ at each $x \in (a, b)$ exist and are equal to each other except on a countable set. Moreover, $\phi'_+(x)$ and $\phi'_-(x)$ are monotone nondecreasing functions and $\phi'_-(x) \leqslant \phi'_+(x)$ for each $x \in (a, b)$.

(v) If ϕ is convex on $(-\infty, \infty)$ and f is integrable on $[0, 1]$, show that $\int \phi(f(t))\,dt \geqslant \phi[\int f(t)\,dt]$.

4. Let $\{a_n\}_{n=1}^{\infty}$ be a set of distinct points in the interval $[a, b]$. Let $(\alpha_n)_{n=1}^{\infty}$ and $(\beta_n)_{n=1}^{\infty}$ be sequences of real numbers such that

$$\sum_{n=1}^{\infty} |\alpha_n| < \infty$$

and

$$\sum_{n=1}^{\infty} |\beta_n| < \infty.$$

Define

$$s_n(x) = \begin{cases} 0 & \text{if } x < a_n \\ \alpha_n & \text{if } x = a_n \\ \beta_n & \text{if } x > a_n \end{cases}$$

Show that

$$s(x) = \sum_{n=1}^{\infty} s_n(x)$$

has a finite derivative a.e., and that $s'(x) = 0$ a.e.
[*Hint*: The function s has finite variation; find each V_a^x by using the numbers $|\alpha_n|$ and $|\beta_n|$. Then use Proposition 6.]

5. (a) Prove that a function f on R^1 has the form

$$f(x) = \int_{-\infty}^{x} \phi(t)\,dt$$

for some integrable function ϕ on R^1 if and only if f is absolutely continuous on $[-K, K]$ for all $K > 0$, $V_{-\infty}^{\infty} f$ is finite and

$$\lim_{x\to-\infty} f(x) = 0.$$

(b) Show that if f and g are functions on R^1 satisfying the conditions of part (a), then

$$\begin{aligned}\int_{R^1} f(t)g'(t)\,dt + \int_{R^1} f'(t)g(t)\,dt &= \int_{R^1} f'(t)\,dt \cdot \int_{R^1} g'(t)\,dt \\ &= \lim_{x\to\infty} f(x) \cdot \lim_{x\to\infty} g(x).\end{aligned}$$

6. Let f be an integrable function over $[a, b]$.

(a) Show that there exists a set $E \subset (a, b)$ such that $\mu([a, b] \cap E^c) = 0$ and

$$\lim_{h\downarrow 0} \int_{x}^{x+h} |f(t) - \alpha|\,dt = \lim_{h\downarrow 0} \int_{x-h}^{x} |f(t) - \alpha|\,dt = |f(x) - \alpha|$$

for all $\alpha \in R^1$ and all $x \in E$. [Hints; Let $\{\beta_n\}_{n=1}^{\infty}$ be any countable dense subset of R^1. The functions g_n defined by $g_n(t) = |f(t) - \beta_n|$ for $n \in N$ are integrable over $[a, b]$. By Proposition 11 there are sets $E_n \subset (a, b)$ such that $\mu(E_n^c \cap [a, b]) = 0$ and

$$\lim_{h\downarrow 0}(1/h) \int_{x}^{x+h} g_n(t)\,dt = \lim_{h\downarrow 0}(1/h) \int_{x-h}^{x} g_n(t)\,dt = g_n(x)$$

for all $x \in E_n$.

Let

$$E = \bigcap_{n=1}^{\infty} E_n.$$

Then $\mu(E^c \cap [a,b]) = 0$. For $\varepsilon > 0$ and $\alpha \in R^1$, pick an n such that $|\beta_n - \alpha| < \varepsilon/3$. Then $||f(t) - \alpha| - |f(t) - \beta_n|| \leqq |\beta_n - \alpha| < \varepsilon/3$ for all $t \in [a,b]$. It follows that

$$\left| \frac{1}{h} \int_x^{x+h} |f(t) - \alpha| \, dt - \frac{1}{h} \int_x^{x+h} |f(t) - \beta_n| \, dt \right| \leqq \frac{1}{h} \int_x^{x+h} \frac{\varepsilon}{3} \, dt = \frac{\varepsilon}{3}$$

and therefore

$$\begin{aligned} &\left| \frac{1}{h} \int_x^{x+h} |f(t) - \alpha| \, dt - |f(x) - \alpha| \right| \\ &\qquad \leqq \left| \frac{1}{h} \int_x^{x+h} \left| f(t) - \alpha \right| \, dt - \frac{1}{h} \int_x^{x+h} \left| f(t) - \beta_n \right| \, dt \right| \\ &\qquad + \left| \frac{1}{h} \int_x^{x+h} g_n(t) \, dt - g_n(x) \right| + |\beta_n - \alpha| \\ &\qquad \leqq \frac{\varepsilon}{3} + \frac{\varepsilon}{3} + \frac{\varepsilon}{3} = \varepsilon \end{aligned}$$

if $x \in E$ and $0 < h < h_0$, where h_0 depends on ε and n. But n depends only on ε and α. Thus we have

$$\lim_{h \downarrow 0} \frac{1}{h} \int_x^{x+h} |f(t) - \alpha| \, dt = |f(x) - \alpha|$$

for all $x \in E$ and $\alpha \in R^1$.]

(b) Show that

$$\lim_{h \downarrow 0} \int_0^h |f(x+t) + f(x-t) - 2f(x)| \, dt = 0$$

for almost all $x \in (a,b)$.
[*Hint*: For fixed $x \in (a,b)$, write

$$\begin{aligned} \frac{1}{h} \int_0^h |f(x+t) + f(x-t) - 2f(x)| \, dt &\leqslant \frac{1}{h} \int_x^{x+h} |f(t) - f(x)| \, dt \\ &\quad + \frac{1}{h} \int_{x-h}^{x} |f(t) - f(x)| \, dt. \end{aligned}$$

Apply part (a) with $\alpha = f(x)$.]

7. (a) Let f be a continuous real-valued function on $[a, b]$. Define

$$E = \{x : a \leqslant x < b,\ D^+ f(x) \leqslant 0\}.$$

Supposing that $f(E)$ contains no interval, show that f is nondecreasing.
[*Hint*: If $f(c) > f(d)$, where $a \leqslant c < d \leqslant b$, choose any $y_0 \in (f(d), f(c))$ and let $x_0 = \sup\{x : c \leqslant x < d,\ f(x) \geqslant y_0\}$. Show that $f(x_0) = y_0$, and hence that $D^+ f(x_0) \leqslant 0$. This implies that $f(E) \supset (f(d), f(c))$, a contradiction to the hypothesis.]

(b) Let f be as in part (a). Suppose that $D_+ f$ is nonnegative a.e. on $[a, b]$. Suppose also the set $B = \{x : a \leqslant x < b,\ D_+ f(x) = -\infty\}$ is countable. Prove that f is nondecreasing.
[*Hint*: Let $A = \{x : x \in (a, b),\ D^+ f(x)$ or $D_+ f(x)$ is negative and finite $\}$. Let ψ be as in Proposition 4 of this chapter for the set A. Let $\sigma(x) = \psi(x) + x$ and let $g = f + \varepsilon\sigma$, where ε is a positive number. Show that $D^+ g(x) = \infty$ for $x \in A$ and that $D^+ g(x)$ is positive for $x \in A^c \cap B^c$. Hence the set $E = \{x : D^+ g(x) \leqslant 0\}$ is contained in the countable set B and $g(E)$ can accordingly contain no interval. By part (a), g is nondecreasing. But ε is arbitrary, so f also must be nondecreasing.]

(c) Let f be as in part (a). Suppose that $f'(x)$ exists and is finite for all but a countable set x in (a, b) and that f' is integrable in $[a, b]$. Show that

$$f(x) - f(a) = \int_a^x f'(t)\, dt$$

for all $a \leqslant x \leqslant b$. This means in particular that f is absolutely continuous.
[*Hint*: Let $g_n = \max\{f', -n\}$ and $f_n(x) = \int_a^x g_n(t)\, dt$.
Then by the Dominated Convergence Theorem

$$\lim_{n \to \infty} f_n(x) = \int_a^x f'(t)\, dt.$$

Show next that

$$D_+(f_n - f)(x) = f'_n(x) - f'(x) = g_n(x) - f'(x) \geqslant 0$$

a.e. on (a, b). Since

$$\frac{f_n(x+h) - f_n(x)}{h} \geqslant \frac{1}{h} \int_x^{x+h} (-n)\, dt = -n,$$

$D_+(f_n - f)(x)$ is greater than $-\infty$ except on the countable set where f' is nonfinite or does not exist. By part (b), $f_n - f$ is nondecreasing. Thus

$$f_n(x) - f(x) \geqslant f_n(a) - f(a) = -f(a)$$

and so

$$\int_a^x f'(t)\, dt = \lim_{n\to\infty} f_n(x) \geqslant f(x) - f(a).$$

Replace f by $-f$ to reverse the last inequality.]

8. (a) Extend Proposition 18 to nonmeasurable sets A by showing that for any $A \subset R^1$, $\tau(x, A) = 1$ for all x in A except a subset of measure zero.
   (b) Deduce that a set $A \subset R^1$ is measurable if and only if $\tau(x, A) = 0$ for almost all x in $R^1 - A$.
9. If f is a real-valued function over R^1, we say that f is *approximately continuous at* x_0 provided that for every $\varepsilon > 0$, x_0 is a point of dispersion for the set $\{x : |f(x) - f(x_0)| \geqslant \varepsilon\}$. Show that f is measurable if and only if it is approximately continuous almost everywhere.
10. Let f be a continuous real-valued function on $[a, b]$. For each $y \in R^1$, let $A_y = \{x \in [a, b] : f(x) = y\}$. Define ν on R^1 by

$$\nu(y) = \begin{cases} \overline{A}_y & \text{if } A_y \text{ is finite,} \\ \infty & \text{if } A_y \text{ is infinite,} \end{cases}$$

where $\overline{A}_y$ is to denote here the cardinal number of the set A_y. The function ν is known as the *Banach indicatrix of f*. Prove that ν is measurable and that

$$\int_{R^1} \nu(y)\, dy = V_a^b f.$$

[*Hint*: Let $\Delta = \{a = x_0 < x_1 < \ldots < x_n = b\}$ be a subdivision of $[a, b]$. We define the mesh of Δ, written $|\Delta|$, as

$$|\Delta| = \max\{x_k - x_{k-1} : 1 \leqslant k \leqslant n\}.$$

Let $\Delta_1 \subset \Delta_2 \subset \ldots$ be a sequence of subdivisions of $[a, b]$ such that $|\Delta_n| \to 0$ as $n \to \infty$. Let $\Delta_n = \{a = x_0^{(n)} < \ldots < x_{m_n}^{(n)} = b\}$. For each n define

$$\nu_n = \sum_{k=1}^{m_n} \chi_{B_{n,k}}, \text{ where}$$

$B_{n,k} = f([x_{k-1}{}^{(n)}, x_k{}^{(n)}])$. Show that $\nu_n(y) \to \nu(y)$ for almost all $y \in R^1$ and then use the Monotone Convergence Theorem.]

11. Let f be a function with domain $[a, b] \subset R^1$ and range $[\alpha, \beta] \subset R^1$. If $\mu(E) = 0$ implies $\mu(f(E)) = 0$ for all $E \subset [a, b]$, then f is said to *satisfy the condition N*.
    (a) Show that $f(E)$ is measurable for every measurable set E if and only if f satisfies the condition N.
    (b) Absolutely continuous functions satisfy the condition N.
    (c) If f is continuous and of finite variation and if it satisfies the condition N, then it is absolutely continuous.
12. Let f and g be real-valued functions on a bounded interval $[a, b]$. Suppose that f is continuous and g is absolutely continuous on $[a, b]$. Show that the Stieltjes integral

$$\int_a^b f(x)\, dg(x)$$

is equal to the Lebesgue integral

$$\int_a^b f(x) g'(x)\, dx.$$

Note that everything remains intact, if we replace for f, the requirement of continuity by the assumption that f is of finite variation over $[a, b]$.

13. Calculate the four Dini derivatives of the following two functions:
    (i) $f(0) = 0, f(x) = x \sin(1/x)$ if $x \neq 0$,
    (ii) $g(0) = 0, g(x) = x\{1 + \sin(\log x)\}$ if $x > 0$, and $g(x) = x + \sqrt{-x} \sin^2(\log -x)$ if $x < 0$.

    [*Hints*: $D^+ f(0) = D^- f(0) = 1, D_+ f(0) = D_- f(0) = -1$; $D^+ g(0) = 2, D_+ g(0) = 0, D^- g(0) = 1, D_- g(0) = -\infty$.]
14. Let f and g be absolutely continuous function. Then $f(g)$ is absolutely continuous if and only if it has finite variation.
15. Let f and g be integrable over R^1 and such that $\int_{-\infty}^{\infty} (f(x)h(x) + g(x)h'(x))\, dx = 0$ for all functions $h : R^1 \to R^1$ such that h and h' are continuous and $h(x) = 0$ if x is outside some finite interval. Show that $g(x) = \int_{-\infty}^{x} f(t)\, dt$ a.e.
16. Show that an absolutely continuous function $f : [a, b] \to R^1$ transforms:
    (a) sets of measure zero into sets of measure zero,
    (b) measurable sets into measurable sets.

17. Let $f \in L^1[a,b]$. Prove that

$$\lim_{h\to 0} \frac{1}{h} \int_x^{x+h} |f(t) - f(x)|\, dt = 0$$

for almost all points x of $[a,b]$.

18. Verify that F has a representation of the form

$$F(x) = C + \int_a^x f(t)\, dt$$

with $f \in L^p[a,b]$ for $p > 1$ and $b - a < \infty$ if and only if for every partition $a = x_0 < x_1 < \ldots < x_n = b$ we have that

$$\sum_{k=0}^{n-1} \frac{|F(x_{k+1}) - F(x_k)|^p}{(x_{k+1} - x_k)^{p-1}} \leqq K,$$

where K is independent of the mode of partition. What can be said about the case $p = 1$?

19. Let $g(x) = x$ for $0 \leqq x \leqq 1$ and $g(x) = 2 - x$ for $0 \leqq x \leqq 2$ and extend the definition of $g(x)$ to all real x by setting $g(x + 2) = g(x)$. Define

$$f(x) = \sum_{n=0}^{\infty} \left(\frac{3}{4}\right)^n g(4^n x).$$

Then f is continuous on R^1 but has no finite derivative at any point.

20. Let f be continuous on $[0, 1]$. Suppose that

$$(*) \qquad \int_0^1 x^n f(x)\, dx = 0$$

for $n = 0, 1, 2, \ldots$. Show that f is identically 0 on $[0, 1]$. If f is merely integrable on $[0, 1]$ and if f satisfies condition (*), verify that $f = 0$ a.e. in $[0, 1]$.

[*Hint*: Look at the discussion in connection with Exercise 34.]

21. Let $F(x) = x^2 \sin(1/x)$. Is it true that

$$\int_{-1}^{1} F'(x)\, dx = F(1) - F(-1)\ ?$$

22. Let $f(x) = x^p \sin(1/x^q)$ for $0 < x \leqq 1$. Show that if $0 < q < p$, then f is absolutely continuous, but if $0 < p \leqq q$, then f is not even of bounded variation.

23. Let functions f and g be defined on $(0, 1]$ by

$$f(x) = x^2|\sin(1/x)|, \quad g(x) = x^{1/2}.$$

Show that f and g are each absolutely continuous; the composite function $f(g)$ is absolutely continuous, but $g(f)$ is not.

24. Show that if f and g are each absolutely continuous and g is monotone, then $f(g)$ is absolutely continuous.

25. Let ω denote the Cantor function defined after Proposition 5 and let us set $\omega(x) = 0$ for $x < 0$ and $\omega(x) = 1$ for $x > 1$. Then $\omega'(x) = 0$ a.e. on R^1. If we have the interval $[a, b]$, we define the corresponding Cantor function in a similar way, namely, we set it equal to $\omega((x - a)/(b - a))$ for $0 \leqq x \leqq 1$. Now, let $I_1, I_2, \ldots$ be the intervals $[0, 1]$, $[0, 1/2]$, $[1/2, 1]$, $[0, 1/4], \ldots, [0, 1/8], \ldots$ and let ω_n be the Cantor function corresponding to I_n. Then

$$f(x) = \sum_{n=1}^{\infty} \omega_n(x)/2^n$$

is continuous and strictly increasing on $I = [0, 1]$; moreover, $f'(x) = 0$ a.e. (by Proposition 6).

N.B. Another example of a strictly increasing continuous function whose derivative is zero almost everywhere can be found on pp. 48-49 of *Functional Analysis* by F. Riesz and B. Sz.-Nagy, Ungar, New York, 1955.

26. Here we sketch another proof for Proposition 3; this method of proof is due to F. Riesz. We begin with a few definitions. Let $g : D \to R_e^1$, where $D \subset R^1$. We then define a function on $\overline{D}$ to R_e^1, called the *limit superior* of g, denoted by g^s, as follows:

$$g^s(x) = \inf_{U \in U(x)} \sup\{g(t) : t \in U\}, \qquad x \in \overline{D},$$

where $U(x)$ denotes the class of all sets $U = D \cap V$, with V being any neighborhood of x. The function $g : D \to R_e^1$ is said to be *upper semicontinuous* at a point $x \in D$ if $g^s(x) = g(x)$.

LEMMA. *Let g be a bounded function on the interval (a, b). We define a subset E of (a, b) as follows:*

$$E = \{x : a < x < b \text{ and for } x \text{ there exists an } x_1 > x \text{ with } g(x_1) > g^s(x)\}.$$

If E is nonempty, it is an open set and relative to its representation

$$\bigcup_{j=1}^{\infty} (a_j, b_j)$$

as union of disjoint open intervals we have that $g(x) \leqq g^s(b_j)$ *for* $a_j < x < b_j$ *and all* j.

Proof. The limit superior g^s is upper semicontinuous on $[a,b]$; hence E is an open set. If E is nonempty, let (a_j, b_j) be one of the open component intervals of E and let $a_j < x_1 < b_j$. We put $x_2 = \sup\{x : x_1 \leqq x \leqq b_j \text{ and } g(x_1) \leqq g^s(x)\}$; since g^s is upper semicontinuous, we get that $g(x_1) \leqq g^s(x_2)$. If in this inequality $x_2 = b_j$, then the lemma is proved. If, however, $x_2 < b_j$, then $x_2 \in E$, thus there is an $x_3 > x_2$ with $g(x_3) > g^s(x_2)$ and so $g^s(x_3) \geqq g(x_3) > g^s(x_2) \geqq g(x_1)$. The assumption $x_3 \leqq b_j$ leads to the contradiction $x_2 \geqq x_3$; hence $x_3 > b_j$. Since $b_j \notin E$, we have that $g^s(b_j) \geqq g(x_3)$; taking into account the inequality further above, we obtain $g(x_1) < g^s(b_j)$. But this is what we have set out to do.

THEOREM. *Let f be a bounded monotone function on an open interval I. Then f has a finite derivative almost everywhere.*

Proof. We proceed in four steps.

(i) We have to show that $D^+f \leqq D_-f$ and $D^-f \leqq D_+f$ hold a.e. Since f is assumed to be monotone, it will be sufficient to verify only one of these two inequalities, say, the first inequality; for, if f is monotone, then the function g defined by $g(x) = -f(-x)$ is monotone in the same sense as f, and we get in addition to $D^+f \leqq D_-f$ also that $D^+g \leqq D_-g$ and this leads to $D^-f \leqq D_+f$ on account of $D^+g = D^-f$ and $D_-g = D_+f$. Why? Now, $\{x : D^+f(x) > D_-f(x)\} = \{x : D^+f(x) > R > r > D_-f(x) : r, R$ positive and rational $\}$ and it will be enough to show that, for $0 < r < R$, the set $\{x : D^+f(x) > R > r > D_-f(x)\} = S_R \cap T_r$ with $S_R = \{x : D^+f(x) > R\}$ and $T_r = \{x : D_-f(x) < r\}$ is a set of measure zero; here we make use of the fact that a countable union of sets of measure zero has also measure zero. In addition, we must show that D^+f is finite almost everywhere.

(ii) Let f be a bounded nondecreasing function on the interval $J = (a,b)$ and suppose that D denotes the points of discontinuity of f, and C denotes the points of continuity of f, on J; we know that D is a countable set. Let $0 < r < R$. We observe:
If $J' = (a', b')$ is an open subinterval of $J = (a,b)$, then the set $J' \cap S_R \cap T_r \cap C$ can be covered by a countable number of pairwise disjoint open intervals J'_i, the sum of whose lenghts is smaller than $(r/R)(b' - a')$, that is,

$$J' \cap S_R \cap T_r \cap C \subset \cup J'_i \text{ and } \sum_i (b'_i - a'_i) \leqq \frac{r}{R}(b' - a').$$

(a) We apply the lemma to $g_1(y) = f(-y) + ry$ defined on $(-b', -a')$ and we obtain the set

$$E_1 = \bigcup_j (-b_j, -a_j).$$

Since $f(-y)$ will be nonincreasing, we obtain that

$$g_1^s(y) = \lim\{g_1(y') : y' \uparrow y\} = g_1(y -)$$

[and in particular that $g_1^s(-b') = g_1((-b') +)$]. Observe that the notation $h(x +)$ and $h(x -)$ denotes limit from the right and limit from the left, respectively. If $x \in J' \cap T_r \cap C$, then there is an $x_1 < x$ with $f(x) - f(x_1) < r(x - x_1)$, or with $x = -y$ and $x_1 = -y_1$ substituted and taking into account the continuity of g_1 at y, we see that $g_1^s(y) = g_1(y) < g_1(y_1)$; by the definition of E_1 this means that $y \in E_1$, or $x \in E_1^*$ provided that E_1^* denotes the set

$$\bigcup_j (a_j, b_j)$$

obtained by reflecting the set E_1 about the zero point. Our result is: $J' \cap T_r \cap C \subset E_1^*$. By the Lemma, $g_1(y) \leqq g_1((-a_j) -)$ for $-b_j < y < -a_j$ and hence $g_1((-b_j) +) \leqq g_1((-a_j) -)$; this implies that

$$r(b_j - a_j) \geqq f(b_j -) - f(a_j +).$$

(b) We now apply the lemma to the function $g_2(x) = f(x) - Rx$ defined on (a_j, b_j) of part (a) above. We obtain as in part (a) a set

$$E_{2j} = \bigcup_k (a_{jk}, b_{jk})$$

with

$$(*) \qquad (a_j, b_j) \cap S_R \cap C \subset E_{2j} \text{ and } R(b_{jk} - a_{jk}) \leqq f(b_{jk} +) - f(a_{jk} +).$$

Remark. In case $b_{jk} = b_j$ we can replace $f(b_{jk} +)$ in (*) by the smaller $f(b_j -)$ because g_2 is only defined on (a_j, b_j) and hence $g_2^s(b_j) = g_2(b_j -)$.

We obtain

$$J' \cap S_R \cap T_r \cap C \subset \bigcup_{j,k} (a_{jk}, b_{jk});$$

upon summation of the above inequalities, making use of the foregoing remark and using the easily seen fact that if (a_i, b_i) is a sequence of pairwise disjoint open intervals in (a, b), then

$$\sum_i (b_i - a_i) \leqq b - a,$$

we readily get that

$$\begin{aligned}\sum_{j,k} (b_{jk} - a_{jk}) &\leqq \frac{1}{R} \sum_{j,k} (f(b_{jk}+) - f(a_{jk}+)) \\ &\leqq \frac{1}{R} \sum_j (f(b_j -) - f(a_j +)) \leqq \frac{r}{R} \sum_j (b_j - a_j) \\ &\leqq \frac{r}{R}(b' - a').\end{aligned}$$

But this is what we wanted to establish.

(iii) If one starts out with the interval $J = (a, b)$, applies the considerations of step (ii), repeats this on the obtained subintervals, etc., one gets upon m applications a covering of $S_R \cap T_r \cap C$ consisting of countably many disjoint open intervals whose total length, i.e., the sum of the individual length of intervals, is $\leqq (r/R)^m$ $(b - a)$. As $m \to \infty$, it follows that $S_R \cap T_r \cap C$ and hence also $S_R \cap T_r \subset S_r \cap T_r \cap C \cup D$ is a set of measure zero.

(iv) If one applies the method discussed in part (b) of step (ii) directly to (a, b) in place of (a_j, b_j), then one obtains in the corresponding set E_2 a covering of $S_R \cap C$ consisting of countably many disjoint open intervals the sum of whose lenghts, as the summation of the corresponding inequalities in (*) shows, is smaller than $(1/R)(f(b -) - f(a +))$. As $R \to +\infty$, we see that $\{x : D^+ f(x) = +\infty\}$ has measure zero for $\{x : D^+ f(x) = +\infty\}$ is contained in $S_R \cap C \cup D$ and the latter in turn is contained in $E_2 \cup D$.

The proof of the theorem is now finished.

N.B In the theorem we assumed that f was bounded; this is not an essential restriction for if $f : (a, b) \to R^1$ is nondecreasing, then (a, b) is a countable union of open subintervals on each of which f is bounded. The same applies in case f is nonincreasing.

27. Let F be a compact set in R^1. Consider the set E_r of points that are at distance r from F, i.e.,

$$E_r = \{x \in R^1 : \inf_{y \in F} |x - y| = r\},$$

where $r > 0$. Show that E_r is a set of measure zero.
[*Hints*: For, if not, E_r contains a point of density; let y be such a point. Let x be a point of F at distance r from y. A neighborhood U of x of radius r cannot contain any point of E_r, since all points of U are distant less than r from y. Now any neighborhood V of y is half in U and so $V \cap E_r^c$ contains an interval at least half as long as V. The existence of such intervals contradicts the assumption that y is a point of density of E_r.]

28. Let f be a real-valued nondecreasing function on the interval $[a, b]$. Show:
    (i) If $x_1, \ldots, x_n$ are arbitrary points in the interior of $[a, b]$, then

$$(f(a+0) - f(a)) + \sum_{k=1}^{n} (f(x_k + 0) - f(x_k - 0)) + (f(b) - f(b-0)) \leqq f(b) - f(a).$$

    (ii) The set of discontinuities of f is at most countable; if $x_1, x_2, \ldots$ are the points of discontinuity of f, then

$$(f(a+0) - f(a)) + \sum_{k=1}^{\infty} (f(x_k + 0) - f(x_k - 0)) + (f(b) - f(b-0)) \leqq f(b) - f(a).$$

    (iii) Let $s(a) = 0$ and

$$s(x) = (f(a+0) - f(a)) + \sum_{x_k < x} (f(x_k + 0) - f(x_k - 0)) + (f(x) - f(x_k - 0))$$

    for $a < x \leqq b$; s is called the *saltus function of f*. Then the difference $h(x) = f(x) - s(x)$ is a nondecreasing continuous function on $[a, b]$.

29. Let f be a real-valued function on the interval $[a, b]$. Show:
    (i) If f is monotone on $[a, b]$, then it is of finite variation on $[a, b]$.
    (ii) If f is of finite variation on $[a, b]$, then it is bounded on $[a, b]$.
    (iii) If f is of finite variation on $[a, b]$ and $a < c < b$, then

$$V_a^b(f) = V_a^c(f) + V_c^b(f).$$

    (iv) The function f is of finite variation if and only if it can be represented as the difference of two nondecreasing functions on $[a, b]$.

(v) If f is of finite variation on $[a, b]$, then it can be represented as the sum of its saltus function and a continuous function of finite variation on $[a, b]$.

30. (i) From an infinite family of uniformly bounded real-valued functions on the interval $[a, b]$, one can select a sequence of functions which converges on any countable subset E of $[a, b]$.

(ii) From an infinite family of nondecreasing real-valued functions on the interval $[a, b]$, one can select a sequence of functions which converges at every point of $[a, b]$ to some nondecreasing function g on $[a, b]$.

(iii) Let H be an infinite family of real-valued functions on $[a, b]$ such that

$$|f(x)| \leqq K_1 < \infty, \quad V_a^b(f) \leqq K_2 < \infty$$

for all $f \in H$ and all $x \in [a, b]$. Then we may select from H a sequence of functions which converges at every point of $[a, b]$ to some function of finite variation on $[a, b]$.

31. A sequence (finite or not) whose terms are complex numbers c_n $(n = \ldots, -1, 0, 1, \ldots)$ is said to be *positive definite*, if for any finite set of complex numbers $\lambda_1, \ldots, \lambda_m$ the inequality

$$\sum_{k,j=1}^{m} c_{k-j} \lambda_k \bar{\lambda}_j \geqq 0$$

holds; $\bar{\lambda}_j$ denotes the complex conjugate of λ_j.
Show: A necessary and sufficient condition for a sequence $(c_n)_{n=-\infty}^{\infty}$ to be positive definite is that it have the representation

$$c_n = \int_{-\pi}^{\pi} e^{inx}\, dg(x) \qquad (n = 0, \pm 1, \pm 2, \ldots),$$

where g denotes a nondecreasing function of finite variation on $[-\pi, \pi]$. (theorem of G. Herglotz).

32. We say that a sequence $(\sigma_j)_{j=1}^{\infty}$ of distinct natural numbers is of *slow growth*, if there exists a natural number M such that

$$\sigma_j < Mj \qquad (j = 1, 2, \ldots).$$

Show: Let $(\sigma_j)_{j=1}^{\infty}$ be a sequence of slow growth. Then there is a natural number m and real numbers $\lambda_1, \ldots, \lambda_m$ such that all integers n, for which the numbers

$$\frac{\lambda_k n}{2\pi} \qquad (k = 1, \ldots, m)$$

differ from integers by not more than 1/4, can be represented in the form

$$n = \sigma_p + \sigma_q - \sigma_r - \sigma_s,$$

where σ_p, σ_q, σ_r, and σ_s are suitable elements of the sequence $(\sigma_j)_{j=1}^{\infty}$. (The foregoing number-theoretic result can be proved by use of the theorem of G. Herglotz, stated in Exercise 31; for details of proof see the author's forthcoming paper, "On sequences of slow growth" in *Portugaliae Mathematica.*)

33. Let f be bounded and measurable on (0,1). Is it true that

$$\lim_{n\to\infty} f\left(x - \frac{1}{n}\right) = f(x)$$

almost everywhere?

34. Let f be an integrable function on $[0, 2\pi]$ and suppose that

$$(*) \qquad \int_0^{2\pi} f(t) \begin{smallmatrix}\cos nt\\ \sin nt\end{smallmatrix} \, dt = 0$$

for $n = 0, 1, 2, \ldots$. Show that $f = 0$ a.e. on $[0, 2\pi]$.
[*Hints*: Let f be continuous on $[0, 2\pi]$ and show that (*) implies $f = 0$ everywhere. Suppose not, i.e., suppose $f(t_0) \neq 0$, that is, $f(t_0) > 0$. Then by continuity it would follow that $f(t) > \varepsilon > 0$ on an interval $I = [t_0 - \delta, t_0 + \delta]$. It will be enough to show that there is a sequence of trigonometric polynomials T_n such that
(i) $T_n(t) \geqq 0$ for $x \in I$,
(ii) T_n tends uniformly to $+\infty$ in every interval J inside I,
(iii) the T_n are uniformly bounded outside I.
To do this we set $T_n = x^n$ with $x(t) = 1 + \cos(t - t_0) - \cos\delta$. Then $x(t) \geqq 1$ in I, $x(t) > 1$ in J and $|x(t)| \leqq 1$ outside I. But

$$0 = \int_0^{2\pi} f(t)T_n(t)\, dt = \int_I f(t)T_n(t)\, dt + \int_{[0,2\pi]-I} f(t)T_n(t)\, dt.$$

It is clear that

$$\int_I f(t)T_n(t)\, dt > \varepsilon\mu(J)\min_{t\in J} T_n(t)$$

which by (ii) tends to $+\infty$ for $n \to \infty$. On the other hand,

$$\int_{[0,2\pi]-I} f(t)T_n(t)\,dt$$

remains bounded in view of (iii). Thus

$$0 = \int_0^{2\pi} f(t)T_n(t)\,dt \to +\infty,$$

a contradiction.

Finally, we assume that f is in $L^1[0, 2\pi]$ and we set

$$F(t) = \int_0^t f(s)\,ds,$$

noting that F is therefore continuous (in fact, absolutely continuous). Condition (*) yields that $F(2\pi) = 0$. It is clear that $F(0) = 0$ as well. We also observe that

$$\int_0^{2\pi} F(t)\cos nt\,dt = -\frac{1}{n}\int_0^{2\pi} f(t)\sin nt\,dt = 0 \text{ and}$$

$$\int_0^{2\pi} F(t)\sin nt\,dt = \frac{1}{n}\int_0^{2\pi} f(t)\cos nt\,dt = 0.$$

This follows from (*) and using integration by parts (which is applicable on account of the absolute continuity of F).

Let us put

$$\alpha = \frac{1}{2\pi}\int_0^{2\pi} F(t)\,dt.$$

Then

$$\int_0^{2\pi} (F(t) - \alpha)\,{\cos nt \atop \sin nt}\,dt = 0$$

for $n = 0, 1, 2, \ldots$ by what we have derived further above. But $F(t) - \alpha$ is a continuous function and so by the first part of the proof $F(t) - \alpha \equiv 0$. Since $F(0) = 0$ we see that $F(t) \equiv 0$ (because F is constant) and so $f = 0$ a.e. in $[0, 2\pi]$ because

$$F(t) = \int_0^t f(s)\,ds.$$

Incidentally, the first part of the proof can be accomplished with less effort when we use the fact that a real-valued continuous function f of period 2π on R^1 can be approximated uniformly by a trigonometric polynomial T, for this means that $f(x) = T(x) + \varepsilon h(x)$, where $|h(x)| \leqq 1$, on $[0, 2\pi]$ for any $\varepsilon > 0$. By condition (*) we then get that

$$\int_0^{2\pi} f^2(x)\,dx \leqq \varepsilon \int_0^{2\pi} |f(x)|\,dx.$$

If $f \neq 0$, then ε would be bounded from below; this however is a contradiction because $\varepsilon > 0$ is arbitrary.]

35. Let $(x_n)_{n=1}^{\infty}$ and $(y_n)_{n=1}^{\infty}$ be two sequences of real numbers. Suppose, moreover, that $y_n \to \infty$ and, for some n_0, $y_{n+1} > y_n$ whenever $n > n_0$. Show:

$$\lim_{n\to\infty} \frac{x_n}{y_n} = \lim_{n\to\infty} \frac{x_n - x_{n-1}}{y_n - y_{n-1}}$$

provided that the limit on the right exists (be it finite or infinite).

36. Read the article "On absolutely continuous functions" by D. E. Varberg, *Amer. Math. Monthly*, vol. 72 (1965), pp. 831–841.

Chapter 5

Abstract Measure and Integration

In this and the next chapter we abstract some of the most important features of Lebesgue measure and integration theory discussed in the first three chapters and also consider some further developments. We commence with some definitions.

By a *σ-algebra* $\mathfrak{B}$ we mean a family of subsets of a given set X which contains the empty set $\varnothing$ and is closed with respect to complements and with respect to countable unions.

A *set function* μ is a mapping which assigns an extended real number to certain sets in a family of sets.

A *measurable space* is a pair $(X, \mathfrak{B})$ consisting of a set X and a σ-algebra $\mathfrak{B}$ of subsets of X. A subset A of X is called *measurable* (or *measurable with respect to* $\mathfrak{B}$) if $A \in \mathfrak{B}$.

By a *measure* μ on a measurable space $(X, \mathfrak{B})$ we mean a nonnegative set function defined for all sets of $\mathfrak{B}$ and satisfying

(i) $\mu(\varnothing) = 0$, and

(ii) $$\mu(\bigcup_{k=1}^{\infty} E_k) = \sum_{k=1}^{\infty} \mu(E_k)$$

for any sequence $(E_k)_{k=1}^{\infty}$ of mutually disjoint measurable (with respect to $\mathfrak{B}$) sets.

By a *measure space* $(X, \mathfrak{B}, \mu)$ we mean a measurable space $(X, \mathfrak{B})$ together with a measure μ defined on $\mathfrak{B}$.

Property (ii) of μ is called *countable additivity*; it implies, of course, *finite additivity*,. i.e.,

$$\mu(\bigcup_{k=1}^{m} E_k) = \sum_{k=1}^{m} \mu(E_k)$$

for mutually disjoint sets E_k belonging to $\mathfrak{B}$ since we can use countable additivity with $E_j = \varnothing$ for $j > m$.

An example of a measure space is $(R^1, \mathfrak{M}, \mu)$, where R^1 is the real line, $\mathfrak{M}$ the Lebesgue measurable sets of the real line, and μ the Lebesgue measure. Another measure space results if we replace R^1 by the interval $[0, 1]$ and $\mathfrak{M}$ by the Lebesgue measurable subsets of $[0, 1]$. A third example is $(R^1, \mathcal{B}, \mu)$, where $\mathcal{B}$ is the class of Borel sets and μ is again the Lebesgue measure.

Let $\nu(E)$ be ∞ for any infinite set E and be equal to the number of elements in E when E is a finite set. Then ν is a countably additive set function and it is easy to see that $(R^1, \mathcal{P}(R^1), \nu)$, where $\mathcal{P}(R^1)$ is the power set of R^1, i.e., the collection of all subsets of R^1, is a measure space; ν is called *counting measure.*

Let X be any uncountable set, $\mathfrak{B}$ the family of those subsets which are either countable or the complement of a countable set. Then $\mathfrak{B}$ is a σ-algebra and we can define a measure on it by setting $\mu(A) = 0$ for each countable set A and $\mu(B) = 1$ for each set B whose complement is countable.

From these definitions we get the following facts about measure.

PROPOSITION 1. *Let $(X, \mathfrak{B}, \mu)$ be a measure space. If A, $B \in \mathfrak{B}$ and $A \subset B$, then $\mu(A) \leqq \mu(B)$.*

Proof. Since $B = A \cup (B - A)$ is a disjoint union, we have $\mu(B) = \mu(A) + \mu(B - A) \geqq \mu(A)$.

PROPOSITION 2. *Let $(X, \mathfrak{B}, \mu)$ be a measure space. Let $(A_n)_{n=1}^{\infty}$ be a sequence of sets in $\mathfrak{B}$ such that $A_1 \subset A_2 \subset \ldots \subset A_n \subset \ldots$. Then*

$$\mu\Big(\bigcup_{n=1}^{\infty} A_n\Big) = \lim_{n\to\infty} \mu(A_n).$$

Proof. Put $A_0 = \varnothing$. Then

$$\bigcup_{n=1}^{\infty} A_n = \bigcup_{n=1}^{\infty} (A_n \cap A_{n-1}^c).$$

By countable additivity,

$$\mu\Big(\bigcup_{n=1}^{\infty} A_n\Big) = \sum_{n=1}^{\infty} \mu(A_n \cap A_{n-1}^c) = \lim_{p\to\infty} \sum_{n=1}^{p} \mu(A_n \cap A_{n-1}^c)$$
$$= \lim_{p\to\infty} \mu\Big(\bigcup_{n=1}^{p} (A_n \cap A_{n-1}^c)\Big) = \lim_{p\to\infty} \mu(A_p).$$

PROPOSITION 3. *Let $(X, \mathfrak{B}, \mu)$ be a measure space. If $(A_n)_{n=1}^{\infty}$ is a sequence of sets in $\mathfrak{B}$ such that $\mu(A_1) < \infty$ and $A_1 \supset A_2 \supset \ldots \supset A_n \supset \ldots$, then*

$$\mu\Big(\bigcap_{n=1}^{\infty} A_n\Big) = \lim_{n\to\infty} \mu(A_n).$$

In particular, if

$$\bigcap_{n=1}^{\infty} A_n = \varnothing,$$

then

$$\lim_{n\to\infty} \mu(A_n) = 0.$$

Proof. The sequence $(A_1 \cap A_n^c)_{n=1}^{\infty}$ is nondecreasing, and all $A_1 \cap A_n^c$ are in $\mathfrak{B}$. Applying Proposition 2, we get

$$\begin{aligned}\mu(A_1) - \lim_{n\to\infty} \mu(A_n) &= \lim_{n\to\infty}(\mu(A_1) - \mu(A_n)) = \lim_{n\to\infty} \mu(A_1 \cap A_n^c) \\ &= \mu(\bigcup_{n=1}^{\infty} (A_1 \cap A_n^c)) = \mu(A_1 \cap (\bigcup_{n=1}^{\infty} A_n^c)) \\ &= \mu(A_1 \cap (\bigcap_{n=1}^{\infty} A_n)^c) = \mu(A_1) - \mu(\bigcap_{n=1}^{\infty} A_n).\end{aligned}$$

Subtracting $\mu(A_1)$ (which can be done since it is finite), we obtain

$$\lim_{n\to\infty} \mu(A_n) = \mu(\bigcap_{n=1}^{\infty} A_n).$$

PROPOSITION 4. *Let* $(X, \mathfrak{B}, \mu)$ *be a measure space and let* $(A_n)_{n=1}^{\infty}$ *be any sequence of sets in* $\mathfrak{B}$. *Then*

$$\text{(i)} \qquad \mu(\bigcup_{k=1}^{\infty} (\bigcap_{n=k}^{\infty} A_n)) \leqq \varliminf_{k\to\infty} \mu(A_k);$$

$$\text{(ii)} \qquad \mu(\bigcap_{k=1}^{\infty} (\bigcup_{n=k}^{\infty} A_n)) \geqq \varlimsup_{k\to\infty} \mu(A_k).$$

If

$$\bigcap_{k=1}^{\infty} (\bigcup_{n=k}^{\infty} A_n) = \bigcup_{k=1}^{\infty} (\bigcap_{n=k}^{\infty} A_n) = B$$

and

$$\mu(\bigcup_{k=1}^{\infty} A_k) < \infty,$$

then

$$\lim_{k\to\infty} \mu(A_k)$$

exists and is equal to $\mu(B)$.

Proof. We only prove (i); (ii) is proved in a similar fashion. Evidently

$$\bigcap_{n=1}^{\infty} A_n \subset \bigcap_{n=2}^{\infty} A_n \subset \bigcap_{n=3}^{\infty} A_n \subset \cdots .$$

Proposition 2 implies that

$$\mu\Big(\bigcup_{k=1}^{\infty} \Big(\bigcap_{n=k}^{\infty} A_n\Big)\Big) = \lim_{k\to\infty} \mu\Big(\bigcap_{n=k}^{\infty} A_n\Big).$$

We also have

$$\mu(A_k) \geqq \mu\Big(\bigcap_{n=k}^{\infty} A_n\Big)$$

for all k. This implies that

$$\lim_{k\to\infty} \mu\Big(\bigcap_{n=k}^{\infty} A_n\Big) \leqq \varliminf_{k\to\infty} \mu(A_k),$$

proving (i).

The proof of the last part of the proposition follows from the inequalities

$$\varlimsup_{k\to\infty} \mu(A_k) \leqq \mu(B) \leqq \varliminf_{k\to\infty} \mu(A_k).$$

PROPOSITION 5. *Let* $(X, \mathfrak{B}, \mu)$ *be a measure space and* $(E_k)_{k=1}^{\infty}$ *be a sequence in sets in* $\mathfrak{B}$. *Then*

$$\mu\Big(\bigcup_{k=1}^{\infty} E_k\Big) \leqq \sum_{k=1}^{\infty} \mu(E_k).$$

Proof. Let

$$G_n = E_n - \Big(\bigcup_{k=1}^{n-1} E_k\Big).$$

Then $G_n \subset E_n$ and the sets G_n are disjoint. Hence $\mu(G_n) \leqq \mu(E_n)$ while

$$\mu\Big(\bigcup_{k=1}^{\infty} E_k\Big) = \sum_{n=1}^{\infty} \mu(G_n) \leqq \sum_{n=1}^{\infty} \mu(E_n).$$

(Q.E.D.)

Let $(X, \mathfrak{B}, \mu)$ be a measure space. The measure μ is called *finite* if $\mu(X) < \infty$; it is called *σ-finite* if there is a sequence $(X_n)_{n=1}^{\infty}$ of sets in $\mathfrak{B}$ such that

$$X = \bigcup_{n=1}^{\infty} X_n$$

and $\mu(X_n) < \infty$ for each n.

Remark. It should be noted that we may take $(X_n)_{n=1}^{\infty}$ to be a pairwise disjoint sequence (this is so on account of Proposition 14 in Chapter 1).

The Lebesgue measure on $[0, 1]$ is an example of a finite measure, while the Lebesgue measure on $(-\infty, \infty)$ is an example of a σ-finite measure. The counting measure on an uncountable set is a measure that is not σ-finite.

A *set* E is said to be of *finite measure* if $E \in \mathfrak{B}$ and $\mu(E) < \infty$. A set E is said to be of *σ-finite measure* if E is the union of a countable collection of measurable sets of finite measure.

Any measurable set contained in a set of σ-finite measure is itself of σ-finite measure, and the union of a countable collection of sets of σ-finite measure is again of σ-finite measure. If μ is σ-finite, then every measurable set is of σ-finite measure.

If $(X, \mathfrak{B}, \mu)$ is a measure space and $Y \in \mathfrak{B}$, then we can form a new measure space $(Y, \mathfrak{B}_Y, \mu_Y)$ by letting $\mathfrak{B}_Y$ consist of those sets of $\mathfrak{B}$ that are contained in Y and defining $\mu_Y(E) = \mu(E)$ for $E \in \mathfrak{B}_Y$. The measure μ_Y is called the *restriction of μ to Y.*

A measure space $(X, \mathfrak{B}, \mu)$ is said to be *complete* if $\mathfrak{B}$ contains all subsets of sets of measure zero, i.e., if $B \in \mathfrak{B}$, $\mu(B) = 0$, and $A \subset B$ imply $A \in \mathfrak{B}$.

We note that Lebesgue measure is complete, while Lebesgue measure restricted to the σ-algebra of Borel sets is not complete.

Each measure space $(X, \mathfrak{B}, \mu)$ can be completed by the addition of subsets of sets of measure zero as we shall see next.

PROPOSITION 6. *If $(X, \mathfrak{B}, \mu)$ is a measure space, then we can find a complete measure space $(X, \mathfrak{B}_0, \mu_0)$ such that*

(*i*) $\mathfrak{B} \subset \mathfrak{B}_0$;

(*ii*) $E \in \mathfrak{B}$ *implies* $\mu(E) = \mu_0(E)$;

(*iii*) $\mathfrak{B}_0$ *consists of all sets of the form* $E \cup F$, *where* $E \in \mathfrak{B}$ *and* F *is a subset of a set* M *in* $\mathfrak{B}$ *with* $\mu(M) = 0$.

Proof. We begin by showing that $\mathfrak{B}_0$ defined by (iii) is a σ-algebra. For convenience we shall assume throughout this proof that the letter E with or without subscripts will denote a set in $\mathfrak{B}$, the letter M with or without subscripts will denote a set in $\mathfrak{B}$ for which $\mu(M) = 0$, and the letter F with or without subscripts will denote a subset of M.

To see that the complement of a set $E \cup F$ in $\mathfrak{B}_0$ is also in $\mathfrak{B}_0$, let $F \subset M$ so that

$$(E \cup F)^c = E^c \cap F^c \supset E^c \cap M^c,$$

$$E^c \cap F^c - E^c \cap M^c = E^c \cap (F^c - M^c) \subset M.$$

Thus, if

$$F_1 = E^c \cap F^c - E^c \cap M^c,$$

then

$$F_1 \subset M \text{ and } (E \cup F)^c = (E^c \cap M^c) \cup F_1.$$

Hence $\mathfrak{B}_0$ contains the complement of every one of its elements.

Let $\{E_n \cup F_n\}_{n=1}^{\infty} \subset \mathfrak{B}_0$ and $F_n \subset M_n$ for $n = 1, 2, \ldots$. Then, since

$$\bigcup_{n=1}^{\infty} (E_n \cup F_n) - \bigcup_{n=1}^{\infty} E_n \subset \bigcup_{n=1}^{\infty} F_n \subset \bigcup_{n=1}^{\infty} M_n = M,$$

it is clear that

$$(\#) \qquad \bigcup_{n=1}^{\infty} (E_n \cup F_n) = \Big(\bigcup_{n=1}^{\infty} E_n\Big) \cup F,$$

where

$$F = \bigcup_{n=1}^{\infty} (E_n \cup F_n) - \bigcup_{n=1}^{\infty} E_n.$$

Thus $\mathfrak{B}_0$ is a σ-algebra. It is obvious that $\mathfrak{B} \subset \mathfrak{B}_0$ because the empty set $\varnothing$ is in $\mathfrak{B}$ and $\mu(\varnothing) = 0$.

We define $\mu_0(E \cup F) = \mu(E)$. To see that μ_0 is uniquely defined on $\mathfrak{B}_0$ we observe the following: if $E_1 \cup F_1 = E_2 \cup F_2$ and $F_1 \subset M_1$, $F_2 \subset M_2$, we let $M = M_1 \cup M_2$ so that $E_1 \cup M = E_2 \cup M$ and thus $\mu(E_1) = \mu(E_1 \cup M) = \mu(E_2)$. Moreover, from formula(#) we see that μ_0 is countably additive.

The completeness of the measure μ_0 is an immediate consequence of the fact that $\mathfrak{B}_0$ contains all subsets of sets of measure zero in $\mathfrak{B}$. This completes the proof of the proposition.

The following propositions are proved in complete analogy to the corresponding propositions in Chapter 2.

PROPOSITION 7. *Let $(X, \mathfrak{B})$ be a measurable space and suppose that f is an extended real-valued function defined on X. Then the following statements are equivalent*:

(*i*) $\{x \in X : f(x) < \alpha\} \in \mathfrak{B}$ *for each* $\alpha \in R^1$.
(*ii*) $\{x \in X : f(x) \leqq \alpha\} \in \mathfrak{B}$ *for each* $\alpha \in R^1$.
(*iii*) $\{x \in X : f(x) > \alpha\} \in \mathfrak{B}$ *for each* $\alpha \in R^1$.
(*iv*) $\{x \in X : f(x) \geqq \alpha\} \in \mathfrak{B}$ *for each* $\alpha \in R^1$.

DEFINITION. *An extended real-valued function f defined on X is called measurable* (*or measurable with respect to* $\mathfrak{B}$) *if one of the four statements of Proposition 7 holds.*

PROPOSITION 8. *If c is a constant and the functions f and g are measurable, then so are the functions $f + c$, cf, $f + g$, fg, and $f \vee g$. Moreover, if $(f_n)_{n=1}^{\infty}$ is a sequence of measurable functions, then*

$$\sup_n f_n, \inf_n f_n, \overline{\lim_{n\to\infty}} f_n \text{ and } \underline{\lim_{n\to\infty}} f_n$$

are all measurable. If $(f_n)_{n=1}^{\infty}$ is a sequence of measurable functions defined on X and

$$\lim_{n\to\infty} f_n(x)$$

exists in R_e^1 for all $x \in X$, then

$$\lim_{n\to\infty} f_n$$

is measurable.

PROPOSITION 9. *If μ is a complete measure and f is a measurable function in $(X, \mathfrak{B}, \mu)$, then $f = g$ a.e. with respect to μ implies that g is also measurable.*

N. B. The foregoing proposition is not true if the measure space is incomplete.

By a *simple function* we mean, as before, a finite linear combination

$$s(x) = \sum_{j=1}^{n} c_j \chi_{E_j}(x) \qquad (c_j \in R^1, \text{ the } E_j\text{'s mutually disjoint})$$

of characteristic functions of measurable sets E_j.

PROPOSITION 10. *Let f be an extended real-valued $\mathfrak{B}$-measurable function defined on X. Then there exists a sequence $(s_n)_{n=1}^{\infty}$ of simple functions defined on X such that $|s_1| \leqq |s_2| \leqq \ldots \leqq |s_n| \leqq \ldots$ and $s_n(x) \to f(x)$ for each $x \in X$. If f is bounded, then the functions s_n can be so chosen that the convergence is uniform. If $f \geqq 0$, the sequence $(s_n)_{n=1}^{\infty}$ can be chosen so that $0 \leqq s_1 \leqq s_2 \leqq \ldots \leqq f$; if f is defined on a σ-finite measure space, then we may choose the functions s_n so that each vanishes outside a set of finite measure. (See Proposition 11 of Chapter 2 and the Remark following it.)*

If E is a measurable set in $(X, \mathfrak{B}, \mu)$ and s a nonnegative simple function in $(X, \mathfrak{B}, \mu)$ of the form

$$s(x) = \sum_{k=1}^{n} c_k \chi_{E_k}(x),$$

we define the *integral of s over E* as

$$\int_E s = \sum_{k=1}^{n} c_k \mu(E_k \cap E).$$

Of course, it is assumed that E and the E_k's are measurable sets in $(X, \mathfrak{B}, \mu)$. It can be seen that $\int_E s$ is independent of the representation of s which we use. Next, we define the integral of a nonnegative function as follows:

Let f be a nonnegative extended real-valued measurable function on a complete measure space $(X, \mathfrak{B}, \mu)$. Then $\int f$ is defined as the supremum of the integrals $\int s$ as s ranges over all simple functions with $0 \leqq s \leqq f$.

We now establish *Fatou's Lemma*.

PROPOSITION 11. *Let $(f_n)_{n=1}^{\infty}$ be a sequence of nonnegative measurable functions that converge μ-a.e. on a set E of $\mathfrak{B}$ to a function f. Then*

$$\int_E f \leqq \varliminf_{n\to\infty} \int_E f_n .$$

Proof. Without loss of generality we may assume that $f_n(x) \to f(x)$ for each $x \in E$. From the definition of $\int_E f$ it is enough to show that, if s is any nonnegative simple function with $s \leqq f$, then

$$\int_E s \leqq \varliminf_{n\to\infty} \int_E f_n .$$

If $\int_E s = \infty$, then there is a measurable set $A \subset E$ with $\mu(A) = \infty$ such that $s > \alpha > 0$ on A. Put

$$A_n = \{x \in E : f_k(x) > \alpha \text{ for all } k \geqq n\}.$$

Then $(A_n)_{n=1}^{\infty}$ is an increasing sequence of measurable sets whose union contains A, since

$$s \leqq \varliminf_n f_n .$$

Hence

$$\lim_n \mu(A_n) = \infty.$$

Since $\int_E f_n \geqq \alpha\mu(A_n)$, we have $\int_E f_n = \infty = \int_E s$.

If $\int_E s < \infty$, then there is a measurable set $A \subset E$ with $\mu(A) < \infty$ such that s vanishes identically on $E - A$. Let M be the maximum of s, let $\varepsilon > 0$ be a given positive real number, and set

$$A_n = \{x \in E : f_k(x) > (1 - \varepsilon)s(x) \text{ for all } k \geqq n\}.$$

Then $(A_n)_{n=1}^{\infty}$ is an increasing sequence of sets whose union contains A, and so $(A - A_n)_{n=1}^{\infty}$ is a decreasing sequence of sets whose intersection is empty. By Proposition 3,

$$\lim_{n\to\infty} \mu(A - A_n) = 0,$$

and so we can find an n such that $\mu(A - A_k) < \varepsilon$ for all $k \geqq n$. Thus, for $k \geqq n$, we have

$$\int_E f_k \geqq \int_{A_k} f_k \geqq (1 - \varepsilon) \int_{A_k} s \geqq (1 - \varepsilon) \int_E s - \int_{A - A_k} s$$
$$\geqq \int_E s - \varepsilon(\int_E s + M).$$

Hence

$$\varliminf_{n\to\infty} \int_E f_n \geqq \int_E s - \varepsilon(\int_E s + M).$$

Since ε is arbitrary,

$$\varliminf_{n\to\infty} \int_E f_n \geqq \int_E s.$$

We now derive the *Monotone Convergence Theorem.*

PROPOSITION 12. *Let $(f_n)_{n=1}^{\infty}$ be a sequence of nonnegative measurable functions that converge μ-a.e. to a function f and suppose that $f_n \leqq f$ for all n. Then*

$$\int f = \lim_{n\to\infty} \int f_n .$$

Proof. Since $f_n \leqq f$, we have $\int f_n \leqq \int f$. Thus, by Proposition 11, we have

$$\int f \leqq \varliminf_{n\to\infty} \int f_n \leqq \varlimsup_{n\to\infty} \int f_n \leqq \int f.$$

Using Proposition 12, it is easy to show that the integral is a linear mapping which preserves order.

PROPOSITION 13. *If f and g are nonnegative measurable functions and a and b are nonnegative real numbers, then*

$$\int (af + bg) = a \int f + b \int g.$$

We also have

$$\int f \geqq 0$$

with equality only if $f = 0$ μ-a.e.

Proof. To show the first statement, we let $(s_n)_{n=1}^{\infty}$ and $(h_n)_{n=1}^{\infty}$ be two sequences of simple functions such that $s_n \uparrow f$ and $h_n \uparrow g$. Then $(as_n + bh_n)_{n=1}^{\infty}$ is a sequence of simple functions such that

$$(as_n + bh_n) \uparrow (af + bg).$$

By Proposition 12, we have that

$$\begin{aligned}\int (af + bg) &= \lim_{n\to\infty} \int (as_n + bh_n) = \lim_{n\to\infty}\left(a \int s_n + b \int h_n\right)\\ &= a \int f + b \int g,\end{aligned}$$

where the linearity of the integral for simple functions is trivial.

To prove the second part of the proposition, we observe that $\int f \geqq 0$ whenever $f \geqq 0$. If $\int f = 0$, let

$$A_n = \left\{x \in X : f(x) \geqq \frac{1}{n}\right\}.$$

Then $f \geqq (1/n)\chi_{A_n}$ and so $\mu(A_n) = \int \chi_{A_n} = 0$. Since the set where $f > 0$ is the union of the sets A_n, it has μ-measure zero.

Remark. Proposition 12 also has the following consequence: If $(f_n)_{n=1}^{\infty}$ is a sequence of nonnegative measurable functions, then

$$\int \sum_{n=1}^{\infty} f_n = \sum_{n=1}^{\infty} \int f_n .$$

A nonnegative function f is called *integrable* (with respect to μ) over a measurable set $E \in \mathfrak{B}$ if f is measurable and

$$\int_E f < \infty.$$

An arbitrary measurable function f: $E \to R_e^1$ with $E \in \mathfrak{B}$ is said to be *integrable over* E if both f^+ and f^-, with

$$f^+(x) = \max\{f(x), 0\} \text{ and } f^-(x) = -\min\{f(x), 0\},$$

are integrable over E; in this case we define

$$\int_E f = \int_E f^+ - \int_E f^- .$$

We now consider the *Dominated Convergence Theorem.*

PROPOSITION 14. *Let g be integrable over $E \in \mathfrak{B}$ and suppose that $(f_n)_{n=1}^{\infty}$ is a sequence of extended real-valued μ-measurable functions such that μ-a.e. on E*

$$|f_n(x)| \leqq g(x)$$

and such that μ-a.e. on E $f_n(x) \to f(x)$. Then f is integrable over E and

$$\int_E f = \lim_{n\to\infty} \int_E f_n .$$

Proof. The verification of the proposition consists in applying Proposition 11 to the sequences $(g + f_n)_{n=1}^{\infty}$ and $(g - f_n)_{n=1}^{\infty}$.

We note the following two consequences of the foregoing proposition:

(*i*) *If f is integrable over E and $(E_n)_{n=1}^{\infty}$ is a disjoint sequence in $\mathfrak{B}$ and*

$$E = \bigcup_{n=1}^{\infty} E_n,$$

then

$$\int_E f = \sum_{n=1}^{\infty} \int_{E_n} f.$$

(*ii*) *If $(f_n)_{n=1}^{\infty}$ are $\mathfrak{B}$-measurable functions of the complete measure space $(X, \mathfrak{B}, \mu)$ such that*

$$\sum_{n=1}^{\infty} |f_n|$$

is integrable, then

$$\sum_{n=1}^{\infty} f_n$$

is integrable and

$$\int_X \sum_{n=1}^{\infty} f_n = \sum_{n=1}^{\infty} \int_X f_n .$$

The following proposition gives some of the standard properties of the integral; the proof is simple and will be omitted.

PROPOSITION 15. *If f and g are integrable functions over $E \in \mathfrak{B}$, then*

(*i*) $\int_E (af + bg) = a \int_E f + b \int_E g$ (*with $a, b \in R^1$*);

(*ii*) *if $|h| \leqq |f|$ and h is measurable on E, then h is integrable over E*;

(*iii*) *if $f \geqq g$ μ-a.e., then*

$$\int_E f \geqq \int_E g.$$

Let $L^1(X, \mathfrak{B}, \mu)$ denote the set of all real-valued functions defined μ-a.e. on X and integrable with respect to μ over X; the measure space $(X, \mathfrak{B}, \mu)$ is assumed complete.

PROPOSITION 16. *Let $f \in L^1(X, \mathfrak{B}, \mu)$. For every $\varepsilon > 0$ there exists a $\delta > 0$ depending only on ε and f such that for all $E \in \mathfrak{B}$ satisfying $\mu(E) < \delta$, we have*

$$\int_E |f| < \varepsilon.$$

Proof. For $n \in N$, let

$$h_n(x) = \begin{cases} |f(x)| \text{ if } |f(x)| \leqq n, \\ n \text{ otherwise.} \end{cases}$$

Then $(h_n)_{n=1}^{\infty}$ is a nondecreasing sequence of $\mathfrak{B}$-measurable functions and

$$\lim_{n \to \infty} h_n = |f|.$$

By Proposition 12 we have

$$\lim_{n \to \infty} \int_X h_n = \int_X \lim_{n \to \infty} h_n = \int_X |f|.$$

We pick n so that

$$\int_X (|f| - h_n) < \frac{\varepsilon}{2}.$$

We set $\delta = \varepsilon/2n$ and choose any $E \in \mathfrak{B}$ such that $\mu(E) < \delta$. We get

$$\int_E h_n \leqq n\mu(E) < \frac{\varepsilon}{2}.$$

It follows that

$$\left| \int_E f \right| \leqq \int_E |f| = \int_E (|f| - h_n) + \int_E h_n < \int_X (|f| - h_n) + \frac{\varepsilon}{2}$$
$$< \frac{\varepsilon}{2} + \frac{\varepsilon}{2} = \varepsilon$$

for all $E \in \mathfrak{B}$ such that $\mu(E) < \delta$.

PROPOSITION 17. *If a nonnegative function $f : E \to R_e^1$ is integrable over $E \in \mathfrak{B}$, then, for every positive number α, the subset*

$$E_\alpha = \{x \in E : f(x) \geqq \alpha\}$$

is of finite measure and the subset

$$D = \{x \in E : f(x) > 0\}$$

is of σ-finite measure.

Proof. Consider the simple function $s : E \to R^1$ defined by

$$s(x) = \begin{cases} \alpha & \text{if } x \in E, \\ 0 & \text{if } x \in E - E_\alpha. \end{cases}$$

Then $0 \leqq s \leqq f$ and hence

$$\int_E f \geqq \int_E s = \alpha\mu(E_\alpha).$$

Since $\int_E f < \infty$, this shows that E_α is of finite measure.

Finally, since

$$D = \bigcup_{n=1}^{\infty} E_{1/n},$$

it follows that D is of σ-finite measure. The proof is finished.

Remark. Much in the same vein we could continue here with the exploration of similarities to the theory discussed in Chapters 2 and 3; instead of doing this, we now turn to a new topic of study which will culminate in the so-called Radon-Nikodym Theorem and which in turn will permit us to extend our knowledge about the structure of L^p spaces.

As a rule we shall continue requiring completeness of the measure space; Proposition 6 attests to the fact that we lose little by doing so, yet we gain so much.

DEFINITION. *By a signed measure on a measurable space $(X, \mathfrak{B})$ we mean an extended real-valued set function ν defined for the sets of $\mathfrak{B}$ which satisfies:*

(*i*) *ν assumes at most one of the values ∞ and $-\infty$;*

(*ii*) *$\nu(\varnothing) = 0$, where $\varnothing$ is the empty set;*

(*iii*) *ν is countably additive, i.e.,*

$$\nu\left(\bigcup_{n=1}^{\infty} E_n\right) = \sum_{n=1}^{\infty} \nu(E_n)$$

for any disjoint sequence $(E_n)_{n=1}^{\infty}$ *in* $\mathfrak{B}$; *here it is understood that the series*

$$\sum_{n=1}^{\infty} \nu(E_n)$$

either converges absolutely or diverges properly (*to* ∞ *or* $-\infty$, *independent of the order of the terms*).

Thus every measure on $\mathfrak{B}$ is also a signed measure on $\mathfrak{B}$, but the converse is false in general.

A signed measure $\nu : \mathfrak{B} \to R_e^1$ is said to be *finite* if $\nu(\mathfrak{B}) \subset R^1$. The set of all finite signed measures on $\mathfrak{B}$ clearly forms a linear space.

Consider a given signed measure $\nu : \mathfrak{B} \to R_e^1$ on a measurable space $(X, \mathfrak{B})$ and recall that the members of $\mathfrak{B}$ are called the measurable subsets of X.

A subset E of X is said to be *positive* with respect to ν if E is measurable, i.e., if $E \in \mathfrak{B}$, and for every measurable subset A of E we have that $\nu(A) \geqq 0$. Similarly, a subset E of X is said to be *negative* with respect to the signed measure ν if E is measurable, and for every measurable subset A of E we have $\nu(A) \leqq 0$. A subset E of X which is both positive and negative with respect to ν is called a *null set* of X.

N.B. A measurable set is a null set if and only if every measurable subset of it has ν measure zero; while every null set must have measure zero, a set of measure zero may well be a union of two sets whose measures are not zero, but are negatives of each other. Similarly, a positive set is not to be confused with a set that merely has positive ν measure.

PROPOSITION 18. *Every measurable subset of a positive subset of* X *is positive, and the union of any countable collection of positive subsets of* X *is positive. Every measurable subset of a negative subset of* X *is negative, and the union of any countable collection of negative subsets of* X *is negative.*

Proof. We shall only prove the first half of the proposition; the proof of the second half is similar to the proof of the first half and will therefore be omitted.

It is obvious that every measurable subset of a positive set must be positive. Now, let E be the union of $(E_n)_{n=1}^{\infty}$, where each set E_n is assumed positive. To prove that the union E of $(E_n)_{n=1}^{\infty}$ is positive, let D denote any measurable subset of E. For each $n \in N$, let

$$D_n = D \cap E_n \cap (X - E_{n-1}) \cap \ldots \cap (X - E_1).$$

Then D_n is a measurable subset of the positive set E_n. Hence we have $\nu(D_n) \geqq 0$. Since the sets D_n $(n \in N)$ are disjoint and D is their union, it follows from the countable additivity of ν that

$$\nu(D) = \sum_{n=1}^{\infty} \nu(D_n) \geqq 0.$$

But this proves that D is positive.

PROPOSITION 19. *Every measurable subset E of X with finite negative measure, that is, $-\infty < \nu(E) < 0$, contains a negative subset D with $\nu(D) < 0$.*

Proof. If E is negative, then we may take $D = E$ and the proposition is proved. Otherwise, E contains subsets of positive measure. Let n_1 denote the smallest natural number such that there is a measurable subset E_1 of E with

$$\nu(E_1) > \frac{1}{n_1}.$$

Since E_1 and $E - E_1$ are disjoint, it follows from ((ii) and (iii) of) the definition of signed measure that

$$\nu(E) = \nu(E_1) + \nu(E - E_1).$$

Since $\nu(E)$ is finite, it follows from ((i) of) the definition of signed measure that both $\nu(E_1)$ and $\nu(E - E_1)$ are finite. Hence we obtain

$$\nu(E - E_1) = \nu(E) - \nu(E_1) < 0,$$

because $\nu(E) < 0$ and $\nu(E_1) > 0$. If $E - E_1$ is negative, then we may take $D = E - E_1$ and the proposition is proved. Otherwise, $E - E_1$ contains subsets of positive measure. Let n_2 denote the smallest natural number such that there is a measurable subset E_2 of $E - E_1$ with

$$\nu(E_2) > \frac{1}{n_2}.$$

Then, just as before, both $\nu(E_2)$ and $\nu(E - (E_1 \cup E_2))$ are finite. Hence we obtain

$$\nu(E - (E_1 \cup E_2)) = \nu(E) - \nu(E_1) - \nu(E_2) < 0.$$

Continuing this process, we get either a negative subset D of E with $\nu(D) < 0$, or a sequence $\{n_j : j \in N\}$ of natural numbers and a sequence $\{E_j : j \in N\}$ of disjoint measurable subsets of E with

$$\frac{1}{n_j} < \nu(E_j) < \infty \text{ for every } j \in N.$$

In the latter case, let

$$D = E - (\bigcup_{j=1}^{\infty} E_j).$$

By countable additivity of ν we have

$$\nu(E) = \nu(D) + \nu(\bigcup_{j=1}^{\infty} E_j) = \nu(D) + \sum_{j=1}^{\infty} \nu(E_j) > \nu(D) + \sum_{j=1}^{\infty} \frac{1}{n_j}.$$

By (i) of the definition of signed measure and by our assumption that $\nu(E)$ is finite, this implies that $\nu(D)$ is finite and that the sequence $\{1/n_j : j \in N\}$ of positive real numbers is summable. Hence we obtain that $n_j \to \infty$ and

$$\nu(D) < \nu(E) - \sum_{j=1}^{\infty} \frac{1}{n_j} < 0.$$

It remains to prove that D is negative. For this purpose, let M denote any measurable subset of D. For each $j \in N$, we have

$$M \subset D \subset E - (\bigcup_{k=1}^{j-1} E_k).$$

By the choice of the natural number n_j, we have

$$\nu(M) \leqq \frac{1}{n_j - 1}.$$

Since this holds for every $j \in N$ and since $n_j \to \infty$, we must have $\nu(M) \leqq 0$. This shows that D is negative and completes the proof.

PROPOSITION 20 (HAHN'S DECOMPOSITION THEOREM). *For an arbitrary signed measure* ν: $\mathfrak{B} \to R_e^1$ *on a* σ*-algebra of subsets of* X, *there exists a positive subset* P *and a negative subset* Q *of* X *with respect to* ν *satisfying*

$$P \cup Q = X \qquad \text{and} \qquad P \cap Q = \varnothing .$$

Proof. In view of (i) of the definition of signed measure, we may assume that ν does not take the value $-\infty$ (by passing to the signed measure $-\nu$ if necessary).

Consider the family $\mathfrak{I}$ of all negative subsets of X and let

$$\lambda = \inf\{\nu(E) : E \in \mathfrak{I}\}.$$

Then there exists a sequence $(E_n)_{n=1}^{\infty}$ in $\mathfrak{I}$ such that

$$\lim_{n\to\infty} \nu(E_n) = \lambda.$$

Let

$$Q = \bigcup_{n=1}^{\infty} E_n .$$

By Proposition 18, Q is a negative subset of X. Hence we have $\lambda \leqq \nu(Q)$. On the other hand, consider the subset $Q - E_n$ of Q. Since Q is negative, we have $\nu(Q - E_n) \leqq 0$. Since ν is countably additive, we obtain

$$\nu(Q) = \nu(E_n) + \nu(Q - E_n) \leqq \nu(E_n).$$

As this is true for every $n \in N$, we must have $\nu(Q) \leqq \lambda$. Consequently we obtain $\nu(Q) = \lambda$ and $\lambda > -\infty$.

We now show that the subset $P = X - Q$ of X is positive. For this purpose, let us assume that P is not positive. Then, by definition, there is a measurable subset E of P with $\nu(E) < 0$. Hence, by Proposition 19, E must contain a negative subset D of X with $\nu(D) < 0$. Since D and Q are disjoint negative subsets of X, it follows from Proposition 18 and countable additivity of ν that $D \cup Q$ is negative and hence

$$\lambda \leqq \nu(D \cup Q) = \nu(D) + \nu(Q) = \nu(D) + \lambda.$$

This implies $\nu(D) \geqq 0$. But this contradicts $\nu(D) < 0$ and completes the proof.

The pair $\{P, Q\}$ in the foregoing proposition is called a *Hahn decomposition* of X with respect to the signed measure $\nu : \mathfrak{B} \to R_e^1$. The positive subset P and the negative subset Q of X are called *positive component* and *negative component*, respectively.

The Hahn decompositions are obviously not unique, for any nonempty null set in one of the components may be moved to the other component without destroying its being a Hahn decomposition. However, these are almost unique in the sense of the following proposition.

PROPOSITION 21. *If $\{P, Q\}$ and $\{P', Q'\}$ are any two Hahn decompositions of X with respect to a signed measure ν: $\mathfrak{B} \to R_e^1$, then we have, for every $E \in \mathfrak{B}$,*

$$\nu(E \cap P) = \nu(E \cap P') \text{ and } \nu(E \cap Q) = \nu(E \cap Q').$$

Proof. Let $E \in \mathfrak{B}$ be arbitrarily given. Since the set $E \cap P \cap Q'$ is a subset of both P and Q', we must have $\nu(E \cap P \cap Q') = 0$. Similarly, we also have $\nu(E \cap P' \cap Q) = 0$. Hence

$$\nu(E \cap P) = \nu(E \cap (P \cup P')) = \nu(E \cap P'),$$

$$\nu(E \cap Q) = \nu(E \cap (Q \cup Q')) = \nu(E \cap Q').$$

To see the last step, observe that $E \cap (P - P')$ is a subset of $E \cap P \cap Q'$ because $E \cap (P - P') \subset E \cap P$ and $E \cap (P - P') \subset E \cap Q'$. Hence $\nu(E \cap (P - P')) = 0$. Similarly, $\nu(E \cap (P' - P)) = 0$. But $E \cap (P - P')$

$= E \cap ((P \cup P') - P') = E \cap (P \cup P') - E \cap P'$ and $E \cap (P' - P) = E \cap ((P' \cup P) - P) = E \cap (P' \cup P) - E \cap P$. Thus, ν being subtractive,

$$0 = \nu(E \cap (P - P')) = \nu(E \cap (P \cup P')) - \nu(E \cap P')$$

and

$$0 = \nu(E \cap (P \cup P')) - \nu(E \cap P).$$

The proof is complete.

On account of the last two propositions, we obtain, for an arbitrarily given signed measure, $\nu : \mathfrak{B} \to R_e^1$, three uniquely defined set functions

$$\nu^+, \nu^-, |\nu| : \mathfrak{B} \to R_e^1$$

as follows. Let $\{P, Q\}$ be any Hahn decomposition of X with respect to ν. Then these functions are defined by setting, for every $E \in \mathfrak{B}$, $\nu^+(E) = \nu(E \cap P)$, $\nu^-(E) = -\nu(E \cap Q)$, $|\nu|(E) = \nu^+(E) + \nu^-(E)$.

One can easily see that the set functions ν^+, ν^-, and $|\nu|$ are in fact measures on the σ-algebra $\mathfrak{B}$. They are called *positive variation*, *negative variation*, and *total variation* of the signed measure ν, respectively. It is clear that, for every $E \in \mathfrak{B}$,

$$\nu(E) = \nu^+(E) - \nu^-(E).$$

Let μ_1 and μ_2 be two measures on a measurable space $(X, \mathfrak{B})$. We say that μ_1 and μ_2 are *mutually singular* and write $\mu_1 \perp \mu_2$, if there exists a set $E \in \mathfrak{B}$ such that $\mu_1(E) = 0$ and $\mu_2(X - E) = 0$. In case of signed measures ν_1 and ν_2, we define $\nu_1 \perp \nu_2$ if $|\nu_1| \perp |\nu_2|$.

By a *Jordan decomposition* of a signed measure $\nu : \mathfrak{B} \to R_e^1$ we mean a pair $\{\nu_1, \nu_2\}$ of mutually singular measures ν_1 and ν_2 satisfying $\nu = \nu_1 - \nu_2$. Obviously, $\{\nu^+, \nu^-\}$ is a Jordan decomposition of ν. As to the uniqueness of this decomposition, we have the following proposition.

PROPOSITION 22. *Every signed measure $\nu : \mathfrak{B} \to R_e^1$ has a unique Jordan decomposition.*

Proof. We only need to prove the uniqueness. For this purpose, let $\{\nu_1, \nu_2\}$ denote any Jordan decomposition of ν. Since $\nu_1 \perp \nu_2$, there exists a set $Q \in \mathfrak{B}$ such that $\nu_1(Q) = 0 = \nu_2(X - Q)$. Let $P = X - Q$. Since ν_1 and ν_2 are measures on $\mathfrak{B}$ and since $\nu = \nu_1 - \nu_2$, it is obvious that P is a positive and Q is a negative set with respect to the signed measure ν. Hence $\{P, Q\}$ is a Hahn decomposition of X with respect to ν. Consequently. we have

$$\nu_1(E) = \nu_1(E \cap P) + \nu_1(E \cap Q) = \nu_1(E \cap P) = \nu(E \cap P)$$
$$= \nu^+(E),$$
$$\nu_2(E) = \nu_2(E \cap P) + \nu_2(E \cap Q) = \nu_2(E \cap P) = \nu(E \cap P)$$
$$= \nu^-(E)$$

for every $E \in \mathfrak{B}$. Hence $\{\nu_1, \nu_2\} = \{\nu^+, \nu^-\}$ and the proof is finished.

Consider a given measure space $(X, \mathfrak{B}, \mu)$. Let f be a nonnegative measurable extended real-valued function on X. Define a function ν: $\mathfrak{B} \to R_e^1$ by taking

$$\nu(E) = \int_E f\, d\mu$$

for every $E \in \mathfrak{B}$. Then ν is countably additive and hence ν is a measure on $\mathfrak{B}$, which we call the *measure defined by f and* μ.

An arbitrary measure ν: $\mathfrak{B} \to R_e^1$ is said to be *absolutely continuous with respect to a measure* $\mu : \mathfrak{B} \to R_e^1$, in symbols $\nu \ll \mu$, if $\nu(E) = 0$ for every set $E \in \mathfrak{B}$ for which $\mu(E) = 0$. In case of signed measures ν and μ we say $\nu \ll \mu$ if $|\nu| \ll |\mu|$.

We see that if ν is a measure defined by a nonnegative measurable function f: $X \to R_e^1$ and μ, then it follows that $\nu \ll \mu$; if $\mu(E) = 0$, then every function f: $E \to R_e^1$ is measurable and integrable with

$$\int_E f\, d\mu = 0.$$

In case μ is a σ-finite measure, then the converse is also true; in fact, we have the so-called *Radon-Nikodym Theorem*.

PROPOSITION 23. *If the measure* μ: $\mathfrak{B} \to R_e^1$ *of a measure space* $(X, \mathfrak{B}, \mu)$ *is* σ-*finite, then, for every measure* ν: $\mathfrak{B} \to R_e^1$ *with* $\nu \ll \mu$, *there exists a nonnegative measurable function* f: $X \to R_e^1$ *such that*

$$\nu(E) = \int_E f\, d\mu$$

holds for every $E \in \mathfrak{B}$. *The function f is almost unique in the sense that if any measurable function* g: $X \to R_e^1$ *has this property, then we have* $f = g$ *a.e. in* X *with respect to* μ.

Proof. To show the existence of f, we first assume that μ is a finite measure.

If μ is finite, $\nu - \xi\mu$ is a signed measure on $\mathfrak{B}$ for every rational number ξ. Let T denote the set of all nonnegative rational numbers. For each $\tau \in T$, select a Hahn decomposition $\{P_\tau, Q_\tau\}$ for $\nu - \tau\mu$. Then P_τ is a positive set for $\nu - \xi\mu$ whenever $\xi \leqq \tau$ and Q_τ is a negative set for $\nu - \xi\mu$ whenever $\xi \geqq \tau$. Let

$$P = \bigcap_{\tau \in T} P_\tau .$$

As the intersection of a countable collection of members of $\mathfrak{B}$, P is in $\mathfrak{B}$. Let E denote any measurable subset of P. For every $\tau \in T$, we have $E \subset P_\tau$ and hence $(\nu - \tau\mu)(E) \geqq 0$. This implies $\tau\mu(E) \leqq \nu(E)$. Since this holds for every $\tau \in T$, we must have either $\mu(E) = 0$ or $\nu(E) = \infty$. Hence $\mu(E) > 0$ implies $\nu(E) = \infty$.

Now we define a nonnegative function f: $\mathfrak{B} \to R_e^1$ by taking

$$f(x) = \inf\{\tau \in T : x \in Q_\tau\}$$

for every $x \in X$. Here we follow the usual convention that the infimum of an empty collection of real numbers is ∞. Hence we have $f(x) = \infty$ if and only if $x \in P$. To prove that f is measurable, let r denote any real number. For every rational number ξ, let

$$S_\xi = \bigcup_{\tau < \xi} Q_\tau .$$

Then, according to the definition of f, we have

$$\{x \in X : f(x) \leqq r\} = \bigcup_{\xi > r} S_\xi .$$

Since this subset of X is in $\mathfrak{B}$, f is measurable.

Next we show that

$$\nu(E) = \int_E f \, d\mu$$

holds for every $E \in \mathfrak{B}$. For this purpose, we first assume that $\mu(E \cap P) > 0$. Since $f(x) = \infty$ for every $x \in P$, it follows that the restriction of f to E is not integrable and hence

$$\int_E f \, d\mu = \infty .$$

(N.B. If a measurable function f: $E \to R_e^1$ is integrable, then it is finite almost everywhere in E, that is, the subsets $f^{-1}(\infty)$ and $f^{-1}(-\infty)$ are of measure 0. For example, let $A = f^{-1}(\infty)$. To show that $\mu(A) = 0$, let n be any natural number. Since $n\chi_A \leqq f^+$, we have $\int_E f^+ \geqq n\mu(A)$. Since f^+ is integrable and since this holds for all $n \in N$, we must have $\mu(A) = 0$.) On the other hand,

$$\mu(E) \geqq \mu(E \cap P) > 0$$

implies

$$\nu(E) \geqq \nu(E \cap P) = \infty = \int_E f \, d\mu .$$

This establishes the equality for the case where $\mu(E \cap P) > 0$.

Now we consider the case where $\mu(E \cap P) = 0$. Since $\nu \ll \mu$, this implies $\nu(E \cap P) = 0$. Let $n \in N$ be arbitrarily given. For each integer $k \geqq 0$, consider the subset

$$E_k = \left\{x \in X : \frac{k-1}{n} < f(x) \leqq \frac{k}{n}\right\}$$

of X. For each rational number $\xi > k/n$, we have

$$E_k \subset \left\{x \in X : f(x) \leqq \frac{k}{n}\right\} \subset S_\xi = \bigcup_{\tau < \xi} Q_\tau .$$

For every $\tau < \xi$, Q_τ is a negative set for the signed measure $\nu - \xi\mu$. By Proposition 18, E_k is also a negative set for $\nu - \xi\mu$. This implies $\nu(E_k) \leqq \xi\mu(E_k)$. Since this is true for every rational number $\xi > k/n$, we must have

$$\nu(E_k) \leqq \frac{k}{n}\mu(E_k).$$

On the other hand, since

$$E_k \subset \left\{x \in X : f(x) > \frac{k-1}{n}\right\} \subset P_{(k-1)/n},$$

it follows from Proposition 18 that E_k is a positive set for $\nu - ((k-1)/n)\mu$. This implies

$$\frac{k-1}{n}\mu(E_k) \leqq \nu(E_k).$$

Combining these two inequalities, we obtain

$$(*) \qquad \frac{k-1}{n}\mu(E_k) \leqq \nu(E_k) \leqq \frac{k}{n}\mu(E_k).$$

Since E is the disjoint union of $E \cap P$ and the sequence $(E_n)_{n=0}^{\infty}$, and since $\mu(E \cap P) = 0 = \nu(E \cap P)$, it follows from the countable additivity of measure and integral that we have

$$\int_E f\,d\mu = \sum_{k=0}^{\infty} \int_{E_k} f\,d\mu, \quad \nu(E) = \sum_{k=0}^{\infty} \nu(E_k).$$

By the definition of E_k, we have

$$(**) \qquad \frac{k-1}{n}\mu(E_k) \leqq \int_{E_k} f\,d\mu \leqq \frac{k}{n}\mu(E_k).$$

From the inequalities (*) and (**) we deduce

$$-\frac{1}{n}\mu(E_k) \leqq \nu(E_k) - \int_{E_k} f\,d\mu \leqq \frac{1}{n}\mu(E_k).$$

Summing over $k = 0, 1, 2, \ldots$, we obtain

$$-\frac{1}{n}\mu(E) \leqq \nu(E) - \int_{E} f\,d\mu \leqq \frac{1}{n}\mu(E).$$

Since $\mu(E)$ is finite and this holds for all $n \in N$, we must have

$$\nu(E) = \int_{E} f\,d\mu.$$

This establishes the existence of f for the case where μ is a finite measure.

Next, we assume that $\mu\colon \mathfrak{B} \to R_e^1$ is σ-finite. Then X is the union of a sequence $(E_n)_{n=1}^{\infty}$ of measurable subsets E_n with $\mu(E_n) < \infty$ for every $n \in N$. For each $n \in N$, let

$$X_n = E_n - \left(\bigcup_{j=1}^{n-1} E_j\right).$$

Then X_n is measurable with $\mu(X_n) < \infty$. By the preceding part of the existence proof, there exists a nonnegative measurable function

$$f_n : X_n \to R_e^1$$

such that

$$\nu(E) = \int_{E} f_n\,d\mu$$

holds for every measurable subset E of X_n. Since X is the disjoint union of the sequence $(X_n)_{n=1}^{\infty}$, we have a well-defined combined function

$$f : X \to R_e^1$$

given by $f(x) = f_n(x)$ for every $x \in X_n$. This function f is obviously measurable. By the countable additivity of measure and integral, one can easily verify that

$$\nu(E) = \int_{E} f\,d\mu$$

holds for every $E \in \mathfrak{B}$. This completes the existence proof.

To establish the almost uniqueness of the function $f\colon X \to R_e^1$, let $g\colon X \to R_e^1$ denote any measurable function satisfying

$$\nu(E) = \int_E g\, d\mu$$

for every $E \in \mathfrak{B}$. For each $n \in N$, we let

$$A_n = \left\{x \in X : f(x) - g(x) \geqq \frac{1}{n}\right\} \in \mathfrak{B}$$

and

$$B_n = \left\{x \in X : g(x) - f(x) \geqq \frac{1}{n}\right\} \in \mathfrak{B}.$$

Then we have

$$\frac{1}{n}\mu(A_n) \leqq \int_{A_n} (f - g)\, d\mu = \int_{A_n} f\, d\mu - \int_{A_n} g\, d\mu = 0.$$

This implies $\mu(A_n) = 0$. Similarly, we also have $\mu(B_n) = 0$. Since

$$D = \{x \in X : f(x) \neq g(x)\} = \bigcup_{n=1}^{\infty} (A_n \cup B_n),$$

it follows that $\mu(D) = 0$. This completes the proof of the proposition.

Remark. The almost unique function $f\colon X \to R_e^1$ in the foregoing proposition is sometimes called the *Radon-Nikodym derivative* of the measure ν with respect to the measure μ; in symbols,

$$f = \frac{d\nu}{d\mu}.$$

PROPOSITION 24 (LEBESGUE'S DECOMPOSITION THEOREM) . *If the measure μ: $\mathfrak{B} \to R_e^1$ of a measure space $(X, \mathfrak{B}, \mu)$ is σ-finite, then, for an arbitrarily given σ-finite measure ν: $\mathfrak{B} \to R_e^1$, there exists a measure $\nu_0 : \mathfrak{B} \to R_e^1$ with $\nu_0 \perp \mu$ and a measure $\nu_1 : \mathfrak{B} \to R_e^1$ with $\nu_1 \ll \mu$ such that $\nu = \nu_0 + \nu_1$. Moreover, the measures ν_0 and ν_1 are unique.*

Proof. Since μ and ν are both σ-finite, so is their sum $\sigma = \mu + \nu$. Obviously we have $\mu \ll \sigma$ and $\nu \ll \sigma$. By the Radon-Nikodym Theorem (see Proposition 23), there exist nonnegative measurable functions $f, g : X \to R_e^1$ satisfying

$$\mu(E) = \int_E f\, d\sigma, \quad \nu(E) = \int_E g\, d\sigma$$

for every $E \in \mathfrak{B}$. Let

$$C = \{x \in X : f(x) > 0\}, \quad D = \{x \in X : f(x) = 0\}.$$

Then $C, D \in \mathfrak{B}$ and X is the disjoint union of C and D. Furthermore, we have

$$\mu(D) = \int_D f\,d\sigma = 0.$$

We define two functions $\nu_0, \nu_1 : \mathfrak{B} \to R_e^1$ by taking for every $E \in \mathfrak{B}$

$$\nu_0(E) = \nu(E \cap D), \qquad \nu_1(E) = \nu(E \cap C).$$

Then ν_0 and ν_1 are measures on $\mathfrak{B}$ and satisfy $\nu = \nu_0 + \nu_1$.

Since $\nu_0(C) = \nu(C \cap D) = \nu(\varnothing) = 0$, we have $\nu_0 \perp \mu$. To prove $\nu_1 \ll \mu$, let $E \in \mathfrak{B}$ with $\mu(E) = 0$ be arbitrarily given. Then we have

$$\int_E f\,d\sigma = \mu(E) = 0.$$

Since f is nonnegative, this implies that $f(x) = 0$ almost everywhere in E with respect to σ. Since $f(x) > 0$ for every $x \in E \cap C$, this implies

$$\nu_1(E) = \nu(E \cap C) \leqq \sigma(E \cap C) = 0.$$

Hence $\nu_1(E) = 0$. This proves that $\nu_1 \ll \mu$ and completes the existence proof.

To show the uniqueness of the measures ν_0 and ν_1, let us consider arbitrary measures $\nu_0', \nu_1' : \mathfrak{B} \to R_e^1$ satisfying

$$\nu = \nu_0' + \nu_1', \quad \nu_0' \perp \mu, \quad \nu_1' \ll \mu.$$

Since $\nu_0 + \nu_1 = \nu = \nu_0' + \nu_1'$, we obtain a signed measure

$$\lambda = \nu_0 - \nu_0' = \nu_1' - \nu_1 : \mathfrak{B} \to R_e^1.$$

It remains to verify that $\lambda = 0$. Since $\nu_0 \perp \mu$ and $\nu_0' \perp \mu$, there exist two measurable subsets A and B such that

$$\mu(A) = 0, \qquad \nu_0(X - A) = 0,$$

$$\mu(B) = 0, \qquad \nu_0'(X - B) = 0.$$

Then we have $\mu(A \cup B) = 0$. To prove $\lambda = 0$, let $E \in \mathfrak{B}$ be arbitrarily given. Then we have $\mu(E \cap (A \cup B)) = 0$. Since $\nu_1 \ll \mu$ and $\nu_1' \ll \mu$, this implies

$$\lambda(E \cap (A \cup B)) = \nu_1'(E \cap (A \cup B)) - \nu_1(E \cap (A \cup B)) = 0.$$

On the other hand, since

$$E - (A \cup B) = (E - A) \cap (E - B)$$

is contained in both $E - A$ and $E - B$, we have

$$\lambda(E - (A \cup B)) = \nu_0(E - (A \cup B)) - \nu_0'(E - (A \cup B)) = 0.$$

We therefore obtain that

$$\lambda(E) = \lambda(E \cap (A \cup B)) + \lambda(E - (A \cup B)) = 0.$$

This shows that $\lambda = 0$, and we have what we wanted to prove.

Remark. The decomposition $\nu = \nu_0 + \nu_1$ in the foregoing proposition is sometimes called the *Lebesgue decomposition* of the σ-finite measure ν: $\mathfrak{B} \to R_e^1$ with respect to the σ-finite measure μ: $\mathfrak{B} \to R_e^1$ into its singular and absolutely continuous components with respect to μ.

In the remainder of this chapter we shall consider two applications of the Radon-Nikodym Theorem; the first of these applications pertains to "change of variable" in an integral and the second application deals with the so-called "conjugate space" of an L^p-space.

Let X and Y be nonempty sets and ψ be a mapping of X into Y. Associated with ψ are several mappings of objects associated with Y into corresponding objects associated with X. For example, the set mapping

$$\Psi(E) = \psi^{-1}(E)$$

is a mapping of subsets of Y into the subsets of X. This mapping preserves unions, intersections, and complements. Ψ is called the *set mapping induced by* ψ. We call ψ a *point mapping*. If $(X, \mathfrak{A})$ and $(Y, \mathfrak{B})$ are measurable spaces, we call the point mapping ψ of X into Y a *measurable transformation* if

$$\psi^{-1}(E) \in \mathfrak{A}$$

for each $E \in \mathfrak{B}$. Thus, ψ is a measurable transformation if and only if Ψ maps $\mathfrak{B}$ into $\mathfrak{A}$.

If f is a real-valued function on $(X, \mathfrak{A})$, then f is a mapping of X into R^1 and f is *measurable with respect to* $\mathfrak{A}$ if and only if f is a measurable transformation of $(X, \mathfrak{A})$ into $(R^1, \mathcal{B})$, where $\mathcal{B}$ is the σ-algebra of Borel sets, i.e., the smallest σ-algebra containing the topology of R^1. A real-valued function of a real variable is Lebesgue measurable if and only if it is a measurable transformation of $(R^1, \mathfrak{M})$ into $(R^1, \mathcal{B})$, where $\mathfrak{M}$ is the class of Lebesgue measurable sets in R^1; it is *Borel measurable* if and only if it is a measurable transformation of $(R^1, \mathcal{B})$ into $(R^1, \mathcal{B})$. In case f takes on infinite values, we merely add the requirement that

$$f^{-1}(\infty) \text{ and } f^{-1}(-\infty)$$

be Borel sets or Lebesgue measurable sets, as the case may be.

Remark. In defining measurability for extended real-valued functions on $(X, \mathfrak{A})$, we can also adopt the simple procedure of defining the class $\mathcal{B}_e$ of Borel sets in R_e^1 directly: We say that a set $B \subset R_e^1$ is a *Borel set in* R_e^1 if it is the union of a set in $\mathcal{B}$ (the class of Borel sets in R^1) with any subset of $R_e^1 - R^1 = \{-\infty, \infty\}$. A function $f: X \to R_e^1$ on $(X, \mathfrak{A})$ is then said to be measurable with respect to $\mathfrak{A}$ if and only if

$$f^{-1}(B) \in \mathfrak{A}$$

for every $B \in \mathcal{B}_e$.

We note the following simple fact: If $f: X \to Y$ is a measurable transformation from $(X, \mathfrak{A})$ into $(Y, \mathfrak{B})$ and $g: Y \to R_e^1$ is measurable with respect to $\mathfrak{B}$ as a function with extended real values, then the composition $g(f)$ is measurable with respect to $\mathfrak{A}$.

Indeed, for any Borel set $B \in \mathcal{B}_e$ we have $\{x : g(f)(x) \in B\} = f^{-1}\{y : g(y) \in B\} = f^{-1}(E)$ for some $E \in \mathfrak{B}$ and is therefore in $\mathfrak{A}$.

Note that the foregoing shows in particular that a Borel measurable function of a Lebesgue measurable function is Lebesgue measurable.

If we start with a measure space $(X, \mathfrak{A}, \mu)$ and f is a measurable transformation from $(X, \mathfrak{A})$ into $(Y, \mathfrak{B})$, it is natural to use f to define a measure ν on $\mathfrak{B}$ by putting

$$(*) \qquad \nu(E) = \mu(f^{-1}(E)) \text{ for } E \in \mathfrak{B}.$$

With this definition of ν it is clear that $(Y, \mathfrak{B}, \nu)$ is a measure space. If (*) holds, we will write

$$\nu = \mu f^{-1}.$$

This permits us to carry out a "change of variable" in an integral as in the next proposition.

PROPOSITION 25. *If f is a measurable transformation from a measure space $(X, \mathfrak{A}, \mu)$ into a measurable space $(Y, \mathfrak{B})$ and $g: Y \to R^1$ is $\mathfrak{B}$-measurable, then*

$$\int_Y g\, d(\mu f^{-1}) = \int_X g(f)\, d\mu$$

in the sense that if either integral exists, so does the other and the two are equal.

Proof. It is enough to consider nonnegative functions $g: Y \to R^1$. Suppose first that $g = \chi_E$, the characteristic function of a set E in $\mathfrak{B}$. Then

$$g(f)(x) = \begin{cases} 1 & \text{if } x \in f^{-1}(E), \\ 0 & \text{if } x \notin f^{-1}(E); \end{cases}$$

hence $g(f)$ is the characteristic function of $f^{-1}(E)$, a set in $\mathfrak{A}$. Thus, in this case, by (*)

$$\int g\,d(\mu f^{-1}) = \mu f^{-1}(E) = \mu(f^{-1}(E)) = \int g(f)\,d\mu.$$

By linearity, the result follows now for nonnegative $\mathfrak{B}$-simple functions g. If $(g_n)_{n=1}^{\infty}$ is an increasing sequence of nonnegative simple functions converging to the measurable function g, then $g_n(f)$ will be an increasing sequence of simple functions converging to $g(f)$ and by the Monotone Convergence Theorem (see Proposition 12) we are in; the proof is finished.

Often in integration, when the variable is changed, one wants to integrate with respect to a new measure $\nu \neq \mu f^{-1}$. One can do so easily when μf^{-1} is absolutely continuous with respect to ν.

PROPOSITION 26. *Given σ-finite measure spaces $(X, \mathfrak{A}, \mu)$ and $(Y, \mathfrak{B}, \nu)$ and a measurable transformation f from $(X, \mathfrak{A})$ into $(Y, \mathfrak{B})$ such that μf^{-1} is absolutely continuous with respect to ν. Let*

$$\psi = \frac{d(\mu f^{-1})}{d\nu}$$

denote the Radon-Nikodym derivative. Then

$$\int g(f)\,d\mu = \int g\psi\,d\nu,$$

for every measurable $g\colon Y \to R_e^1$ in the sense that, if either integral exists, so does the other and the two are equal.

Proof. By Proposition 25 we have

$$\int_X g(f)\,d\mu = \int_Y g\,d(\mu f^{-1}).$$

Since μf^{-1} is absolutely continuous with respect to ν, there exists a measurable ψ such that for every $E \in \mathfrak{A}$

$$\int_E \psi\,d\nu = (\mu f^{-1}(E)).$$

If g is the characteristic function of a measurable set E, it now follows that

$$\int g\,d(\mu f^{-1}) = (\mu f^{-1})(E) = \int g\psi\,d\nu$$

and the required result now follows by successive extension to functions g which are (1) nonnegative, simple; (2) nonnegative, measurable; (3) measurable. The proof is finished.

The following is a direct consequence of the foregoing proposition.

PROPOSITION 27. *If $h: R^1 \to R_e^1$ is a nonnegative Lebesgue integrable function and*

$$H(x) = \int_{-\infty}^{x} h(t)\,dt,$$

then

$$\int_a^b g(x)\,dx = \int_\alpha^\beta g(H(t))h(t)\,dt,$$

where $a = H(\alpha)$ and $b = H(\beta)$.

Proof. We shall show that

$$\int_a^b g(x)\,dx = \int_\alpha^\beta g(H(t))\,d\mu_H = \int_\alpha^\beta g(H(t))h(t)\,dt,$$

where μ_H is the measure obtained by setting

$$\mu_H(E) = \int_E h\,d\mu$$

for any Lebesgue measurable set E in R^1.

Under the conditions of the proposition, we consider the set mapping induced by $H: R^1 \to R^1$ from the Lebesgue measure space $(R^1, \mathfrak{M}, \mu)$ to $(R^1, \mathfrak{B}, \mu_H)$. Proposition 25 then gives that

$$\int_a^b g(x)\,dx = \int_\alpha^\beta g(H(t))\,d\mu_H.$$

But the measure μ_H is absolutely continuous with respect to Lebesgue measure μ and h is a possible definition of the Radon-Nikodym derivative $d\mu_H/d\mu$ and so

$$\int_\alpha^\beta g(H(t))\,d\mu_H = \int_\alpha^\beta g(H(t))h(t)\,dt$$

follows from Proposition 26; the proof is finished.

If $(X, \mathfrak{B}, \mu)$ is a σ-finite complete measure space, we denote by $L^p(X, \mathfrak{B}, \mu)$ the space of all measurable functions on X for which $\int |f|^p \, d\mu < \infty$, considering two functions in $L^p(X, \mathfrak{B}, \mu)$ to be equivalent if they are equal μ-a.e. As before we define $L^\infty(X, \mathfrak{B}, \mu)$ to be the space of bounded measurable functions and we set

$$\|f\|_p = \{\int |f|^p \, d\mu\}^{1/p}$$

for $1 \leqq p < \infty$ and

$$\|f\|_\infty = \text{ess sup} |f|$$

for $p = \infty$. The space $L^\infty(X, \mathfrak{B}, \mu)$ depends on the choice of μ to determine the norm and the class of equivalent functions; but this only requires knowing what the sets of μ-measure zero are.

The following facts can be obtained in complete analogy to the proofs used in Chapter 3.

For $1 \leqq p \leqq \infty$, the spaces $L^p(X, \mathfrak{B}, \mu)$ are complete normed linear spaces and if $f \in L^p(X, \mathfrak{B}, \mu)$ and $g \in L^q(X, \mathfrak{B}, \mu)$, with $q = p/(p-1)$, then $fg \in L^1(X, \mathfrak{B}, \mu)$ and

$$\int |fg| \, d\mu \leqq \|f\|_p \|g\|_q.$$

For $1 \leqq p < \infty$ for every $f \in L^p(X, \mathfrak{B}, \mu)$ one can find a simple function s vanishing outside a set of finite measure such that $s \in L^p(X, \mathfrak{B}, \mu)$ and $\|f - s\|_p < \varepsilon$ for any given $\varepsilon > 0$.

Let S be a normed linear space over R^1. A mapping F of the space S into R^1 such that

$$F(\alpha f + \beta g) = \alpha F(f) + \beta F(g)$$

for all $\alpha, \beta \in R^1$ and $f, g \in S$ is called a *linear functional* on S. We say that the linear functional F is *bounded* if there is a constant M such that

$$|F(f)| \leqq M\|f\|$$

for all f in S. The smallest constant M for which this inequality is true is called the *norm* of the functional F. Thus

$$\|F\| = \sup \frac{|F(f)|}{\|f\|},$$

as f ranges over all nonzero elements of the space X.

If $g \in L^q(X, \mathfrak{B}, \mu)$, we can define a bounded linear funcional F on $L^p(X, \mathfrak{B}, \mu)$ by setting

$$F(f) = \int fg\, d\mu.$$

By Hőlder's Inequality

$$\|F\| \leqq \|g\|_q.$$

Actually, $\|F\| = \|g\|_q$, as we can easily see. We do this here for the case $1 < p < \infty$ only, and we set

$$f = |g|^{q/p} \operatorname{sign} g,$$

where the *signum* of a real number x is defined by

$$\operatorname{sign} x = \begin{cases} x/|x| & \text{if } x \neq 0, \\ 0 & \text{if } x = 0. \end{cases}$$

Then $|f|^p = |g|^q = fg$. Hence $f \in L^p(X, \mathfrak{B}, \mu)$ and $\|f\|_p = (\|g\|_q)^{q/p}$. But

$$F(f) = \int fg = \int |g|^q = (\|g\|_q)^q = \|g\|_q \|f\|_p$$

and so $\|F\|$ is at least as large as $\|g\|_q$.

Our aim is to establish that every bounded linear functional on $L^p(X, \mathfrak{B}, \mu)$ is of the form

$$F(f) = \int fg\, d\mu$$

with $g \in L^q(X, \mathfrak{B}, \mu)$.

PROPOSITION 28. *If $(X, \mathfrak{B}, \mu)$ is a finite, complete measure space and if g is an integrable function such that for some constant M,*

$$\left|\int gs\, d\mu\right| \leqq M\|s\|_p$$

for all simple functions s in $L^p(X, \mathfrak{B}, \mu)$, then $g \in L^q(X, \mathfrak{B}, \mu)$.

Proof. We shall prove this for $p > 1$ only; the case when $p = 1$ is left to the reader to fill in.

Let $(\psi_n)_{n=1}^{\infty}$ be a sequence of nonnegative simple functions such that $\psi_n \uparrow |g|^q$. We put

$$s_n = (\psi_n)^{1/p} \operatorname{sign} g.$$

Then s_n is a simple function and

$$\|s_n\|_p = (\int \psi_n \, d\mu)^{1/p}.$$

Since

$$s_n g \geqq |s_n||\psi_n|^{1/q} = |\psi_n|^{(1/p)+(1/q)} = |\psi_n| = \psi_n,$$

we have

$$\int \psi_n \, d\mu \leqq \int s_n g \, d\mu \leqq M\|s_n\|_p \leqq M(\int \psi_n \, d\mu)^{1/p}.$$

But $1 - (1/p) = (1/q)$ and so

$$(\int \psi_n \, d\mu)^{1/q} \leqq M, \text{ or } \int \psi_n \, d\mu \leqq M^q;$$

therefore, by the Monotone Convergence Theorem (see Proposition 12),

$$\int |g|^q \, d\mu \leqq M^q.$$

This completes the proof.

PROPOSITION 29. *Suppose that $(E_n)_{n=1}^{\infty}$ is a sequence of disjoint sets in $\mathfrak{B}$ and, for each $n \in N$, f_n is a function in $L^p(X, \mathfrak{B}, \mu)$, with $1 \leqq p < \infty$, which vanishes outside E_n. Let*

$$f = \sum_{n=1}^{\infty} f_n.$$

Then $f \in L^p(X, \mathfrak{B}, \mu)$ if and only if

$$\sum_{n=1}^{\infty} \|f_n\|_p^p < \infty.$$

In this case

$$f = \sum_{n=1}^{\infty} f_n$$

is in $L^p(X, \mathfrak{B}, \mu)$; that is,

$$\|f - \sum_{j=1}^{n} f_j\|_p \to 0$$

and

$$\|f\|_p^p = \sum_{n=1}^{\infty} \|f_n\|_p^p.$$

The proof of this proposition is left as an exercise.

We now take up the *Riesz Representation Theorem* for L^p-spaces.

PROPOSITION 30. *Let F be a bounded linear functional on $L^p(X, \mathfrak{B}, \mu)$, with $1 \leqq p < \infty$ and μ a σ-finite measure. Then there is a unique element g in $L^q(X, \mathfrak{B}, \mu)$, where $q = p/(p-1)$, such that*

$$F(f) = \int fg\, d\mu$$

and we also have $\|F\| = \|g\|_q$.

Proof. First we take up the case when μ is a finite measure.

If μ is finite, then every bounded measurable function is in $L^p(X, \mathfrak{B}, \mu)$. We define a set function ν on $\mathfrak{B}$ by

$$\nu(E) = F(\chi_E).$$

If E is the union of a sequence $(E_n)_{n=1}^{\infty}$ of disjoint sets in $\mathfrak{B}$, we let

$$c_n = \operatorname{sign} F(\chi_{E_n}),$$

and set

$$f = \sum_{n=1}^{\infty} c_n \chi_{E_n}.$$

By Proposition 29 and by the boundedness of F we then have

$$\sum_{n=1}^{\infty} |\nu(E_n)| = F(f) < \infty$$

and

$$\sum_{n=1}^{\infty} \nu(E_n) = F(\chi_E) = \nu(E).$$

This shows that ν is a signed measure and so by Proposition 23 there is a measurable function g such that for each $E \in \mathfrak{B}$ we have

$$\nu(E) = \int_E g\, d\mu.$$

Since ν is always finite, g is integrable.

If s is a simple function, the linearity of F and of the integral yield that

$$F(s) = \int sg\, d\mu.$$

Since $F(s)$ is bounded by $\|F\|\|s\|_p$, we have by Proposition 28 that $g \in L^q(X, \mathfrak{B}, \mu)$. Let G be the bounded linear functional defined on $L^p(X, \mathfrak{B}, \mu)$ by

$$G(f) = \int fg \, d\mu.$$

Then $G - F$ is a bounded linear functional that vanishes on the dense subspace of all simple functions in $L^p(X, \mathfrak{B}, \mu)$, and it follows that $G - F = 0$. Thus, for all $f \in L^p(X, \mathfrak{B}, \mu)$

$$F(f) = \int fg \, d\mu$$

and it is easily seen that $\|F\| = \|G\| = \|g\|_q$.

The function g must determine a unique element of $L^q(X, \mathfrak{B}, \mu)$, for if g_1 and g_2 determine the same functional F, then $g_1 - g_2$ must determine the zero functional and therefore $\|g_1 - g_2\|_q = 0$ or $g_1 = g_2$ μ-a.e.

We now assume that μ is a σ-finite measure and that $(X_n)_{n=1}^{\infty}$ is an increasing sequence of μ-measurable sets of finite measure whose union is X. From what we have proved already, we know that for each $n \in N$ there exists a function g_n in $L^q(X, \mathfrak{B}, \mu)$, which vanishes outside X_n such that $F(f) = \int fg_n \, d\mu$ for all $f \in L^p(X, \mathfrak{B}, \mu)$ that vanish outside X_n. We also have that $\|g_n\|_q \leqq \|F\|$. Since any function g_n with this property is uniquely determined on X_n up to sets of μ-measure zero and since g_{n+1} also has this property, we may assume $g_{n+1} = g_n$ on X_n. For $x \in X_n$ we put $g(x) = g_n(x)$. Then g is a well-defined measurable function and $|g_n| \uparrow |g|$. By the Monotone Convergence Theorem (see Proposition 12)

$$\int |g|^q \, d\mu = \lim_{n \to \infty} \int |g_n|^q \, d\mu \leqq \|F\|^q$$

and $g \in L^q(X, \mathfrak{B}, \mu)$.

If $f \in L^p(X, \mathfrak{B}, \mu)$, let $f_n = f$ on X_n and $f_n = 0$ outside X_n. Then $f_n \to f$ pointwise and in $L^p(X, \mathfrak{B}, \mu)$. Since $|fg|$ is integrable and $|f_n g| \leqq |fg|$, the Dominated Convergence Theorem (see Proposition 14) shows that

$$\int fg \, d\mu = \lim_{n \to \infty} \int f_n g \, d\mu = \lim_{n \to \infty} \int f_n g_n \, d\mu = \lim_{n \to \infty} F(f_n) = F(f)$$

and the proof is finished.

Remark. It is clear that the set of all bounded linear functionals on a normed linear space S over R^1 form in a natural manner again a normed linear space. This new normed linear space is called the *conjugate space* of S. Proposition 30 identifies the conjugate spaces of L^p-spaces - at least up to isomorphism.

EXERCISES

1. In a normed linear space $(S, \| \ \|)$ a series

$$\sum_{n=1}^{\infty} s_n$$

with $s_n \in S$ for all $n \in N$ is said to be *convergent* if the sequence of partial sums of the series converges in S; the series is said to be *absolutely convergent* if the numerical series

$$\sum_{n=1}^{\infty} \|s_n\|$$

is convergent.
   (a) Prove that a normed linear space is complete if and only if every absolutely convergent series is convergent.
   (b) Prove Proposition 29.
2. Verify Proposition 28 for the case when $p = 1$.
3. Show that in Proposition 30 the requirement of σ-finiteness can be dropped if we consider only the case when $p > 1$.
4. Suppose that μ and ν are σ-finite measures on a measurable space $(X, \mathfrak{B})$ such that $\mu \ll \nu$ and $\nu \ll \mu$. Prove that

$$\frac{d\nu}{d\mu} \neq 0 \text{ a.e.}$$

and

$$\frac{d\mu}{d\nu} = 1 \Big/ \frac{d\nu}{d\mu} \text{ a.e.}$$

noting that μ and ν have exactly the same sets of zero measures.
5. Let μ_0, μ_1, and μ_2 be σ-finite measures on $(X, \mathfrak{B})$ such that

$$\mu_2 \ll \mu_1 \text{ and } \mu_1 \ll \mu_0 .$$

Show that
   (i) $\mu_2 \ll \mu_0$ and
   (ii) $$\frac{d\mu_2}{d\mu_o} = \frac{d\mu_2}{d\mu_1}\frac{d\mu_1}{d\mu_0} \quad (\mu_0 - \text{a.e.}).$$
6. For any signed measure μ on $(X, \mathfrak{B})$, prove that

$$-\mu^-(E) \leqq \mu(E) \leqq \mu^+(E)$$

and

$$|\mu(E)| \leqq |\mu|(E)$$

hold for every $E \in \mathfrak{B}$.

7. Let ν and μ be signed measures on $(X, \mathfrak{B})$. Show that
   (i) $\nu \ll \mu \Leftrightarrow \nu^+ \ll \mu$ and $\nu^- \ll \mu \Leftrightarrow |\nu| \ll \mu$.
   (ii) $\nu \ll \mu$ and $\nu \perp \mu$ imply $\nu = 0$.
   [*Hints* for part (ii): $\nu \perp \mu$ implies that there is a set $A \in \mathfrak{B}$ such that $|\nu|(A) = 0 = |\mu|(X - A)$; $\nu \ll \mu$ implies that for all $B \in \mathfrak{B}$ such that $|\mu|(B) = 0$ we have $|\nu|(B) = 0$. Pick any $E \in \mathfrak{B}$. Since $A \cap E \subset A$, we have $|\nu|(A \cap E) \leqq |\nu|(A) = 0$. But $E = (A \cap E) \cup (E - A)$ is a disjoint union, we get from $E - A \subset X - A$ that $|\mu|(E - A) \leqq |\mu|(X - A) = 0$ and hence $|\nu|(E - A) = 0$ by the assumption that $\nu \ll \mu$. But $|\nu|(E) = |\nu|(A \cap E) + |\nu|(E - A) = 0$ and hence $|\nu|(E) = 0$ for all $E \in \mathfrak{B}$. But $0 \leqq |\nu(E)| \leqq |\nu|(E) = 0$ for all $E \in \mathfrak{B}$ and so $\nu = 0$.]
8. Let μ and ν be signed measures on $(X, \mathfrak{B})$. Show that the following two conditions are equivalent:
   (i) $\mu \perp \nu$;
   (ii) $\mu^+ \perp \nu^+, \mu^+ \perp \nu^-, \mu^- \perp \nu^+, \mu^- \perp \nu^-$.
9. Let μ, ν, and ω be signed measures on $(X, \mathfrak{B})$ such that $\nu + \omega$ is defined, and let α be a real number. Show that
   (i) $\nu \ll \mu$ and $\omega \ll \mu$ imply $\alpha\nu \ll \mu$ and $(\nu + \omega) \ll \mu$;
   (ii) $\nu \perp \mu$ and $\omega \perp \mu$ imply $\alpha\nu \perp \mu$ and $(\nu + \omega) \perp \mu$.
   [*Hints*: For part (i) take $E \in \mathfrak{B}$ such that $|\mu|(E) = 0$. Then $(\alpha\nu)(E) = \alpha\nu(E) = \alpha 0 = 0$ and $(\nu + \omega)(E) = \nu(E) + \omega(E) = 0$. For part (ii), choose A and B in $\mathfrak{B}$ such that $|\mu|(A) = |\mu|(B) = 0$, $|\nu|(A^c) = 0$ and $|\omega|(B^c) = 0$. Let $C = A \cup B$. Then $|\nu + \omega|(C^c) \leqq (|\nu| + |\omega|)(C^c) \leqq |\nu|(A^c) + |\omega|(B^c) = 0$ and $|\mu|(C) \leqq |\mu|(A) + |\mu|(B) = 0$. Thus $(\nu + \omega) \perp \mu$. It is obvious that $\alpha\nu \perp \mu$.]
10. Let $(X, \mathfrak{B})$ be a measurable space. A *measurable dissection* of a set $E \in \mathfrak{B}$ is any finite pairwise disjoint family $\{E_1, E_2, \ldots, E_n\}$ such that

$$\bigcup_{k=1}^{n} E_k = E.$$

Suppose that ν is a signed measure on $(X, \mathfrak{B})$. Prove that

$$|\nu|(E) = \sup\{\sum_{k=1}^{n} |\nu(E_k)| : \{E_1, \ldots, E_n\} \text{ is a measurable dissection of } E\}$$

for every $E \in \mathfrak{B}$.

[*Hints*: Let β denote the right side of the equation. Then

$$\sum_{k=1}^{n} |\nu(E_k)| = \sum_{k=1}^{n} |\nu^+(E_k) - \nu^-(E_k)| \leqq \sum_{k=1}^{n} (\nu^+(E_k) + \nu^-(E_k))$$
$$= \sum_{k=1}^{n} |\nu|(E_k) = |\nu|(E)$$

for every measurable dissection $\{E_1, \ldots, E_n\}$ of E, and so $\beta \leqq |\nu|(E)$. To show the reverse inequality, let $\{P, Q\}$ be a Hahn decomposition of X for ν and consider the dissection $\{E \cap P, E \cap Q\}$. We obtain $\beta \geqq |\nu(E \cap P)| + |\nu(E \cap Q)| = \nu^+(E) + \nu^-(E) = |\nu|(E)$.]

11. Let $(X, \mathfrak{B}, \mu)$ be a measure space and $f\colon X \to R_e^1$ be any μ-integrable function. Define $\nu\colon \mathfrak{B} \to R^1$ by

$$\nu(E) = \int_E f d\mu.$$

Show that ν is a signed measure on $(X, \mathfrak{B})$. Let $f^+ = f \vee 0$, $f^- = -(f \wedge 0)$. Show that

$$\nu^+(E) = \int_E f^+ d\mu$$
$$\nu^-(E) = \int_E f^- d\mu$$
$$|\nu|(E) = \int_E |f| d\mu$$

for every $E \in \mathfrak{B}$. The set functions ν^+, ν^-, and $|\nu|$ are finite measures on $(X, \mathfrak{B})$.

12. Let μ and ν be two measures with $\nu \ll \mu$. Prove that

$$\int f d\nu = \int f \frac{d\nu}{d\mu} d\mu$$

holds for every measurable function $f\colon X \to R_e^1$ which is ν-integrable.

13. Generalize the Radon-Nikodym Theorem (see Proposition 23) to the case where $\nu\colon \mathfrak{B} \to R_e^1$ is a signed measure.

14. Let μ and ν be σ-finite measures on a measurable space $(X, \mathfrak{B})$. Show that $\nu \perp \mu$ if and only if there exists no σ-finite measure ω on $(X, \mathfrak{B})$ such that $\omega \neq 0$, $\omega \leqq \nu$, and $\omega \ll \mu$.
[*Hints*: If $\nu \perp \mu$, then $\nu = 0 + \nu$ is the unique Lebesgue decomposition of ν given in Proposition 24. Suppose that ω is a σ-finite measure on $(X, \mathfrak{B})$ such that $\omega \leqq \nu$ and $\omega \ll \mu$. Then there is a σ-finite measure τ on $(X, \mathfrak{B})$ such that $\nu = \omega + \tau$; define

$$\tau(E) = \lim_{n\to\infty} (\nu(E \cap A_n) - \omega(E \cap A_n)),$$

where $A_1 \subset A_2 \subset \ldots$, $\omega(A_n) < \infty$, and

$$X = \bigcup_{n=1}^{\infty} A_n.$$

Thus $\nu = \omega + \tau_1 + \tau_2$, where $\tau_1 \ll \mu$ and $\tau_2 \perp \mu$. Since $\omega + \tau_1 \ll \mu$, $\nu = \omega + \tau_1 + \tau_2$ is a Lebesgue decomposition of ν. Therefore $\omega = 0$. Next, suppose that ν and μ are not mutually singular. Using Proposition 24, write $\nu = \nu_1 + \nu_2$, where $\nu_1 \ll \mu$ and $\nu_2 \perp \mu$. Then $\nu_1 \neq 0$ and $\nu_1 \leqq \nu$; hence ν_1 is the desired ω.]

15. Prove Proposition 10.
16. Let $(f_n)_{n=1}^{\infty}$ be a sequence of nonnegative, extended real-valued, $\mathfrak{B}$-measurable functions on X. Show that

$$\int_X \left(\sum_{n=1}^{\infty} f_n\right) d\mu = \sum_{n=1}^{\infty} \int_X f_n \, d\mu.$$

17. Let $(f_k)_{k=1}^{\infty}$ be a nondecreasing sequence of extended real-valued, $\mathfrak{B}$-measurable functions on X such that, for some k, $\int_X f_k^- \, d\mu < \infty$. Show that

$$\lim_{k\to\infty} \int_X f_k \, d\mu = \int_X \left(\lim_{k\to\infty} f_k\right) d\mu.$$

[*Hints*: One may assume without loss of generality that $\int_X f_1^- \, d\mu < \infty$, and that no f_k assumes the value $-\infty$. If any $\int_X f_k \, d\mu$ is equal to ∞, the result is trivial. Otherwise, we define

$$g_k(x) = \begin{cases} f_{k+1}(x) - f_k(x) & \text{if } f_k(x) < \infty, \\ \infty & \text{otherwise.} \end{cases}$$

Then

$$\lim_{n\to\infty} f_n = \lim_{n\to\infty} \left(f_1 + \sum_{k=1}^{n-1} g_k\right) = f_1 + \sum_{n=1}^{\infty} g_k,$$

and so by Problem 16,

$$\int_X \left(\lim_{n\to\infty} f_n\right) d\mu = \int_X f_1 \, d\mu + \sum_{k=1}^{\infty} \left(\int_X f_{k+1} \, d\mu - \int_X f_k \, d\mu\right)$$
$$= \lim_{n\to\infty} \int_X f_n \, d\mu.]$$

18. Let $(f_n)_{n=1}^{\infty}$ be a sequence of nonnegative, extended real-valued, $\mathfrak{B}$-measurable functions on X. Show that

$$\int_X \underline{\lim}_{n\to\infty} f_n \, d\mu \leqq \underline{\lim}_{n\to\infty} \int_X f_n \, d\mu.$$

19. Let $(f_n)_{n=1}^{\infty}$ be a sequence of extended real-valued, $\mathfrak{B}$-measurable functions each defined a.e. on X. Suppose that there is a nonnegative μ-integrable function g such that for each n, $|f_n(x)| \leqq g(x)$ holds μ-a.e. on X. Show that

   (i)

$$\int_X \underline{\lim}_{n\to\infty} f_n \, d\mu \leqq \underline{\lim}_{n\to\infty} \int_X f_n \, d\mu$$

   and

$$\int_X \overline{\lim}_{n\to\infty} f_n \, d\mu \geqq \overline{\lim}_{n\to\infty} \int_X f_n \, d\mu;$$

   (ii) if

$$\lim_{n\to\infty} f_n(x)$$

   exists for μ-almost all $x \in X$, then

$$\lim_{n\to\infty} \int_X f_n \, d\mu$$

   exists and

$$\int_X \lim_{n\to\infty} f_n \, d\mu = \lim_{n\to\infty} \int_X f_n \, d\mu.$$

20. Let $(X, \mathfrak{B}, \mu)$ be a complete measure space and let $(f_n)_{n=1}^{\infty}$ be a sequence of $\mathfrak{B}$-measurable functions each of which is defined μ-a.e. on X. Suppose that f is defined μ-a.e. on X and that

$$\lim_{n\to\infty} f_n(x) = f(x)$$

   μ-a.e. on X. Show that f is μ-measurable.
   [*Hints*: Let

$$A = (\text{dom } f) \cap (\bigcap_{n=1}^{\infty} \text{dom } f_n) \cap \{x \in X : f_n(x) \to f(x)\}.$$

   Then $A \in \mathfrak{B}$ and $\mu(A^c) = 0$. For each $n \in N$, define

$$g_n(x) = \begin{cases} f_n(x) & \text{if } x \in A, \\ 0 & \text{if } x \in A^c, \end{cases}$$

and define

$$g(x) = \begin{cases} f(x) & \text{if } x \in A, \\ 0 & \text{if } x \in A^c. \end{cases}$$

By Proposition 9, g_n is $\mathfrak{B}$-measurable for each $n \in N$. It is clear that $g_n(x) \to g(x)$ for all $x \in X$. By Proposition 8 we see that g is $\mathfrak{B}$-measurable. Therefore, by Proposition 9, f must be $\mathfrak{B}$-measurable.]

21. Prove the following statements:
    (i) Let $(X, \mathfrak{B}, \mu)$ be a finite measure space and let f and $(f_n)_{n=1}^{\infty}$ be $\mathfrak{B}$-measurable functions that are defined and finite μ-a.e. on X. Suppose that $f_n \to f$ μ-a.e. on X. Then for each pair of positive real δ and ε, there exists a set $J \in \mathfrak{B}$ and an integer $n_0 \in N$ such that $\mu(J^c) < \varepsilon$ and $|f(x) - f_n(x)| < \delta$ for all $x \in J$ and $n \geqq n_0$.
    (ii) Let $(X, \mathfrak{B}, \mu)$, f, and $(f_n)_{n=1}^{\infty}$ be as in part (i). Then $f_n \to f$ in measure.
    (iii) Let $(X, \mathfrak{B}, \mu)$, f and $(f_n)_{n=1}^{\infty}$ be as in part (i). Then for each $\varepsilon > 0$ there exists a set $A \in \mathfrak{B}$ such that $\mu(A^c) < \varepsilon$ and $f_n \to f$ uniformly on A.

    [*Hint*: See Propositions 15, 16, and 17 in Chapter 2.]
22. Let $(X, \mathfrak{B}, \mu)$ be a measure space and let f and $(f_n)_{n=1}^{\infty}$ be $\mathfrak{B}$-measurable functions such that $f_n \to f$ in measure. Show that there exists a subsequence $(f_{n_k})_{k=1}^{\infty}$ such that $f_{n_k} \to f$ μ-a.e.
    [*Hint*: See Proposition 16 in Chapter 3.]
23. A *Dynkin system* $\mathfrak{D}$ of subsets of a set X is a class of sets containing X which is closed under proper differences (i.e., if $E, D \in \mathfrak{D}$ and $D \subset E$, then $E - D \in \mathfrak{D}$) and countable disjoint unions. Show:
    (i) A Dynkin system $\mathfrak{D}$ is a σ-algebra if it is closed under finite intersections.
    (ii) If $\mathcal{E}$ is a system of sets closed under finite intersections, then the σ-algebra generated by $\mathcal{E}$ and the Dynkin system generated by $\mathcal{E}$ (i.e., the smallest Dynkin system containing $\mathcal{E}$) are identical.
24. Prove that a finite measure ν on a σ-algebra $\mathfrak{B}$ is absolutely continuous with respect to a measure μ on $\mathfrak{B}$ if and only if for every $\varepsilon > 0$ there exists a $\delta > 0$ such that for all $E \in \mathfrak{B}$

$$\mu(E) \leqq \delta \quad \text{implies} \quad \nu(E) \leqq \varepsilon.$$

25. Let ν and μ be two measures on a σ-algebra $\mathfrak{B}$ of subsets of X. If, for any positive number ϵ, there exist disjoint sets $A_\epsilon \in \mathfrak{B}$, $B_\epsilon \in \mathfrak{B}$, such that $X = A_\epsilon \cup B_\epsilon$ and $\nu(A_\epsilon) < \epsilon$, $\mu(B_\epsilon) < \epsilon$, then ν and μ are mutually singular.

26. Let μ be a σ-finite measure on a σ-algebra $\mathfrak{B}$. Prove the existence of a finite measure ν on $\mathfrak{B}$ having the same null sets as μ.

Chapter 6

Outer Measure and Product Measure

Let X be a nonempty set and let $\mathcal{P}(X)$ denote the power set of X, i.e., the set of all subsets of X. A set function μ^* defined on $\mathcal{P}(X)$ is called an *outer measure* if the following relations hold:

(i) $0 \leqq \mu^*(A) \leqq \infty$ for all $A \subset X$;
(ii) $\mu^*(\varnothing) = 0$, where $\varnothing$ denotes the empty set;
(iii) μ^* is *monotone*, i.e., $\mu^*(A) \leqq \mu^*(B)$ if $A \subset B \subset X$;
(iv) μ^* is *countably subadditive*,, i.e.,

$$\mu^*(\bigcup_{n=1}^{\infty} A_n) \leqq \sum_{n=1}^{\infty} \mu^*(A_n)$$

for all sequences $(A_n)_{n=1}^{\infty}$ of subsets of X.

We note that the "counting measure" is an outer measure but that the Lebesgue measure on the real line R^1 is not (because it is not defined for all subsets of R^1).

An outer measure μ^* in a set X is said to be *finite* if $\mu^*(X)$ is finite; it is said to be *σ-finite* if there is a sequence $(X_n)_{n=1}^{\infty}$ of subsets of X satisfying the requirement that

$$X = \bigcup_{n=1}^{\infty} X_n \text{ with } \mu^*(X_n) < \infty$$

for every $n \in N$, where N denotes the set of natural numbers.

Let X be a nonempty set and $\mathfrak{A}$ a (Boolean) algebra of subsets of X, i.e., $\mathfrak{A}$ contains $\varnothing$ and X and is closed under the operations of complementation, (finite) union and (finite) intersection. A set function μ defined only on $\mathfrak{A}$ is called *finitely additive measure on an algebra* if

(i) $0 \leqq \mu(A) \leqq \infty$ for all $A \in \mathfrak{A}$;
(ii) $\mu(\varnothing) = 0$;
(iii) $\mu(A \cup B) = \mu(A) + \mu(B)$ if $A, B \in \mathfrak{A}$ and the sets A and B are disjoint.

A finitely additive measure μ on an algebra such that

(iv) $$\mu(\bigcup_{n=1}^{\infty} A_n) = \sum_{n=1}^{\infty} \mu(A_n)$$

for all pairwise disjoint sequences $(A_n)_{n=1}^{\infty}$ such that $A_n \in \mathfrak{A}$ and

$$\bigcup_{n=1}^{\infty} A_n \in \mathfrak{A}$$

is called a *countably additive measure on an algebra*, or simply a *measure on an algebra*. If $\mathfrak{A}$ is a σ-algebra of subsets of X, i.e., if $\mathfrak{A}$ is a (Boolean) algebra closed under the operations of countable union and countable intersection, and μ is a (countably additive) measure on algebra, then μ is a *measure*; in other words, a "measure on an algebra" is a "measure" if and only if its domain of definition $\mathfrak{A}$ is a σ-algebra.

Let X be a nonempty set and μ^* an outer measure on $\mathcal{P}(X)$. A set A of $\mathcal{P}(X)$ is said to be μ^*-*measurable* if

$$\mu^*(S) = \mu^*(S \cap A) + \mu^*(S \cap A^c),$$

with c denoting complementation in X, holds for every set S in $\mathcal{P}(X)$. It is clear that we can replace $S \cap A^c$ by $S - A$ in the formula above.

Evidently, A is μ^*-measurable if

$$\mu^*(S) \geqq \mu^*(S \cap A) + \mu^*(S \cap A^c)$$

holds for all S in $\mathcal{P}(X)$ since the inequality in the other direction is trivially true by the subadditivity of outer measure.

PROPOSITION 1. *The family $\mathfrak{B}$ of all μ^*-measurable subsets of X is a σ-algebra and the restriction $\bar{\mu}$ of μ^* to $\mathfrak{B}$ is a complete measure on $\mathfrak{B}$.*

Proof. It is clear that the empty set, the whole space X, and the complement of any μ^*-measurable set are μ^*-measurable. Let $S \in \mathcal{P}(X)$. We shall show now that the intersection of two μ^*-measurable sets A and B is also a μ^*-measurable set. Let $S \subseteqq X$. Since A is μ^*-measurable

(i) $\mu^*(S \cap B) = \mu^*(S \cap B \cap A) + \mu^*(S \cap B \cap A^c)$,

and since B is μ^*-measurable,

(ii) $\mu^*(S) = \mu^*(S \cap B) + \mu^*(S \cap B^c)$,

$\mu^*(S \cap (A \cap B)^c) = \mu^*(S \cap (A \cap B)^c \cap B) + \mu^*(S \cap (A \cap B)^c \cap B^c)$,

(iii) $\mu^*(S \cap (A \cap B)^c) = \mu^*(S \cap B \cap A^c) + \mu^*(S \cap B^c)$

because $S \cap (A \cap B)^c \cap B = S \cap B \cap A^c$ and $S \cap (A \cap B)^c \cap B^c = S \cap B^c$.

From (i) and (ii) it follows that

$$\mu^*(S) = \mu^*(S \cap B \cap A) + \mu^*(S \cap B \cap A^c) + \mu^*(S \cap B^c),$$

and from (iii) that

$$\mu^*(S) = \mu^*(S \cap B \cap A) + \mu^*(S \cap (A \cap B)^c).$$

Thus $A \cap B$ is a μ^*-measurable. Since $A \cup B = (A^c \cap B^c)^c$, we conclude that $\mathfrak{B}$ is an algebra.

If A_1 and A_2 are disjoint μ^*-measurable sets, it follows by replacing S by $S \cap (A_1 \cup A_2)$ in the defining relation

$$\mu^*(S) = \mu^*(S \cap A) + \mu^*(S \cap A^c)$$

that

$$\mu^*(S \cap (A_1 \cup A_2)) = \mu^*(S \cap A_1) + \mu^*(S \cap A_2).$$

By induction we see that if A is the union of a finite sequence $(A_n)_{n=1}^k$ of disjoint μ^*-measurable sets, then

$$\mu^*(S \cap A) = \sum_{n=1}^{k} \mu^*(S \cap A_n)$$

for all $S \subseteqq X$.

Since the μ^*-measurable sets form an algebra by what we have shown already, it will be enough to show that the union A of any sequence $(A_n)_{n=1}^\infty$ of disjoint μ^*-measurable sets is again a μ^*-measurable set in order to see that the μ^*-measurable sets in fact form a σ-algebra. Indeed, if $(B_n)_{n=1}^\infty$ is any sequence of μ^*-measurable subsets of X, then

$$\bigcup_{n=1}^{\infty} B_n = B_1 \cup (B_2 \cap B_1^c) \cup (B_3 \cap B_2^c \cap B_1^c) \cup \ldots$$
$$\cup (B_n \cap B_{n-1}^c \cap \ldots \cap B_1^c) \cup \ldots .$$

But each set of the form

$$A_n = (B_n \cap B_{n-1}^c \cap \ldots \cap B_1^c)$$

is μ^*-measurable. Furthermore, the sets A_n are pairwise disjoint and

$$\bigcup_{n=1}^{\infty} A_n = \bigcup_{n=1}^{\infty} B_n .$$

Let A be the union of any sequence $(A_n)_{n=1}^\infty$ of disjoint μ^*-measurable sets. It follows from what we have established already that, for $S \subseteqq X$,

$$\begin{aligned}\mu^*(S) &= \mu^*(S \cap \bigcup_{n=1}^{k} A_n) + \mu^*(S \cap (\bigcup_{n=1}^{k} A_n)^c) \\ &= \sum_{n=1}^{k} \mu^*(S \cap A_n) + \mu^*(S \cap (\bigcup_{n=1}^{k} A_n)^c) \\ &\geqq \sum_{n=1}^{k} \mu^*(S \cap A_n) + \mu^*(S \cap A^c).\end{aligned}$$

Thus

$$\begin{aligned}\mu^*(S \cap A) + \mu^*(S \cap A^c) &\geqq \mu^*(S) \\ &\geqq \sum_{n=1}^{\infty} \mu^*(S \cap A_n) + \mu^*(S \cap A^c) \\ &\geqq \mu^*(S \cap A) + \mu^*(S \cap A^c).\end{aligned}$$

This proves that A is μ^*-measurable and also shows (replacing S by $S \cap A$) that

$$\mu^*(S \cap A) = \sum_{n=1}^{\infty} \mu^*(S \cap A_n);$$

hence $\mathfrak{B}$ is a σ-algebra and μ^* is seen to be countably additive on $\mathfrak{B}$.

Now we show that every subset A of X such that $\mu^*(A) = 0$ is μ^*-measurable and $\mu^*(T) = \mu^*(T \cap A^c)$ for all $T \subseteqq X$.

Let $T \in \mathcal{P}(X)$. Then $\mu^*(T \cap A) = 0$ since $T \cap A \subset A$. Also, we have

$$\mu^*(T) \leqq \mu^*(T \cap A) + \mu^*(T \cap A^c) = \mu^*(T \cap A^c) \leqq \mu^*(T).$$

It follows that

$$\mu^*(T) = \mu^*(T \cap A) + \mu^*(T \cap A^c) = \mu^*(T \cap A^c);$$

the first equality shows that A is μ^*-measurable.

The completeness of $\bar{\mu}$ is now easily established. Let $A \in \mathfrak{B}$ with $\mu^*(A) = 0$ and $E \subset A$ be arbitrarily given. Then

$$0 \leqq \mu^*(E) \leqq \mu^*(A) = \bar{\mu}(A) = 0$$

implying $\mu^*(E) = 0$. But $\mu^*(E) = 0$ implies that $E \in \mathfrak{B}$ by what we have shown above.

The proof is complete.

Let μ be a measure on an algebra with domain of definition $\mathfrak{A}$, a (Boolean) algebra of subsets of X. We shall show that μ can be extended to a measure whose domain of definition is a σ-algebra $\mathfrak{B}$ containing $\mathfrak{A}$. We shall do this by using μ to construct an outer measure μ^* and show that the measure $\bar{\mu}$ induced by μ^* (see Proposition 1) is the desired extension of μ. The technique by which we construct μ^* from μ is similar to the one by which we constructed Lebesgue outer measure in Chapter 1 from the length of intervals; for each $E \subseteqq X$ we let

$$\mu^*(E) = \inf \sum_{n=1}^{\infty} \mu(E_n),$$

where the infimum is taken over all sequences $(E_n)_{n=1}^{\infty}$ of sets of $\mathfrak{A}$ whose union contains E.

PROPOSITION 2. *Let μ be a measure of an algebra with domain of definition $\mathfrak{A}$, a (Boolean) algebra of subsets of a set X. For each $E \subseteqq X$, let*

$$\mu^*(E) = \inf \sum_{n=1}^{\infty} \mu(E_n),$$

where the infimum is taken over all sequences $(E_n)_{n=1}^{\infty}$ of sets in $\mathfrak{A}$ whose union contains E. Then μ^* is an outer measure and every set in $\mathfrak{A}$ is μ^*-measurable. Moreover, $\mu^*(E) = \mu(E)$ for E in $\mathfrak{A}$. We shall call this μ^* the outer measure induced by μ.

Proof. Evidently μ^* is a monotone, nonnegative set function defined on $\mathcal{P}(X)$ and the properties $\mu^*(\varnothing) = 0$ and $\mu^*(A) \leqq \mu^*(B)$, for $A \subseteqq B$, and $A, B \in \mathfrak{A}$, are clear.

Let E be the union of an arbitrary sequence $(E_n)_{n=1}^{\infty}$ of sets in X. Let $\varepsilon > 0$ and, for each $n = 1, 2, \ldots$, let the sequence $(E_{m,n})_{m,n=1}^{\infty}$ have the properties

$$E_{m,n} \in \mathfrak{A},$$

$$E_n \subseteqq \bigcup_{m=1}^{\infty} E_{m,n},$$

$$\sum_{m=1}^{\infty} \mu(E_{m,n}) \leqq \mu^*(E_n) + \frac{\varepsilon}{2^{n+1}}.$$

Then

$$\bigcup_{m,n=1}^{\infty} E_{m,n} \supseteqq E,$$

and thus

$$\mu^*(E) \leqq \sum_{m,n=1}^{\infty} \mu(E_{m,n}) \leqq \sum_{n=1} \mu^*(E_n) + \varepsilon,$$

which proves, since $\varepsilon > 0$ is arbitrary, that μ^* is countably subadditive. Thus μ^* is an outer measure defined on $\mathcal{P}(X)$.

We now let $E \in \mathfrak{A}$. Since $E \subseteqq E$, it follows that $\mu(E) \geqq \mu^*(E)$. If $E_n \in \mathfrak{A}$ for all $n \in N$ and

$$E \subseteqq \bigcup_{n=1}^{\infty} E_n,$$

then the sets

$$A_1 = E_1, \quad A_n = E_n \cap (\bigcup_{j<n} E_j)^c, \; n > 1,$$

are disjoint sets in $\mathfrak{A}$ with

$$\bigcup_{n=1}^{\infty} A_n = \bigcup_{n=1}^{\infty} E_n.$$

Thus

$$\begin{aligned}\mu(E) &= \mu(E \cap \bigcup_{n=1}^{\infty} A_n) = \mu(\bigcup_{n=1}^{\infty} (E \cap A_n)) = \sum_{n=1}^{\infty} \mu(E \cap A_n) \\ &\leqq \sum_{n=1}^{\infty} \mu(A_n) \leqq \sum_{n=1}^{\infty} \mu(E_n),\end{aligned}$$

which shows that $\mu(E) \leqq \mu^*(E)$. Thus $\mu(E) = \mu^*(E)$ if E is in $\mathfrak{A}$.

To see that every set E in $\mathfrak{A}$ is μ^*-measurable, let S be an arbitrary member of $\mathcal{P}(X)$. We have to verify that

$$\mu^*(S) \geqq \mu^*(S \cap E) + \mu^*(S \cap E^c).$$

For $\varepsilon > 0$ there are sets $E_n \in \mathfrak{A}$, $n = 1, 2, \ldots$, with

$$S \subseteqq \bigcup_{n=1}^{\infty} E_n$$

and

$$\sum_{n=1}^{\infty} \mu(E_n) \leqq \mu^*(S) + \varepsilon.$$

But, since

$$S \cap E \subseteqq \bigcup_{n=1}^{\infty} (E \cap E_n)$$

and

$$S \cap E^c \subseteqq \bigcup_{n=1}^{\infty} (E_n \cap E^c),$$

$$\begin{aligned}\varepsilon + \mu^*(S) &\geqq \sum_{n=1}^{\infty} \mu(E_n) = \sum_{n=1}^{\infty} \{\mu(E \cap E_n) + \mu(E_n \cap E^c)\} \\ &\geqq \mu^*(S \cap E) + \mu^*(S \cap E^c).\end{aligned}$$

The proof is finished.

Let $\mathfrak{A}$ be a (Boolean) algebra of subsets of a set X. Let μ be a measure on an algebra with domain of definition $\mathfrak{A}$. Then μ is said to be *σ-finite* on $\mathfrak{A}$ if X is the union of a sequence $(E_n)_{n=1}^{\infty}$ of sets in $\mathfrak{A}$ such that $\mu(E_n) < \infty$, $n = 1, 2, \ldots$, and μ is said to be *finite* on $\mathfrak{A}$ if $\mu(X) < \infty$.

PROPOSITION 3. (CARATHÉODORY'S EXTENSION THEOREM). *Let μ be a measure on an algebra with domain of definition $\mathfrak{A}$ and μ^* the outer measure induced by μ. Then the restriction $\bar{\mu}$ of μ^* to the μ^*-measurable sets is an extension of μ to a σ-algebra containing $\mathfrak{A}$. If μ is σ-finite, then $\bar{\mu}$ is the only measure on the smallest σ-algebra $\mathfrak{A}_0$ containing $\mathfrak{A}$ which is an extension of μ.*

Proof. That $\bar{\mu}$ is an extension of μ from $\mathfrak{A}$ to a measure on a σ-algebra containing $\mathfrak{A}$ follows from Propositions 2 and 1; it can easily be seen that $\bar{\mu}$ is finite or σ-finite whenever μ is.

Let μ be σ-finite on $\mathfrak{A}$ and let $\mathfrak{A}_0$ denote the smallest σ-algebra containing $\mathfrak{A}$. We want to show the uniqueness of $\bar{\mu}$. Suppose that $\bar{\mu}_1$ is some other measure on $\mathfrak{A}_0$ which agrees with μ on $\mathfrak{A}$. Since μ is σ-finite on $\mathfrak{A}$, to prove the uniqueness of the extension, it will be enough to show that $\mu^*(E) = \bar{\mu}_1(E)$ for every set E in $\mathfrak{A}_0$ contained in a set F in $\mathfrak{A}$ for which $\mu(F) < \infty$. Let $E_n \in \mathfrak{A}$ and

$$E \subseteqq \bigcup_{n=1}^{\infty} E_n .$$

Then, since

$$\bar{\mu}_1(E) \leqq \sum_{n=1}^{\infty} \bar{\mu}_1(E_n) = \sum_{n=1}^{\infty} \mu(E_n),$$

it follows that

$$\bar{\mu}_1(E) \leqq \mu^*(E).$$

Similarly,

$$\bar{\mu}_1(F - E) \leqq \mu^*(F - E).$$

Since

$$\bar{\mu}_1(E) + \bar{\mu}_1(F - E) = \bar{\mu}_1(F) \leqq \mu^*(F) = \mu^*(E) + \mu^*(F - E),$$

we get that $\bar{\mu}_1(E) = \mu^*(E)$.

The proof is complete.

Remarks. It is clear that the foregoing extension procedure of C. Carathéodory not only extends μ to a measure on the smallest σ-algebra $\mathfrak{A}_0$ of $\mathfrak{A}$ but also completes the measure. Therefore, a side product of the foregoing extension procedure is the fact, proved already in Proposition 6 of Chapter 5, that every measure space extends to a complete measure space.

The smallest σ-algebra $\mathfrak{A}_0$ containing $\mathfrak{A}$ is, of course, the intersection of all σ-algebras containing $\mathfrak{A}$; that there are σ-algebras containing $\mathfrak{A}$ is clear: $\mathcal{P}(X)$ is one such. $\mathfrak{A}_0$ is sometimes called the σ-algebra *generated by* $\mathfrak{A}$.

Let $\mathfrak{A}$ be a (Boolean) algebra of subsets of a set X. Let $\mathfrak{A}_\sigma$ denote those sets which are countable unions of sets of $\mathfrak{A}$ and let $\mathfrak{A}_{\sigma\delta}$ denote those sets which are countable intersections of sets in $\mathfrak{A}_\sigma$.

PROPOSITION 4. *Let μ be a measure on an algebra with domain of definition $\mathfrak{A}$ and let μ^* be the outer measure induced by μ. Then for any set $E \in \mathcal{P}(X)$ and for any $\varepsilon > 0$, there is a set $A \in \mathfrak{A}_\sigma$ with $E \subset A$ and*

$$\mu^*(A) \leqq \mu^*(E) + \varepsilon.$$

There is also a set $B \in \mathfrak{A}_{\sigma\delta}$ with $E \subset B$ and

$$\mu^*(E) = \mu^*(B).$$

Proof. By the definition of μ^* there is a sequence $(A_k)_{k=1}^\infty$ in $\mathfrak{A}$ such that

$$E \subseteqq \bigcup_{k=1}^\infty A_k$$

and

$$\sum_{k=1}^\infty \mu(A_k) \leqq \mu^*(E) + \varepsilon.$$

Taking

$$A = \bigcup_{k=1}^\infty A_k,$$

we get

$$\mu^*(A) \leqq \sum_{k=1}^\infty \mu^*(A_k) = \sum_{k=1}^\infty \mu(A_k).$$

To verify the second part of the proposition, we observe that for each $n \in N$ there is a set A_n in $\mathfrak{A}_\sigma$ such that $E \subset A_n$ and

$$\mu^*(A_n) < \mu^*(E) + \frac{1}{n}.$$

Let

$$B = \bigcap_{n=1}^{\infty} A_n.$$

Then $B \in \mathfrak{A}_{\sigma\delta}$ and $E \subset B$. Since $B \subset A_n$,

$$\mu^*(B) \leqq \mu^*(A_n) < \mu^*(E) + \frac{1}{n}.$$

But n is arbitrary and so $\mu^*(B) \leqq \mu^*(E)$. On the other hand, $E \subset B$ implies $\mu^*(B) \geqq \mu^*(E)$. Hence $\mu^*(B) = \mu^*(E)$ and the proof is finished.

Remark. In Proposition 3 we established that if μ is σ-finite on $\mathfrak{A}$, then $\bar{\mu}$ is the only measure on the smallest σ-algebra $\mathfrak{A}_0$ containing $\mathfrak{A}$ which is an extension of $\mathfrak{A}$. Proposition 4 lends itself to proving the foregoing statement in yet another way; we shall do this now.

Let $\bar{\mu}_1$ be some measure on $\mathfrak{A}_0$ which agrees with μ on $\mathfrak{A}$. Since each member of $\mathfrak{A}_\sigma$ can be expressed as a disjoint countable union of sets in $\mathfrak{A}$, it can easily be seen that $\bar{\mu}_1$ must agree with $\bar{\mu}$ on $\mathfrak{A}_\sigma$. Let B be any set in $\mathfrak{A}_0$ with $\mu^*(B) < \infty$. Then, by Proposition 4, there is a set $A \in \mathfrak{A}_\sigma$ such that $B \subset A$ and $\mu^*(A) \leqq \mu^*(B) + \varepsilon$. Since $B \subset A$,

$$\bar{\mu}_1(B) \leqq \bar{\mu}_1(A) = \mu^*(A) \leqq \mu^*(B) + \varepsilon.$$

But ε is arbitrary and so

$$\bar{\mu}_1(B) \leqq \mu^*(B)$$

for all $B \in \mathfrak{A}_0$.

Since the class of μ^*-measurable sets is a σ-algebra containing $\mathfrak{A}$, each B in $\mathfrak{A}_0$ must be μ^*-measurable because $\mathfrak{A}_0$ is the smallest σ-algebra containing $\mathfrak{A}$. If B is μ^*-measurable and A is in $\mathfrak{A}_\sigma$ with $B \subset A$ and $\mu^*(A) \leqq \mu^*(B) + \varepsilon$, then

$$\mu^*(A) = \mu^*(B) + \mu^*(A - B)$$

and so

$$\mu^*(A - B) \leqq \varepsilon \text{ if } \mu^*(B) < \infty.$$

Hence

$$\mu^*(B) \leqq \mu^*(A) = \bar{\mu}_1(A) = \bar{\mu}_1(B) + \bar{\mu}_1(A - B) \leqq \bar{\mu}_1(B) + \varepsilon.$$

But ε is arbitrary and so $\mu^*(B) \leqq \bar{\mu}_1(B)$. This shows that $\mu^*(B) = \bar{\mu}_1(B)$.

Since μ is σ-finite on $\mathfrak{A}$, let $(X_n)_{n=1}^{\infty}$ be a countable disjoint collection of sets in $\mathfrak{A}$ with

$$X = \bigcup_{n=1}^{\infty} X_n \text{ and } \mu(X_n) < \infty \text{ for all } n \in N.$$

If B is any set in $\mathfrak{A}_0$, then

$$B = \bigcup_{n=1}^{\infty} (X_n \cap B)$$

and this is a disjoint union of sets in $\mathfrak{A}_0$. We therefore have

$$\bar{\mu}_1(B) = \sum_{n=1}^{\infty} \bar{\mu}_1(X_n \cap B)$$

and

$$\bar{\mu}(B) = \sum_{n=1}^{\infty} \bar{\mu}(X_n \cap B).$$

But $\mu^*(X_n \cap B) < \infty$, and so we obtain that

$$\bar{\mu}(X_n \cap B) = \bar{\mu}_1(X_n \cap B);$$

this means that $\bar{\mu} = \bar{\mu}_1$ on $\mathfrak{A}_0$ and we obtain what we have set out to do.

In preparation for our discussion of product measure, it is convenient to start with a set function on a collection Λ of subsets of a set X having less structure than a (Boolean) algebra of sets.

We say that a class Λ of subsets of X such that

(i) $\varnothing \in \Lambda$;

(ii) if $A, B \in \Lambda$, then $A \cap B \in \Lambda$;

(iii) if $A, B \in \Lambda$, then

$$A - B = \bigcup_{j=1}^{n} E_j,$$

where $\{E_1, \ldots, E_n\}$ is a finite sequence of mutually disjoint sets in Λ, is a *semi-ring*.

By a *ring* of subsets of X we mean any nonempty class $\mathscr{R}$ of subsets of X such that

$$\text{if } A, B \in \mathscr{R}, \text{ then } A \cap B \in \mathscr{R} \text{ and } A \triangle B \in \mathscr{R}.$$

Here $A \triangle B$ denotes the *symmetric difference* $(A - B) \cup (B - A)$ of the sets A and B.

Since $\varnothing = A \triangle A$, $A \cup B = (A \triangle B) \triangle (A \cap B)$ and $A - B = A \triangle (A \cap B)$ we see that a ring is a class of sets closed under the operations of union, intersection, and difference and the empty set $\varnothing$ is in $\mathscr{R}$. Thus a ring is also a semi-ring. It is also clear that any class $\mathfrak{A}$ of subsets of X which is a ring and contains X is a (Boolean) algebra because it is closed under the operation of taking the complement.

PROPOSITION 5. *The smallest ring $\mathscr{R}(\Lambda)$ containing the semi-ring Λ consists of the sets that can be represented in the form*

$$E = \bigcup_{k=1}^{n} A_k$$

of a finite disjoint union of sets of Λ.

Proof. The ring $\mathscr{R}(\Lambda)$ certainly must contain all sets of this form, since it has to be closed under finite unions.

To verify that the system Ω of sets of this type forms a ring, suppose

$$A = \bigcup_{k=1}^{n} A_k, \quad B = \bigcup_{k=1}^{m} B_k$$

and let $C_{ij} = A_i \cap B_j \in \Lambda$. It is clear that the sets C_{ij} are mutually disjoint and

$$A \cap B = \bigcup_{i=1}^{n} \bigcup_{j=1}^{m} C_{ij}.$$

This shows that the system Ω is closed under intersections.

It remains to verify that the system Ω is also closed under the operation of taking the symmetric difference. But from the definition of a semi-ring, an inductive argument shows that

$$A_i = \bigcup_{j=1}^{m} C_{ij} \cup \bigcup_{k=1}^{r_i} D_{ik}, \ (i = 1, \ldots, n),$$

$$B_j = \bigcup_{i=1}^{n} C_{ij} \cup \bigcup_{k=1}^{s_j} E_{kj}, \ (j = 1, \ldots, m),$$

where the finite sequences $\{D_{ik}\}$ $(k = 1, \ldots, r_i)$ and $\{E_{kj}\}$ $(k = 1, \ldots, s_j)$ consist of disjoint sets in Λ. Thus

$$A \triangle B = \bigcup_{i=1}^{n} (\bigcup_{k=1}^{r_i} D_{ik}) \cup \bigcup_{j=1}^{m} (\bigcup_{k=1}^{s_j} E_{kj})$$

and so the system Ω is seen to be a ring. The proof is finished.

Let Λ be a semi-ring. A set function $\mu : \Lambda \to R_e^1$ is said to be (finitely) *additive* if

(i) $\mu(\varnothing) = 0$,

(ii) for every family $E_1, E_2, \ldots, E_n$ of disjoint sets of Λ such that

$$\bigcup_{i=1}^{n} E_i \in \Lambda$$

we have

$$\mu(\bigcup_{i=1}^{n} E_i) = \sum_{i=1}^{n} \mu(E_i).$$

The function $\mu : \Lambda \to R_e^1$ is said to be *countably additive* if

(i) $\mu(\varnothing) = 0$,

(ii) for any mutually disjoint sequence $(E_i)_{i=1}^{\infty}$ of sets of Λ such that

$$E = \bigcup_{i=1}^{\infty} E_i \in \Lambda$$

we have

$$\mu(E) = \sum_{i=1}^{\infty} \mu(E_i).$$

Any nonnegative, countably additive set function μ on a semi-ring Λ we call a *measure on a semi-ring.*

PROPOSITION 6. *If μ is a nonnegative, additive set function defined on a semi-ring Λ, then there is a unique additive set function ν defined on the smallest ring $\mathcal{R} = \mathcal{R}(\Lambda)$ containing Λ such that ν is an extension of μ. The extension ν of μ is nonnegative on $\mathcal{R}$; it is called the extension of μ from Λ to $\mathcal{R}(\Lambda)$.*

Proof. Let $A \in \mathcal{R} = \mathcal{R}(\Lambda)$. Then, by Proposition 5,

$$A = \bigcup_{k=1}^{n} E_k,$$

where the sets E_k are mutually disjoint and $E_k \in \Lambda$. We define

$$(*) \qquad \nu(A) = \sum_{k=1}^{n} \mu(E_k).$$

To see that ν is well-defined on $\mathcal{R}$, we must verify that (*) gives the same result for any two decompositions of A into disjoint subsets in Λ.

Suppose

$$A = \bigcup_{k=1}^{n} E_k = \bigcup_{j=1}^{m} F_j,$$

where $F_j \in \Lambda$ and the F_j's are disjoint. We put $H_{kj} = E_k \cap F_j$. Since Λ is a semi-ring, the sets $H_{kj} \in \Lambda$ are disjoint and

$$E_k = \bigcup_{j=1}^{m} H_{kj} \ (k = 1, 2, \ldots, n),$$

$$F_j = \bigcup_{k=1}^{n} H_{kj} \ (j = 1, 2, \ldots, m).$$

Since μ is additive on Λ,

$$\sum_{k=1}^{n} \mu(E_k) = \sum_{k=1}^{n} (\sum_{j=1}^{m} \mu(H_{kj})) = \sum_{j=1}^{m} (\sum_{k=1}^{n} \mu(H_{kj})) = \sum_{j=1}^{m} \mu(F_j)$$

and it therefore makes no difference which decomposition of A is used in (*) to define $\nu(A)$.

If A and B are disjoint sets of $\mathscr{R}$, and

$$A = \bigcup_{k=1}^{n} E_k, \ B = \bigcup_{j=1}^{m} F_j,$$

then $A \cup B$ is a set of $\mathscr{R}$ with a possible decomposition into disjoint subsets of Λ given by

$$A \cup B = \bigcup_{k=1}^{n} E_k \cup \bigcup_{j=1}^{m} F_j.$$

Thus

$$\nu(A \cup B) = \sum_{k=1}^{n} \mu(E_k) + \sum_{j=1}^{m} \mu(F_j) = \nu(A) + \nu(B),$$

because ν is uniquely determined by (*). Since $\mathscr{R}$ is a ring, ν is finitely additive on $\mathscr{R}$. It is also clear that ν is nonnegative.

Finally, let ν_1 be any extension of μ from Λ to $\mathscr{R} = \mathscr{R}(\Lambda)$ which is additive. If $A \in \mathscr{R}$ and

$$A = \bigcup_{k=1}^{n} E_k$$

is a disjoint decomposition into disjoint sets of Λ,

$$\begin{aligned} \nu_1(A) &= \sum_{k=1}^{n} \nu_1(E_k) && \text{(since } \nu_1 \text{ is additive)} \\ &= \sum_{k=1}^{n} \mu(E_k) && \text{(since } \nu_1 \text{ is an extension)} \\ &= \nu(A) && \text{(by (*))}. \end{aligned}$$

Thus, ν is the unique additive extension of μ from Λ to $\mathscr{R}(\Lambda)$ and the proof is complete.

PROPOSITION 7. *If μ is a measure on a semi-ring Λ, then the unique extension of μ to the smallest ring $\mathscr{R} = \mathscr{R}(\Lambda)$ containing Λ is a measure on the ring $\mathscr{R} = \mathscr{R}(\Lambda)$.*

Proof. In view of Proposition 6, it will be enough to show that the unique additive extension ν of μ from Λ to $\mathscr{R}$ is countably additive on $\mathscr{R}$. We do this now.

Let $E \in \mathscr{R}$, $E_k \in \mathscr{R}$ for $k = 1, 2, \ldots$, and let the E_k's be pairwise disjoint and

$$E = \bigcup_{k=1}^{\infty} E_k.$$

We put

$$E = \bigcup_{j=1}^{n} A_j \qquad (A_j \text{ disjoint sets in } \Lambda),$$

$$E_k = \bigcup_{i=1}^{n_k} B_{ki} \qquad (B_{ki} \text{ disjoint sets in } \Lambda).$$

We let

$$C_{jki} = A_j \cap B_{ki}$$

($j = 1, 2, \ldots, n; k = 1, 2, \ldots; i = 1, 2, \ldots, n_k$). Evidently the C_{jki}'s form a mutually disjoint family of sets in Λ and

$$A_j = \bigcup_{k=1}^{\infty} \bigcup_{i=1}^{n_k} C_{jki}, \quad B_{ki} = \bigcup_{j=1}^{n} C_{jki}$$

are disjoint decomposition into sets of Λ. Since μ is additive on Λ,

$$\mu(B_{ki}) = \sum_{j=1}^{n} \mu(C_{jki})$$

and since μ is countably additive on Λ,

$$\mu(A_j) = \sum_{k=1}^{\infty} \sum_{i=1}^{n_k} \mu(C_{jki}).$$

Since we are dealing with series of nonnegative terms, we have

$$\begin{aligned} \nu(E) &= \sum_{j=1}^{n} \mu(A_j) = \sum_{j=1}^{n} \Big(\sum_{k=1}^{\infty} \sum_{i=1}^{n_k} \mu(C_{jki})\Big) = \sum_{k=1}^{\infty} \Big(\sum_{i=1}^{n_k} \sum_{j=1}^{n} \mu(C_{jki})\Big) \\ &= \sum_{k=1}^{\infty} \sum_{i=1}^{n_k} \mu(B_{ki}) = \sum_{k=1}^{\infty} \nu(E_k). \end{aligned}$$

This completes the proof.

We say that a class Γ of subsets of a set X such that

(i) $\varnothing \in \Gamma$;
(ii) if $A, B \in \Gamma$, then $A \cap B \in \Gamma$;
(iii) if $A \in \Gamma$, then

$$A^c = X - A = \bigcup_{j=1}^{n} A_j,$$

where the A_j's are mutually disjoint sets in Γ, is a *semi-algebra*.

If Γ is any semi-algebra of subsets of X, then the class $\mathfrak{A}$ of sets consisting of the empty set and all finite disjoint unions of sets in Γ is a (Boolean) algebra of sets. It is readily seen that $\mathfrak{A}$ is in fact the smallest (Boolean) algebra of subsets of X containing Γ. We shall therefore call $\mathfrak{A}$ the *algebra generated by* Γ.

Recalling the proof of Proposition 5, it is clear, for example, that $\mathfrak{A}$ is closed under the operation of finite intersection. To see that $\mathfrak{A}$ is also closed under complementation, we take any $A \in \mathfrak{A}$. Then

$$A = \bigcup_{k=1}^{n} A_k$$

where the A_k's are disjoint sets and $A_k \in \Gamma$ for $k = 1, 2, \ldots, n$. But

$$A^c = \bigcap_{k=1}^{n} A_k^c$$

and, since $A_k \in \Gamma$, we have that $A_k^c \in \mathfrak{A}$, and so, since $\mathfrak{A}$ is closed under finite intersection, $A^c \in \mathfrak{A}$.

If μ is a set function on Γ, it is natural, in view of our discussion about semi-rings, to define a set function ν on $\mathfrak{A}$ by setting

$$\nu(A) = \sum_{i=1}^{n} \mu(A_i)$$

whenever A is the disjoint union of the sets A_i in Γ.

PROPOSITION 8. *Let Γ be a semi-algebra of sets and μ a nonnegative set function defined on Γ with $\mu(\varnothing) = 0$. Then μ has a unique extension to a measure on the algebra $\mathfrak{A}$ generated by Γ, provided that the following two conditions are satisfied*:

(*i*) *If a set C in Γ is the union of a finite disjoint collection $\{C_1, \ldots, C_n\}$ of sets in Γ, then*

$$\mu(C) = \sum_{i=1}^{n} \mu(C_i).$$

(*ii*) *If a set C in Γ is the union of a countable disjoint family $\{C_i\}_{i=1}^{\infty}$ of sets in Γ, then*

$$\mu(C) \leqq \sum_{i=1}^{\infty} \mu(C_i).$$

Remark. In view of our discussion about semi-rings, it does not seem necessary to go into details as far as a proof for the foregoing proposition is concerned, except to observe that condition (i) implies: if A is the union of each of two disjoint finite collections $\{C_1, \ldots, C_n\}$ and $\{D_1, \ldots, D_m\}$, then

$$\sum_{k=1}^{n} \mu(C_k) = \sum_{j=1}^{m} \mu(D_j).$$

Moreover, condition (ii) implies that the unique extension ν of μ from Γ to $\mathfrak{A}$ is countably additive on $\mathfrak{A}$ because finite additivity and monotonicity already imply the reverse inequality.

We now turn to the construction of product measure.

Let $(X, \mathfrak{A}, \mu)$ and $(Y, \mathfrak{B}, \nu)$ be two complete measure spaces, and consider the Cartesian product $X \times Y$ of X and Y. If $A \subset X$ and $B \subset Y$, we call $A \times B$ a *rectangle* and A and B its *sides*. If $A \in \mathfrak{A}$ and $B \in \mathfrak{B}$, we call $A \times B$ a *measurable rectangle*. The class $\mathscr{R}$ of all measurable rectangles forms a semi-algebra, since

$$(A \times B) \cap (C \times D) = (A \cap C) \times (B \cap D)$$

and

$$(A \times B)^c = (A^c \times B) \cup (A \times B^c) \cup (A^c \times B^c).$$

It is clear that for any two nonempty rectangles $E_1 = A_1 \times B_1$ and $E_2 = A_2 \times B_2$ we have

$$E_1 = E_2 \text{ if and only if } A_1 = A_2 \text{ and } B_1 = B_2;$$

$$E_1 \subset E_2 \text{ if and only if } A_1 \subset A_2 \text{ and } B_1 \subset B_2;$$

$$E_1 - E_2 = [(A_1 - A_2) \times B_1] \cup [(A_1 \cap A_2) \times (B_1 - B_2)].$$

A rectangle is empty if and only if at least one of its sides is empty.

If $A \times B$ is a measurable rectangle, we define

$$\lambda(A \times B) = \mu(A) \cdot \nu(B).$$

PROPOSITION 9. *Let $\{A_j \times B_j\}_{j=1}^{\infty}$ be a countable disjoint collection of measurable rectangles whose union is a measurable rectangle $A \times B$. Then*

$$\lambda(A \times B) = \sum_{j=1}^{\infty} \lambda(A_j \times B_j).$$

Proof. We fix a point $x \in A$. Then, for each $y \in B$, the point (x, y) belongs to a unique rectangle $A_j \times B_j$. Thus B is the disjoint union of those B_j such that x is in the corresponding A_j. Hence, for every $x \in A$,

$$\sum_{j=1}^{\infty} \nu(B_j) \cdot \chi_{A_j}(x) = \nu(B) \cdot \chi_A(x)$$

because ν is countably additive and it follows by a consequence of the Monotone Convergence Theorem that

$$\sum_{j=1}^{\infty} \int \nu(B_j) \cdot \chi_{A_j} \, d\mu = \int \nu(B) \cdot \chi_A \, d\mu,$$

or

$$\sum_{j=1}^{\infty} \nu(B_j) \cdot \mu(A_j) = \nu(B) \cdot \mu(A).$$

This completes the proof.

Let $\mathcal{R}_0$ be the set of all finite disjoint unions of measurable rectangles in $X \times Y$. Then $\mathcal{R}_0$ is the smallest (Boolean) algebra containing $\mathcal{R}$, the semi-algebra of all measurable rectangles in $X \times Y$. Proposition 9 shows that λ satisfies the conditions of Proposition 8 and hence has a unique extension λ_0 to a measure on the algebra $\mathcal{R}_0$; λ_0 is taken as

$$\lambda_0(E) = \sum_{j=1}^{n} \lambda(A_j \times B_j)$$

for every finite disjoint union

$$E = \bigcup_{j=1}^{n} (A_j \times B_j)$$

of measurable rectangles in $X \times Y$. Proposition 3 in turn can be used to extend λ_0 to a complete measure on a σ-algebra $\mathfrak{S}$ containing $\mathcal{R}_0$. This extended measure we call *product measure of* μ *and* ν and denote it by $\mu \times \nu$. If μ and ν are finite (or σ-finite), so is $\mu \times \nu$.

If $\mathfrak{A}$ and $\mathfrak{B}$ are σ-algebras on X and Y, respectively, then the smallest σ-algebra containing the measurable rectangles is denoted by $\mathfrak{A} \times \mathfrak{B}$ (note that $\mathfrak{A} \times \mathfrak{B}$ is *not* the Cartesian product of $\mathfrak{A}$ and $\mathfrak{B}$). Thus, $\mu \times \nu$ is defined on a σ-algebra containing $\mathfrak{A} \times \mathfrak{B}$ and since $\mu \times \nu$ is obtained by use of Proposition 3, it is complete; however, $\mu \times \nu$ may well be an incomplete measure on $\mathfrak{A} \times \mathfrak{B}$, even if μ and ν are complete measures on $\mathfrak{A}$ and $\mathfrak{B}$, respectively.

If $(X, \mathfrak{A}, \mu)$ and $(Y, \mathfrak{B}, \nu)$ are two σ-finite measure spaces, then $\mu \times \nu$ is the only measure on $\mathfrak{A} \times \mathfrak{B}$ which assigns the value $\mu(A) \cdot \nu(B)$ to each measurable rectangle. In general, a measure on $\mathfrak{A} \times \mathfrak{B}$ with this property need not be unique, if we do not have σ-finiteness.

We now consider the structure of the sets that are measurable with respect to $\mu \times \nu$.

If E is any subset of $X \times Y$ and x is a point of X, we define the *x-cross section* E_x by

$$E_x = \{y \in Y : (x, y) \in E\};$$

similarly, for $y \in Y$, we define the *y-cross section* E^y by

$$E^y = \{x \in X : (x,y) \in E\}.$$

The characteristic function of E_x is related to the characteristic function of E by

$$\chi_{E_x}(y) = \chi_E(x,y).$$

We also note that

$$(E^c)_x = (E_x)^c$$

and

$$(\bigcup_{\alpha \in I} E_\alpha)_x = \bigcup_{\alpha \in I} (E_\alpha)_x$$

for any collection $\{E_\alpha : \alpha \in I\}$.

PROPOSITION 10. *Let x be a point of X and E a set in $\mathcal{R}_{\sigma\delta}$. Then E_x is a measurable subset of Y.*

Proof. If E is in the class $\mathcal{R}$ of measurable rectangles, the proposition is trivially true. Suppose that $E \in \mathcal{R}_\sigma$. Let

$$E = \bigcup_{j=1}^{\infty} E_j,$$

where each E_j is a measurable rectangle. Then

$$\chi_{E_x}(y) = \chi_E(x,y) = \sup_j \chi_{E_j}(x,y) = \sup_j \chi_{(E_j)_x}(y).$$

Since each E_j is a measurable rectangle,

$$\chi_{(E_j)_x}(y)$$

is a measurable function of y and so

$$\chi_{E_x}$$

must also be measurable, whence E_x is measurable.

Finally, assume that $E \in \mathcal{R}_{\sigma\delta}$, i.e., take

$$E = \bigcap_{j=1}^{\infty} E_j$$

with $E_j \in \mathcal{R}_\sigma$. Then

$$\chi_{E_x}(y) = \chi_E(x,y) = \inf_j \chi_{E_j}(x,y) = \inf_j \chi_{(E_j)_x}(y),$$

and we see that

$$\chi_{E_x}$$

is measurable. Thus E_x is measurable for any $E \in \mathscr{R}_{\sigma\delta}$, and the proof is complete.

PROPOSITION 11. *Let E be a set in $\mathscr{R}_{\sigma\delta}$ with $(\mu \times \nu)(E) < \infty$. Then the function g defined by*

$$g(x) = \nu(E_x)$$

is a measurable function of x and

$$\int g\, d\mu = (\mu \times \nu)(E).$$

Proof. The proposition is clear if E is a measurable rectangle. To proceed, we first note that any set in $\mathscr{R}_\sigma$ can be represented as a disjoint union of measurable rectangles. Let $(E_j)_{j=1}^{\infty}$ be a disjoint sequence of measurable rectangles and let

$$E = \bigcup_{j=1}^{\infty} E_j.$$

Put

$$g_j(x) = \nu[(E_j)_x].$$

Then each g_j is a nonnegative measurable function and

$$g = \sum_{j=1}^{\infty} g_j.$$

Thus g is measurable and, by a consequence of the Monotone Convergence Theorem, we get

$$\int g\, d\mu = \sum_{j=1}^{\infty} \int g_j\, d\mu = \sum_{j=1}^{\infty} (\mu \times \nu)(E_j) = (\mu \times \nu)(E).$$

Thus, the proposition is true when $E \in \mathscr{R}_\sigma$.

Let E be a set of finite measure in $\mathscr{R}_{\sigma\delta}$. Then there is a sequence $(E_j)_{j=1}^{\infty}$ of sets in $\mathscr{R}_\sigma$ such that $E_{j+1} \subset E_j$ and

$$E = \bigcap_{j=1}^{\infty} E_j.$$

By Proposition 4, we may take $(\mu \times \nu)(E_1) < \infty$. Let

$$g_j(x) = \nu[(E_j)_x].$$

Then

$$g(x) = \lim_{j\to\infty} g_j(x),$$

and so g is measurable. Since

$$\int g_1\, d\mu = (\mu \times \nu)(E_1) < \infty,$$

we have $g_1(x) < \infty$ for almost all x. For an x with $g_1(x) < \infty$, we have that $((E_j)_x)_{j=1}^{\infty}$ is a decreasing sequence of measurable sets of finite measure whose intersection is E_x. Thus, by Proposition 2 of Chapter 5, we have

$$g(x) = \nu(E_x) = \lim_{j\to\infty} \nu[(E_j)_x] = \lim_{j\to\infty} g_j(x).$$

Hence $g_j \to g$ a.e., and so g is measurable. Since $0 \leqq g_j \leqq g_1$, the Dominated Convergence Theorem implies that

$$\int g\, d\mu = \lim_{j\to\infty} \int g_j\, d\mu = \lim_{j\to\infty} (\mu \times \nu)(E_j) = (\mu \times \nu)(E),$$

the last equality following from Proposition 2 of Chapter 5. The proof is finished.

PROPOSITION 12. *Let E be a set such that $(\mu \times \nu)(E) = 0$. Then for almost all x we have $\nu(E_x) = 0$.*

Proof. By Proposition 4 there is a set F in $\mathcal{R}_{\sigma\delta}$ such that $E \subset F$ and $(\mu \times \nu)(F) = 0$. It follows from Proposition 11 that for almost all x we have $\nu(F_x) = 0$. But $E_x \subset F_x$, and so $\nu(E_x) = 0$ for almost all x because ν is assumed complete.

PROPOSITION 13. *Let E be a measurable subset of $X \times Y$ such that $(\mu \times \nu)(E)$ is finite. Then for almost all x the set E_x is a measurable subset of Y. The function g defined by*

$$g(x) = \nu(E_x)$$

is a measurable function defined for almost all x and

$$\int g\, d\mu = (\mu \times \nu)(E).$$

Proof. By Proposition 4 there is a set F in $\mathcal{R}_{\sigma\delta}$ such that $E \subset F$ and $(\mu \times \nu)(F) = (\mu \times \nu)(E)$. Let $G = F - E$. Since E and F are measurable, so is G, and

$$(\mu \times \nu)(F) = (\mu \times \nu)(E) + (\mu \times \nu)(G).$$

Since $(\mu \times \nu)(E)$ is finite and equal to $(\mu \times \nu)(F)$, it follows that $(\mu \times \nu)(G) = 0$. By Proposition 12, $\nu(G_x) = 0$ for almost all x. Thus

$$g(x) = \nu(E_x) = \nu(F_x) \text{ a.e.},$$

and so g is measurable by Proposition 11. Again, by Proposition 11,

$$\int g \, d\mu = (\mu \times \nu)(F) = (\mu \times \nu)(E),$$

and the proof is finished.

PROPOSITION 14 (FUBINI'S THEOREM). *Let $(X, \mathfrak{A}, \mu)$ and $(Y, \mathfrak{B}, \nu)$ be two complete measure spaces and f an integrable function on $X \times Y$. Then*

(*i*) *for almost all x, the function f_x defined by*

$$f_x(y) = f(x, y)$$

is an integrable function on Y;

(*i'*) *for almost all y, the function f_y defined by*

$$f_y(x) = f(x, y)$$

is an integrable function on X;

(*ii*) *$\int_Y f(x, y) \, d\nu(y)$ is an integrable function on X;*

(*ii'*) *$\int_X f(x, y) \, d\mu(x)$ is an integrable function on Y;*

(*iii*) *$\int_X [\int_Y f \, d\nu] \, d\mu = \int_{X \times Y} f \, d(\mu \times \nu) = \int_Y [\int_X f \, d\mu] \, d\nu$.*

Proof. Since there is symmetry between x and y, it is enough to prove (i), (ii), and the first half of (iii). Moreover, it will be sufficient to consider the case when f is nonnegative; if the conclusion of the proposition holds for each of two functions, it also holds for their difference. Now, by Proposition 13, Fubini's Theorem holds if f is the characteristic function of a measurable set of finite measure; hence Fubini's Theorem must also be true if f is a simple function that vanishes outside a set of finite measure. But for each nonnegative integrable function f there is an increasing sequence $(s_n)_{n=1}^{\infty}$ of nonnegative simple functions such that $s_n \uparrow f$, and since each s_n is integrable and simple, it must vanish outside a set of finite measure; actually, for each pair of nonnegative integers (n, k) we can take

$$E_{n,k} = \left\{ x : \frac{k}{2^n} \leqq f(x) < \frac{k+1}{2^n} \right\}$$

and put

$$s_n = \frac{1}{2^n} \sum_{k=0}^{2^{2n}} k \chi_{E_{n,k}}$$

in order to obtain an approximating sequence of the required kind. Thus f_x is the limit of the increasing sequence $((s_n)_x)_{n=1}^{\infty}$ and is measurable. By the Monotone Convergence Theorem

$$\int_Y f(x,y)\,d\nu(y) = \lim_{n\to\infty} \int_Y s_n(x,y)\,d\nu(y),$$

and so this integral is a measurable function of x. Again by the Monotone Convergence Theorem

$$\int_X \left[\int_Y f\,d\nu\right] d\mu = \lim_{n\to\infty} \int_X \left[\int_Y s_n\,d\nu\right] d\mu = \lim_{n\to\infty} \int_{X\times Y} s_n\,d(\mu\times\nu)$$
$$= \int_{X\times Y} f\,d(\mu\times\nu).$$

The proof is complete.

PROPOSITION 15 (TONELLI'S THEOREM). *Let $(X,\mathfrak{A},\mu)$ and $(Y,\mathfrak{B},\nu)$ be two complete σ-finite measure spaces, and let f be a nonnegative measurable function on $X\times Y$. Then*

(*i*) *for almost all x, the function f_x defined by*

$$f_x(y) = f(x,y)$$

is a measurable function on Y;

(*i'*) *for almost all y, the function f_y defined by*

$$f_y(x) = f(x,y)$$

is a measurable function on X;

(*ii*) *$\int_Y f(x,y)\,d\nu(y)$ is a measurable function on X;*

(*ii'*) *$\int_X f(x,y)\,d\mu(x)$ is a measurable function on Y;*

(*iii*) *$\int_X [\int_Y f\,d\nu]\,d\mu = \int_{X\times Y} f\,d(\mu\times\nu) = \int_Y [\int_X f\,d\mu]\,d\nu$.*

Proof. For a nonnegative measurable function f the only point in the proof of Fubini's Theorem where the integrability was used was to infer the existence of an increasing sequence $(s_n)_{n=1}^{\infty}$ of simple functions each vanishing outside a set of finite measure such that $s_n \uparrow f$. But if μ and ν are σ-finite, then so is $\mu\times\nu$, and any nonnegative measurable function on $X\times Y$ can be so approximated. The proof is finished.

PROPOSITION 16. *Let $(X,\mathfrak{A},\mu)$ and $(Y,\mathfrak{B},\nu)$ be two complete σ-finite measure spaces and let f be a measurable function on $X\times Y$ such that*

$$\int_X \left[\int_Y |f|\,d\nu\right] d\mu < \infty.$$

Then f is an integrable function on $X \times Y$ and

$$\int_{X\times Y} f\,d(\mu \times \nu) = \int_X \left[\int_Y f\,d\nu\right] d\mu = \int_Y \left[\int_X f\,d\mu\right] d\nu.$$

Proof. By our hypothesis and Tonelli's Theorem (see Proposition 15) we get that

$$\int_{X\times Y} |f|\,d(\mu \times \nu) = \int_X \left[\int_Y |f|\,d\nu\right] d\mu < \infty.$$

This shows that f is an integrable function on $X \times Y$. To complete the proof, we only need to apply Fubini's Theorem (see Proposition 14).

Discussion (*relating to Proposition 16*). In Proposition 16 we could have used in place of the condition

$$\int_X \left[\int_Y |f|\,d\nu\right] d\mu < \infty$$

either one of the following two conditions

$$\int_Y \left[\int_X |f|\,d\mu\right] d\nu < \infty, \quad \int_{X\times Y} |f|\,d(\mu \times \nu) < \infty$$

to conclude that the iterated integrals of f are equal to the double integral of f. It is also clear that if one of the foregoing three integrals is finite, then they all must be finite and equal. It can also be seen that if

$$\int_X \left[\int_Y |f|\,d\nu\right] d\mu = \infty$$

in Proposition 16, then f cannot be integrable on $X \times Y$.

If f is measurable, but *not* integrable, on $X \times Y$, it may happen that both iterated integrals exist and are different. On the other hand, if it happens that the two iterated integrals are equal and finite, it still does not follow that f is integrable on $X \times Y$. We shall now illustrate this.

Example 1. Let $I = (0,1) \times (0,1)$ and

$$f(x,y) = \frac{x^2 - y^2}{(x^2 + y^2)^2} \qquad \text{if } (x,y) \in I.$$

Clearly, f is continuous and therefore measurable in I.

Suppose x is fixed and $0 < x < 1$; then

$$f(x,y) = \frac{\partial}{\partial y} F(x,y),$$

where

$$F(x,y) = \frac{y}{x^2 + y^2},$$

and since $f(x,y)$ is bounded for $0 \leqq y \leqq 1$, we have

$$\int_0^1 f(x,y)\, dy = F(x,1) - F(x,0) = \frac{1}{x^2+1}.$$

Hence

$$\int_0^1 \left[\int_0^1 f(x,y)\, dy \right] dx = \int_0^1 \frac{dx}{x^2+1} = \operatorname{arc\,tan} x \Big|_0^1 = \frac{\pi}{4}.$$

Since $f(x,y) = -f(y,x)$, it follows that

$$\int_0^1 \left[\int_0^1 f(x,y)\, dx \right] dy = -\frac{\pi}{4}.$$

Since the two iterated integrals are not equal, f cannot be integrable over I; since f is known to be measurable in I, we must have that

$$\int_I |f(x,y)|\, d(x,y) = \infty,$$

i.e., by Proposition 15,

$$\int_0^1 \left[\int_0^1 |f(x,y)|\, dy \right] dx = \infty.$$

This can easily be verified directly; for, if $0 < x < 1$, it follows as before that

$$\int_0^x f(x,y)\, dy = F(x,x) - F(x,0) = \frac{1}{2x}$$

and so

$$\int_0^1 |f(x,y)|\, dy \geqq \frac{1}{2x},$$

which means that

$$\int_0^1 \left[\int_0^1 |f(x,y)|\,dy\right] dx \geqq \frac{1}{2}\int_0^1 \frac{1}{x}\,dx = \infty.$$

Example 2. Let f be defined in the plane R^2 such that

$$f(x,y) = \frac{xy}{(x^2+y^2)^2} \quad \text{when } x^2+y^2>0.$$

We shall see that f is not integrable over R^2, although

$$\int_{-\infty}^{\infty}\left[\int_{-\infty}^{\infty} f(x,y)\,dy\right] dx = \int_{-\infty}^{\infty}\left[\int_{-\infty}^{\infty} f(x,y)\,dx\right] dy = 0.$$

It is easy to see that, if $x \neq 0$, then

$$\int_0^{\infty} f(x,y)\,dy = \frac{1}{2x} \quad \text{and} \quad \int_{-\infty}^{0} f(x,y)\,dy = -\frac{1}{2x},$$

so that

$$\int_{-\infty}^{\infty} f(x,y)\,dy = 0 \quad \text{while} \quad \int_{-\infty}^{\infty} |f(x,y)|\,dy = \frac{1}{|x|},$$

all integrals in question existing in the Cauchy-Riemann sense. A similar argument shows that

$$\int_{-\infty}^{\infty} f(x,y)\,dx = 0 \quad \text{if } y \neq 0,$$

and so the two iterated integrals are equal and zero. At the same time

$$\int_{-\infty}^{\infty}\left[\int_{-\infty}^{\infty} |f(x,y)|\,dy\right] dx = \int_{-\infty}^{\infty} \frac{dx}{|x|} = \infty$$

and this means, by Proposition 15, that f is not integrable over R^2.

From Proposition 16 one can readily deduce the "formula for integration by parts" for the Lebesgue integral in R^1; we have already met the "formula for integration by parts" in Chapter 4 and there the proof for it was based on the theory of absolutely continuous functions.

Essentially, we have to show that if f and g are integrable over the interval (a,b) and

$$F(t) = \int_a^t f(x)\,dx \quad \text{and} \quad G(t) = \int_a^t g(x)\,dx,$$

then we have

$$\int_a^b (f(x)G(x) + g(x)F(x))\,dx = F(b)G(b).$$

To prove this, let $S = (a, b) \times (a, b)$ and define

$$h(x, t) = \begin{cases} g(t) & \text{if } t < x, \\ 0 & \text{if } t \geqq x. \end{cases}$$

Then

$$\int_a^b f(x)G(x)\,dx = \int_a^b \left[\int_a^b f(x)h(x, t)\,dt\right] dx.$$

It is easy to see that $f(x)h(x, t)$ is measurable in S and that

$$\int_a^b \left[\int_a^b |f(x)h(x, t)|\,dt\right] dx < \infty.$$

By Proposition 16, it then follows that

$$\int_a^b f(x)G(x)\,dx = \int_a^b \left[\int_a^b f(x)h(x, t)\,dx\right] dt = \int_a^b \left[\int_t^b f(x)g(t)\,dx\right] dt$$

and hence

$$\begin{aligned}\int_a^b (f(x)G(x) + g(x)F(x))\,dx &= \int_a^b g(t)(F(b) - F(t))\,dt \\ &\quad + \int_a^b g(t)F(t)\,dt \\ &= F(b)\int_a^b g(t)\,dt = F(b)G(b).\end{aligned}$$

Addendum

In the following we shall briefly discuss another approach to the construction of product measure. Since most measure spaces that occur in classical analysis are σ-finite, we shall restrict our discussion to this case.

Let $(X, \mathfrak{A})$ and $(Y, \mathfrak{B})$ be any two measurable spaces. As before, for $A \in \mathfrak{A}$ and $B \in \mathfrak{B}$, the set $A \times B \subset X \times Y$ we call a *measurable rectangle* and the smallest σ-algebra of subsets of $X \times Y$ containing all measurable rectangles we denote by $\mathfrak{A} \times \mathfrak{B}$ and call it the *product σ-algebra*. For $E \subset X \times Y$ and $x \in X$, we let

$$E_x = \{y \in Y : (x,y) \in E\}$$

and call it the *x-cross section of E*; the *y-cross section of E* we define as the set

$$E^y = \{x \in X : (x,y) \in E\}$$

for $y \in Y$. Analogously we define the *x-* and *y-cross section of a function f* on $X \times Y$ as

$$f_x(y) = f(x,y) \text{ for a fixed } x \in X$$

and

$$f^y(x) = f(x,y) \text{ for a fixed } y \in Y,$$

respectively.

LEMMA 1. *Let $(X, \mathfrak{A})$ and $(Y, \mathfrak{B})$ be any two measurable spaces. Then the class Ω of all finite, pairwise disjoint unions of measurable rectangles is a (Boolean) algebra of subsets of $X \times Y$.*

Proof. If $A \times B$ and $C \times D$ are two measurable rectangles, then $(A \times B) \cap (C \times D) = (A \cap C) \times (B \cap D)$. Hence, if

$$E = \bigcup_{k=1}^{m} (A_k \times B_k) \text{ and } F = \bigcup_{j=1}^{n} (C_j \times D_j)$$

are in Ω, where these are pairwise disjoint unions of measurable rectangles, then

$$E \cap F = \bigcup_{k=1}^{m} \bigcup_{j=1}^{n} ((A_k \cap C_j) \times (B_k \cap D_j)),$$

and $E \cap F$ is in Ω.

Finally, if $A \times B$ is a measurable rectangle, then

$$(A \times B)^c = (A^c \times B) \cup (X \times B^c),$$

which is the union of two disjoint measurable rectangles. For

$$E = \bigcup_{k=1}^{m} (A_k \times B_k) \in \Omega, \text{ we have } E^c = \bigcap_{k=1}^{m} (A_k \times B_k)^c,$$

which is a finite intersection of sets in Ω, and therefore $E^c \in \Omega$. The proof is finished.

LEMMA 2. *Let* $(X, \mathfrak{A})$ *and* $(Y, \mathfrak{B})$ *be any two measurable spaces. If* $E \in \mathfrak{A} \times \mathfrak{B}$, *then*

(*i*) $E_x \in \mathfrak{B}$ *for all* $x \in X$, *and*
(*ii*) $E^y \in \mathfrak{A}$ *for all* $y \in Y$.

Proof. It will suffice to prove part (i) only; the proof of part (ii) is similar to the proof of part (i).

Let

$$\Lambda = \{E \in \mathfrak{A} \times \mathfrak{B} : E_x \in \mathfrak{B} \text{ for all } x \in X\}.$$

For any measurable rectangle $A \times B$ and $x \in X$, we have

$$(A \times B)_x = \begin{cases} B \text{ if } x \in A, \\ \varnothing \text{ if } x \notin A. \end{cases}$$

Thus Λ contains all measurable rectangles. To complete the proof of part (i), we only have to show that Λ is a σ-algebra. If $(E_n)_{n=1}^{\infty}$ is a sequence in Λ, then it can be seen that

$$\Big(\bigcup_{n=1}^{\infty} E_n\Big)_x = \bigcup_{n=1}^{\infty} (E_n)_x \qquad \text{for all } x \in X;$$

thus Λ is closed under the operation of countable unions. For $E \in \Lambda$, we have $(E^c)_x = (E_x)^c \in \mathfrak{B}$ for all $x \in X$, and so Λ is also closed under the operation of complementation. The proof is finished.

LEMMA 3. *Let* $(X, \mathfrak{A})$ *and* $(Y, \mathfrak{B})$ *be any measurable spaces, and let* f *be an extended real-valued* $\mathfrak{A} \times \mathfrak{B}$ *-measurable function on* $X \times Y$. *Then*

(*i*) f_x *is* $\mathfrak{B}$*-measurable for all* $x \in X$, *and*
(*ii*) f^y *is* $\mathfrak{A}$*-measurable for all* $y \in Y$.

Proof. Suppose that $f = \chi_E$ for some $E \in \mathfrak{A} \times \mathfrak{B}$. Since the statements:

$$f_x(y) = 1; \quad (x,y) \in E; \quad \chi_{E_x}(y) = 1$$

are mutually equivalent, it follows that

$$f_x = \chi_{E_x}$$

for all $x \in X$. Thus (i) follows from Lemma 2, part (i) in the case that

$$f = \chi_E.$$

From this we see that (i) also must hold if f is a simple function. The general case follows from the foregoing by applying Propositions 10 and 8 of Chapter 5. Part (ii) of the lemma is proved in an entirely similar way.

LEMMA 4. *Let S be any set and let Ω be a (Boolean) algebra of subsets of S. Let $\mathfrak{S}(\Omega)$ denote the smallest σ-algebra containing Ω. Then $\mathfrak{S}(\Omega)$ is the smallest class $\mathfrak{K}$ of subsets of S that contains Ω and satisfies the following two conditions*:

(*i*) *If $E_n \in \mathfrak{K}$ and $E_n \subset E_{n+1}$ for $n = 1, 2, \ldots$, then*

$$\bigcup_{n=1}^{\infty} E_n \in \mathfrak{K}.$$

(*ii*) *If $F_n \in \mathfrak{K}$ and $F_n \supset F_{n+1}$ for $n = 1, 2, \ldots$, then*

$$\bigcap_{n=1}^{\infty} F_n \in \mathfrak{K}.$$

Thus, if the (Boolean) algebra Ω satisfies (i) and (ii), then Ω is a σ-algebra. Classes that satisfy (i) and (ii) are called monotone classes.

Proof. It is clear that the class $\mathfrak{K}$ exists because $\mathscr{P}(S)$ is a monotone class and the intersection of all monotone classes containing Ω is again a monotone class; this intersection is $\mathfrak{K}$.

Since any σ-algebra is a monotone class and $\mathfrak{S}(\Omega) \supset \Omega$, we get that $\mathfrak{S}(\Omega) \supset \mathfrak{K} \supset \Omega$. To complete the proof it is enough to prove that $\mathfrak{K}$ is a σ-algebra. For a monotone sequence $(H_n)_{n=1}^{\infty}$, we write $\lim H_n$ for the set

$$\bigcup_{n=1}^{\infty} H_n$$

or

$$\bigcap_{n=1}^{\infty} H_n$$

according as $(H_n)_{n=1}^{\infty}$ is an increasing or a decreasing sequence. If $E \in \mathfrak{K}$, we write

$$\mathfrak{K}_E = \{F \in \mathfrak{K} : F \cap E^c \in \mathfrak{K}, E \cap F^c \in \mathfrak{K}, E \cup F \in \mathfrak{K}\}.$$

It is clear that $F \in \mathfrak{K}_E$ if and only if $E \in \mathfrak{K}_F$. It is also clear that if $(F_n)_{n=1}^{\infty}$ is a monotone sequence in $\mathfrak{K}_E$, then

$$(\lim F_n) \cap E^c = \lim(F_n \cap E^c) \in \mathfrak{K},$$

$$E \cap (\lim F_n)^c = E \cap (\lim F_n^c) = \lim(E \cap F_n^c) \in \mathfrak{K},$$

and

$$E \cup (\lim F_n) = \lim(E \cup F_n) \in \mathfrak{K},$$

because $\mathfrak{K}$ is a monotone class. Thus $\mathfrak{K}_E$ is a monotone class for every $E \in \mathfrak{K}$.

If $E, F \in \Omega$, then E belongs to $\mathfrak{K}_F$ and F belongs to $\mathfrak{K}_E$ because Ω is a (Boolean) algebra. Thus $\Omega \subset \mathfrak{K}_E$ for all $E \in \Omega$. Since $\mathfrak{K}$ is the smallest monotone class containing Ω, we see that $\mathfrak{K} \subset \mathfrak{K}_E$ for all $E \in \Omega$. Thus for any $F \in \mathfrak{K}$ and $E \in \Omega$ we have $F \in \mathfrak{K}_E$; hence $E \in \mathfrak{K}_F$. This shows that $\Omega \subset \mathfrak{K}_F$ for all $F \in \mathfrak{K}$. Since each $\mathfrak{K}_F$ is a monotone class, it follows that $\mathfrak{K} \subset \mathfrak{K}_F$ for all $F \in \mathfrak{K}$. The definition of $\mathfrak{K}_F$ now shows that $\mathfrak{K}$ is a (Boolean) algebra.

To see that $\mathfrak{K}$ is a σ-algebra, we let $(F_n)_{n=1}^{\infty}$ be a sequence in $\mathfrak{K}$. For each $n \in N$, we put $G_n = F_1 \cup F_2 \cup \ldots \cup F_n$. Then $(G_n)_{n=1}^{\infty}$ is an increasing sequence in $\mathfrak{K}$ because $\mathfrak{K}$ is a (Boolean) algebra. Therefore

$$\bigcup_{n=1}^{\infty} F_n = \bigcup_{n=1}^{\infty} G_n = \lim G_n \in \mathfrak{K}.$$

The proof is finished.

Remark. If $(X, \mathfrak{A})$ and $(Y, \mathfrak{B})$ are measurable spaces, then Lemmas 1 and 4 show that $\mathfrak{A} \times \mathfrak{B}$ is the smallest monotone class of subsets of $X \times Y$ that contains all finite disjoint unions of measurable rectangles.

LEMMA 5. *Let $(X, \mathfrak{A}, \mu)$ and $(Y, \mathfrak{B}, \nu)$ be σ-finite measure spaces and let $E \in \mathfrak{A} \times \mathfrak{B}$. Then*

(*i*) *the function $x \to \nu(E_x)$ on X is $\mathfrak{A}$-measurable*;

(*ii*) *the function $x \to \mu(E^y)$ on Y is $\mathfrak{B}$-measurable*;

(*iii*) $\int_X \nu(E_x)\, d\mu(x) = \int_Y \mu(E^y)\, d\nu(y)$.

Proof. Let $\mathfrak{K}$ be the class of all sets $E \in \mathfrak{A} \times \mathfrak{B}$ such that (i), (ii), and (iii) hold for E. We will prove that $\mathfrak{K}$ is a monotone class that contains all finite disjoint unions of measurable rectangles. By the Remark preceding the lemma we will then have that $\mathfrak{K} = \mathfrak{A} \times \mathfrak{B}$.

Let $E = A \times B$ be a measurable rectangle. Since

$$E_x = \begin{cases} B \text{ if } x \in A, \\ \varnothing \text{ if } x \in A^c, \end{cases}$$

and

$$E^y = \begin{cases} A \text{ if } y \in B, \\ \varnothing \text{ if } y \in B^c, \end{cases}$$

we get that

$$\nu(E_x) = \nu(b)\chi_A(x) \quad \text{and} \quad \mu(E^y) = \mu(A)\chi_B(y);$$

thus (i) and (ii) hold for this E. We may also write

$$\int_X \nu(E_x)\, d\mu(x) = \int_X \mu(B)\chi_A\, d\mu = \nu(B) \cdot \mu(A) = \int_Y \mu(A)\chi_B\, d\nu$$
$$= \int_Y \mu(E^y)\, d\nu(y).$$

Hence (iii) holds for this E, and therefore $E \in \mathfrak{R}$. Thus $\mathfrak{R}$ contains all measurable rectangles.

Let $\{E_1, E_2, \ldots, E_p\}$ be a finite, pairwise disjoint subclass of $\mathfrak{R}$. Since

$$(\bigcup_{n=1}^{p} E_n)_x = \bigcup_{n=1}^{p} (E_n)_x \quad \text{and} \quad (\bigcup_{n=1}^{p} E_n)^y = \bigcup_{n=1}^{p} (E_n)^y$$

for all $x \in X$ and all $y \in Y$, we get that

$$(\bigcup_{n=1}^{p} E_n) \in \mathfrak{R}.$$

This shows that $\mathfrak{R}$ contains all finite disjoint unions of measurable rectangles.

Next, let $(E_n)_{n=1}^{\infty}$ be an increasing sequence in $\mathfrak{R}$ and let

$$E = \bigcup_{n=1}^{\infty} E_n = \lim E_n.$$

Then, by Proposition 2 of Chapter 5,

$$\nu(E_x) = \lim_{n\to\infty} \nu((E_n)_x)$$

for all $x \in X$, and so Proposition 8 of Chapter 5 implies that (i) holds for E. By the Monotone Convergence Theorem (see Proposition 12 or Problem 17 in Chapter 5) we therefore get that

$$\int_X \nu(E_x)\, d\mu(x) = \lim_{n\to\infty} \int_X \nu((E_n)_x)\, d\mu(x) = \lim_{n\to\infty} \int_Y \mu((E_n)^y)\, d\nu(y)$$
$$= \int_Y \mu(E^y)\, d\nu(y).$$

Therefore, E is in $\mathfrak{R}$, and so $\mathfrak{R}$ is closed under the operation of taking unions of increasing sequences.

It remains to show that $\mathfrak{R}$ is closed under the formation of intersections of decreasing sequences. Let $(F_n)_{n=1}^{\infty}$ be a decreasing sequence in $\mathfrak{R}$ such that $F_1 \subset A \times B$ for some measurable rectangle $A \times B$ for which $\mu(A) < \infty$ and $\nu(B) < \infty$. We let

$$F = \bigcap_{n=1}^{\infty} F_n.$$

For each $x \in X$, we have $(F_1)_x \subset B$, and so $\nu((F_1)_x) < \infty$. It follows from Proposition 3 of Chapter 5 that

$$\nu(F_x) = \lim_{n\to\infty} \nu((F_n)_x)$$

for all $x \in X$. Therefore, the function $x \to \nu(F_x)$ is the pointwise limit of a sequence of $\mathfrak{A}$-measurable functions, and so is $\mathfrak{A}$-measurable by Proposition 8 of Chapter 5; hence (i) holds for F. In the same manner we see that (ii) holds for F. Since

$$\int_X \nu((F_1)_x)\, d\mu(x) = \int_X \nu(B)\chi_A\, d\mu < \infty$$

and

$$\int_Y \mu((F_1)^y)\, d\nu(y) \leqq \int_Y \mu(A)\chi_B\, d\nu < \infty,$$

the Dominated Convergence Theorem (see Proposition 14 or Problem 19 of Chapter 5) implies that

$$\int_X \nu(F_x)\, d\mu(x) = \lim_{n\to\infty} \int_X \nu((F_n)_x)\, d\mu(x) = \lim_{n\to\infty} \int_Y \mu((F_n)^y)\, d\nu(y)$$
$$= \int_Y \mu(F^y)\, d\nu(y),$$

implying that (iii) holds for F.

Using the σ-finiteness assumption, we choose increasing sequences $(A_k)_{k=1}^{\infty} \subset \mathfrak{A}$ and $(B_k)_{k=1}^{\infty} \subset \mathfrak{B}$ such that $\mu(A_k) < \infty$ and $\nu(B_k) < \infty$ for all k, and such that

$$X = \bigcup_{k=1}^{\infty} A_k \quad \text{and} \quad Y = \bigcup_{k=1}^{\infty} B_k.$$

Let

$$\mathcal{E} = \{E \in \mathfrak{A} \times \mathfrak{B} : E \cap (A_k \times B_k) \in \mathfrak{R} \text{ for all } k \in N\}.$$

Since the class Ω of all finite disjoint unions of measurable rectangles is a (Boolean) algebra by Lemma 1, and since $\Omega \subset \mathfrak{R}$, we have that $\Omega \subset \mathcal{E}$. If $(E_n)_{n=1}^{\infty}$ is an increasing sequence in $\mathcal{E}$, then

$$\left(\bigcup_{n=1}^{\infty} E_n\right) \cap (A_k \times B_k) = \bigcup_{n=1}^{\infty} (E_n \cap (A_k \times B_k)) \in \mathfrak{R}$$

since $\mathfrak{K}$ is closed under the formation of unions of increasing sequences. Thus $\mathcal{E}$ is closed under limits of increasing sequences. If $(E_n)_{n=1}^{\infty}$ is a decreasing sequence in $\mathcal{E}$, then

$$\left(\bigcap_{n=1}^{\infty} E_n\right) \cap (A_k \times B_k) = \bigcap_{n=1}^{\infty} (E_n \cap (A_k \times B_k)) \in \mathfrak{K},$$

as proved in the foregoing paragraph. Hence $\mathcal{E}$ is also closed under the formation of intersections of decreasing sequences. By the Remark following Lemma 4, $\mathcal{E} = \mathfrak{A} \times \mathfrak{B}$. Now let $(F_n)_{n=1}^{\infty}$ be any decreasing sequence in $\mathfrak{K}$ and let

$$F = \bigcap_{n=1}^{\infty} F_n .$$

As $F \in \mathcal{E}$, we have that $F \cap (A_k \times B_k) \in \mathfrak{K}$ for all $k \in N$. But $\mathfrak{K}$ is closed under the formation of unions of increasing sequences, and so

$$F = \bigcup_{k=1}^{\infty} (F \cap (A_k \times B_k)) \in \mathfrak{K}.$$

Hence $\mathfrak{K}$ is a montone class and the proof is finished.

Definition. *Let $(X, \mathfrak{A}, \mu)$ and $(Y, \mathfrak{B}, \nu)$ be two σ-finite measure spaces. For $E \in \mathfrak{A} \times \mathfrak{B}$, we define*

$$(\mu \times \nu)(E) = \underset{X}{K} \nu(E_x)\, d\mu(x).$$

By part (iii) of Lemma 5, we also have

$$(\mu \times \nu)(E) = \int_Y \mu(E^y)\, d\nu(y).$$

Lemma 6. *The set function $\mu \times \nu$ is a σ-finite measure on $\mathfrak{A} \times \mathfrak{B}$.*

Proof. Let $(E_n)_{n=1}^{\infty}$ be a disjoint sequence of sets in $\mathfrak{A} \times \mathfrak{B}$. By consequence (ii) of Proposition 14 in Chapter 5 (or, alternately, by Problem 16 of Chapter 5) we get

$$\begin{aligned}(\mu \times \nu)\left(\bigcup_{n=1}^{\infty} E_n\right) &= \int_X \nu\left(\left(\bigcup_{n=1}^{\infty} E_n\right)_x\right) d\mu(x) = \int_X \sum_{n=1}^{\infty} \nu((E_n)_x)\, d\mu(x) \\ &= \sum_{n=1}^{\infty} \int_X \nu((E_n)_x)\, d\mu(x) = \sum_{n=1}^{\infty} (\mu \times \nu)(E_n).\end{aligned}$$

Thus $\mu \times \nu$ is countably additive. It is clear that $\mu \times \nu \geqq 0$ and $(\mu \times \nu)(\varnothing) = 0$; therefore $\mu \times \nu$ is a measure on $\mathfrak{A} \times \mathfrak{B}$. To see that $\mu \times \nu$ is σ-finite, we let $(A_k)_{k=1}^{\infty}$ and $(B_k)_{k=1}^{\infty}$ be as in the proof of Lemma 5. Then

$$X \times Y = \bigcup_{k=1}^{\infty} (A_k \times B_k)$$

and $(\mu \times \nu)(A_k \times B_k) = \mu(A_k) \cdot \nu(B_k) < \infty$ for all $k \in N$, completing the proof.

Let X be an arbitrary set and Ω a (Boolean) algebra of subsets of X and let μ be a measure on an algebra with domain of definition Ω. We define a set function μ^* on $\mathcal{P}(X)$ as follows: for $E \subset X$, we let

$$\mu^*(E) = \inf\{\sum_{n=1}^{\infty} \mu(E_n) : E \subset \bigcup_{n=1}^{\infty} E_n \text{ and } E_1, \ldots, E_n, \ldots \in \Omega\}.$$

As we have seen earlier in this chapter, the following results can be deduced:

(a) μ^* is an outer measure on $\mathcal{P}(X)$;
(b) μ^* is equal to μ on the (Boolean) algebra Ω;
(c) all elements of Ω are measurable with respect to μ^*;
(d) μ can be extended to a measure defined on a σ-algebra of subsets of X that contain Ω.

Letting X, Ω, μ, and μ^* be as above, we denote by $\mathfrak{S} = \mathfrak{S}(\Omega)$ the smallest σ-algebra of subsets of X that contains the (Boolean) algebra Ω. Suppose that $(X, \mathfrak{S}, \nu)$ is a measure space such that $\nu(A) = \mu(A)$ for all $A \in \Omega$. The following results obtain:

(i) If $B \in \mathfrak{S}$, then $\mu^*(B) \geqq \nu(B)$. [*Hint*: Let Ω_σ denote the class of all countable unions of sets in Ω. If

$$A = \bigcup_{n=1}^{\infty} A_n \in \Omega_\sigma, \text{ where } (A_n)_{n=1}^{\infty} \text{ is in } \Omega,$$

then

$$A = A_1 \cup \bigcup_{n=2}^{\infty} (A_n \cap A_1^c \cap A_2^c \cap \ldots \cap A_{n-1}^c)$$

is a disjoint union of sets in Ω, and so $\nu(A) = \mu^*(A)$. Hence

$$\begin{aligned}\mu^*(B) &= \inf\{\mu^*(A) : B \subset A \in \Omega_\sigma\} = \inf\{\nu(A) : B \subset A \in \Omega_\sigma\} \\ &\geqq \nu(B).]\end{aligned}$$

(ii) If $F \in \mathfrak{S}$ and $\mu^*(F) < \infty$, then $\nu(F) = \mu^*(F)$.
[*Hint*: For $\varepsilon > 0$, choose $A \in \Omega_\sigma$ such that $F \subset A$ and $\mu^*(A) < \mu^*(F) + \varepsilon$. Then use part (i) to show that

$$\begin{aligned}\mu^*(F) &\leqq \mu^*(A) = \nu(A) = \nu(F) + \nu(A \cap F^c) \\ &\leqq \nu(F) + \mu^*(A \cap F^c) < \nu(F) + \varepsilon.]\end{aligned}$$

(iii) If there is a sequence $(F_n)_{n=1}^{\infty} \subset \Omega$ such that $\mu(F_n) < \infty$ for all n and

$$X = \bigcup_{n=1}^{\infty} F_n,$$

then $\nu(E) = \mu^*(E)$ for all $E \in \mathfrak{S}$, that is, the extension of μ to $\mathfrak{S}$ is unique. [*Hint*: We may suppose that $F_n \cap F_m = \varnothing$ for $n \neq m$. Then by (ii) we have

$$\nu(E) = \sum_{n=1}^{\infty} \nu(E \cap F_n) = \sum_{n=1}^{\infty} \mu^*(E \cap F_n) + \mu^*(E).]$$

(iv) If the σ-finiteness assumption in (iii) fails, the μ may have more than one extension to $\mathfrak{S}$. [*Hint*: Let $X = [0, 1)$ and let Ω be the (Boolean) algebra of all finite unions of intervals of the form $[a, b) \subset [0, 1)$. Define μ on Ω by $\mu(\varnothing) = 0$ and $\mu(A) = \infty$ if $A \neq \varnothing$. Then there are exactly 2^c measures on the Borel sets of $[0, 1)$ that agree with μ on Ω.]

Returning now to the subject of product measure, we see from a computation in the proof of Lemma 5 that $(\mu \times \nu)(A \times B) = \mu(A) \cdot \nu(B)$ for all measurable rectangles $A \times B$ (providing we agree to $0 \cdot \infty = 0$.) If ω is any measure on $\mathfrak{A} \times \mathfrak{B}$ such that $\omega(A \times B) = \mu(A) \cdot \nu(B)$ for all measurable rectangles $A \times B$, then $(\mu \times \nu)(E) = \omega(E)$ for all E in the (Boolean) algebra Ω of all finite disjoint unions of measurable rectangles. Since $\mu \times \nu$ is σ-finite and $\mathfrak{S}(\Omega) = \mathfrak{A} \times \mathfrak{B}$, it follows from part (iii) above that $(\mu \times \nu)(E) = \omega(E)$ for all $E \in \mathfrak{A} \times \mathfrak{B}$. Therefore, the product measure $\mu \times \nu$ is uniquely determined by the requirement that it be a measure on $\mathfrak{A} \times \mathfrak{B}$ and that $(\mu \times \nu)(A \times B) = \mu(A) \cdot \nu(B)$.

THEOREM 1. *Let $(X, \mathfrak{A}, \mu)$ and $(Y, \mathfrak{B}, \nu)$ be two σ-finite measure spaces and let $(X \times Y, \mathfrak{A} \times \mathfrak{B}, \mu \times \nu)$ be the product measure space constructed above. If f is a nonnegative, extended real-valued, $\mathfrak{A} \times \mathfrak{B}$-measurable function on $X \times Y$, then*

(i) *the function $x \to f(x,y)$ is $\mathfrak{A}$-measurable for each $y \in Y$;*
(i′) *the function $y \to f(x,y)$ is $\mathfrak{B}$-measurable for each $x \in X$;*
(ii) *the function $y \to \int_X f(x,y)\,d\mu(x)$ is $\mathfrak{B}$-measurable;*
(ii′) *the function $x \to \int_Y f(x,y)\,d\nu(y)$ is $\mathfrak{A}$-measurable; and*
(iii) *the equalities*

$$\int_{X \times Y} f(x,y)\,d(\mu \times \nu)(x,y) = \int_Y \left[\int_X f(x,y)\,d\mu(x)\right] d\nu(y)$$

$$= \int_X \left[\int_Y f(x,y)\,d\nu(y)\right] d\mu(x)$$

hold.

Proof. Conclusions (i) and (i′) are just part (i) and (ii) of Lemma 3. To prove (ii), (ii′), and (iii), we first suppose that $f = \chi_E$ for some $E \in \mathfrak{A} \times \mathfrak{B}$. Clearly,

$$\int_X \chi_E(x,y)\,d\mu(x) = \int_X \chi_{E^y}(x)\,d\mu(x) = \mu(E^y)$$

for every $y \in Y$ and

$$\int_Y \chi_E(x,y)\,d\nu(y) = \int_Y \chi_{E_x}(y)\,d\nu(y) = \nu(E_x)$$

for every $x \in X$. Thus (ii), (ii′), and (iii) follow for χ_E at once from Lemma 5 and the definition of $\mu \times \nu$. For a simple $\mathfrak{A} \times \mathfrak{B}$-measurable function f, the assertions (ii), (ii′), and (iii) are clear from the linearity of the integrals.

Let f be an arbitrary nonnegative, extended real-valued $\mathfrak{A} \times \mathfrak{B}$-measurable function on $X \times Y$. Let $(s_n)_{n=1}^{\infty}$ be an increasing sequence of nonnegative, real-valued, $\mathfrak{A} \times \mathfrak{B}$-measurable simple functions on $X \times Y$ such that $s_n(x,y) \to f(x,y)$ for all $(x,y) \in X \times Y$ (see Proposition 10 in Chapter 5). By the Monotone Convergence Theorem (see Proposition 12 or Problem 17 in Chapter 5) we get that

$$\int_X f(x,y)\,d\mu(x) = \lim_{n\to\infty} \int_X s_n(x,y)\,d\mu(x).$$

This shows that the function in (ii) is the pointwise limit of a sequence of $\mathfrak{B}$-measurable functions, and so (ii) follows from Proposition 8 of Chapter 5. The verification of (ii′) is similar to the proof of (ii). To show (iii), we use the Monotone Convergence Theorem again and write

$$\begin{aligned}
\int_{X\times Y} f(x,y)\,d(\mu\times\nu)(x,y) &= \lim_{n\to\infty} \int_{X\times Y} s_n(x,y)\,d(\mu\times\nu)(x,y)\\
&= \lim_{n\to\infty} \int_Y \left[\int_X s_n(x,y)\,d\mu(x)\right] d\nu(y)\\
&= \int_Y \left[\lim_{n\to\infty} \int_X s_n(x,y)\,d\mu(x)\right] d\nu(y)\\
&= \int_Y \left[\int_X f(x,y)\,d\mu(x)\right] d\nu(y).
\end{aligned}$$

A similar calculation proves that

$$\int_{X\times Y} f(x,y)\,d(\mu\times\nu)(x,y) = \int_X \left[\int_Y f(x,y)\,d\mu(y)\right] d\nu(x)$$

and the proof is finished.

THEOREM 2. *Let* $(X, \mathfrak{A}, \mu)$ *and* $(Y, \mathfrak{B}, \nu)$ *be two* σ*-finite measure spaces and let* $(X \times Y, \mathfrak{A} \times \mathfrak{B}, \mu \times \nu)$ *be the product measure space constructed above. Let* f *be an extended real-valued* $\mathfrak{A} \times \mathfrak{B}$*-measurable function on* $X \times Y$ *and suppose that any one of the three*

$$\int_{X\times Y} |f(x,y)|\, d(\mu \times \nu)(x,y), \quad \int_Y \left[\int_X |f(x,y)|\, d\mu(x)\right] d\nu(y),$$

and

$$\int_X \left[\int_Y |f(x,y)|\, d\nu(y)\right] d\mu(x)$$

is finite (note that if one of these integrals is finite, then by Theorem 1, part (iii) all are finite and equal). Then

(i) *the function* $x \to f(x,y)$ *is in* $L^1(X, \mathfrak{A}, \mu)$ *for* ν*-almost all* $y \in Y$;
(i′) *the function* $y \to f(x,y)$ *is in* $L^1(Y, \mathfrak{B}, \nu)$ *for* μ*-almost all* $x \in X$;
(ii) *the function* $y \to \int_X f(x,y)\, d\mu(x)$ *is in* $L^1(Y, \mathfrak{B}, \nu)$ *(it should be noted that this function is defined only for those* $y \in Y$ *for which* $x \to f(x,y)$ *is in* $L^1(X, \mathfrak{A}, \mu)$*);*
(ii′) *the function* $x \to \int_Y f(x,y)\, d\nu(y)$ *is in* $L^1(X, \mathfrak{A}, \mu)$ *(a remark similar to the one under (ii) applies here as well);*
(iii) *the equalities*

$$\int_{X\times Y} f(x,y)\, d(\mu \times \nu)(x,y) = \int_Y \left[\int_X f(x,y)\, d\mu(x)\right] d\nu(y)$$
$$= \int_X \left[\int_Y f(x,y)\, d\nu(y)\right] d\mu(x)$$

hold.

Proof. Our hypothesis and Theorem 1 show that

$$(*) \qquad \int_{X\times Y} |f|\, d(\mu \times \nu) = \int_Y \int_X |f|\, d\mu\, d\nu = \int_X \int_Y |f|\, d\nu\, d\mu < \infty.$$

Thus $f \in L^1(X \times Y, \mathfrak{A} \times \mathfrak{B}, \mu \times \nu)$. We write $f = f_1 - f_2$, where $f_1 = f^+$ and $f_2 = f^-$. Then f_j is nonnegative and in $L^1(X \times Y, \mathfrak{A} \times \mathfrak{B}, \mu \times \nu)$ and $f_j \leqq |f|$, for $j = 1, 2$. The functions described in (i) and (i′) are measurable by Lemma 3. From (*) we have

$$\int_X |f_j(x,y)|\, d\mu(x) \leqq \int_X |f(x,y)|\, d\mu(x) < \infty$$

for ν-almost all y; it follows that $f_j^y \in L^1(X, \mathfrak{A}, \mu)$ for ν-almost all y ($j = 1, 2$). Thus (i) holds. Assertion (i′) is proved in a similar way. On account of (i), the function in (ii) is defined for ν-almost all $y \in Y$, and its $\mathfrak{B}$-measurability follows by applying Theorem 1, part (ii) to each f_j and taking linear combinations. Hence for ν-almost all y, we get

$$\left|\int_X f(x,y)\, d\mu(x)\right| \leqq \int_X |f(x,y)|\, d\mu(x);$$

it is clear that (ii) now follows by applying (*). The verification of (ii′) is similar. Applying part (iii) of Theorem 1 to each f_j and then taking the linear combinations we get (iii); this last step is legitimate since the integrals in question are linear on the various L^1-spaces. The proof is finished.

Since we have not used Carathéodory's extension theorem (see Proposition 3) in the construction of the product measure $\mu \times \nu$ here, we cannot expect the measure space $(X \times Y, \mathfrak{A} \times \mathfrak{B}, \mu \times \nu)$ to be complete; in fact, the product measure space $(X \times Y, \mathfrak{A} \times \mathfrak{B}, \mu \times \nu)$ can be incomplete even if $(X, \mathfrak{A}, \mu)$ and $(Y, \mathfrak{B}, \nu)$ are complete as the following example illustrates: $(R^2, \mathfrak{M} \times \mathfrak{M}, \mu \times \mu)$ is an incomplete measure space. (Hint: Let $(X, \mathfrak{A}, \mu)$ and $(Y, \mathfrak{B}, \nu)$ be σ-finite measure spaces. Suppose that there exists a set $A \subset X$ such that $A \notin \mathfrak{A}$ and suppose that there exists a nonempty set $B \in \mathfrak{B}$ such that $\nu(B) = 0$. Show that $(X \times Y, \mathfrak{A} \times \mathfrak{B}, \mu \times \nu)$ is incomplete.)

As in Proposition 6 of Chapter 5, let $(X \times Y, (\mathfrak{A} \times \mathfrak{B})_0, (\mu \times \nu)_0)$ denote the completion of the measure space $(X \times Y, \mathfrak{A} \times \mathfrak{B}, \mu \times \nu)$. In the following we shall see that Theorems 1 and 2 can be extended to this completed product space.

LEMMA 7. *Let $(X, \mathfrak{A}, \mu)$ and $(Y, \mathfrak{B}, \nu)$ be two complete, σ-finite measure spaces. Let $E \in \mathfrak{A} \times \mathfrak{B}$ be such that $(\mu \times \nu)(E) = 0$ and assume that $F \subset E$. Then*

(*i*) $\mu(F^y) = 0$ ν-a.e. *and*

(*ii*) $\nu(F_x) = 0$ μ-a.e.

Proof. Because of symmetry we shall only verify (ii). By Theorem 1 we see that

$$0 = (\mu \times \nu)(E) = \int_X \left[\int_Y \chi_E\, d\nu\right] d\mu = \int_X \nu(E_x)\, d\mu(x).$$

Since $\nu(E_x) \geqq 0$ for all $x \in X$, it follows that $\nu(E_x) = 0$ μ-a.e. It is clear that $F_x \subset E_x$ for all x; thus $F_x \in \mathfrak{B}$ and $\nu(F_x) = 0$ for μ-almost all x. The proof is finished.

THEOREM 3. *Let $(X, \mathfrak{A}, \mu)$ and $(Y, \mathfrak{B}, \nu)$ be complete, σ-finite measure spaces and let f be a nonnegative, extended real-valued $(\mathfrak{A} \times \mathfrak{B})_0$-measurable function on $X \times Y$. Then*

(i) *the function* $x \to f(x,y)$ *is* $\mathfrak{A}$*-measurable for* ν*-almost all* $y \in Y$;
(i′) *the function* $y \to f(x,y)$ *is* $\mathfrak{B}$*-measurable for* μ*-almost all* $x \in X$;
(ii) *the function* $y \to \int_X f(x,y)\,d\mu(x)$ *is* $\mathfrak{B}$*-measurable*;
(ii′) *the function* $x \to \int_Y f(x,y)\,d\nu(y)$ *is* $\mathfrak{A}$*-measurable*;
(iii) *the equalities*

$$\int_{X\times Y} f(x,y)\,d(\mu\times\nu)_0(x,y) = \int_Y \left[\int_X f(x,y)\,d\mu(x)\right] d\nu(y)$$
$$= \int_X \left[\int_Y f(x,y)\,d\nu(y)\right] d\mu(x)$$

hold.

Proof. Let $H \in (\mathfrak{A}\times\mathfrak{B})_0$. By Proposition 6 in Chapter 5, H has the form $E \cup F$, where $E \in \mathfrak{A}\times\mathfrak{B}$ and $F \subset M$ for some $M \in \mathfrak{A}\times\mathfrak{B}$ such that $(\mu\times\nu)(M) = 0$. For each $x \in X$, we have $H_x = E_x \cup F_x$, and so it follows by Lemma 7 and part (i) of Lemma 2 that $H_x \in \mathfrak{B}$ for μ-almost all $x \in X$. This proves (i) for the case that $f = \chi_H$; (i′) is similar.

From what we have seen already, it is clear that $\nu(H_x) = \nu(E_x)$ for μ-almost all $x \in X$. Since

$$\int_Y \chi_H(x,y)d\nu(y) = \nu(H_x),$$

part (i) of Lemma 5 shows that the function

$$x \to \int_Y \chi_H(x,y)\,d\nu(y)$$

is equal μ-a.e. to the $\mathfrak{A}$-measurable function $x \to \nu(E_x)$. This proves (ii′) for $f = \chi_H$; (ii) is similar.

To prove (iii) for χ_H, we observe that

$$\int_{X\times Y} \chi_H\,d(\mu\times\nu)_0 = (\mu\times\nu)_0(H) = (\mu\times\nu)(E) = \int_Y \left[\int_X \chi_E\,d\mu\right] d\nu$$
$$= \int_Y \left[\int_X \chi_E\,d\nu\right] d\mu.$$

The second equality in (iii) is similarly established. The remainder of the proof is like that of Theorem 1.

Remark. Theorem 2 is also valid for $(\mathfrak{A}\times\mathfrak{B})_0$-measurable functions; this result can be deduced from Theorem 3 in the same way that Theorem 2 was deduced from Theorem 1.

EXERCISES

1. Let Ω be an arbitrary class of subsets of a given set X. We form successively:
   (i) the class Ω_1 consisting of the empty set $\varnothing$, X and the $A \subset X$ such that A or A^c belongs to Ω;
   (ii) the class Ω_2 of finite intersections of subsets of X in Ω_1;
   (iii) the class Ω_3 of finite unions of pairwise disjoint subsets belonging to Ω_2.

   Show that Ω_3 is the smallest (Boolean) algebra containing Ω.
2. If $\mathfrak{A}$ is a collection of sets which is closed under finite unions and finite intersections, show that
   (i) $\mathfrak{A}_\sigma$ is closed under countable unions and finite intersections;
   (ii) each set in $\mathfrak{A}_{\sigma\delta}$ is the intersection of a decreasing sequence of sets in $\mathfrak{A}_\sigma$.
3. Using Proposition 4, verify the following statement: If μ is a σ-finite measure on an algebra with domain of definition $\mathfrak{A}$, and μ^* is the outer measure generated by μ, then a set E is μ^*-measurable if and only if E is the set-theoretic difference $A - B$ of a set A in $\mathfrak{A}_{\sigma\delta}$ and a set B with $\mu^*(B) = 0$. Moreover, each set B with $\mu^*(B) = 0$ is contained in a set C in $\mathfrak{A}_{\sigma\delta}$ with $\mu^*(C) = 0$.
4. Let Ω be a (Boolean) algebra of subsets of X and let $\mathfrak{S} = \mathfrak{S}(\Omega)$ denote the smallest σ-algebra of subsets of X containing Ω. Let μ be a measure on an algebra with domain of definition Ω and let μ^* be the outer measure generated by μ. Suppose that $(X, \mathfrak{S}, \nu)$ is a measure space such that $\nu(A) = \mu(A)$ for all $A \in \Omega$. Show that:
   (i) if $B \in \mathfrak{S}$, then $\mu^*(B) \geqq \nu(B)$;
   (ii) if $F \in \mathfrak{S}$ and $\mu^*(F) < \infty$, then $\nu(F) = \mu^*(F)$;
   (iii) if there is a sequence $(F_n)_{n=1}^{\infty} \subset \Omega$ such that $\mu(F_n) < \infty$ for all $n \in N$ and
   $$X = \bigcup_{n=1}^{\infty} F_n,$$
   then $\nu(E) = \mu^*(E)$ for all $E \in \mathfrak{S}$, i.e., the extension of μ to $\mathfrak{S}$ is unique;
   (iv) if the σ-finiteness assumption in (iii) fails, then μ may have more than one extension to $\mathfrak{S}$.
5. If product measure is constructed as in the Addendum, prove that $(R^2, \mathfrak{M} \times \mathfrak{M}, \mu \times \mu)$ is an incomplete measure space (note that $\mathfrak{M}$ stands for the class of Lebesgue measurable sets of the real line).
6. (i) Let F be a monotone increasing function continuous on the right. If

$$(a,b] \subset \bigcup_{j=1}^{\infty} (a_j, b_j],$$

show that

$$F(b) - F(a) \leqq \sum_{j=1}^{\infty} (F(b_j) - F(a_j)).$$

(ii) Let Γ be the class of all half-open intervals of the form $(a,b]$. Observing that Γ is a semi-algebra, show that the set function μ with $\mu(a,b] = F(b) - F(a)$ admits a unique extension ν to a measure on an algebra, where the domain of definition of ν is the algebra generated by Γ.

7. Some properties of the Gamma function:

(i) Show that the function $e^{-x}x^{p-1}$ is integrable over $[0,\infty)$ for $p > 0$, but not for $p \leqq 0$.
[*Hints*: Since

$$\int_0^1 e^{-1}x^{p-1}\,dx \leqq \int_0^1 e^{-x}x^{p-1}\,dx \leqq \int_0^1 x^{p-1}\,dx,$$

the integral in the middle is finite for $p > 0$ and infinite for $p \leqq 0$. Moreover, given the real number p, there is a positive constant M such that

$$e^{-(1/2)x}x^{p-1} < M \quad \text{for } x \geqq 1$$

and so

$$e^{-x}x^{p-1} < Me^{-(1/2)x} \quad \text{for } x \geqq 1.]$$

DEFINITION. *For $p > 0$, the Gamma function Γ is defined by*

$$\Gamma(p) = \int_0^{\infty} e^{-x}x^{p-1}\,dx.$$

(ii) Verify that, for $p > 0$, we have $\Gamma(p+1) = p\Gamma(p)$.
[*Hints*: Let $p > 0$, $x \geqq 0$, $f(x) = px^{p-1}$ and $g(x) = e^{-x}$, so

$$F(x) = \int_0^x f(t)\,dt \quad \text{and} \quad G(x) = \int_0^x g(t)\,dt,$$

and then use integration by parts.]

DEFINITION. *Since the function $x^{p-1}(1-x)^{q-1}$ is integrable on $[0,1]$ when $p > 0$ and $q > 0$, we define, for $p > 0$, $q > 0$, the Beta function B by*

$$B(p,q) = \int_0^1 x^{p-1}(1-x)^{q-1}\,dx.$$

(*Observe that* $B(p,q)$ *is the limit for* $h \downarrow 0$ *of the Riemann integral*

$$\int_h^{1-h} x^{p-1}(1-x)^{q-1}\,dx$$

and so $B(p,q) = B(q,p)$ *by the rules of Riemann integration.*)

(iii) Prove that, for $p > 0$, $q > 0$, the Gamma and the Beta functions are connected by the formula

$$\Gamma(p)\Gamma(q) = \Gamma(p+q)B(p,q).$$

[*Hints*: Let c by any finite real number. Then

$$\int_a^b f(x)\,dx = \int_{a+c}^{b+c} f(x-c)\,dx$$

and hence, for any $y > 0$,

$$e^{-y}\Gamma(p) = \int_0^\infty e^{-x-y}x^{p-1}\,dx = \int_y^\infty e^{-x}(x-y)^{p-1}\,dx$$
$$= \int_0^\infty f(x,y)\,dx,$$

where

$$f(x,y) = \begin{cases} e^{-x}(x-y)^{p-1} & \text{for } 0 \leqq y < x, \\ 0 & \text{for } 0 \leqq x \leqq y. \end{cases}$$

Hence

$$\int_0^\infty \left[\int_0^\infty y^{q-1}f(x,y)\,dx\right] dy = \int_0^\infty \Gamma(p)e^{-y}y^{q-1}\,dy = \Gamma(p)\Gamma(q).$$

But f is seen to be measurable on $[0,\infty) \times [0,\infty)$, and $y^{p-1}f(x,y)$ is therefore nonnegative and measurable on $[0,\infty) \times [0,\infty)$. By Proposition 15,

$$\Gamma(p)\Gamma(q) = \int_0^\infty \left[\int_0^\infty y^{q-1}f(x,y)\,dx\right] dy = \int_0^\infty \left[\int_0^\infty y^{q-1}f(x,y)\,dy\right] dx.$$

By the rules of Riemann integration,

$$\int_0^\infty y^{q-1} f(x,y)\, dy = e^{-x} \int_0^x y^{q-1}(x-y)^{p-1}\, dy$$

$$= e^{-x} x^{p+q-1} \int_0^1 t^{q-1}(1-t)^{p-1}\, dt$$

$$= e^{-x} x^{p+q-1} B(p,q);$$

we have made the substitution $t = y/x$. Thus

$$\Gamma(p)\Gamma(q) = B(p,q) \int_0^\infty e^{-x} x^{p+q-1}\, dx = \Gamma(p+q)B(p,q).]$$

(iv) Show that, if $\alpha > 0$ and f is integrable over $[0,b]$ with $0 < b < \infty$, then

$$h(x) = \int_0^x (x-t)^{\alpha-1} f(t)\, dt$$

exists as a finite number for almost all $x \in [0,b]$ and h is integrable over $[0,b]$.

[*Hints*: Let $g(v) = v^{\alpha-1}$ for $0 \leqq v \leqq b$, and $g(v) = 0$ for all other values of v. We have to scrutinize the existence and integrability of

$$h(x) = \int_0^b g(x-t) f(t)\, dt$$

on $[0,b]$. But $F(x,t) = g(x-t)|f(t)|$ is nonnegative and measurable on $[-\infty,\infty] \times [0,b]$ and so

$$\int_{-\infty}^{\infty} \int_0^b F(x,t) d(x,t) = \int_0^b \left[|f(t)| \int_{-\infty}^{\infty} g(x-t)\, dx \right] dt$$

$$= \int_0^b |f(t)|\, dt \int_{-\infty}^{\infty} g(x)\, dx < \infty.$$

This shows, by Proposition 16, that $g(x-t)f(t)$ is integrable over $[0,b] \times [0,b]$ and so, by Proposition 15,

$$\infty > \int_0^b \int_0^b g(x-t) f(t)\, d(x,t) = \int_0^b \left[\int_0^b g(x-t) f(t)\, dt \right] dx$$

$$= \int_0^b \left[\int_0^b (x-t)^{\alpha-1} f(t)\, dt \right] dx.$$

It follows that $h(x)$ exists as a finite number for almost all x in $[0, b]$ and that h is integrable on $[0, b]$.]

DEFINITION. *If $\alpha > 0$, and f is integrable over $[0, b]$ with $0 < b < \infty$, the fractional integral of order α of f is defined on $[0, b]$ by*

$$I_\alpha f(x) = \frac{1}{\Gamma(\alpha)} \int_0^x (x - t)^{\alpha - 1} f(t)\, dt.$$

(v) If $\alpha > 0$, $\beta > 0$, show that the fractional integral of order β of the fractional integral of order α of f equals the fractional integral of order $\alpha + \beta$ of f.
[*Hints*: The definition implies that

$$(*) \qquad \Gamma(\alpha)\Gamma(\beta) I_\beta I_\alpha f(x) = \int_0^x \left[(x - t)^{\beta - 1} \int_0^t (t - v)^{\alpha - 1} f(v)\, dv \right] dt.$$

From

$$\begin{aligned}
\int_v^x (x - t)^{\beta - 1} (t - v)^{\alpha - 1}\, dt &= \int_0^{x - v} (x - v - t)^{\beta - 1} t^{\alpha - 1}\, dt \\
&= (x - v)^{\alpha + \beta - 1} \int_0^1 (1 - t)^{\beta - 1} t^{\alpha - 1}\, dt \\
&= (x - v)^{\alpha + \beta - 1} B(\alpha, \beta) \\
&= (x - v)^{\alpha + \beta - 1} \frac{\Gamma(\alpha)\Gamma(\beta)}{\Gamma(\alpha + \beta)}
\end{aligned}$$

we get that

$$\begin{aligned}
&\int_0^x \left[(x - t)^{\beta - 1} \int_0^t (t - v)^{\alpha - 1} |f(v)|\, dv \right] dt \\
&\quad = \int_0^x \left[|f(v)| \int_v^x (x - t)^{\beta - 1} (t - v)^{\alpha - 1}\, dt \right] dv \\
&\quad = \frac{\Gamma(\alpha)\Gamma(\beta)}{\Gamma(\alpha + \beta)} \int_0^x (x - v)^{\alpha + \beta - 1} |f(v)|\, dv \\
&\quad = \Gamma(\alpha)\Gamma(\beta) I_{\alpha + \beta} |f|(x)
\end{aligned}$$

and this is finite for almost all x in $[0, b]$. We can therefore invert the order of integration in $(*)$ for these values of x; we get

$$\Gamma(\alpha)\Gamma(\beta)I_\beta I_\alpha f(x) = \int_0^x \left[f(v) \int_v^x (x-t)^{\beta-1}(t-v)^{\alpha-1}\,dt \right] dv$$

$$= \frac{\Gamma(\alpha)\Gamma(\beta)}{\Gamma(\alpha+\beta)} \int_0^x (x-v)^{\alpha+\beta-1} f(v)\,dv$$

$$= \Gamma(\alpha)\Gamma(\beta) I_{\alpha+\beta} f(x),$$

so that $I_\beta I_\alpha f(x) = I_{\alpha+\beta} f(x)$ a.e. on $[0, b]$.]

8. Let $f : [0, \infty) \to R^1$ be integrable over $[0, \infty)$. Since $e^{-a} \leqq 1$ when $a \geqq 0$, the function $e^{-xt} f(t)$ is also integrable over $[0, \infty)$ for every $x \geqq 0$. We define the function $\hat{f}$ by

$$\hat{f}(x) = \int_0^\infty e^{-xt} f(t)\,dt$$

for $0 \leqq x < \infty$; $\hat{f}$ is called the *Laplace transform of f.*

(i) If f and g are integrable over $[0, \infty)$, then the integral $\int_0^x f(x - t) g(t)\,dt$ exists for almost all $x \in [0, \infty)$ and is an integrable function on $[0, \infty)$. [*Hints*: Define $f(t) = 0 = g(t)$ for $-\infty < t < 0$. For each real t we have, by a change of variable $u = x - t$, $\int_{-\infty}^\infty |f(x-t)|\,dx = \int_{-\infty}^\infty |f(u)|\,du$, and hence

$$\int_{-\infty}^\infty \{\int_{-\infty}^\infty |f(x-t)g(t)|\,dx\}\,dt = \int_{-\infty}^\infty |g(t)|\,dt \int_{-\infty}^\infty |f(u)|\,du$$

$$= \int_0^\infty |g(t)|\,dt \cdot \int_0^\infty |f(u)|\,du.$$

The last quantity is finite, since $f, g \in L^1([0, \infty])$. Thus $f(x - t)g(t)$ is in $L^1(R^2)$ by Proposition 16. By Fubini's Theorem $\int_{-\infty}^\infty f(x-t)g(t)\,dt$ exists for almost all x and is integrable; but $g(t) = 0$ for $t < 0$ and $f(x - t) = 0$ for $t > x$.]

(ii) If f and g are as in part (i) and

$$h(x) = \int_0^x f(x-t)g(t)\,dt,$$

show that

$$\hat{h} = \hat{f}\hat{g}.$$

9. Let f and g be real-valued integrable functions on R^1, i.e., $f, g \in L^1(R^1, \mathfrak{M}, \mu)$, where μ denotes Lebesgue measure and $\mathfrak{M}$ the class of all μ-measurable sets in R^1. For brevity we denote this measure space in this and in the subsequent problems by $L^1(R^1)$.

Show that for almost all $x \in R^1$, the function $y \to f(x-y)g(y)$ is in $L^1(R^1)$; for all such x define

$$f*g(x) = \int_{-\infty}^{\infty} f(x-y)g(y)\,dy$$

(the function $f*g$ is called the *convolution of f and g*). Verify that

$$f*g \in L^1(R^1) \text{ and } \|f*g\|_1 \leqq \|f\|_1 \cdot \|g\|_1 .$$

10. Assume that $1 < p < \infty$. Let $f \in L^1(R^1)$ and $g \in L^p(R^1)$.

    Show that for almost all $x \in R^1$, the function $y \to f(x-y)g(y)$ and the function $y \to f(y)g(x-y)$ are in $L^1(R^1)$; for all such x define

$$f*g(x) = \int_{R^1} f(x-y)g(y)\,dy$$

and

$$g*f(x) = \int_{R^1} g(x-y)f(y)\,dy.$$

Verify that $f*g = g*f$ a.e., $f*g \in L^p(R^1)$ and $\|f*g\|_p \leqq \|f\|_1 \cdot \|g\|_p$.

11. Let $1 \leqq p < \infty, f \in L^p(R^1)$, $g \in L^q(R^1)$; here $q = p/(p-1)$ if $p > 0$, and $q = \infty$ if $p = 1$. Define $f*g$ on R^1 by

$$f*g(x) = \int_{R^1} f(x-y)g(y)\,dy.$$

Show that $f * g$ is uniformly continuous on R^1 and

$$\sup_{x \in R^1} |f*g(x)| \leqq \|f\|_p \cdot \|g\|_q .$$

[*Hints*: Let $f \in L^p(R^1)$ and let f_t denote the translate of the function f, i.e., $f_t(x) = f(x+t)$. Then

$$\lim_{t \downarrow 0} \|f_t - f\|_p = \lim_{t \uparrow 0} \|f_t - f\|_p = 0,$$

by Problem 23 in Chapter 3. Hence, there is a $\delta > 0$ such that $x, t \in R^1$ and $|x-t| < \delta$ imply $\|f_x - f_t\|_p \cdot \|g\|_q < \varepsilon$. Then $|x-t| < \delta$ implies

$$|f * g(x) - f * g(t)| \leqq \int_{R^1} |f(x-y) - f(t-y)| \cdot |g(y)|\,dy$$
$$\leqq \|f_x - f_t\|_p \cdot \|g\|_q < \varepsilon.]$$

12. Show that the convolution product (defined in Problem 9) has no multiplicative unit, i.e., there is no $h \in L^1(R^1)$ such that $h*f = f$ a.e. for all $f \in L^1(R^1)$.
[*Hints*: Suppose the contrary. Then, by the absolute continuity of the Lebesgue integral, there exists a $\delta > 0$ such that

$$\int_{-2\delta}^{2\delta} |h(t)|\, dt < 1.$$

Let $f = \chi_{[-\delta,\delta]}$. Then $f \in L^1(R^1)$ and so for almost all $x \in R^1$ we have

$$f(x) = h*f(x) = \int_{R^1} h(x-y)f(y)\,dy = \int_{-\delta}^{\delta} h(x-y)\,dy$$

$$= \int_{x-\delta}^{x+\delta} h(t)\,dt.$$

Since $\mu([-\delta,\delta]) = 2\delta > 0$, there must be an $x \in [-\delta,\delta]$ such that

$$1 = f(x) = \int_{x-\delta}^{x+\delta} h(t)\,dt.$$

Since $[x-\delta, x+\delta] \subset [-2\delta, 2\delta]$, our choice of δ implies that

$$1 = \left|\int_{x-\delta}^{x+\delta} h(t)\,dt\right| \leqq \int_{x-\delta}^{x+\delta} |h(t)|\,dt \leqq \int_{-2\delta}^{2\delta} |h(t)|\,dt < 1,$$

which is a contradiction.]

13. Let $\mathfrak{A}$ be an algebra of subsets of a set X and let μ be a finitely additive measure of the algebra $\mathfrak{A}$. If X is a topological space, μ is said to be regular if given $\varepsilon > 0$, $A \in \mathfrak{A}$, we can find $E, F \in \mathfrak{A}$ with $\overline{E}$ compact, $\overline{E} \subset A \subset \mathring{F}$ ($\mathring{F}$ denotes the interior of F and $\overline{F}$ the closure of F), $\mu(A - E) < \varepsilon$ and $\mu(F - A) < \varepsilon$. Show that if μ is a regular finitely additive measure on the algebra $\mathfrak{A}$, then μ is a countably additive measure on $\mathfrak{A}$.

14. Finite unions of half-open intervals $[a, b) \subset [A, B)$ form an algebra Σ of subsets of $[A, B)$. Let w be a nondecreasing function on $[A, B)$ and define

$$\mu_w([a,b)) = w(b) - w(a);$$

μ_w gives rise to a finitely additive measure on Σ. If w is not left-continuous, we can change our definition μ_w to read

$$\mu_w([a, b)) = w(b - 0) - w(a - 0).$$

Is μ_w a countably additive measure on Σ? How would one go about constructing Lebesgue-Stieltjes measure?

15. Let

$$h(x) = \int_a^x f(y)\, dy,$$

with $f \geqq 0$ and $f \in L^1[a, b]$. If g is a bounded measurable function on $[h(a), h(b)]$, show that

$$\int_{h(a)}^{h(b)} g(t)\, dt = \int_a^b g(h(x)) f(x)\, dx.$$

16. Reading Assignment: Study Birkoff's ergodic theorem (Reference: P.R. Halmos, *Lectures on Ergodic Theory*, Chelsea Publishing Co., New York, 1956, pp. 18-21).
17. Compute

$$\int_0^1 \int_0^1 f(x,y)\, dx\, dy, \quad \int_0^1 \int_0^1 f(x,y)\, dy\, dx, \quad \int_0^1 \int_0^1 |f(x,y)|\, dx\, dy,$$

and

$$\int_0^1 \int_0^1 |f(x,y)|\, dy\, dx$$

for the functions

(a)
$$f(x,y) = \begin{cases} (x - \tfrac{1}{2})^{-3} & \text{for } 0 < y < |x - \tfrac{1}{2}|, \\ 0 & \text{otherwise,} \end{cases}$$

(b)
$$f(x,y) = \frac{x - y}{(x^2 + y^2)^{3/2}},$$

(c)
$$f(x,y) = (1 - xy)^{-p}, \text{ where } p > 0,$$

and compare with the Fubini Theorem.

18. Show that there exists no real solution to the equation

$$u(x) = 1 + \int_x^1 u(t)u(t-x)\,dt, \quad 0 \leqq x \leqq 1.$$

[*Hint*: Integrate both sides.]

19. In this exercise we consider one of the four proofs which Gauss gave of the Fundamental Theorem of Algebra:
If $f(x) = x^n + a_1 x^{n-1} + \ldots + a_n \quad (n > 0)$, where $a_1, \ldots, a_n$ are real or complex numbers, then f has at least one real or complex root.

We put $x = r(\cos\theta + i\sin\theta)$; then $x^k = r^k(\cos k\theta + i\sin k\theta)$, hence $f(x) = P + iQ$, where

$$P = r^n \cos n\theta + \ldots, \quad Q = r^n \sin n\theta + \ldots$$

and all other terms of P and Q contain only smaller powers of r and terms not containing r will be constant.

The fundamental theorem of algebra will be proved if we can show that $P^2 + Q^2$ is zero for a certain pair of values r and θ. We introduce the function

$$U = \operatorname{arc\,tan} \frac{P}{Q}.$$

Then

$$\frac{\partial U}{\partial r} = \frac{(\partial P/\partial r)Q - P(\partial Q/\partial r)}{P^2 + Q^2}, \quad \frac{\partial U}{\partial \theta} = \frac{(\partial P/\partial \theta)Q - P(\partial Q/\partial \theta)}{P^2 + Q^2};$$

hence

$$\frac{\partial^2 U}{\partial r \partial \theta} = \frac{H(r,\theta)}{(P^2 + Q^2)^2}.$$

Here $H(r,\theta)$ is a continuous function of its variables whose exact form is of no interest to us.

Finally, we need the double integrals

$$I_1 = \int_0^R \int_0^{2\pi} \frac{\partial^2 U}{\partial r \partial \theta}\, d\theta\, dr \quad \text{and} \quad I_2 = \int_0^{2\pi} \int_0^R \frac{\partial^2 U}{\partial r \partial \theta}\, dr\, d\theta;$$

R is a positive constant whose value we shall determine later on.

If the function $P^2 + Q^2$ would not become zero anywhere, then the integrand would be continuous and this would, of course, imply that

$I_1 = I_2$. But we shall see that $I_1 \neq I_2$ when R becomes large. This means however that the function $P^2 + Q^2$ must become zero at some point in the interior of the circle $x^2 + y^2 = R^2$, proving the fundamental theorem of algebra. We compute the interior integral in I_1 and obtain

$$\int_0^{2\pi} \frac{\partial^2 U}{\partial r \partial \theta} d\theta = \frac{\partial U}{\partial r}\bigg|_0^{2\pi} = 0$$

since clearly $\partial U/\partial r$ is a function dependent on θ and having period 2π. From this it follows that $I_1 = 0$. We next consider the integral I_2. Here

$$\int_0^{R} \frac{\partial^2 U}{\partial r \partial \theta} dr = \frac{\partial U}{\partial \theta}\bigg|_0^{R}.$$

For what follows it is important to consider the highest powers of r in the numerator and the denominator of $\partial U/\partial \theta$. Since

$$\frac{\partial P}{\partial \theta} = -nr^n \sin n\theta + \dots$$

$$\frac{\partial Q}{\partial \theta} = nr^n \cos n\theta + \dots$$

we get

$$\frac{\partial P}{\partial \theta} Q - P \frac{\partial Q}{\partial \theta} = -nr^{2n} + \dots .$$

But

$$P^2 + Q^2 = r^{2n} + \dots ,$$

and so, finally,

$$\frac{\partial U}{\partial \theta} = \frac{-nr^{2n} + \dots}{r^{2n} + \dots}.$$

Since the remaining terms in the numerator and the denominator are made up of smaller powers of r whose coefficients are bounded functions of θ, we have not only that

$$\lim_{r \to \infty} \frac{\partial U}{\partial \theta} = -n,$$

but even that this convergence to $-n$ is uniform in θ, i.e., given any $\varepsilon > 0$, there is a real number $M = M(\varepsilon) > 0$ not dependent on θ such that $|\partial U/\partial \theta + n| < \varepsilon$ for all θ whenever $r > M$. Since $\partial U/\partial \theta = 0$ for

$r = 0$ (note that in this case we have $\partial P/\partial\theta = \partial Q/\partial\theta = 0$), the interior integral of I_2 leads to the value $\partial U/\partial\theta$ for $r = R$. In case $R \to \infty$, this value tends to $-n$ uniformly in θ. Hence we obtain

$$\lim_{R\to\infty} I_2 = -2\pi n.$$

We see therefore that the integral I_2 is negative for sufficiently large R. Thus, the equality $I_1 = I_2$ cannot be fulfilled and this completes the proof.

20. Prove the following Generalized Minkowski Inequality

$$\left\{\int_a^b \left|\int_c^d g(x,y)\,dy\right|^p dx\right\}^{1/p} \leqq \int_c^d \left\{\int_a^b |g(x,y)|^p\,dx\right\}^{1/p} dy$$

for $p \geqq 1$, by passing from functions which assume only a finite number of values (first only two values) to an arbitrary function $g(x,y)$ measurable on the rectangular region E $(a < x < b, c < y < d)$ for which the term on the right-hand side of the inequality has meaning. Observe that a special case of this, for $p = 1$, manifests itself in Proposition 16, namely, if

$$\int_c^d \left\{\int_a^b |g(x,y)|\,dx\right\} dy < \infty,$$

then

$$\int_E g(P)\,dP = \int_a^b \left\{\int_c^d g(x,y)\,dy\right\} dx = \int_c^d \left\{\int_a^b g(x,y)\,dx\right\} dy.$$

21. Let f be defined on R^1 and suppose that f is integrable in Lebesgue's sense over $[a,b]$ for arbitrary finite a and b. For $h > 0$, let f_h with

$$f_h(t) = \frac{1}{h}\int_{t-\frac{1}{2}h}^{t+\frac{1}{2}h} f(u)\,du = \frac{1}{h}\int_{-\frac{1}{2}h}^{\frac{1}{2}h} f(t+v)\,dv$$

denote the Steklov function with increment h. From Chapter 4 we know that f_h has almost everywhere the derivative

$$f_h'(t) = \frac{1}{h}\{f(t+\tfrac{1}{2}h) - f(t-\tfrac{1}{2}h)\}.$$

Show: If $f \in L^p(-\infty,\infty)$ with $p \geqq 1$, then f_h belongs to the same space, where $\|f_h\|_p \leqq \|f\|_p$; this follows from the Generalized Minkowski Inequality (see Problem 20). On the other hand, using Hölder's Inequality we find that the Steklov function is always bounded:

$$|f_h(t)| \leqq h^{-1/p} \|f\|_p \qquad (-\infty < t < \infty).$$

Combining both these facts we get for arbitrary $r \geqq p$ ($\geqq 1$) :

$$\left\{ \int_{-\infty}^{\infty} |f_h(t)|^r \, dt \right\}^{1/r} \leqq h^{(1/r)-(1/q)} \left\{ \int_{-\infty}^{\infty} |f(t)|^p \, dt \right\}^{1/p}.$$

Verify also that $\|f_h - f\|_p \to 0$ as $h \to 0$.

(We should point out that an interesting application of the foregoing exercise is a characterization of relatively compact sets in L^p; see, for example, §9 in the book by L.A. Lyusternik and V.I. Sobolev: *Elements of Functional Analysis*, Ungar, New York, 1955.)

22. Let f and g be continuous functions over $[a, b]$ and put

$$B = \iint_S [f(x)g(y) - f(y)g(x)]^2 \, dx \, dy$$

with $S = [a, b] \times [a, b]$. Prove that

$$\left(\int_a^b f(x)g(x) \, dx\right)^2 \leqq \left(\int_a^b f^2(x) \, dx\right)\left(\int_a^b g^2(x) \, dx\right).$$

Generalize this inequality.

23. Show that the Lebesgue integral of a nonnegative, measurable function over R^1 equals the Lebesgue measure in R^2 of the "ordinate set", that is,

$$\int_E f = \mu\{(x, y) : x \in E, \ 0 \leqq y \leqq f(x)\}.$$

24. In Exercise 12 we have seen that the convolution product has no multiplicative unit. However, we can construct an "approximate unit" as follows: Let $f \in L^1(R^1)$ and, for $n = 1, 2, \ldots$ and $-\infty < t < \infty$, let

$$h_n(t) = (1 - \cos nt)/(\pi n t^2).$$

Show that $\lim_{n \to \infty} \|h_n * f - f\|_1 = 0$.

Chapter 7

Topological and Metric Spaces

This chapter is devoted to certain topics in the theory of topological and metric spaces which are of special interest to us.

A family of subsets τ of a set X is said to be a *topology* in X if

(i) τ contains the empty set $\varnothing$ and the set X;

(ii) τ contains the union of every one of its subfamilies;

(iii) τ contains the intersection of every one of its finite subfamilies.

The pair (X,τ) is called a *topological space* and the sets in τ are called the *open sets* of (X,τ). For brevity we sometimes write X instead of (X,τ).

If τ and τ_0 are two topologies in X, τ is called *stronger* than τ_0 and τ_0 is said to be *weaker* than τ if $\tau \supset \tau_0$.

Let (X,τ) be a topological space. A *neighborhood of the point* $x \in X$ is any open subset U of X such that $x \in U$; a *neighborhood of the set* $A \subset X$ is an open set containing A. If A is a subset of X, then a point x is a *limit point*, or a *point of accumulation, of* A provided every neighborhood of x contains at least one point $y \neq x$, with $y \in A$. The *interior* of a set in X is the union of its open subsets; the interior of a set A we shall denote by $\mathring{A}$. It is clear that $\mathring{A}$ is the largest open set contained in A. A set is said to be *closed* if its complement, i.e., $A^c = X - A$, is open.

The following is readily seen: The intersection of any family of closed sets is closed and the union of any finite family of closed sets is closed, and $\varnothing$ and X are closed. If $\mathcal{G}$ is any family of subsets of X having the properties of the foregoing statement and τ is the family of complements of members of $\mathcal{G}$, then (X,τ) is a topological space, and $\mathcal{G}$ is the family of closed sets of this topology.

By the *closure* of a set A in a topological space (X,τ) we mean the intersection of all closed sets containing A or, equivalently, the smallest closed set of (X,τ) which contains A; we denote the closure of A by $\overline{A}$.

It is clear that a set A is closed in (X,τ) if and only if $A = \overline{A}$ and that a set A is open in (X,τ) if and only if $A = \mathring{A}$. Evidently, a set is closed if and only if it contains all of its limit points, and a set is open if and only if it contains a neighborhood of each of its points.

Let (X,τ) be a topological space. A family $\beta \subset \tau$ is called a *base* for the topology τ if every set in τ is the union of some subfamily of β; a family $\alpha \subset \tau$ is called a *subbase* for the topology τ if every set in τ is a union of finite intersections of sets in α. It can be seen that α is a subbase for the topology τ if and only if τ is the weakest topology containing α.

PROPOSITION 1. *Let β be a family of subsets of X and τ be the family of all unions of subfamilies of β, i.e., $\tau = \{\cup \alpha : \alpha \subset \beta\}$. Then (X,τ) is a topological space, and β is a base for τ, if and only if*

(*i*) $\cup \beta = X$ *and*

(*ii*) $U, V \in \beta$ *and* $x \in U \cap V$ *imply that there exists* $W \in \beta$ *such that* $x \in W \subset U \cap V$.

Proof. Suppose that τ is a topology for X. Then $X \in \tau$, so there exists $\alpha \subset \beta$ such that $X = \cup \alpha \subset \cup \beta \subset X$. Thus (i) is true. Next let U, V be sets in β and let $x \in U \cap V$. Then $U \cap V$ is in τ, so there is some $\omega \subset \beta$ such that $U \cap V = \cup \omega$. Thus we have $x \in W \subset U \cap V$ for some $W \in \omega$, proving (ii).

Conversely, suppose that (i) and (ii) hold. We have to verify that τ is a topology. Let $\{U_t\}_{t \in T}$ be any subfamily of τ. Then, by the definition of τ, for every t there exists $\alpha_t \subset \beta$ such that $U_t = \cup \alpha_t$. [Here the axiom of choice is used to pick just one α_t for each $t \in T$.] Let $\alpha = \cup \{\alpha_t : t \in T\}$. We have that $\alpha \subset \beta$ and $\cup \alpha = \cup \{U_t : t \in T\}$; thus τ is closed under the formation of arbitrary unions. Let $U, V \in \tau$. Then there are subfamilies $\{U_t\}_{t \in T}$ and $\{V_s\}_{s \in S}$ of β such that $U = \cup \{U_t : t \in T\}$ and $V = \cup \{V_s : s \in S\}$. Thus for each $x \in U \cap V$, there exist $t \in T$ and $s \in S$ such that $x \in U_t \cap V_s$ and therefore, by (ii), there is a W_x in β such that $x \in W_x \subset U_t \cap V_s \subset U \cap V$. Let $\alpha = \{W_x : x \in U \cap V\}$. Then $\alpha \subset \beta$ and $U \cap V = \cup \alpha \in \tau$. Thus τ is closed under the formation of finite intersections. By (i), X is in τ and, since $\varnothing \subset \beta$, we have $\varnothing = \cup \varnothing \in \tau$. This proves that τ is a topology for X. It is evident that β is a base for τ. The proof is finished.

If $Y \subset X$, and τ is a topology for X, then the topology

$$\tau_Y = \{A : A = B \cap Y, B \in \tau\}$$

is called the *relative topology* of Y induced by τ and the set Y with this topology is called a *subspace of* (X,τ). A subset of Y is said to be *relatively open* if it is in τ_Y; *relatively closed* if its complement relative to Y is relatively open. Other terms like *relative closure* of a set are defined analogously. A topological space is said to be *connected* if it is not the union of two nonempty disjoint closed sets; for example, R^1 with the usual topology is connected.

A topological space (X,τ) is said to have a *countable base* if there is a base for the topology τ of X which is a countable family. A subset D of a topological space (X,τ) is said to be *dense in* X if $\overline{D} = X$. A topological space (X,τ) is said to be *separable* if X contains a countable dense subset.

PROPOSITION 2 (LINDELÖF). *Let* (X,τ) *be a topological space, and let* τ *have a countable base* β. *Then every family* $\sigma \subset \tau$ *contains a countable subfamily* σ_0 *with* $\cup\, \sigma = \cup\, \sigma_0$.

Proof. Let $B_1, B_2, \ldots$ be an enumeration of β. Let β_0 be the family of elements of β which are contained in some subset of σ. If $B_n \in \beta_0$, let C_n be some set in σ which contains B_n, and let σ_0 be the family of all these C_n. Then it can be seen that

$$\cup\, \sigma \supset \cup\, \sigma_0 \supset \cup\, \beta_0 .$$

Since β is a basis, if $p \in A \in \sigma$, there is a $B_n \in \beta$ such that $p \in B_n \subset A$, and so $\cup\, \beta_0 \supset \cup\, \sigma$. The proof is finished.

PROPOSITION 3. *Any topological space* (X,τ) *with a countable base is separable.*

Proof. Let β be a countable base of (X,τ). For each nonempty $B \in \beta$ let $x_B \in B$. Then the set $D = \{x_B : B \in \beta\}$ is countable and dense. The proof is finished.

If (X,τ) and (Y,τ_1) are topological spaces, and $f\colon X \to Y$, then f is said to be *continuous at a point* $x \in X$ if for each neighborhood V of $f(x)$ there exists a neighborhood U of x such that $f(U) \subset V$ and f is said to be *continuous on* X if f is continuous at each point of X. If f is a continuous one-to-one map of X onto Y, and if the inverse function f^{-1} is also continuous, then f is called a *homeomorphism* and the spaces (X,τ) and (Y,τ_1) are said to be *homeomorphic*.

PROPOSITION 4. *Let* (X,τ) *and* (Y,τ_1) *be topological spaces and let* f *be a function from* X *into* Y. *Then* f *is continuous on* X *if and only if* $f^{-1}(V) = \{x : f(x) \in V\}$ *is open in* X *whenever* V *is open in* Y.

Proof. Assume that f is continuous on X and let V be open in Y. We wish to show that $f^{-1}(V)$ is open in X. For $x \in f^{-1}(V)$ there exists a neighborhood U_x of x such that $f(U_x) \subset V$, i.e., $U_x \subset f^{-1}(V)$, because f is continuous at x. It follows that

$$f^{-1}(V) = \cup\, \{U_x : x \in F^{-1}(V)\}$$

which is a union of open sets, so that $f^{-1}(V)$ is open. Thus $f^{-1}(V) \in \tau$ for every $V \in \tau_1$.

Conversely, suppose that $f^{-1}(V) \in \tau$ whenever $V \in \tau_1$. Let $x \in X$ and let V be a neighborhood of $f(x)$. Then $f^{-1}(V)$ is a neighborhood of x and $f(f^{-1}(V)) \subset V$. Thus f is continuous at x. Since x is an arbitrary point of X, f is seen to be continuous on X. The proof is complete.

Remark. An immediate corollary of the foregoing proposition is that f is continuous on X if and only if the inverse image under f of each closed set in Y is a closed set in X.

PROPOSITION 5. *Let (X,τ) and (Y,τ_1) be topological spaces and let f be a function from X into Y. Suppose that α is a subbase for the topology τ_1 of Y and such that $f^{-1}(A) \in \tau$ for every $A \in \alpha$. Then f is continuous on X.*

Proof. Let β be the family of all sets of the form

$$B = \bigcap_{k=1}^{n} A_k,$$

where $\{A_1, \ldots, A_n\}$ is a finite subfamily of α. Then β is a base for the topology τ_1 of Y and the set

$$f^{-1}(B) = \bigcap_{k=1}^{n} f^{-1}(A_k),$$

being a finite intersection of open sets, is open for every $B \in \beta$. Now, let $V \in \tau_1$. Then $V = \cup \{B_t : t \in T\}$ for some family $\{B_t\}_{t \in T} \subset \beta$. Thus

$$f^{-1}(V) = f^{-1}\left(\bigcup_{t \in T} B_t\right) = \bigcup_{t \in T} f^{-1}(B_t)$$

which, being a union of open sets, is open in X. The proof is finished.

PROPOSITION 6. *Let X, Y, Z be topological spaces, and let f: $X \to Y$ and g: $Y \to Z$ be continuous. Then the composite function $g \circ f$ is continuous.*

Proof. We shall show: For $x \in X$, if f is continuous at x and g is continuous at $f(x)$, then $g \circ f$ is continuous at x.

Let W be any neighborhood of $(g \circ f)(x) = g(f(x))$. Then there is a neighborhood V of $f(x)$ with $g(V) \subset W$. Since f is continuous at x, there is a neighborhood U of x such that $f(U) \subset V$. Thus we have obtained a neighborhood U of x such that

$$(g \circ f)(U) = g(f(U)) \subset g(V) \subset W$$

and the proof is finished.

PROPOSITION 7. *Let (X,τ) and (Y,τ_1) be topological spaces and let f be a continuous function from X into Y. Let $E \subset X$. Then the function f with its domain of definition restricted to E is a continuous function from E (with its relative topology) into Y.*

Proof. Let $x \in E$ and let V be a neighborhood of $f(x)$. Then there is a neighborhood U (with $U \in \tau$) of x such that $f(U) \subset V$. But then $U \cap E$

is a neighborhood of x in the relative topology on E and $f(U \cap E) \subset f(U) \subset V$. This ends the proof.

Remark. We note in passing that if f and g are continuous real-valued functions on a topological space (X,τ), then the functions given by the expressions

$$|f(x)|, \quad f(x) + g(x), \quad cf(x) \quad (\text{with } c \in R^1),$$
$$\max\{f(x), g(x)\}, \quad \min\{f(x), g(x)\}$$

are continuous functions on X as well.

Consider the following set of "separation axioms" for a topological space (X,τ):

T_1 : Given two distinct points x and y, there is an open set which contains y but not x.

T_2 : Given two distinct points x and y, there are disjoint open sets U_1 and U_2 such that $x \in U_1$ and $y \in U_2$.

T_3 : In addition to T_1, given a closed set F and a point x not in F, there are disjoint open sets U_1 and U_2 such that $x \in U_1$ and $F \subset U_2$.

T_4 : In addition to T_1, given two disjoint closed sets F_1 and F_2, there are disjoint open sets U_1 and U_2 such that $F_1 \subset U_1$ and $F_2 \subset U_2$.

Remark. Condition T_1 is equivalent to the statement that each set consisting of a single point is closed. Indeed, if each set $\{x\}$ is closed, given two distinct points x and y, we take U to be the complement of $\{x\}$ in X. Then U is an open set containing y but not x. On the other hand, suppose that condition T_1 holds. Each $y \in X - \{x\}$ is contained in an open set $U \subset X - \{x\}$. Thus the set $X - \{x\}$ is the union of the open sets contained in it and therefore must be open and so $\{x\}$ is closed.

In view of the foregoing remark we see that condition T_{i+1} implies condition T_i for $i = 1, 2, 3$.

A topological space which satisfies condition T_2 is said to be a *Hausdorff* space, one which satisfies T_3 is said to be a *regular* space, and one which satisfies T_4 is called a *normal* space.

PROPOSITION 8 (URYSOHN'S LEMMA). *Let A and B be disjoint closed sets in a normal topological space X. Then there is a continuous real-valued function f defined on X such that* $0 \leqq f(x) \leqq 1, f(A) = 0, f(B) = 1$.

Proof. Let D denote the set of all nonnegative dyadic rational numbers, that is, the set of all numbers of the form

$$\frac{a}{2^q},$$

where a and q are nonnegative integers.

We begin by constructing an indexed family $\Lambda = \{U_t : t \in D\}$ of open sets in X such that, for any two distinct $s, t \in D$, $s < t$ implies

$$\overline{U}_s \subset U_t.$$

For this purpose, we let $U_t = X$ for every $t > 1$ in D and we put $U_1 = X - B$. Since X is normal and A, B are disjoint, there exist two open sets G and H of X such that $A \subset G$, $B \subset H$, and $G \cap H = \varnothing$. We then let $U_0 = G$. This implies that

$$\overline{U}_0 \subset X - H \subset X - B = U_1.$$

Next, let $t \in D$ and $0 < t < 1$. Then t can be uniquely written in the form

$$t = \frac{2m+1}{2^n}.$$

We will construct U_t by induction on n. Let

$$\alpha = \frac{2m}{2^n} = \frac{m}{2^{n-1}},$$

$$\beta = \frac{2m+2}{2^n} = \frac{m+1}{2^{n-1}}.$$

Then $\alpha < t < \beta$ and, by the inductive hypothesis, the open sets U_α and U_β have already been constructed in such a way that

$$\overline{U}_\alpha \subset U_\beta.$$

Hence $\overline{U}_\alpha$ and $X - U_\beta$ are two disjoint closed sets in X. Since X is normal, there exist two open sets V and W of X such that

$$\overline{U}_\alpha \subset V, \quad X - U_\beta \subset W, \quad V \cap W = \varnothing.$$

Letting $U_t = V$, we obtain

$$\overline{U}_\alpha \subset U_t, \quad \overline{U}_t \subset X - W \subset U_\beta.$$

This completes the inductive construction of the indexed family $\Lambda = \{U_t : t \in D\}$ of open sets in X. From the construction it is clear that Λ covers X.

We now define a real-valued function $f\colon X \to [0, 1]$ by taking

$$f(x) = \inf\{t : x \in U_t\}$$

for each $x \in X$. Then we have $f(A) = 0$, $f(B) = 1$. It only remains to verify the continuity of this function f on X. We do this now.

For each real number $a \in [0, 1]$, let

$$L_a = \{t \in [0, 1] : t < a\}$$

and

$$R_a = \{t \in [0, 1] : t > a\}.$$

Then the collection $\{L_a, R_a : a \in [0, 1]\}$ forms a subbasis of the topology of $[0, 1]$. By Proposition 5 it is enough to prove that, for each $a \in [0, 1]$, $f^{-1}(L_a)$ and $f^{-1}(R_a)$ are open in X.

To see that the set $f^{-1}(L_a) = \{x \in X : f(x) < a\}$ is open, we proceed as follows. Since $f(x) = \inf\{t : x \in U_t\}$ and since the infimum is less than a if and only if some member of $\{t : x \in U_t\}$ is less than a, the set $f^{-1}(L_a)$ consists of all points $x \in X$ such that $x \in U_t$ for some $t < a$. Hence

$$f^{-1}(L_a) = \cup \{U_t : t \in D \text{ and } t < a\}.$$

This implies that $f^{-1}(L_a)$ is an open set of X.

To see that $f^{-1}(R_a)$ is open as well, it is enough to verify that

$$f^{-1}([0, 1] - R_a) = \{x \in X : f(x) \leqq a\}$$

is closed. Since $f(x) = \inf\{t : x \in U_t\}$, we see that $f(x) \leqq a$ if and only if $x \in U_t$ for every $t > a$ in D. Thus

$$f^{-1}([0, 1] - R_a) = \cap \{U_t : t \in D \text{ and } t > a\}.$$

We now show that

$$f^{-1}([0, 1] - R_a) = \cap \{\overline{U}_t : t \in D \text{ and } t > a\}.$$

Let r be any member of D with $r > a$. Since D is dense in $[0, 1]$, there is an $s \in D$ such that $a < s < r$ and hence

$$\overline{U}_s \subset U_r.$$

Since $s > a$, we get that

$$\cap \{\overline{U}_t : t \in D \text{ and } t > a\} \subset \overline{U}_s \subset U_r.$$

Since r is an arbitrary member of D with $r > 0$, this implies that

$$\cap \{\overline{U}_t : t \in D \text{ and } t > a\} \subset f^{-1}([0, 1] - R_a).$$

But the inclusion

$$f^{-1}([0, 1] - R_a) \subset \cap \{\overline{U}_t : t \in D \text{ and } t > a\}$$

is obvious and so the proof is finished.

Remark. If X is a topological space and for each pair of disjoint closed sets A, B in X there exists a continuous function $f: X \to [0, 1]$ such that $f(A) = 0$ and $f(B) = 1$, then X must be normal. Indeed, the sets $U = \{x : f(x) < 1/2\}$ and $V = \{x : f(x) > 1/2\}$ are disjoint open sets which $U \supset A$ and $V \supset B$.

PROPOSITION 9 (THE TIETZE EXTENSION THEOREM). *If f is a bounded real-valued continuous function defined on a closed set A of a normal space X, there is a continuous real-valued function F defined on X with $F(x) = f(x)$ for x in A, and*

$$\sup_{x \in X} |F(x)| = \sup_{x \in A} |f(x)|.$$

Proof. We assume that f is not identically zero on A; if $f = 0$ on A, the proposition is obvious.

Let

$$f_0(x) = f(x), \quad \alpha_0 = \sup_{x \in A} |f_0(x)|,$$

$$A_0 = \{x \in A : f_0(x) \leqq -\alpha_0/3\} \text{ and}$$

$$B_0 = \{x \in A : f_0(x) \geqq \alpha_0/3\}.$$

Then A_0 and B_0 are disjoint closed sets. Using the foregoing proposition, we can find a function F_0 defined on all of X such that $F_0(A_0) = -\alpha_0/3$, $F_0(B_0) = \alpha_0/3$, $-\alpha_0/3 \leqq F_0(x) \leqq \alpha_0/3$. Let $f_1(x) = f_0(x) - F_0(x)$ for $x \in A$. Then f_1 is continuous and $\alpha_1 \leqq (2/3)\alpha_0$, where

$$\alpha_1 = \sup_{x \in A} |f_1(x)|.$$

By applying to the pair f_1, α_1 the procedure used on f_0, α_0, and then continuing inductively, we obtain a sequence F_j, $j = 1, 2, \ldots$ of real-valued continuous functions on X such that

$$|f(x) - \sum_{j=0}^{n} F_j(x)| \leqq \left(\frac{2}{3}\right)^{n+1} \alpha_0, \qquad x \in A,$$

and

$$\sup_{x \in X} |F_n(x)| \leqq \frac{1}{3}\left(\frac{2}{3}\right)^{n} \alpha_0 .$$

This shows that the series

$$\sum_{n=0}^{\infty} F_n(x)$$

converges and defines a function F on X which coincides with f on A. To see that F is continuous, let $\varepsilon > 0$ and fix n so that $2\alpha_0(2/3)^{n+1} < \varepsilon/2$. Then

$$\begin{aligned} |F(x) - F(y)| &\leqq |F(x) - \sum_{j=0}^{n} F_j(x)| \\ &\quad + |\sum_{j=0}^{n} F_j(x) - \sum_{j=0}^{n} F_j(y)| + |\sum_{j=0}^{n} F_j(y) - F(y)| \\ &\leqq 2\alpha_0\left(\frac{2}{3}\right)^{n+1} + \sum_{j=0}^{n} |F_j(x) - F_j(y)| \\ &< \frac{\varepsilon}{2} + \sum_{j=0}^{n} |F_j(x) - F_j(y)|. \end{aligned}$$

But there is a neighborhood V of x such that

$$\sum_{j=0}^{n} |F_j(x) - F_j(y)| < \frac{\varepsilon}{2}, \qquad y \in V,$$

(observe the Remark following Proposition 7) and thus $|F(x) - F(y)| < \varepsilon$ for $y \in V$. This proves the proposition.

PROPOSITION 10. *A real-valued continuous function f defined on a closed subset of a normal space X has a real-valued continuous extension defined on all of X.*

Proof. We need only consider the case where f is unbounded. If f is real-valued and continuous on the closed set A in the normal space X, the bounded function arc tan $f(x)$ has a continuous extension $g(x)$ defined on X. The closed sets A and

$$B = \left\{x : g(x) = \frac{\pi}{2}\right\}$$

are disjoint and hence there is a continuous function h with $0 \leqq h(x) \leqq 1$, which vanishes on B, and has the constant value 1 on A. The function tan $h(x)g(x)$ is therefore a continuous extension of f and the proof is complete.

Remarks. Let X be the closed unit interval $[0, 1]$, A the set $(0, 1]$ and f the function defined on A by $f(x) = \sin(1/x)$. X is clearly normal but F is not a closed subset of X and f cannot be extended continuously to X in any manner whatsoever.

We have used Urysohn's Lemma to prove Tietze's Extension Theorem. This process can be reversed and Urysohn's Lemma can be deduced from Tietze's Extension Theorem.

Let (X,τ) be a topological space. A *cover* of X is any family α of subsets of X such that $\cup\ \alpha = X$. A cover in which each member is an open set is said to be an *open cover*. A subfamily of a cover which is also a cover is called a *subcover*.

A topological space (X,τ) is said to be a *compact* space if each open cover of X admits a finite subcover.

A family of sets is said to have the *finite intersection property* if each finite subfamily has nonempty intersection.

PROPOSITION 11. *A topological space (X,τ) is compact if and only if each family of closed subsets of X having the finite intersection property has nonempty intersection.*

Proof. Λ is an open cover of X if and only if $\mathcal{G} = \{U^c : U \in \Lambda\}$ is a family of closed sets with empty intersection. Hence every open cover has a finite subcover if every family of closed sets having empty intersection has a finite subfamily with empty intersection. This completes the proof.

PROPOSITION 12. *Let (X,τ) be a Hausdorff space and let A be a subspace of X that is compact in its relative topology τ_A. Then A is a closed subset of (X,τ).*

Proof. We show that A^c is open. Let $y \in A^c$. For each $x \in A$ we choose disjoint open sets U_x and V_x such that $x \in U_x$, $y \in V_x$. Then

$$\{U_x \cap A : x \in A\}$$

is an open cover of A, in its relative topology; so there exists a finite set $\{x_1, \ldots, x_n\} \subset A$ such that

$$A \subset \bigcup_{j=1}^{n} U_x .$$

Let

$$V = \bigcap_{j=1}^{n} V_{x_j} .$$

Then V is a neighborhood of y and $V \cap A = \varnothing$, which means that $V \subset A^c$ and the proposition is proved.

PROPOSITION 13. *If* (X,τ) *is a compact space and* A *is a closed subset of* X, *then* A *is a compact subspace of* X.

Proof. Let $\mathcal{G}$ be any family of closed (in the relative topology) subsets of A having the finite intersection property. Then each member of $\mathcal{G}$ is closed in X and, by Proposition 11, $\cap\, \mathcal{G} \neq \varnothing$. Therefore A is compact by Proposition 11. This ends the proof.

PROPOSITION 14. *A compact Hausdorff space is normal.*

Proof. Let A be a closed subset of the compact Hausdorff space X, and let $p \notin A$. Then, if $q \in A$, there is a neighborhood U_q of q and a neighborhood V_q of p such that $U_q \cap V_q = \varnothing$. Since A is compact, a finite set $U_{q_1}, \ldots, U_{q_n}$ covers A and

$$(U_{q_1} \cup \ldots \cup U_{q_n}) \cap (V_{q_1} \cap \ldots \cap V_{q_n}) = \varnothing .$$

Thus any compact set A and any point $p \notin A$ have disjoint neighborhoods. Let A and B be closed and disjoint. Then if $p \in A$, there is a neighborhood U_p of p, and a neighborhood V_p of B, such that $U_p \cap V_p = \varnothing$. Then a finite U_{p_i} will cover A, and the sets

$$U_{p_1} \cup U_{p_2} \cup \ldots \cup U_{p_m}$$

and

$$V_{p_1} \cap V_{p_2} \cap \ldots \cap V_{p_m}$$

are disjoint neighborhoods of A and B, respectively. This proves the proposition.

A topological space X is said to be *locally compact* if each point $x \in X$ has a neighborhood U such that $\overline{U}$ is compact.

Let X be a locally compact Hausdorff space and let x_∞ be an object not in X, and form the set $X_\infty = X \cup \{x_\infty\}$. We define a topology on X_∞ by specifying the following as open sets:

(i) the open subsets of X, regarded as subsets of X_∞,
(ii) the complements in X_∞ of the compact subspaces of X, and
(iii) the full set X_∞.

Using Proposition 12, it is easy to show that this class of sets actually is a topology on X_∞, and also that the given topology on X equals its relative topology as a subspace of X_∞.

PROPOSITION 15. *The topological space X_∞ associated with the locally compact Hausdorff space X in the manner described above is a compact Hausdorff space.*

Proof. First we show that X_∞ is compact. Let $\{G_t : t \in T\}$ be an open cover of X_∞. We must find a finite subcover. If X_∞ occurs among the G_t's, then $\{G_t : t \in T\}$ clearly has a finite subcover, namely $\{X_\infty\}$. We may therefore assume that each G_t is a set of type (i) or type (ii). At least one G_t, say G_{t_0}, must contain the point x_∞, and this set is necessarily of type (ii). Its complement $G_{t_0}^c$ is thus a compact subspace of X which is contained in the union of some class of open subsets of X of the form $G_t \cap X$, so it is contained in the union of some finite subclass of these sets, say $\{G_1 \cap X, G_2 \cap X, \ldots, G_n \cap X\}$. It can be seen therefore that the class $\{G_{t_0}, G_1, G_2, \ldots, G_n\}$ is a finite subcover of the original open cover of X_∞ and so X_∞ is compact.

Next, we show that X_∞ is Hausdorff. Since X is Hausdorff, any pair of distinct points in X both of which are situated in X can be separated by open subsets of X, and thus can be separated by open subsets of X_∞ of type (i). It is therefore enough to show that any point x in X and the point x_∞ can be separated by open subsets of X_∞. X is locally compact, so x has a neighborhood G whose closure $\overline{G}$ in X is compact. Evidently G and $\overline{G}^c$ are disjoint open subsets of X_∞ such that $x \in G$ and $x_\infty \in \overline{G}^c$, so X_∞ is compact. This proves the proposition.

The compact Hausdorff space X_∞ associated with the locally compact Hausdorff space X in the manner described above is called the *one-point compactification* of X, and the point x_∞ is called the *point at infinity*.

The one-point compactification is useful in simplifying proofs of theorems about locally compact Hausdorff spaces. As a case in point, we consider the following locally compact version of Urysohn's Lemma.

PROPOSITION 16. *Let X be a locally compact Hausdorff space, let A be a compact subspace of X, and let U be an open set such that $A \subset U$. Then there exists a continuous function from X into $[0, 1]$ such that $f(x) = 1$ for all $x \in A$ and $f(x) = 0$ for all $x \in U^c$.*

Proof. To see the validity of the proposition we only need to apply

Urysohn's Lemma (see Proposition 8) to the closed sets C and $X_\infty - U$ of the compact Hausdorff space X_∞ which is normal by Proposition 14. This ends the proof.

Remark. A Hausdorff space is said to be *completely regular* if, for every point y in X and every closed set F not containing y, there is a continuous $f: X \to [0, 1]$ such that $f(y) = 0$ and, for x in F, $f(x) = 1$. Proposition 16 shows that a locally compact Hausdorff space is completely regular. Indeed, take $U = \{y\}^c$ and $C = F$ and everything is clear.

PROPOSITION 17. *Let (X,τ) and (Y, τ_1) be topological spaces and f be a function from X into Y. Suppose that X is compact and that f is continuous on X. Then $f(X)$ is a compact subspace of Y.*

Proof. Let $\mathcal{S}$ be an open cover of $f(X)$. Then $\{f^{-1}(S) : S \in \mathcal{S}\}$ is an open cover of X and so there exist $S_1, \ldots, S_n \in \mathcal{S}$ such that

$$X = \bigcup_{k=1}^{n} f^{-1}(S_k) = f^{-1}\Big(\bigcup_{k=1}^{n} S_k\Big).$$

It follows therefore that

$$f(X) \subset \bigcup_{k=1}^{n} S_k.$$

The proof is finished.

A corollary of the foregoing proposition is that every real-valued continuous function on a compact space attains its supremum and infimum.

PROPOSITION 18. *A continuous one-to-one function from a compact space to a Hausdorff space is a homeomorphism.*

Proof. Let X be a compact space, Y a Hausdorff space, and f a one-to-one continuous function on X, with $f(X) = Y$. By Proposition 13, a closed set A in X is a compact subspace and its continuous image $f(A)$ is compact (by Proposition 17), and, since Y is Hausdorff, $f(A)$ is closed (by Proposition 12). Thus f^{-1} is continuous by the Remark following Proposition 4, and the proof is finished.

PROPOSITION 19. *If a set X is a Hausdorff space with respect to a topology τ_1 and a compact space with respect to a topology τ_2 and if $\tau_1 \subset \tau_2$, then $\tau_1 = \tau_2$.*

Proof. Since $\tau_1 \subset \tau_2$, the identity map $f(x) = x$ of the compact space (X, τ_2) onto the Hausdorff space (X, τ_1) is continuous and so, by Proposition 18, f is a homeomorphism. This means however that $\tau_1 = \tau_2$ and the proof is complete.

We now digress briefly into set theory for some terminology and facts.

The *ordered pair* (x,y) is the set $\{\{x\},\{x,y\}\}$. If X and Y are sets, the *Cartesian product of* X *and* Y is the set $X \times Y$ of all ordered pairs (x,y) such that $x \in X$ and $y \in Y$. A *relation* is any set of ordered pairs.

Let X be a set. A *partial ordering on* X is any relation $\leqq \subset X \times Y$ satisfying

(i) $x \leqq x$ (reflexive);
(ii) $x \leqq y$ and $y \leqq x$ imply $x = y$ (antisymmetric);
(iii) $x \leqq y$ and $y \leqq z$ imply $x \leqq z$ (transitive).

If $\leqq$ also satisfies

(iv) $x, y \in X$ implies $x \leqq y$ or $y \leqq x$ (trichotomy), then $\leqq$ is called a *linear ordering of* X.

If $\leqq$ is a linear ordering such that

(v) $\varnothing \neq Y \subset X$ implies the existence of an element $y \in Y$ such that $y \leqq x$ for each $x \in Y$ (y is the *smallest element* of Y), then $\leqq$ is called a *well-ordering on* X.

A *partially ordered set* is an ordered pair $(X,\leqq)$ where X is a set and $\leqq$ is a partial ordering on X. If $\leqq$ is a linear ordering (resp. well-ordering), then $(X,\leqq)$ is called a *linearly ordered* (resp. *well-ordered*) set.

Let $(X,\leqq)$ be a partially ordered set and $Y \subset X$. An element $u \in X$ is called an *upper bound for* Y if $x \leqq u$ for each $x \in Y$. An element $m \in X$ is called a *maximal element of* Y if $x \in X$ and $m \leqq x$ implies $x \leqq m$. Similarly we define *lower bound* and *minimal element*. An upper bound v of Y is said to be a *least upper bound*, or *supremum, of* Y if every upper bound u of Y has the property that $v \leqq u$. Similarly we define *greatest lower bound*, or *infimum*. A *chain* in X is any subset C of X such that C is linearly ordered under the given order relation $\leqq$ on X.

A partially ordered set $(X,\leqq)$ is said to be a *lattice* if every pair x, y of elements of X has a supremum and infimum, denoted by $x \vee y$ and $x \wedge y$, respectively. Let $(X,\leqq)$ be a lattice and Y a nonempty subset of X with the property that if x and y are in Y, then $x \vee y$ and $x \wedge y$ are also in Y; then Y is called a *sublattice* of X; note that the operations $\vee$ and $\wedge$ are relative to the partial ordering $\leqq$ in $(X,\leqq)$. For example, the configuration (see figure on next page) is a lattice X, but the configuration Y with the points c and d removed from X is not a sublattice of X.

Let $\{A_t\}_{t\in T}$ be any family of sets. The *Cartesian product* of this family, written

$$\bigtimes_{t\in T} A_t,$$

is the set of all functions x having domain of definition T such that $x_t = x(t) \in A_t$ for each $t \in T$. Each such function x is called a *choice function* for the family $\{A_t\}_{t \in T}$. For

$$x \in \mathop{\times}_{t \in T} A_t$$

and $t \in T$, the value $x_t \in A_t$ is called the t-th *coordinate of* x.

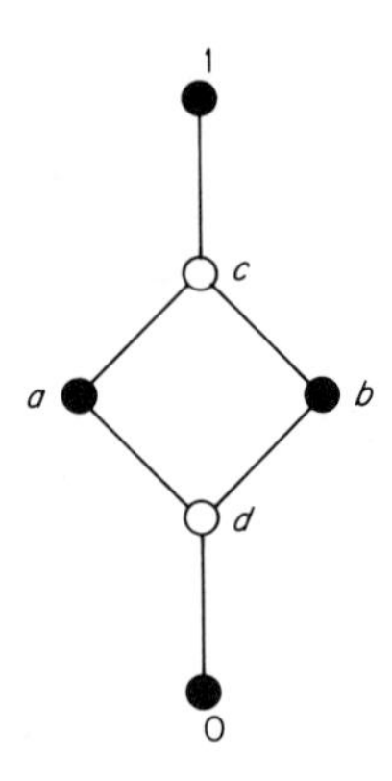

Axiom of Choice. *The Cartesian product of any nonempty family of nonempty sets is a nonempty set, i.e., if $\{A_t\}_{t \in T}$ is a family of sets such that $T \neq \varnothing$ and $A_t \neq \varnothing$ for each $t \in T$, then there exists at least one choice function for the family $\{A_t\}_{t \in T}$.*

Well-ordering Theorem. *Every set can be well-ordered, i.e., if X is a set, then there is some well-ordering on X.*

Zorn's Lemma. *Every nonempty partially ordered set in which each chain has an upper bound has a maximal element.*

One can show that the Axiom of Choice, the Well-ordering Theorem and Zorn's Lemma are pairwise equivalent; we shall not go into this matter, however.

For later use we note the following consequence of the Well-ordering Theorem:

There is an uncountable set X which is well-ordered by the relation $\leqq$ in such a way that

(i) there is a greatest element Ω in X, i.e., $\Omega \in X$ and $x \leqq \Omega$ for all $x \in X$, and
(ii) if $x \in X$ and $x \neq \Omega$, then the set $\{y \in X : y \leqq x\}$ is countable.

It can be shown that this set X is unique in the sense that, if Y is any other well-ordered set with the same properties, then there is an one-to-one order preserving correspondence between X and Y. The greatest element Ω in X is called the *smallest nondenumerable ordinal.* In defining the Baire functions (see Section 2 of Chapter 9 for example) the smallest nondenumerable ordinal does enter.

[The student keen on learning the basics of set theory will find the following two books very interesting and helpful: P. R. Halmos, *Naive Set Theory*, Van Nostrand, 1960; L. E. Sigler, *Exercises in Set Theory*, Van Nostrand, 1966.]

We now turn to an application of Zorn's Lemma.

PROPOSITION 20 (THE THEOREM OF ALEXANDER). *Let* (X,τ) *be a topological space and let* β *be any subbase for the topology* τ *of* X. *Then the following two statements are equivalent*:

(*i*) *The space* (X,τ) *is compact.*
(*ii*) *Every cover of* X *by a subfamily of* β *admits a finite subcover.*

Proof. It is clear that (i) implies (ii). To show the converse, suppose that (ii) holds and (i) fails. Consider the family $\mathfrak{R}$ of all open covers of X without finite subcovers. The family $\mathfrak{R}$ is partially ordered by inclusion and the union of a nonempty chain in $\mathfrak{R}$ is a cover in $\mathfrak{R}$. Zorn's Lemma implies that $\mathfrak{R}$ contains a maximal cover $\mathfrak{S}$. That is, $\mathfrak{S}$ is an open cover of X, $\mathfrak{S}$ has no finite subcover, and if U is any open set not in $\mathfrak{S}$, then $\mathfrak{S} \cup \{U\}$ admits a finite subcover. Let $\mathfrak{N} = \mathfrak{S} \cap \beta$. Then no finite subfamily of $\mathfrak{N}$ covers X, and so (ii) implies that $\mathfrak{N}$ is not a cover of X. Let x be a point in $X \cap \{\cup \mathfrak{N}\}^c$, and select a set V in the cover $\mathfrak{S}$ that contains x. Since β is a subbase, there are sets $S_1, \ldots, S_n$ in β such that

$$x \in \bigcap_{j=1}^{n} S_j \subset V.$$

Since $x \notin (\cup \mathfrak{N})$, no S_j is in $\mathfrak{S}$. Since $\mathfrak{S}$ is maximal, there exists for each j a set A_j which is the union of a finite number of sets in $\mathfrak{S}$ such that

$$S_j \cup A_j = X.$$

Hence

$$V \cup \bigcup_{j=1}^{n} A_j \supset (\bigcap_{j=1}^{n} S_j) \cup (\bigcup_{j=1}^{n} A_j) = X,$$

and therefore X is a union of finitely many sets from $\mathfrak{S}$. This contradicts the choice of $\mathfrak{S}$, however, and the proof is finished.

Let $\{X_\iota\}_{\iota \in T}$ be a nonempty family of topological spaces and let

$$X = \bigtimes_{\iota \in T} X\mathrm{v}_\iota$$

be the Cartesian product of this family, i.e., the set of all functions x having domain of definition T such that $x_t = x(t) \in X_t$ for each $t \in T$. For each $t \in T$ define π_t on X by $\pi_t(x) = x_t$. The function π_t is called the *projection of X onto X_t*. We define the *product topology on the set X* by using as a subbase the family of all sets of the form

$$\pi_t^{-1}(U_t),$$

where t runs through T and U_t runs through the open sets of X_t. A base for the product topology is therefore the family of all finite intersections of inverse projections of open sets. A base for the product topology is the family of all sets of the form

$$\bigtimes_{t \in T} U_t,$$

where U_t is open in X_t for each $t \in T$ and $U_t = X_t$ for all but a finite number of the t's.

PROPOSITION 21 (TYCHONOFF'S THEOREM). *Let $\{X_t\}_{t \in T}$ be a nonempty family of compact topological spaces. Then the Cartesian product X of these spaces is compact in the product topology.*

Proof. By Proposition 20 it is enough to consider open covers of X by subbasic open sets as described further above. Let Λ be any cover of X by subbasic open sets. For each $t \in T$, let Λ_t denote the family of all open sets $U \subset X_t$ such that $\pi_t^{-1}(U) \in \Lambda$. We claim that $\cup\, \Lambda_t = X_t$ for some $t \in T$. If this were not the case, there would be a point $x \in X$ such that for every $t \in T$, $\pi_t(x) = x_t \in X_t \cap (\cup\, \Lambda_t)^c$; hence $x \notin \pi_t^{-1}(U)$ for all $\pi_t^{-1}(U) \in \Lambda_t$. That is, Λ would not be a cover of X. Hence we can choose a $t_0 \in T$ such that $\cup\, \Lambda_{t_0} = X_{t_0}$. Since X_{t_0} is compact, there is a finite family $\{U_1, \ldots, U_n\} \subset \Lambda_{t_0}$ such that $X_{t_0} = U_1 \cup U_2 \cup \ldots \cup U_n$. Clearly $\{\pi_{t_0}^{-1}(U_j) : j = 1, \ldots n\}$ is a finite subcover of Λ for X. The proof is finished.

Let X be a set, and let ρ be a real-valued function on $X \times X$ such that for all $x, y, z \in X$ we have

(i) $\rho(x,y) \geqq 0$,
(ii) $\rho(x,y) = 0$ if and only if $x = y$,
(iii) $\rho(x,y) = \rho(y,x)$, and
(iv) $\rho(x,y) \leqq \rho(x,z) + \rho(z,y)$ [the *triangle inequality*].

Then ρ is called a *metric for X* and $\rho(x,y)$ is called the *distance from x to y*. The pair (X,ρ) is called a *metric space*.

Let (X,ρ) be a metric space. For $\varepsilon > 0$ and $x \in X$, we call the set

$$S(x,\varepsilon) = \{y \in X : \rho(x,y) < \varepsilon\}$$

the *ε-neighborhood of* x or the *open sphere of radius* ε *centered at* x.

PROPOSITION 22. *Let* (X,ρ) *be a metric space and let*

$$\beta = \{S(x,\varepsilon) : \varepsilon > 0, x \in X\}.$$

Then β *is a base for a topology* τ *for* X. *We call this topology the metric topology.*

Proof. We show that β satisfies properties (i) and (ii) of Proposition 1. Property (i) of Proposition 1 clearly holds. To see that property (ii) holds as well, we take $S(x,\varepsilon)$ and $S(y,\delta)$ in β and we let $z \in S(x,\varepsilon) \cap S(y,\delta)$. Then $\rho(x,z) < \varepsilon$ and $\rho(y,z) < \delta$. Let

$$\gamma = \min\{\varepsilon - \rho(x,z), \delta - \rho(y,z)\}.$$

Thus γ is positive, and for $v \in S(z,\gamma)$, we have $\rho(x,v) \leqq \rho(x,z) + \rho(z,v) < (\varepsilon - \gamma) + \gamma = \varepsilon$ and $\rho(y,v) \leqq \rho(y,z) + \rho(z,v) < (\delta - \gamma) + \gamma = \delta$. This shows that $S(z,\gamma) \subset S(x,\varepsilon) \cap S(y,\delta)$ and the proof is finished.

PROPOSITION 23. *A separable metric space* (X,ρ) *has a countable base.*

Proof. Let (X,ρ) be a separable metric space containing a countable dense subset D. Let

$$\beta = \{S(x,r) : x \in D, r > 0, r \text{ rational}\}.$$

Then β is countable. To verify that β is a base, let U be open and let $z \in U$. Then there is an $\varepsilon > 0$ such that $S(z,\varepsilon) \subset U$. Since D is dense in (X,ρ) there is an

$$x \in S\left(z,\frac{\varepsilon}{3}\right) \cap D.$$

We now pick a rational number r such that $(1/3)\varepsilon < r < (2/3)\varepsilon$. For $y \in S(x,r)$ we then have

$$\rho(y,z) \leqq \rho(x,y) + \rho(x,z) < r + \frac{1}{3}\varepsilon < \varepsilon$$

so that

$$S(x,r) \subset S(z,\varepsilon) \subset U.$$

But $z \in S(x,r)$ because $\rho(x,z) < (1/3)\varepsilon < r$. Thus U is a union of members of β. The proof is complete.

A sequence $(x_n)_{n=1}^{\infty}$ in a topological space is said to *converge to a point* x in the space if for each neighborhood U of x there exists a natural number n_0 such that $x_n \in U$ whenever $n \geqq n_0$.

A sequence $(x_n)_{n=1}^{\infty}$ in a metric space (X,ρ) is said to be a *Cauchy sequence* if for each $\varepsilon > 0$ there exists a natural number n_0 such that

$\rho(x_n, x_m) < \varepsilon$ whenever $m, n \geqq n_0$. A metric space (X,ρ) is said to be *complete* if each Cauchy sequence in (X,ρ) converges to a point of (X,ρ).

Let X be a nonempty topological space and let $\mathfrak{C}$ denote the set of all real-valued bounded continuous functions on X. It is easy to see that

$$\rho(f,g) = \sup\{|f(x) - g(x)| : x \in X\}$$

is a metric for $\mathfrak{C}$; in fact, this ρ is the metric of uniform convergence.

PROPOSITION 24. *$(\mathfrak{C},\rho)$ is a complete metric space.*

Proof. Let $(f_n)_{n=1}^{\infty}$ be a Cauchy sequence in $(\mathfrak{C},\rho)$, i.e.,

$$\lim_{n,m\to\infty}[\sup\{|f_n(x) - f_m(x)| : x \in X\}] = 0. \tag{1}$$

For fixed $x \in X$, (1) implies

$$\lim_{n,m\to\infty}|f_n(x) - f_m(x)| = 0,$$

and so $(f_n(x))_{n=1}^{\infty}$ is a Cauchy sequence in R^1. But R^1 is complete under its usual metric topology and so

$$\lim_{n\to\infty} f_n(x)$$

exists; we denote it by $f(x)$. The mapping $x \to f(x)$ is thus a function from X into R^1. We now show that

$$\lim_{n\to\infty} \rho(f,f_n) = 0. \tag{2}$$

Let $\varepsilon > 0$ be arbitrary and let the natural number p (depending only on ε) be so large that

$$\rho(f_n,f_m) < \frac{\varepsilon}{3} \tag{3}$$

for all $m, n \geqq p$. Take a fixed but arbitrary $x \in X$ and choose m (depending on both x and ε) so large that $m \geqq p$ and also

$$|f_m(x) - f(x)| < \frac{\varepsilon}{3}. \tag{4}$$

Combining (3) and (4) we obtain that

$$|f_n(x) - f(x)| \leqq |f_m(x) - f_n(x)| + |f_m(x) - f(x)| < \frac{\varepsilon}{3} + \frac{\varepsilon}{3} = \frac{2}{3}\varepsilon \tag{5}$$

for all $n \geqq p$. But p is independent of x and x is arbitrary in (5); hence we may take the supremum in (5) and get, for $n \geqq p$

$$\rho(f, f_n) \leqq \frac{2}{3}\varepsilon < \varepsilon.$$

This means that (2) holds.

We now verify that $f \in (\mathfrak{C}, \rho)$. Pick n so that $\rho(f, f_n) < 1$. Then

$$|f(x)| < |f_n(x)| + 1$$

for all $x \in X$. Thus f is bounded and its bound does not exceed $1 + \sup\{|f_n(x)| : x \in X\}$. To show that f is continuous, let x be any point in X, let $\varepsilon > 0$ be arbitrary and let n be so large that $\rho(f, f_n) < \varepsilon/2$. Let U be a neighborhood of x such that

$$|f_n(y) - f_n(x)| < \frac{\varepsilon}{3}$$

for all $y \in U$. If $y \in U$, we thus have that

$$\begin{aligned} |f(y) - f(x)| &\leqq |f(y) - f_n(y)| + |f_n(y) - f_n(x)| + |f_n(x) - f(x)| \\ &\leqq \rho(f, f_n) + |f_n(y) - f_n(x)| + \rho(f, f_n) < \frac{\varepsilon}{3} + \frac{\varepsilon}{3} + \frac{\varepsilon}{3} \\ &= \varepsilon. \end{aligned}$$

But this means that f is continous at x. The proof is finished.

PROPOSITION 25. *A subset A of a metric space (X, ρ) is closed if and only if whenever $(x_n)_{n=1}^{\infty}$ is a sequence with $x_n \in A$, $n = 1, 2, \ldots$, and $(x_n)_{n=1}^{\infty}$ has limit x in X, we have $x \in A$.*

Proof. Assume that A is closed and let $(x_n)_{n=1}^{\infty}$ be a sequence with $x_n \in A$, $n = 1, 2, \ldots$, for which a limit x in X exists. If x were in A^c, then A^c would be a neighborhood of x, and so all but a finite number of the terms of the sequence would be situated in A^c; but this is a contradiction.

Conversely, assume that A is not closed. Then A has a limit point x such that $x \notin A$. For each $n \in N$, select

$$x_n \in A \cap S\left(x, \frac{1}{n}\right).$$

Then $x_n \in A$ for $n \in N$, x_n converges to x and $x \notin A$. This completes the proof.

A subset A of a metric space (X, ρ) is said to be *bounded* if there exists $p \in X$ and $c \in R^1$ such that $\rho(p, x) \leqq c$ for all $x \in A$. If A is a nonempty bounded set in (X, ρ) we call the number

$$\sup\{\rho(x, y) : x, y \in A\}$$

the *diameter of A* and denote it by $\delta(A)$.

PROPOSITION 26 (G. CANTOR). *Let (X,ρ) be a metric space. Then (X,ρ) is complete if and only if whenever $(A_n)_{n=1}^{\infty}$ is a decreasing sequence of nonempty closed subsets of (X,ρ), i.e.,*

$$A_1 \supset A_2 \supset \dots,$$

such that

$$\lim_{n\to\infty} \delta(A_n) = 0,$$

we have

$$\bigcap_{n=1}^{\infty} A_n = \{x\}$$

for some $x \in X$.

Proof. Assume that $(A_n)_{n=1}^{\infty}$ is a decreasing sequence of nonempty closed subsets of (X,ρ) such that

$$\lim_{n\to\infty} \delta(A_n) = 0.$$

For each $n \in N$ let $x_n \in A_n$. Then $m \geqq n$ implies $\rho(x_m, x_n) \leqq \delta(A_n) \to 0$, so $(x_n)_{n=1}^{\infty}$ is a Cauchy sequence. Let

$$x = \lim_{n\to\infty} x_n.$$

For each m, $x_n \in A_m$ for all large n, and A_m is closed, so $x \in A_m$. Thus

$$x \in \bigcap_{n=1}^{\infty} A_n.$$

If

$$y \in \bigcap_{n=1}^{\infty} A_n,$$

then $\rho(x,y) \leqq \delta(A_n)$ for every n. Therefore $\rho(x,y) = 0$. Hence

$$\bigcap_{n=1}^{\infty} A_n = \{x\}.$$

To show the converse, suppose that (X,ρ) has the decreasing closed set property. Let $(x_n)_{n=1}^{\infty}$ be a Cauchy sequence in (X,ρ). For each $n \in N$, let

$$A_n = \overline{\{x_m : m \geqq n\}}.$$

Then $(A_n)_{n=1}^{\infty}$ is a decreasing sequence of closed sets and since $(x_n)_{n=1}^{\infty}$ is a Cauchy sequence,

$$\lim_{n\to\infty} \delta(A_n) = 0.$$

Let

$$\bigcap_{n=1}^{\infty} A_n = \{x\}.$$

If $\varepsilon > 0$, then there is an $n_0 \in N$ such that $\delta(A_{n_0}) < \varepsilon$. But $x \in A_{n_0}$ and thus $n \geqq n_0$ implies that $\rho(x_n, x) < \varepsilon$. The proof is finished.

PROPOSITION 27. *Let X be a metric space and Y be a topological space. Suppose that f is a function from X into Y. Then, for $x \in X$, f is continuous at x if and only if $f(x_n) \to f(x)$ whenever $(x_n)_{n=1}^{\infty}$ is a sequence in X such that $x_n \to x$.*

Proof. Suppose that $f(x_n) \to f(x)$ whenever $x_n \to x$ and assume that f is not continuous at x. Then there exists a neighborhood V of $f(x)$ such that $f(U) \subset V$ for no neighborhood U of x. For each $n \in N$, choose $x_n \in S(x, 1/n)$ such that $f(x_n) \notin V$. Then $x_n \to x$ but $f(x_n) \nrightarrow f(x)$. This contradiction establishes that f is continuous at x.

To prove the converse, suppose that f is continuous at x and let $(x_n)_{n=1}^{\infty}$ be any sequence in X such that $x_n \to x$. Let V be any neighborhood of $f(x)$. Then there exists a neighborhood U of x such that $f(U) \subset V$. Since $x_n \to x$, there exists $n_0 \in N$ such that $n \geqq n_0$ implies $x_n \in U$. Then $n \geqq n_0$ implies $f(x_n) \subset f(U) \subset V$. Consequently $f(x_n) \to f(x)$ and the proof is finished.

PROPOSITION 28(BANACH'S FIXED POINT THEOREM). *Let (X, ρ) be a nonempty complete metric space. Suppose that f is a function from X into X such that for some constant $c \in (0, 1)$ we have*

$$\rho(f(x), f(y)) \leqq c\rho(x, y)$$

for all $x, y \in X$. Then there exists a unique point $p \in X$ such that $f(p) = p$. The point p is called a fixed point and f a contraction map.

Proof. The continuity of f is obvious. Further, f cannot have more than one fixed point: $f(p) = p$, $f(p') = p'$ and $\rho(p, p') > 0$ yield the contradiction $\rho(p, p') = \rho(f(p), f(p')) \leqq c\rho(p, p') < \rho(p, p')$.

To prove that f has a fixed point, we first choose any $x \in X$ and show that the sequence

$$x, f(x) = fx, \ f(f(x)) = f^2 x, \ldots$$

of iterates is a Cauchy sequence. In fact, note that

$$\rho(fx, f^2 x) \leqq c\rho(x, fx),$$

$$\rho(f^2 x, f^3 x) \leqq c\rho(fx, f^2 x) \leqq c^2 \rho(x, fx),$$

so that, by induction,

$$\rho(f^n x, f^{n+1} x) \leqq c^n \rho(x, fx).$$

It follows that for any n and any $m > n$ we have

$$\begin{aligned}\rho(f^n x, f^m x) &\leqq \rho(f^n x, f^{n+1} x) + \ldots + \rho(f^{m-1} x, f^m x)\\ &\leqq (c^n + \ldots + c^{m-1})\rho(x, fx).\end{aligned}$$

and so (because $0 < c < 1$)

$$\rho(f^n x, f^m x) \leqq \frac{c^n}{1 - c} \rho(x, fx)$$

for all $m > n$; therefore, since $c^n \to 0$ as $n \to \infty$, we conclude that the sequence $(f^n x)_{n=1}^{\infty}$ is indeed a Cauchy sequence. But (X, ρ) is complete and so the sequence $(f^n x)_{n=1}^{\infty}$ converges to some $p \in X$. That $f(p) = p$ is seen as follows. On the one hand,

$$\lim_{n \to \infty} f^n x = p$$

and the continuity of f implies that

$$\lim_{n \to \infty} f(f^n x) = f(p);$$

on the other hand, the sequence $(f(f^n x))_{n=1}^{\infty} = (f^{n+1} x)_{n=1}^{\infty}$ is a subsequence of the Cauchy sequence $(f^n x)_{n=1}^{\infty}$ and therefore must converge to p. Thus $p = fp = f(p)$ and the proposition is proved.

Let (X, τ) be a topological space. A set $A \subset X$ is called *nowhere dense* if the interior of the closure of A is empty. A set $F \subset X$ is said to be of *first category* if F is a countable union of nowhere dense sets. Subsets of X which are not of first category are said to be of *second category*.

If A is a nowhere dense subset of a topological space (X, τ) then $X - \overline{A}$ is dense in (X, τ). Indeed, suppose that $X - \overline{A}$ is not dense, i.e., there is a point $p \in X$ and an open set G such that $p \in G$ and $G \cap (X - \overline{A}) = \varnothing$. Then $p \in G \subset \overline{A}$ and so p belongs to the interior of the closure of A, but this is a contradiction.

If G is an open subset of a metric space (X, ρ) and A is a nowhere dense set in (X, ρ), then there is an open ε-sphere $S(p, \varepsilon)$ such that $S(p, \varepsilon) \subset G$ and $S(p, \varepsilon) \cap A = \varnothing$. Indeed, let $H = G \cap (X - \overline{A})$. Then $H \subset G$ and $H \cap A = \varnothing$. Since G and $X - \overline{A}$ are open, H is open; since G is open and $X - \overline{A}$ is dense, H is nonempty. Hence there is an $\varepsilon > 0$ such that $S(p, \varepsilon) \subset H$; moreover, $S(p, \varepsilon) \subset G$ and $S(p, \varepsilon) \cap A = \varnothing$.

PROPOSITION 29 (BAIRE CATEGORY THEOREM). *A complete metric space X is of second category (as a subset of itself).*

Proof. We assume the contrary, i.e., we suppose that X is a complete metric space and

$$X = \bigcup_{n=1}^{\infty} M_n,$$

where the M_n are nowhere dense sets in X. Let S_0 be an open sphere of radius 1. Since M_1 is nowhere dense, there exists an open sphere S_1 of radius $r_1 < 1/2$ such that its closure $\overline{S}_1$ is disjoint from M_1. Since M_2 is nowhere dense there exists in S_1 an open sphere S_2 of radius $r_2 < 1/2^2$ such that its closure $\overline{S}_2$ is disjoint from M_2. Continuing this process, we obtain a sequence of nested closed spheres

$$\overline{S}_1 \supset \overline{S}_2 \supset \overline{S}_3 \supset \ldots$$

and Proposition 26 is applicable. Now let x_0 be the point which is common to all closed spheres $\overline{S}_n (n = 1, 2, 3, \ldots)$. By our construction x_0 will not belong to any of the sets M_n; but this contradicts the assumption that

$$X = \bigcup_{n=1}^{\infty} M_n$$

and the proof is finished.

PROPOSITION 30. *Every incomplete metric space (X, ρ) can be extended to a complete metric space.*

Proof. Let $\tilde{X}$ denote the set of all Cauchy sequences $\tilde{x} = \{x_1, x_2, \ldots\}$, $x_n \in X$. Two Cauchy sequences

$$\tilde{x} = \{x_1, x_2, \ldots\} \text{ and } \tilde{y} = \{y_1, y_2, \ldots\}$$

will be looked upon as "equal" (in symbols $\tilde{x} \sim \tilde{y}$) if and only if

$$\lim_{n\to\infty} \rho(x_n, y_n) = 0.$$

(It is clear that $\sim$ is an equivalence relation on $\tilde{X}$, i.e., if $\tilde{x}$, $\tilde{y}$, $\tilde{z} \in X$, then (i) $\tilde{x} \sim \tilde{x}$, (ii) $\tilde{x} \sim \tilde{y}$ implies $\tilde{y} \sim \tilde{x}$ and (iii) $\tilde{x} \sim \tilde{y}$ and $\tilde{y} \sim \tilde{z}$ imply $\tilde{x} \sim \tilde{z}$ [transitive].) We introduce into $\tilde{X}$ a metric $\tilde{\rho}$ by letting

$$\tilde{\rho}(\tilde{x}, \tilde{y}) = \lim_{n\to\infty} \rho(x_n, y_n),$$

where $\tilde{x} = \{x_1, x_2, \ldots\}$ and $\tilde{y} = \{y_1, y_2, \ldots\}$ (noting that $\tilde{\rho}(\tilde{x}, \tilde{y})$ is independent of the representation of $\tilde{x}$ and $\tilde{y}$). This limit exists; since

$$\rho(x_m, y_m) \leqq \rho(x_m, x_n) + \rho(x_n, y_n) + \rho(y_n, y_m),$$

we have that

$$\rho(x_m, y_m) - \rho(x_n, y_n) \leqq \rho(x_m, x_n) + \rho(y_n, y_m)$$

and so the numbers $\rho(x_n, y_n)$ do form a Cauchy sequence in R^1. But R^1 is complete.

It is simple to see that $(\tilde{X}, \tilde{\rho})$ is a metric space. This metric space contains in particular all sequences of the form

$$\tilde{x} = \{x, x, \ldots\}$$

and the mapping $x \to \{x, x, \ldots\}$ is an isometry from X into $\tilde{X}$, i.e., it preserves distance. We therefore will no longer make a distinction between the element x and the sequence $\{x, x, \ldots\}$. We may thus look upon X as a subspace of $\tilde{X}$. In particular we may speak of the distance between a point $\tilde{x} \in \tilde{X}$ aMd a point $y \in X$ by setting

$$\tilde{\rho}(\tilde{x}, y) = \lim_{n\to\infty} \rho(x_n, y) \text{ for } \tilde{x} = \{x_1, x_2, \ldots\}.$$

Now, if $\tilde{x} = \{x_1, x_2, \ldots\}$ is an arbitrary Cauchy sequence in X, i.e., if for any $\varepsilon > 0$ there is a positive integer $N(\varepsilon)$ such that $m, n > N(\varepsilon)$ implies

$$\rho(x_m, x_n) < \varepsilon,$$

then for $m \to \infty$, we get, for $n > N(\varepsilon)$,

$$(*) \qquad \tilde{\rho}(\tilde{x}, x_n) \leqq \varepsilon$$

so that

$$\lim_{n\to\infty} \tilde{\rho}(\tilde{x}, x_n) = 0$$

holds. This means: For every $\tilde{x} \in \tilde{X}$ and every $\varepsilon > 0$ there exists an $x \in X$ such that $\tilde{\rho}(\tilde{x}, x) < \varepsilon$; in other words, X is dense in $\tilde{X}$.

To see that $(\tilde{X}, \tilde{\rho})$ is complete: Let $\{\tilde{x}_1, \tilde{x}_2, \ldots\}$ be a Cauchy sequence in $\tilde{X}$. As we have just seen, there exists for every $\tilde{x}_n$ an $x_n \in X$ such that

$$(**) \qquad \tilde{\rho}(\tilde{x}_n, x_n) < \frac{1}{n}$$

holds. Thus $\tilde{x} = \{x_1, x_2, \ldots\}$ is a Cauchy sequence and by (*) therefore

$$\lim_{n\to\infty} \tilde{\rho}(\tilde{x}, x_n) = 0.$$

Taking into account (**), we therefore get that

$$\lim_{n\to\infty} \tilde{\rho}(\tilde{x}_n, \tilde{x}) = 0,$$

i.e., $\tilde{x}_n$ converges to $\tilde{x}$.

We also note: The space $\tilde{X}$ is minimal among all the complete metric spaces which contain X in the sense that if Y is a complete metric space containing X, then there is an isometric map from $\tilde{X}$ into Y under which all points $x \in X$ remain fixed. This mapping is simply obtained by assigning to each Cauchy sequence $\tilde{x} = \{x_1, x_2, \ldots\}$ its limit in Y. The proof is finished.

A topological space (X,τ) is said to be *metrizable* if there exists a metric ρ on X which defines the topology τ of X.

PROPOSITION 31 (URYSOHN METRIZATION THEOREM). *A regular topological space with a countable base is metrizable. In particular, a compact Hausdorff space is metrizable if and only if it has a countable base.*

Proof. A compact metric space X is separable. Indeed, for every natural number n there exist finitely many open spheres $S(x_j^{(n)}, 1/n), j = 1, 2, \ldots, N(n)$, $x_j^{(n)} \in X$, which cover X and the at most countably many $x_j^{(n)}$ form a dense set in X. As we have seen earlier, a separable metric space has a countable base. On the other hand, a compact Hausdorff space is normal (Proposition 14) and therefore regular, so the second statement in the proposition at hand follows from the first.

Let X be a regular space with a countable base $\beta = \{U_1, U_2, \ldots\}$. We first show that X is normal. Let A and B be disjoint closed sets in X. Since X is regular there is for each point $x \in A$ a set $U \in \beta$ such that

$$x \in U \subset \overline{U} \subset B^c = X - B.$$

Thus, if $V_1, V_2, \ldots$ is a subsequence of $\{U_1, U_2, \ldots\}$ consisting of all U_n with $\overline{U}_n \subset B^c$ we have

$$A \subset \bigcup_{n=1}^{\infty} V_n.$$

Similarly, if $W_1, W_2, \ldots$ is a subsequence of $\{U_1, U_2, \ldots\}$ consisting of those U_n with $\overline{U}_n \subset A^c$, then

$$B \subset \bigcup_{n=1}^{\infty} W_n.$$

We set $Y_1 = V_1$, $Z_1 = W_1 - \overline{Y}_1$ and inductively define

$$Y_n = V_n - \bigcup_{j=1}^{n-1} \overline{Z}_j,$$

$$Z_n = W_n - \bigcup_{j=1}^{n} \overline{Y}_j.$$

Thus

$$Y = \bigcup_{n=1}^{\infty} Y_n$$

and

$$Z = \bigcup_{n=1}^{\infty} Z_n$$

are open sets. To see that $Y \cap Z = \emptyset$ we verify that $Y_n \cap Z_m = \emptyset$ for all $n, m \geqq 1$. If $n \leqq m$, then $Z_m \subset W_m - \overline{Y}_n \subset W_m - Y_n$ and so $Y_n \cap Z_m = \emptyset$. If $n > m$, then $Y_n \subset V_n - \overline{Z}_m \subset V_n - Z_m$ and so $Y_n \cap Z_m = \emptyset$. To show that $A \subset Y$, let $x \in A$ and pick m with $x \in V_m$. Since $\overline{Z}_n \subset \overline{W}_n \subset A^c$ for $n \geqq 1$ and since $x \in A \cap V_m$, we have that $x \in Y_m$ which gives that $A \subset Y$. Similarly one shows that $B \subset Z$ and thus X is seen to be normal.

With $\beta = \{U_1, U_2, \ldots\}$ a countable base for the open sets τ of X, we see that if $p \in U_m$, there is a U_n such that $p \in U_n \subset \overline{U}_n \subset U_m$. This means that there exist pairs (U_n, U_m) of sets selected from the base β with the property that $\overline{U}_n \subset U_m$; as β is a countable family of sets, there can only be a countable number of such pairs. Let

$$(U_{n_1}, U_{m_1}), \ldots, (U_{n_k}, U_{m_k}), \ldots$$

be an enumeration of such pairs. For each $k = 1, 2, \ldots$ there exists a continuous function f_k with $0 \leqq f_k(x) \leqq 1$, $f_k(\overline{U}_{n_k}) = 0$, $f_k(U^c_{m_k}) = 1$, by Urysohn's Lemma (see Proposition 8). We define ρ on $X \times X$ by

$$\rho(x,y) = \sum_{k=1}^{\infty} \frac{1}{2^k} |f_k(x) - f_k(y)|, \quad x, y \in X.$$

It is clear that ρ satisfies (i), (iii) and (iv) of the definition of a metric. Suppose that $\rho(x,y) = 0$ for some pair $x \neq y$; then $f_k(x) = f_k(y)$, $k = 1, 2, \ldots$. On the other hand, there exists a set $U_{m(x)}$ from the base β such that $x \in U_{m(x)}$, $y \notin U_{m(x)}$. By the regularity of the space there is some set $U_{n(x)}$ from the base β with

$$x \in U_{n(x)} \subset \overline{U}_{n(x)} \subset U_{m(x)},$$

so that $(U_{n(x)}, U_{m(x)})$ is one of the pairs listed above. Hence $f_k(x) \neq f_k(y)$ for some choice of k and this contradiction shows that ρ is in fact a metric on X.

Let x be given and $\varepsilon > 0$. Then $\rho(x, \cdot)$ is seen to be a continuous function on X. Hence there is a set U_m from the base β with $x \in U_m$ such that $y \in U_m$ implies $\rho(x,y) < \varepsilon$. This shows that the identity mapping of X with the given topology τ onto X with the metric topology defined by ρ is a continuous function. On the other hand, if $x \in U_m$, there is a U_n such that

$$x_n \in U_n \subset \overline{U}_n \subset U_m.$$

Hence (U_n, U_m) occurs in the enumeration of the pairs, say as the k-th term. Then, if $\rho(x,y) < 2^{-k}$, we have $|f_k(y)| < 1$ and so $y \in U_m$. Thus

$S(x, 2^{-k}) \subset U_m$; this shows that the identity mapping of X with the metric topology onto (X, τ) is continuous. Thus the identity is a homeomorphism and the space (X, τ) is metrizable. The proof is complete.

Let f be a mapping from a metric space (X, ρ) into a metric space (Y, d). We say that f is *uniformly continuous* on (X, ρ) if for every $\varepsilon > 0$ there exists $\delta > 0$ such that

$$d(f(p), f(q)) < \varepsilon$$

for all p and q in X for which $\rho(p, q) < \delta$.

PROPOSITION 32. *Let f be a continuous mapping of a compact metric space (X, ρ) into a metric space (Y, d). Then f is uniformly continuous on (X, ρ).*

Proof. Let $\varepsilon > 0$ be given. Since f is continuous, we can associate to each point $p \in X$ a positive number $\psi(p)$ such that

$$(*) \qquad q \in X,\ \rho(p, q) < \psi(p) \text{ implies } d(f(p), f(q)) < \frac{\varepsilon}{2}.$$

Let $J(p)$ be the set of all $q \in X$ for which

$$\rho(p, q) < \frac{1}{2}\psi(p).$$

Since $p \in J(p)$, the collection of all sets $J(p)$ is an open cover of X; and since (X, ρ) is compact, there is a finite set of points $p_1, \ldots, p_n$ in X such that

$$(**) \qquad X \subset J(p_1) \cup \ldots \cup J(p_n).$$

We put

$$\delta = \frac{1}{2} \min\{\psi(p_1), \ldots, \psi(p_n)\}.$$

It is clear that $\delta > 0$.

Now let p and q be points of X such that $\rho(p, q) < \delta$. By (**), there is an integer m, $1 \leqq m \leqq n$, such that $p \in J(p_m)$; hence

$$\rho(p, p_m) < \frac{1}{2}\psi(p_m),$$

and we also have that

$$\rho(p, p_m) \leqq \rho(p, q) + \rho(p, p_m) < \delta + \frac{1}{2}\psi(p_m) \leqq \psi(p_m).$$

Finally, (*) shows that therefore

$$d(f(p), f(q)) \leqq d(f(p), f(p_m)) + d(f(q), f(p_m)) < \varepsilon.$$

The proof is finished.

PROPOSITION 33 (PRINCIPLE OF EXTENSION BY CONTINUITY). *Let* (X, ρ) *and* (Y, d) *be metric spaces, and let* (Y, d) *be complete. If* $f\colon A \to Y$ *is uniformly continuous on a dense subset* A *of* (X, ρ), *then* f *has a unique continuous extension* $F\colon X \to Y$ *with* F *uniformly continuous on* (X, ρ).

Proof. If $x \in X$, there is an approximating sequence of points $(a_n)_{n=1}^{\infty}$ with $a_n \in A$ for $n = 1, 2, \ldots$ such that

$$\lim_{n\to\infty} a_n = x.$$

Since $(a_n)_{n=1}^{\infty}$ is convergent, it is a Cauchy sequence; since f is uniformly continuous, the sequence $(f(a_n))_{n=1}^{\infty}$ is a Cauchy sequence in (Y, d). But (Y, d) is complete and so there exists a point $F(x) \in Y$ with

$$\lim_{n\to\infty} f(a_n) = F(x).$$

To see that $F(x)$ is well-defined, we verify that $F(x)$ depends only on x and not on the particular sequence $a_n \to x$. Let $(b_n)_{n=1}^{\infty}$ be another sequence in A with

$$\lim_{n\to\infty} b_n = x.$$

Then clearly $\rho(a_n, b_n) \to 0$ with $n \to \infty$ and, since f is uniformly continuous, $d(f(a_n), f(b_n)) \to 0$ as $n \to \infty$ and so

$$\lim_{n\to\infty} f(b_n) = F(x).$$

It is now easy to see that $\rho(x, x') < \delta$ implies $d(F(x), F(x')) \leqq \varepsilon$ and from this the uniform continuity of F on (X, ρ) follows. The uniqueness of F is trivial and the proof is finished.

A topological space Y is said to be *sequentially compact* if every sequence in Y has a convergent subsequence.

A sequence $(f_n)_{n=1}^{\infty}$ of functions from a set X into a topological space Y is said to *converge at a point* $x \in X$ if the sequence of points $(f_n(x))_{n=1}^{\infty}$ is convergent in Y; $(f_n)_{n=1}^{\infty}$ is said to be *(pointwise) convergent* if it converges at every point of X; $(f_n)_{n=1}^{\infty}$ is said to *converge (pointwise)* to a function f: $X \to Y$ if for each $x \in X$, the sequence $(f_n(x))_{n=1}^{\infty}$ in Y converges to the point $f(x)$.

PROPOSITION 34. *If a sequence* $(f_n)_{n=1}^{\infty}$ *of functions* $f_n : D \to Y$ *from a countable set* D *into a topological space* Y *is such that, for every* $x \in D$, *the subspace*

$$F_x = \overline{\{f_n(x) : n \in N\}}$$

of Y is sequentially compact, then $(f_n)_{n=1}^{\infty}$ has a convergent subsequence.

Proof. Let $D = \{x_1, x_2, \ldots\}$. Since

$$\overline{\{f_n(x_1) : n \in N\}}$$

is sequentially compact, we can pick a subsequence $(f_{1,n})_{n=1}^{\infty}$ of $(f_n)_{n=1}^{\infty}$ such that $(f_{1,n}(x_1))_{n=1}^{\infty}$ converges. We can next pick a subsequence $(f_{2,n})_{n=1}^{\infty}$ of $(f_{1,n})_{n=1}^{\infty}$ such that $(f_{2,n}(x_2))_{n=1}^{\infty}$ converges. Continuing this process we obtain a subsequence $(f_{m,n})_{n=1}^{\infty}$ such that $(f_{m,n}(x_m))_{n=1}^{\infty}$ converges. Considering the "diagonal" sequence $(f_{n,n})_{n=1}^{\infty}$ we see that it converges for every point of the set D. The proof is finished.

A family $\mathcal{G}$ of functions $f: X \to Y$ from a topological space X into a metric space (Y, ρ) is said to be *equicontinuous* at a point p of X if for every $\varepsilon > 0$, there is a neighborhood U of p in X such that

$$\rho(f(p), f(x)) < \varepsilon$$

holds for every $f \in \mathcal{G}$ and every $x \in U$. The family is said to be *equicontinuous* if it is equicontinuous at every point of X.

PROPOSITION 35. *If an equicontinuous sequence $(f_n)_{n=1}^{\infty}$ of functions $f_n : X \to Y$ from a topological space X into a complete metric space (Y, d) is convergent at every point of a dense subset D of X, then $(f_n)_{n=1}^{\infty}$ converges at every point of X. Moreover, if $f: X \to Y$ is the function to which $(f_n)_{n=1}^{\infty}$ converges, then f is continuous on X.*

Proof. Consider an arbitrary point $p \in X$ and $\varepsilon > 0$. Since $(f_n)_{n=1}^{\infty}$ is equicontinuous, there is a neighborhood U of p in X such that $\rho(f_n(p), f_n(x)) < \varepsilon/3$ holds for every $n \in N$ and every $x \in U$. Since D is dense in X, there exists a point x in $D \cap U$. Since $(f_n)_{n=1}^{\infty}$ is convergent at x, $(f_n(x))_{n=1}^{\infty}$ is a Cauchy sequence and so there is a natural number k such that

$$\rho(f_i(x), f_j(x)) < \frac{\varepsilon}{3}$$

holds whenever $i, j > k$. Then

$$\begin{aligned}\rho(f_i(p), f_j(p)) &\leqq \rho(f_i(p), f_i(x)) + \rho(f_i(x), f_j(x)) + \rho(f_j(p), f_j(x)) \\ &< \frac{\varepsilon}{3} + \frac{\varepsilon}{3} + \frac{\varepsilon}{3} = \varepsilon\end{aligned}$$

for all natural numbers $i > k$ and $j > k$. By the completeness of (Y, ρ) we see that $(f_n)_{n=1}^{\infty}$ is convergent at the point $p \in X$. Since p is arbitrary, we have shown the convergence of $(f_n)_{n=1}^{\infty}$ at every point of X.

Finally, let $f: X \to Y$ be the function to which $(f_n)_{n=1}^{\infty}$ converges. To see that f is continuous at the point $p \in X$, let x be any point in the neighborhood U of p selected above. Since $(f_n)_{n=1}^{\infty}$ converges to f at these two points p and x, we have a natural number k such that

$$\rho(f(p), f_n(p)) < \frac{\varepsilon}{3}, \quad \rho(f(x), f_n(x)) < \frac{\varepsilon}{3}$$

hold for all natural numbers $n > k$. Choose any natural number $n > k$. Then

$$\begin{aligned} \rho(f(p), f(x)) &\leqq \rho(f(p), f_n(p)) + \rho(f_n(p), f_n(x)) + \rho(f(x), f_n(x)) \\ &< \frac{\varepsilon}{3} + \frac{\varepsilon}{3} + \frac{\varepsilon}{3} = \varepsilon. \end{aligned}$$

Since x is an arbitrary point of the neighborhood U of p, this establishes that f is continuous at p. Since p is any point of X, we have shown that f is continuous on X. The proof is finished.

PROPOSITION 36. *If $(f_n)_{n=1}^{\infty}$ is an equicontinuous sequence of functions $f: X \to Y$ from a compact topological space X into a metric space (Y, ρ) which converges to a function $f: X \to Y$, then $(f_n)_{n=1}^{\infty}$ converges uniformly to f, i.e., for each $\varepsilon > 0$ there is a natural number k such that*

$$\rho(f(x), f_n(x)) < \varepsilon$$

holds for every $n > k$ and every point $x \in X$.

Proof. Let $\varepsilon > 0$ be given. Since $(f_n)_{n=1}^{\infty}$ is equicontinuous, there exists for each point $p \in X$ an open neighborhood U_p of p in X such that

$$\rho(f_n(p), f_n(x)) < \frac{\varepsilon}{3}$$

holds for every $n \in N$ and every $x \in U_p$. Since the metric

$$\rho : Y \times Y \to R^1$$

is continuous (see, for example, Exercise 15 of Chapter 1), this implies that

$$\rho(f(p), f_n(x)) < \frac{\varepsilon}{3}$$

holds for every $x \in U_p$.

Since X is compact, the open cover $U = \{U_p : p \in X\}$ of X has a finite subcover

$$X = U_{p_1} \cup \ldots \cup U_{p_k}.$$

Since $(f_n)_{n=1}^{\infty}$ converges to f at the points $p_1, \ldots, p_k$ of X, there exists a natural number m such that

$$\rho(f(p_i), f_n(p_i)) < \frac{\varepsilon}{3}$$

holds for every $n > m$ and every $i = 1, 2, \ldots, k$.

To see that $(f_n)_{n=1}^{\infty}$ converges uniformly to f, let x denote any point of X. Then there exists an integer i with $1 \leqq i \leqq k$ and $x \in U_p$. Thus

$$\rho(f(x), f_n(x)) \leqq \rho(f(p_i), f(x)) + \rho(f(p_i), f_n(p_i)) + \rho(f_n(p_i), f_n(x))$$
$$< \frac{\varepsilon}{3} + \frac{\varepsilon}{3} + \frac{\varepsilon}{3} = \varepsilon$$

for every natural number $n > m$. This proves the uniform convergence and the proof is complete.

Combining the foregoing three propositions we obtain Proposition 37.

PROPOSITION 37 (ARZELÀ-ASCOLI THEOREM). *Let $\mathcal{G}$ be an equicontinuous family of functions from a separable topological space X into a complete metric space Y. If a sequence $(f_n)_{n=1}^{\infty}$ in $\mathcal{G}$ is such that, for each $x \in X$, the subset*

$$F_x = \overline{\{f_n(x) : n \in N\}}$$

is compact, then $(f_n)_{n=1}^{\infty}$ has a convergent subsequence $(f_{n_k})_{k=1}^{\infty}$ and the function f: $X \to Y$ to which $(f_{n_k})_{k=1}^{\infty}$ converges is continuous on X. Moreover, the convergence is uniform on every compact subspace of X.

Proof. Since X is separable, it has a countable dense subset D. The subset F_x of Y is closed and compact for every $x \in X$ and hence for every $x \in D$. But a closed and compact subset of any metric space is sequentially compact. [Indeed. Let A be a closed and compact subset of a metric space and suppose that A is not sequentially compact, i.e., some sequence $(a_n)_{n=1}^{\infty}$ contains no convergent subsequence. Hence each point in A has a neighborhood containing at most a finite number of a_n. Since a finite number of these neighborhoods cover A, the sequence $(a_n)_{n=1}^{\infty}$ consists of only a finite number of distinct points of A and, therefore, certainly does have a convergent subsequence. This contradiction shows that A is sequentially compact.] By Proposition 34, $(f_n)_{n=1}^{\infty}$ has a subsequence $(f_{n_k})_{k=1}^{\infty}$ which is convergent at every $x \in D$. By Proposition 35, $(f_{n_k})_{k=1}^{\infty}$ is convergent and f: $X \to Y$ to which $(f_{n_k})_{k=1}^{\infty}$ converges is continuous. The last assertion of the proposition is a direct consequence of Proposition 36. The proof is complete.

Remark. The theory of metric spaces lends itself quite naturally to a variety of interesting applications to analysis. In this connection the interested reader might want to consult Chapter 8 of *Metric Spaces* by E. T. Copson, Cambridge University Press, Cambridge, 1968.

EXERCISES

1. Show that there is no real-valued function f on R^1 which is continuous at each rational point but discontinuous at each irrational point by verifying the following claims:
   (i) The set of all irrational points of R^1 is of second category.
   [*Hints*: The set of rational points of R^1 is of first category since it is the countable union of one-point sets and any one-point set of R^1 is closed and nowhere dense. If the set of irrational points of R^1 were of first category, then R^1 itself would have to be of first category as the union of two sets of first category; but R^1 is a complete metric space and is therefore of second category and we have reached a contradiction.]
   (ii) The set of all irrational points of R^1 is not of type F_σ.
   [*Hints*: Let A denote the set of all irrational points. If A were of type F_σ, then

$$A = \bigcup_{n=1}^{\infty} F_n,$$

   where each F_n is closed. But each F_n contains only irrational points. Hence F_n contains no nonempty open interval. Thus each F_n is closed and nowhere dense. This implies that A is of first category, contradicting part (i) above.]
   (iii) Let f: $R^1 \to R^1$ and let D denote the set of points of R^1 at which f is not continuous. Then D is of type F_σ.
   [*Hint*: See Addendum to Chapter 3.]
2. Show the existence of continuous real-valued functions on [0, 1] which are nowhere differentiable.
   (*Outline of proof*: Let C denote the set of all continuous functions f: $[0, 1] \to R^1$ endowed with the metric of uniform convergence

$$\rho(f, g) = \max\{|f(x) - g(x)| : x \in [0, 1]\}$$

   with $f, g \in C$. Then C is a complete metric space. Let E_n denote the set of functions f in C such that for some x in $[0, 1 - 1/n]$ the inequality $|f(x + h) - f(x)| \leqq nh$ holds for all $0 < h < 1 - x$. To see that E_n is closed, consider any f in the closure of E_n, and let $(f_k)_{k=1}^{\infty}$ be a sequence in E_n that converges to f. There is a corresponding sequence of numbers $(x_k)_{k=1}^{\infty}$ such that, for each k,

$$0 \leqq x_k \leqq 1 - \frac{1}{n}$$

and

$$|f_k(x_k + h) - f_k(x_k)| \leqq nh \qquad \text{for all } 0 < h < 1 - x_k.$$

We assume also that

$$x_k \to x, \text{ for some } 0 \leqq x \leqq 1 - \frac{1}{n},$$

since this condition will be fulfilled if we replace $(f_k)_{k=1}^{\infty}$ by a suitably chosen subsequence. If $0 < h < 1 - x$, the inequality $0 < h < 1 - x_k$ holds for all sufficiently large k, and then

$$\begin{aligned}
|f(x + h) - f(x)| \leqq{} & |f(x + h) - f(x_k + h)| \\
& + |f(x_k + h) - f_k(x_k + h)| \\
& + |f_k(x_k + h) - f_k(x_k)| + |f_k(x_k) - f(x_k)| \\
& + |f(x_k) - f(x)| \\
\leqq{} & |f(x + h) - f(x_k + h)| + \rho(f, f_k) + nh \\
& + \rho(f_k, f) + |f(x_k) - f(x)|.
\end{aligned}$$

Letting $k \to \infty$, and using the fact that f is continuous at x and $x + h$, we get that

$$|f(x + h) - f(x)| \leqq nh \qquad \text{for all } 0 < h < 1 - x.$$

Thus f belongs to E_n.

Any $f \in C$ is uniformly continuous on $[0, 1]$ and may therefore be approximated uniformly and arbitrarily closely by a polygonal function, i.e., a piecewise linear continuous function, say g. To show that E_n is nowhere dense in C, it will be enough to show that for any such function g, and $\varepsilon > 0$, there is a function v in $C - E_n$ such that $\rho(g, v) \leqq \varepsilon$. Let M be the maximum value of the slopes of the line segments that make up the graph of g, and select an integer m such that $m\varepsilon > n + M$. Let s denote the "sawtooth" function

$$s(x) = \min\{x - [x], [x] + 1 - x\}$$

(observe that $s(x)$ is the distance from x to the nearest integer), and put

$$v(x) = g(x) + \varepsilon s(mx).$$

Then at each point of $[0, 1)$ the function v has a one-sided derivative on the right numerically greater than n. This is clear, since $\varepsilon s(mx)$ has everywhere in $[0, 1)$ a right derivative equal to $\pm\varepsilon m$, and g has a right derivative with absolute value at most equal to M. Therefore $v \in C - E_n$. Since $\rho(g, v) = \varepsilon/2$, it follows that E_n is nowhere dense in C. Hence the set

$$E = \bigcup_{n=1}^{\infty} E_n$$

is of first category in C. This is the set of all continuous functions that have bounded right difference quotients at some point of $[0, 1)$.

Similarly, the set of functions that have bounded left difference quotients at some point of $(0, 1]$ is of first category; indeed, we can deduce this from what we have already studied above by considering the isometry of C induced by the substitution of $1 - x$ for x. The union of these two sets includes all functions in C that have a one-sided derivative somewhere in $[0, 1]$.

By similar reasoning one can show: If $\mathfrak{D} = \{f \in C : D^+f(x)$ and $D_-f(x)$ are both finite for some $x \in [0, 1]\}$, then $\mathfrak{D}$ is of first category in the complete metric space C. Thus the set of all continuous real-valued functions on $[0, 1]$ which have at least one infinite right derivative at every point of $[0, 1)$ is dense in C. See, for example, S. Banach, *Studia Math.* **3** (1931), 174-179.)

Historical Note: The investigation of the relationship between continuity and differentiability has a long history. According to Weierstrass (see Weierstrass, K. "Über continuirliche Functionen eines reellen Arguments, die für keinen Werth des letzteren einen bestimmten Differentialquotienten besitzen", Königl. Akad. Wiss. (1872), Mathematische Werke II, 71-74), Riemann told his students in 1861 that the continuous function

$$\sum_{n=1}^{\infty} \frac{\sin n^2 x}{n^2}$$

is nowhere differentiable. As Weierstrass was not able to prove it (and, in fact, until now, it seems to have been neither proved nor disproved), he gave in 1872 his famous example

$$\sum_{n=1}^{\infty} a^n \cos b^n x$$

where b is an odd integer $\geqq 3$, and a is a positive number such that $a < 1$ and $ab > 1 + 3\pi/2$; f is a continuous function which is nowhere differentiable. Later on, Weierstrass's result was improved by G. H. Hardy; see Hardy, G. H. "Weierstrass's non-differentiable function", Trans. Amer. Math. Soc. **17**(1916), 301-325.

3. Uniform Boundedness Principle: Let F be any family of continuous, real-valued functions on a complete metric space X such that for each $x \in X$, $|f(x)| \leqq M_x$ for all $f \in F$. Prove that there is some constant M and a nonempty open set U in X such that $|f(x)| \leqq M$ for each $x \in U$ and each $f \in F$.
   [*Hints*: Let $E_n = \{x \in X : |f(x)| \leqq n \text{ for each } f \in F\}$. Show that E_n is closed. By hypothesis, we have for each $x \in X$, a natural number n_x such that $|f(x)| \leqq n_x$ for all $f \in F$. This implies $x \in E_{n_x}$. Hence

$$X = \bigcup_{n=1}^{\infty} E_n .$$

   But X is of second category. Hence there is a natural number n such that E_n is not nowhere dense. Thus the open set U = interior of the closure of E_n is nonempty. Since E_n is closed, $U = \mathring{E}_n$. Consequently $|f(x)| \leqq n$ holds for every $f \in F$ and every $x \in U$.]
4. Let X be a topological space and $A \subset X$. Show that the following statements are equivalent:
   (i) A is nowhere dense.
   (ii) Every nonempty open set U of X contains a nonempty open set V which is disjoint from A.
   (iii) $X - \overline{A}$ is dense in X.
5. Let A be a set of first category in a compact Hausdorff space X. Show that $X - A$ is dense in X.
6. Let $(f_n)_{n=1}^{\infty}$ be a sequence of continuous real-valued functions defined on a Hausdorff space X. Suppose that a finite limit:

$$\lim_{n \to \infty} f_n(x) = f(x)$$

   exists at every point x of X. Show that the set of points at which the function f is discontinuous is a set of first category.
7. Let X be a locally compact Hausdorff space. Let $x \in X$ and let U be a neighborhood of x. Then there exists a neighborhood V of x such that $\overline{V}$ is compact and $\overline{V} \subset U$. Show this.
8. Let X be as in Exercise 7 and A be a compact subspace of X. Suppose that U is an open subset of X such that $A \subset U$. Show that there exists an open $V \subset X$ such that $\overline{V}$ is compact and $A \subset V \subset \overline{V} \subset U$.
9. Let X be as in Exercise 7. Prove that X is of second category (as a subset of itself).
10. Verify that any Cauchy sequence in a metric space is bounded.
11. Prove that a subset of a metric space is compact if and only if it is closed and sequentially compact.
12. A subset K of a metric space is said to be *totally bounded* if for every $\varepsilon > 0$ it is possible to cover K by a finite number of open spheres

$S(k_i, \varepsilon)$, $i = 1, \ldots, n$, with centers k_i in K. Show:
For a metric space X the following statements are equivalent:
(i) X is complete.
(ii) Every closed and totally bounded subspace of X is compact.

13. Prove: If K is a set in a metric space X, the following statements are equivalent:
(a) K is sequentially compact.
(b) $\overline{K}$ is compact.
(c) K is totally bounded and $\overline{K}$ is complete.
Moreover, a compact metric space is complete and separable.

14. Let $\{X_t : t \in T\}$ be a nonempty family of Hausdorff spaces. Show that the Cartesian product X of these spaces is Hausdorff (in the product topology).

15. Prove: A continuous image of a connected space is connected; a Cartesian product
$$\bigtimes_{t \in T} X_t$$
is connected if and only if every X_t is connected.

16. Partition of unity: Let X be a normal space, F a closed subset of X and $U_1, \ldots, U_n$ be open sets such that
$$\bigcup_{k=1}^{n} U_k \supset F.$$
Show that there exist continuous functions $h_1, \ldots, h_n$ on X with values in $[0, 1]$ such that

(i)
$$\sum_{k=1}^{n} h_k(x) = 1$$
for all $x \in F$ and
(ii) $h_k(U_k^c) = 0$ for $k = 1, \ldots, n$.
[*Hints*: Suppose
$$\bigcup_{k=1}^{n} U_k = X;$$
we show that there are closed sets $A_1, \ldots, A_n$ such that
$$\bigcup_{k=1}^{n} A_k = X$$
and $A_k \subset U_k$ for $k = 1, \ldots, n$. The proof is by induction and is clear for $n = 1$. Suppose that $U_1 \cup U_2 = X$. Then U_1^c and U_2^c are disjoint closed sets and there exist disjoint open sets V_1 and V_2 such that

$U_1^c \subset V_1$ and $U_2^c \subset V_2$. Putting $A_1 = V_1^c$ and $A_2 = V_2^c$, we obtain the present assertion for $n = 2$. Suppose that the assertion is true for $n - 1$ and that

$$\bigcup_{k=1}^{n} U_k = X.$$

Since

$$X = (\bigcup_{k=1}^{n-1} U_k) \cup U_n,$$

there are closed sets A and A_n such that

$$A \subset \bigcup_{k=1}^{n-1} U_k,$$

$A_n \subset U_n$, and $A \cup A_n = X$. For $k = 1, \ldots, n - 1$, define $V_k = U_k \cap A$. Then

$$\bigcup_{k=1}^{n-1} V_k = A$$

and by the induction hypothesis applied to A, there are sets $A_1, \ldots, A_{n-1}$, closed in A, such that

$$\bigcup_{k=1}^{n-1} A_k = A$$

and $A_k \subset V_k$ $(k = 1, \ldots, n - 1)$. Evidently the A_k's are also closed in X, each A_k is contained in U_k, and

$$\bigcup_{k=1}^{n} A_k = X.$$

We now take up the case when $F = X$. Suppose

$$\bigcup_{k=1}^{n} U_k = X$$

and assume that $A_1, \ldots, A_n$ are determined as above. By Urysohn's lemma, there are continuous functions f_k on X such that $f_k(X) \subset [0, 1]$, $f_k(A_k) = 1$, and $f_k(U_k^c) = 0$ $(k = 1, \ldots, n)$. For $k = 1, \ldots, n$, let

$$h_k(x) = \frac{f_k(x)}{\sum_{j=1}^{n} f_j(x)} \qquad \text{for } x \in X.$$

Since

$$\sum_{j=1}^{n} f_j(x) \geqq 1$$

for all x, it is clear that each h_k is continuous and (i) and (ii) are satisfied.

Finally, suppose that F is a closed subset of X and that

$$\bigcup_{k=1}^{n} U_k \supset F.$$

Define $U_0 = F^c$ and observe that

$$\bigcup_{k=0}^{n} U_k = X.$$

By what we have shown already, there are continuous functions $h_0, h_1, \ldots, h_n$ on X with values in $[0, 1]$ such that

$$\sum_{k=0}^{n} h_k(x) = 1$$

for all $x \in X$ and $h_k(U_k^c) = 0$ $(k = 0, 1, \ldots, n)$. Since $h_0(F) = 0$, we have

$$\sum_{k=1}^{n} h_k(x) = 1$$

for $x \in F$. This means that $h_1 \ldots, h_n$ indeed satisfy what is required of them.]

17. Let X be a topological space and let $A \subset X$. A point $p \in A$ is called an *isolated point of* A if it is not a limit point of A, i.e., if there exists a neighborhood U of p such that $U \cap A = \{p\}$. The set A is called *perfect* if it is closed and has no isolated points, i.e., if A is equal to the set of its own limit points.
Show: If X is a complete metric space and A is a nonempty perfect subset of X, then the cardinality of A is larger than or equal to the cardinal number c.
18. Verify that the Cantor ternary set P is compact, nowhere dense and perfect in $[0, 1]$.
19. Prove: If X is a topological space with countable base β and A is a closed subset of X, then X contains a perfect subset P and a countable subset C such that $A = P \cup C$ (Cantor-Bendixson).
[*Hints*: A point $x \in X$ is called a *condensation point of* a subset A of X if $U \cap A$ is uncountable for each neighborhood U of x. Let

$$P = \{x \in X : x \text{ is a condensation point of } A\}$$

and let $C = A \cap P^c$. Since each condensation point is clearly a limit point, we see that $P \subset A$. Evidently $A = P \cup C$. Since no point of C is a condensation point of A, each $x \in C$ has a neighborhood $V_x \in \beta$

such that $A \cap V_x$ is countable. But β is countable, so $C \subset U\{A \cap V_x : x \in C\}$, and C is countable.

Let $x \in P$ and let U be a neighborhood of x. Then $U \cap A$ is uncountable and $U \cap C$ is countable, hence

$$U \cap P = (U \cap A) \cap (U \cap C)^c$$

is uncountable, and so x is a limit point of P. Thus P has no isolated points. To see that P is closed, let $x \in P^c$. Then x has a neighborhood V such that $V \cap A$ is countable. If there is a $y \in V \cap P$, then V is a neighborhood of y and y is a condensation point of A and so $V \cap A$ is uncountable. It follows that V and P are disjoint and so x is not a limit point of P. Therefore P contains all of its limit points, i.e., P is closed. Thus P must be perfect.]

20. Let $(f_n)_{n=1}^{\infty}$ be a differentiable sequence of functions $f_n : [0, 1] \to R^1$ such that the sequence $(f'_n(x_0))_{n=1}^{\infty}$ converges for some x_0 in $[0, 1]$. Show that if the derived sequence $(f'_n)_{n=1}^{\infty}$ is uniformly bounded on $[0, 1]$, then $(f_n)_{n=1}^{\infty}$ contains a uniformly convergent subsequence.
21. Use the fixed point theorem for complete metric spaces to find a constructive solution in R^1 of the equation $x = q \sin x + a$ with a and q constant and $0 < q < 1$.
22. If f: $X \to X$ satisfies $\rho(f(p), f(q)) < \rho(p, q)$ for any two distinct points of a compact metric space (X, ρ), does f have a fixed point? (Yes.) Does the function $f(x) = \pi/2 + x - \text{arc tan } x$ for every $x \in R^1$ have a fixed point? (No.)
23. Let $f(x, y)$ be a real-valued continuous function over a rectangle D with center (x_0, y_0) and suppose that

$$|f(x, y_2) - f(x, y_1)| \leqq K|y_2 - y_1|,$$

where K is a positive constant, whenever (x, y_1) and (x, y_2) belong to D. Show that there is an interval $J = [x_0 - a, x_0 + a]$ and a function g in the variable x defined and continuously differentiable in J and satisfying the conditions $g(x_0) = y_0$ and $g' = f(x, g(x))$. Furthermore, there is only one function g satisfying all these conditions (Cauchy-Picard).
24. Let (a_{ij}) be an $n \times n$ matrix of real numbers such that

$$\sum_{j=1}^{n} |a_{ij}| < 1$$

for all i. For arbitrary reals $b_1, \ldots, b_n$, show that the system of equations

$$\gamma_i - \sum_{j=1}^{n} a_{ij}\gamma_j = b_i$$

has exactly one solution $x_0 = (\gamma_1^0, \ldots, \gamma_n^0)$ and that this solution can be found by an iteration process where one starts with an arbitrary n-tuple of real numbers.

25. Show: If $((X_n, d_n))_{n=1}^{\infty}$ is a sequence of metric spaces, then a metric for the product space

$$X = \bigtimes_{n \in N} X_n$$

is given by the formula

$$d(x,y) = \sum_{n=1}^{\infty} \frac{1}{2^n} \frac{d_n(pr_n x, pr_n y)}{1 + d_n(pr_n x, pr_n y)},$$

where pr_n denotes the projection of X onto X_n. If each of the spaces (X_n, d_n) is complete, then the product space X with the metric d is also complete.

26. Let (X,ρ) be a metric space. For any nonempty set $E \subset X$, define

$$\rho_E = \inf\{\rho(x,y) : y \in E\}.$$

Then ρ_E is uniformly continuous on X (see Exercise 15 of Chapter 1). If A and B are disjoint nonempty closed subsets of X, what is the relevance of the function

$$f(x) = \frac{\rho_A(x)}{\rho_A(x) + \rho_B(x)}$$

to Urysohn's lemma (see Proposition 8)?

27. By a *directed system* we mean a set A together with a relation $<$ satisfying the following conditions:
   (a) $\alpha < \alpha$ for each $\alpha \in A$.
   (b) If $\alpha < \beta$ and $\beta < \gamma$, then $\alpha < \gamma$.
   (c) If $\alpha, \beta \in A$, there is a $\gamma \in A$ with $\alpha < \gamma$ and $\beta < \gamma$.

   A *net* is a mapping of a directed system into a topological space X. A point $x \in X$ is said to be the *limit* of a net $\langle x_\alpha \rangle$ if for each open set U containing x there is an $\alpha_0 \in A$ such that $x_\alpha \in U$ for all $\alpha > \alpha_0$; here x_α denotes the value of the net at α. A point x is said to be a *cluster point* of $\langle x_\alpha \rangle$ if given $\alpha \in A$ there is a $\beta > \alpha$ such that $x_\beta \in U$.

   Nets provide a generalization of sequences. That sequences are not sufficient for the purposes of analysis is already evident in classical integration theory; consider the following example.

   The collection $\mathcal{P}$ of all finite partitions of the closed interval $[a,b]$ into closed subintervals is a directed system, when ordered by the relation $T_1 < T_2$ if and only if T_2 refines T_1. If f is any real-valued function on $[a, b]$, we can define a net $P_L : \mathcal{P} \to R^1$ by letting $P_L(T)$ be the lower Riemann sum of f over T; likewise, we can define $P_U : \mathcal{P} \to R^1$ by letting $P_U(T)$ be the upper Riemann sum of f over T. Convergence of both of these nets to the number c simply means $\int_a^b f(x)\,dx = c$.

Prove:

(i) A point x is a point of closure of a set E if and only if it is the limit of a net $\langle x_\alpha \rangle$ from E.

(ii) X is Hausdorff if and only if every net in X has at most one limit point.

(iii) A function f from a topological space X into a topological space Y is continuous at $x \in X$ if and only if for each net $\langle x_\alpha \rangle$ which converges to x the net $\langle f(x_\alpha) \rangle$ converges to $f(x)$.

28. A real-valued function f, defined on a topological space X, is said to be lower semi-continuous (resp. upper semi-continuous) at a point $a \in X$, if for each $h < f(a)$ (resp. each $k > f(a)$) there is a neighborhood V of a such that $h < f(x)$ (resp. $k > f(x)$) for each $x \in V$.

A real-valued function f is said to be lower semi-continuous (resp. upper semi-continuous) on X if it is lower semi-continuous (resp. upper semi-continuous) at every point of X. Show:

(i) If f is lower semi-continuous at a point, then $-f$ is upper semi-continuous at this point and conversely.

(ii) A real-valued function f on a topological space X is lower semi-continuous if and only if, for each finite real number k the set $f^{-1}((k, +\infty])$ is open in X.

(iii) A subset A of a topological space X is open (resp. closed) in X if and only if the characteristic function of A is lower (resp. upper) semi-continuous on X.

(iv) If f is lower semi-continuous on a compact space X, then there exists $x_0 \in X$ such that $f(x_0) = \inf\{f(x) : x \in X\}$.

(v) If D is a nonempty set of lower semi-continuous functions on a topological space X, then $\sup\{f : f \in D\}$ is a lower semi-continuous function on X.

(vi) Let f be a lower semi-continuous function on a metric space X. Then f is the pointwise limit of a nondecreasing sequence of continuous functions on X.

(vii) Let X be a metric space and let u and v be finite-valued upper and lower semi-continuous functions on X, respectively, such that $u \leqq v$ everywhere on X. Then there is a continuous f on X such that $u \leqq f \leqq v$ everywhere on X.

Chapter 8

The Method of P. J. Daniell

Let L be a family of functions mapping a set X into the real line R^1 such that

(i) L is a linear space over R^1 and
(ii) for each $f \in L$, the function $f^+ \in L$, where

$$f^+(x) = \max\{0, f(x)\}.$$

If we define

$$(f \vee g)(x) = \max\{f(x), g(x)\}$$

and

$$(f \wedge g)(x) = \min\{f(x), g(x)\},$$

the relations

$$f^+ = f \vee 0,$$

$$f \vee g = (f - g) \vee 0 + g,$$

and

$$f \wedge g = f + g - (f \vee g)$$

show that, if f and g belong to L, then $f \vee g$ and $f \wedge g$ will belong to L also. Evidently,

$$|f| = f^+ + (-f)^+$$

and therefore, if $f \in L$, then $|f| \in L$. On the other hand, if $|f| \in L$, then $f \in L$, since

$$f^+ = \tfrac{1}{2}(f + |f|).$$

Any family L of functions $f : X \to R^1$ satisfying the conditions (i) and (ii) is called a *linear lattice* of functions. If I is a linear functional on L

(considered as a linear space over the reals), we say that I is *positive* if $I(f) \geqq 0$ for each nonnegative function f in L. If I is a positive linear functional and $f \leqq g$ with f and g in L, then $I(f) \leqq I(g)$, because I is linear and $g - f \geqq 0$. A positive linear functional I on a linear lattice L is called a *Daniell integral* if one of the following equivalent conditions is satisfied:

(D) : If $(f_n)_{n=1}^{\infty}$ is any sequence of functions in L, which decreases to zero at each point $x \in X$, then

$$\lim_{n\to\infty} I(f_n) = 0.$$

(D*) : If $(f_n)_{n=1}^{\infty}$ is any increasing sequence of functions in L, and if f is a function in L satisfying

$$f(x) \leqq \lim_{n\to\infty} f_n(x)$$

for all $x \in X$, then

$$I(f) \leqq \lim_{n\to\infty} I(f_n).$$

(D**): If $(f_n)_{n=1}^{\infty}$ is any sequence of nonnegative functions in L, and if f is a function in L such that

$$f(x) \leqq \sum_{n=1}^{\infty} f_n(x)$$

for all $x \in X$, then

$$I(f) \leqq \sum_{n=1}^{\infty} I(f_n).$$

Examples of Daniell integrals:

1. Let $X = [a, b]$ and L be the linear lattice $C(X)$ of all continuous functions $f : X \to R^1$. Then the functional $I : L \to R^1$ defined by the Riemann integral

$$I(f) = \int_a^b f(x)\,dx$$

for every $f \in L = C(X)$ is a Daniell integral.

2. Let $(X, \mathfrak{B}, \mu)$ be any complete measure space and L be the linear lattice of all measurable functions $f : X \to R^1$ which are integrable with respect to μ. Then the functional $I : L \to R^1$ defined by

$$I(f) = \int f\,d\mu$$

for every $f \in L$ is a Daniell integral.

Remark. If I is a Daniell integral and $(f_n)_{n=1}^{\infty}$ an increasing sequence in L such that

$$f(x) = \lim_{n\to\infty} f_n(x),$$

$x \in X$, defines a function in L, then

$$I(f) = \lim_{n\to\infty} I(f_n).$$

Indeed, $f_n \leqq f$ for all $n \in N$, and so $I(f) \geqq I(f_n)$ because I is positive. Using condition (D*), we get the required equality.

Suppose that I is a Daniell integral on a linear lattice L. Let L^+ denote the set of all functions $f : X \to R_e^1$ which are limits of monotone increasing functions of L. L^+ is not a linear space but

$$\alpha, \beta \geqq 0 \text{ and } f, g \in L^+ \text{ imply } \alpha f + \beta g \in L^+.$$

Thus, if $(f_n)_{n=1}^{\infty}$ is an increasing sequence in L, $(I(f_n))_{n=1}^{\infty}$ is an increasing sequence in R^1 which has a unique limit in $R^1 \cup \{\infty\}$ and we can define I on L^+ by

$$I(\lim_{n\to\infty} f_n) = \lim_{n\to\infty} I(f_n).$$

This definition is proper because if $(f_n)_{n=1}^{\infty}$ and $(g_n)_{n=1}^{\infty}$ are two increasing sequences each converging to h in L^+, then condition (D*) gives, for fixed n,

$$f_n \leqq h = \lim_{n\to\infty} g_n \text{ implies } I(f_n) \leqq \lim_{n\to\infty} I(g_n)$$

so that

$$\lim_{n\to\infty} I(f_n) \leqq \lim_{n\to\infty} I(g_n);$$

in an analogous manner we can derive the opposite inequality

$$\lim_{n\to\infty} I(g_n) \leqq \lim_{n\to\infty} I(f_n).$$

We are therefore indeed justified in taking

$$I(h) = \lim_{n\to\infty} I(h_n)$$

for every $h \in L^+$, where $(h_n)_{n=1}^{\infty}$ is an increasing approximating sequence in L of the function h, convergence being in the sense of pointwise convergence.

It is clear that I is linear on L^+ in the sense that if $\alpha, \beta \geqq 0$ and $f, g \in L^+$, then $I(\alpha f + \beta g) = \alpha I(f) + \beta I(g)$. It is also clear that L^+ is a lattice, because if $f_n \uparrow f$ and $g_n \uparrow g$, then $f_n \vee g_n \uparrow f \vee g$ and $f_n \wedge g_n \uparrow f \wedge g$.

PROPOSITION 1. *A nonnegative function f belongs to L^+ if and only if there is a sequence $(f_k)_{k=1}^{\infty}$ of nonnegative functions in L such that*

$$f = \sum_{k=1}^{\infty} f_k.$$

In this case

$$I(f) = \sum_{k=1}^{\infty} I(f_k).$$

Proof. The "if" part of the proposition is obvious. To show the "only if" part, we let f be nonnegative and $g_n \uparrow f$ with $g_n \in L$. By replacing g_n by $g_n \vee 0$, we may assume that each g_n is nonnegative. We put $f_1 = g_1, f_n = g_n - g_{n-1}$ for $n > 1$. Then

$$f = \sum_{k=1}^{\infty} f_k,$$

and

$$\begin{aligned} I(f) &= \lim_{n\to\infty} I(g_n) = \lim_{n\to\infty} I\Big(\sum_{k=1}^{n} f_k\Big) = \lim_{n\to\infty} \sum_{k=1}^{n} I(f_k) \\ &= \sum_{k=1}^{\infty} I(f_k). \end{aligned}$$

The proof is finished.

PROPOSITION 2. *If $(f_n)_{n=1}^{\infty}$ is a sequence of nonnegative functions in L^+, then the function*

$$f = \sum_{n=1}^{\infty} f_n$$

is in L^+ and

$$I(f) = \sum_{n=1}^{\infty} I(f_n).$$

Proof. By Proposition 1, for each $n \in N$ there is a sequence $(f_{n,k})_{k=1}^{\infty}$ of nonnegative functions in L such that

$$f_n = \sum_{k=1}^{\infty} f_{n,k}.$$

Hence

$$f = \sum_{n,k} f_{n,k}.$$

But the set of pairs of integers is countable and so f is the sum of a sequence of nonnegative functions in L and so must be in L^+. Since all terms are nonnegative, the order of summation is immaterial and

$$I(f) = \sum_{n=1}^{\infty} (\sum_{k=1}^{\infty} I(f_{n,k})).$$

This completes the proof.

For an arbitrary extended real-valued function $f: X \to R_e^1$ we define its *upper Daniell integral* $I^*(f)$ by

$$I^*(f) = \inf \{I(g) : g \geqq f, g \in L^+\}$$

with the understanding that the infimum of an empty set in R_e^1 is $+\infty$. On the other hand, we define the *lower Daniell integral* $I_*(f)$ *of* f by

$$I_*(f) = -I^*(-f).$$

We say that a function $f: X \to R_e^1$ is *integrable* (with respect to I) if $I^*(f) = I_*(f)$ and is finite. The class of integrable functions we shall denote by $L^1 = L^1(I, L)$. For $f \in L^1$ we call the common finite value of $I^*(f)$ and $I_*(f)$ the *integral of* f and we denote it by $J(f)$. We shall show in Proposition 4 that L^1 is a linear lattice and that J is a Daniell integral on L^1.

The properties of the upper Daniell integral I^* and the lower Daniell integral I_* are given in the following proposition.

PROPOSITION 3. *For an arbitrary given Daniell integral* $I: L \to R^1$ *on a linear lattice* L *of real-valued functions on a set* X, *the upper and lower Daniell integrals* I^*, I_* *have the following properties*:

(*i*) *For any two functions* $f, g: X \to R_e^1$, *we have*

$$I^*(f+g) \leqq I^*(f) + I^*(g),$$

provided the right-hand side of the inequality is defined.

(*ii*) *For any function* $f: X \to R_e^1$ *and any nonnegative real number* α, *we have*

$$I^*(\alpha f) = \alpha I^*(f).$$

(*iii*) *For any two functions* $f, g: X \to R_e^1$ *with* $f \leqq g$, *we have*

$$I^*(f) \leqq I^*(g), \; I_*(f) \leqq I_*(g).$$

(*iv*) *For every function* $f: X \to R_e^1$, *we have*

$$I_* \leqq I_*(f).$$

(*v*) *If a function* $f : X \to R_e^1$ *is in* L^+, *then we have*

$$I_*(f) = I^*(f) = I(f).$$

(*vi*) *If a function* $f : X \to R_e^1$ *is the pointwise sum of a sequence* $(f_n : X \to R_e^1)_{n=1}^{\infty}$ *of nonnegative functions, then*

$$I^*(f) \leqq \sum_{n=1}^{\infty} I^*(f_n).$$

Proof. Parts (i), (ii), and (iii) follow immediately from the definition. It should be noted in connection with (i) that we may put $(f + g)(x) = +\infty$ at those points x for which one of the values $f(x)$ and $g(x)$ is $+\infty$ and the other is $-\infty$ so that (i) is true whatever the value in R_e^1 chosen for $(f + g)(x)$ at such points x. (Incidentally, we shall see later that an integrable function must take finite values at "almost all" points.)

Since $0 = I(0) = I(f - f) \leqq I^*(f) + I^*(-f)$ by (i), it follows that $I_*(f) = -I^*(-f) \leqq I^*(f)$ and this establishes part (iv).

To verify (v), we take $f \in L^+$. Then, by the definition of the upper Daniell integral, $I^*(f) \leqq I(f)$. If $g \in L^+$ and $f \leqq g$, then $I(f) \leqq I(g)$ and so $I(f) \leqq I^*(f)$, whence $I^*(f) = I(f)$. If $g \in L$, then $-g \in L \subset L^+$ so that $I_*(g) = I(g)$. But each f in L^+ is the pointwise limit of an increasing sequence $(g_n)_{n=1}^{\infty}$ in L. Since $f \geqq g_n$, we have $I_*(f) \geqq I_*(g_n) = I(g_n)$, and so

$$I_*(f) \geqq \lim_{n\to\infty} I(g_n) = I(f).$$

By (iii), $I_*(f) \leqq I^*(f)$, and so $I_*(f) = I(f) = I^*(f)$ for all $f \in L^+$.

To see that part (vi) is true, we note that there is nothing to prove in case $I^*(f) = +\infty$ for some n, or if the series

$$\sum_{n=1}^{\infty} I^*(f_n)$$

diverges. Otherwise, given $\varepsilon > 0$, for each $n \in N$, choose $g_n \geqq f_n$, $g_n \in L^+$ such that

$$I^*(f_n) > I(g_n) - \frac{\varepsilon}{2^n}.$$

Then

$$g = \sum_{n=1}^{\infty} g_n \in L^+$$

by Proposition 2, $g \geqq f$ and

$$I^*(f) \leqq I(g) = \sum_{n=1}^{\infty} I(g_n) < \varepsilon + \sum_{n=1}^{\infty} I^*(f_n).$$

Since ε is an arbitrary positive number, the proof is finished.

PROPOSITION 4. *Let I be a Daniell integral on a linear lattice of functions on X to R^1. Then the construction defining a functional J on the set L^1 determines a Daniell integral on a linear lattice L^1 which extends I. Moreover, if $(f_n)_{n=1}^{\infty}$ is an increasing sequence of functions in L^1 and*

$$f(x) = \lim_{n\to\infty} f_n(x)$$

for all $x \in X$, then $f \in L^1$ if and only if

$$\lim_{n\to\infty} J(f_n) < \infty.$$

In this case

$$J(f) = \lim_{n\to\infty} J(f_n).$$

Proof. Part (v) of Proposition 3 shows that $L^1 \supset L$ and that J is an extension of I. Now, if $g \in L^1$, then $\alpha g \in L^1$ for α in R^1 because $\alpha \geqq 0$ implies $I^*(\alpha f) = \alpha I^*(f) = \alpha I_*(f) = I_*(\alpha f)$ and $\alpha < 0$ implies $I^*(\alpha f) = \alpha I_*(f) = \alpha I^*(f) = I_*(\alpha f)$. Moreover, if f and g are both in L^1, then, using part (i) of Proposition 3, we get

$$-I_*(f+g) = I^*(-f-g) \leqq -J(f) - J(g)$$

and so

$$I_*(f+g) \geqq J(f) + J(g) \geqq I^*(f+g);$$

using part (iv) of Proposition 3, we conclude that $f + g \in L^1$ and $J(f+g) = J(f) + J(g)$. Hence L^1 is a real linear space and J is a linear functional on L^1.

To verify that L^1 is a linear lattice, it will be enough to show that $f \in L^1$ implies $f^+ \in L^1$. For a fixed f in L^1 and any $\varepsilon > 0$, we select functions g and h in L^+ such that $-h \leqq f \leqq g$ and

$$I(g) < J(f) + \varepsilon < \infty, \quad I(h) \leqq J(f) + \varepsilon < \infty.$$

But $g = (g \vee 0) + (g \wedge 0)$ and $g \wedge 0 \in L^+$; thus

$$I(g \vee 0) \leqq I(g) - I(g \wedge 0) < \infty.$$

Therefore $g^+ = g \vee 0 \in L^+$ and $I(g^+) < \infty$. Similarly,

$$-h^- = h \wedge 0 \in L^+ \text{ and } -h \leqq f^+ \leqq g^+.$$

Since $-h \leqq g$, $g^+ - h^- \leqq g + h$. Consequently

$$I(g^+) + I(-h^-) \leqq I(g) + I(h) < 2\varepsilon.$$

But

$$-I(-h^-) \leqq I_*(f^+) \leqq I^*(f^+) \leqq I(g^+)$$

and so $I^*(f^+) - I_*(f^+) < 2\varepsilon$. Since ε is an arbitrary positive number and $J(g^+)$ is finite, we have the required relation that $f^+ \in L^1$.

Finally, suppose that $(f_n)_{n=1}^{\infty}$ is an increasing sequence of functions in L^1 and let f be the pointwise limit of $(f_n)_{n=1}^{\infty}$ on X. Then, if

$$\lim_{n\to\infty} J(f_n) = \infty,$$

and $g \leqq f$, $g \in L^1$, it is clear that

$$J(g) \leqq \lim_{n\to\infty} J(f_n)$$

since $J(g)$ is finite. On the other hand, if

$$\lim_{n\to\infty} J(f_n)$$

is finite, we put $h = f - f_1$. Then $h \geqq 0$ and

$$h = \sum_{n=1}^{\infty} (f_{n+1} - f_n).$$

By part (vi) of Proposition 3,

$$I^*(h) \leqq \sum_{n=1}^{\infty} (J(f_{n+1}) - J(f_n)) = \lim_{n\to\infty} J(f_n) - J(f_1);$$

we obtain that

$$I^*(f) = I^*(f_1 + h) \leqq I^*(f_1) + I^*(h) \leqq \lim_{n\to\infty} J(f_n).$$

But $f_n \leqq f$ and so

$$I_*(f) \geqq \lim_{n\to\infty} J(f_n).$$

We therefore conclude that

$$I^*(f) = I_*(f) = \lim_{n\to\infty} J(f_n).$$

This means that, if

$$\lim_{n\to\infty} J(f_n) < \infty,$$

then $f \in L^1$, and

$$J(f) = \lim_{n\to\infty} J(f_n).$$

The positive functional J therefore satisfies condition (D*) and must be a Daniell integral on L^1. The proof is finished.

Remark. The last part of Proposition 4 was an analogue of the Monotone Convergence Theorem; the following proposition is an analogue of Fatou's Lemma.

PROPOSITION 5. *Let $(f_n)_{n=1}^{\infty}$ be a sequence of nonnegative functions in L^1. Then the function*

$$\inf_{n\in N} f_n$$

is in L^1. Moreover, if

$$\underline{\lim}_{n\to\infty} J(f_n) < \infty,$$

then the function

$$\underline{\lim}_{n\to\infty} f_n$$

is in L^1; in this case

$$J(\underline{\lim}_{n\to\infty} f_n) \leqq \underline{\lim}_{n\to\infty} J(f_n).$$

Proof. Let $g_n = f_1 \wedge f_2 \wedge \dots \wedge f_n$. Then $(g_n)_{n=1}^{\infty}$ is a sequence of nonnegative functions in L^1 which decreases to

$$g = \inf_{n\in N} f_n.$$

Thus $-g_n \uparrow -g$, and since $J(-g_n) \leqq 0$, we must have that $g \in L^1$ by Proposition 4.

To complete the proof, we let

$$h_m = \inf_{n\geqq m} f_n.$$

Then $(h_m)_{m=1}^{\infty}$ is a sequence of nonnegative functions in L^1 which increases to

$$\underline{\lim}_{n\to\infty} f_n.$$

Since $h_m \leqq f_n$ for $m \leqq n$,

$$\lim_{m\to\infty} J(h_m) \leqq \varliminf_{n\to\infty} J(f_n) < \infty.$$

Hence

$$\varliminf_{n\to\infty} f_n \in L^1$$

and

$$J(\varliminf_{n\to\infty} f_n) \leqq \varliminf_{n\to\infty} J(f_n)$$

by Proposition 4, completing the proof.

We shall next consider an analogue of the Dominated Convergence Theorem.

PROPOSITION 6. *Let $(f_n)_{n=1}^{\infty}$ be a sequence of functions in L^1 and suppose that there exists a function g in L^1 such that for all $n \in N$ we have $|f_n| \leqq g$. Then, if*

$$f(x) = \lim_{n\to\infty} f_n(x)$$

for all $x \in X$, we have $f \in L^1$ and

$$J(f) = \lim_{n\to\infty} J(f_n).$$

Proof. The functions $f_n + g$ are nonnegative and

$$J(f_n + g) \leqq 2J(g).$$

Hence, by Proposition 5, we have $f + g \in L^1$ and

$$J(f + g) \leqq \varliminf_{n\to\infty} J(f_n + g) = J(g) + \varliminf_{n\to\infty} J(f_n).$$

Therefore we have that

$$J(f) \leqq \varliminf_{n\to\infty} J(f_n).$$

Since the functions $g - f_n$ are also nonnegative, we have

$$J(g - f) \leqq \varliminf_{n\to\infty} J(g - f_n) = J(g) - \varlimsup_{n\to\infty} J(f_n).$$

Hence

$$\varlimsup_{n\to\infty} J(f_n) \leqq J(f),$$

and therefore

$$\lim_{n\to\infty} J(f_n)$$

exists and is equal to $J(f)$. The proof is finished.

Let $(L^+)^-$ denote the class of those functions on X which are the pointwise limit of a decreasing sequence $(f_n)_{n=1}^\infty$ of functions in L^+ with $I(f_n) < \infty$ and

$$\lim_{n\to\infty} I(f_n) > -\infty.$$

PROPOSITION 7. *If f is any real-valued function on X with $I^*(f)$ finite, then there is a $g \in (L^+)^-$ such that $f \leqq g$ and $I^*(f) = J(g)$.*

Proof. By Proposition 4 applied to the sequence $(-f_n)_{n=1}^\infty$, we get that $(L^+)^- \subset L^1$. If f is any function on X such that $I^*(f)$ is finite, then, given any $n \in N$, we can find $h_n \in L^+$ such that

$$f \leqq h_n \text{ and } I(h_n) \leqq I^*(f) + \frac{1}{n}.$$

Putting $g_n = h_1 \wedge h_2 \wedge \dots \wedge h_n$, we have $f \leqq g_n \leqq h_n$ and so $(g_n)_{n=1}^\infty$ is a decreasing sequence of functions in L^+ with

$$I^*(f) \leqq I(g_n) \leqq I^*(f) + \frac{1}{n}.$$

Hence the function

$$g = \lim_{n\to\infty} g_n$$

is in $(L^+)^-$, while $f \leqq g$ and $I^*(f) = J(g)$, as was to be shown.

A function f on X is called a *null function* if $f \in L^1$ and $J(|f|) = 0$. If f is a null function and $|g| \leqq f$, then

$$0 \leqq I_*(|g|) \leqq I^*(|g|) \leqq J(f) = 0;$$

hence $g \in L^1$, and g is a null function as well.

PROPOSITION 8. *A function f on X is in L^1 if and only if f is the difference $g - h$ of a function g in $(L^+)^-$ and a nonnegative null function h. A function h is a null function if and only if there is a null function w in $(L^+)^-$ such that $|h| \leqq w$.*

Proof. If $f = g - h$, then f is the difference of two functions in L^1 and therefore must itself be in L^1.

If $f \in L^1$, then Proposition 7 asserts the existence of $g \in (L^+)^-$ such that $f \leqq g$ and $J(f) = J(g)$. Hence $h = g - f$ is a nonnegative function and $J(h) = 0$ implying that h is a null function. If h is a null function, then, by Proposition 7, there is a function $w \in (L^+)^-$ with $|h| \leqq w$ and $J(w) = J(|h|) = 0$. The proposition is proved.

PROPOSITION 9. *Suppose that I is a Daniell integral on a linear lattice L and I' is an extension of I to a Daniell integral on a linear lattice $L' \supset L$. If I and I' are extended to give Daniell integrals over L^1 and $(L')^1$, then $(L')^1 \supset L^1$ and J' is an extension of J.*

Proof. By applying Proposition 4 twice we see that $(L^+)^- \subset (L')^1$ and that $J(f) = J'(f)$ for $(L^+)^-$. Hence, by the second part of Proposition 8, each function which is null with respect to J must also be null with respect to J'. By the first part of Proposition 8, every function f in L^1 must be in $(L')^1$, and $J(f) = J'(f)$, completing the proof.

Remark. The foregoing proposition shows that the extension J to L^1 of a Daniell integral I on L is unique.

A nonnegative function f on X is said to be *measurable* (with respect to J) if $g \wedge f$ belongs to L^1 for each g in L^1. We say that a set A in X is *measurable* (with respect to J) if its characteristic function χ_A is measurable; a set A in X is called *integrable* (with respect to J) if $\chi_A \in L^1$. It is clear that a measurable subset of an integrable set is itself integrable.

PROPOSITION 10. *If f and g are nonnegative measurable functions, then so are the functions $f \wedge g$ and $f \vee g$. If $(f_n)_{n=1}^{\infty}$ is a sequence of nonnegative measurable functions which converge pointwise to a function f, then f is measurable.*

Proof. If f and g are nonnegative measurable functions and $h \in L^1$, then $h \wedge (f \wedge g) = (h \wedge f) \wedge (h \wedge g)$ and $h \wedge (f \vee g) = (h \wedge f) \vee (h \wedge g)$. Hence the measurability of $f \wedge g$ and $f \vee g$ follow from the fact that L^1 is a lattice. If $(f_n)_{n=1}^{\infty}$ is a sequence of nonnegative measurable functions converging to f pointwise on X and if g is a function in L^1, then $(f_n \wedge g)_{n=1}^{\infty}$ is a sequence of functions in L^1 converging pointwise to $f \wedge g$. Since $|f_n \wedge g| \leqq |g|$, we have that $f \wedge g \in L^1$ by Proposition 6. The proof is finished.

PROPOSITION 11. *A nonnegative function f on X is measurable with respect to J if $u \wedge f \in L^1$ for each $u \in L$.*

Proof. If $u \in L$, then $u \wedge f \in L^1$, and Proposition 4 implies that $g \wedge f \in L^1$ for $g \in L^+$ and $J(g) < \infty$. By Proposition 6, $g \wedge f \in L^1$ for all $g \in (L^+)^-$. If h is any function in L^1, then $h = g - w$, where $g \in (L^+)^-$ and w is a nonnegative null function, by Proposition 8. Since $0 \leqq g \wedge f$

$-h \wedge f \leqq w$, the function $h \wedge f$ differs from the integrable function $g \wedge f$ by a null function. Thus $h \wedge f$ is in L^1, proving that f is measurable.

PROPOSITION 12. *If A and B are measurable sets with respect to J, then so are their union, intersection, and difference. If $(A_n)_{n=1}^{\infty}$ is a sequence of measurable sets with respect to J, then the sets*

$$\bigcap_{n=1}^{\infty} A_n \text{ and } \bigcup_{n=1}^{\infty} A_n$$

will also be measurable. If the function 1 is measurable with respect to J, then the class $\mathfrak{A}$ of measurable sets with respect to J is a σ-algebra.

Proof. Since

$$\chi_{A \cap B} = \chi_A \wedge \chi_B$$

and

$$\chi_{A \cup B} = \chi_A \vee \chi_B,$$

the measurability of $A \cap B$ and $A \cup B$ follow from the measurability of the sets A and B. If $g \in L^1$, we have

$$g \wedge \chi_{A-B} = g \wedge \chi_A - g \wedge \chi_{A \cap B} + g \wedge 0,$$

and the measurability of $A - B$ follows from that of A and B. Setting

$$A = \bigcup_{n=1}^{\infty} A_n,$$

we get

$$\chi_A = \lim_{n\to\infty} (\chi_{A_1} \vee \cdots \vee \chi_{A_n})$$

and the measurability of A is seen to be a consequence of Proposition 10. A similar argument applies in the case

$$\bigcap_{n=1}^{\infty} A_n.$$

If the constant function 1 is a measurable function, then the set X is a measurable set and the complement of a measurable set will be measurable.

PROPOSITION 13. *If 1 is a measurable function with respect to J and f a nonnegative integrable function, i.e., $f \in L^1$, then for each real number α the set*

$$E_\alpha = \{x \in X : f(x) > \alpha\}$$

is measurable with respect to J.

Proof. If $\alpha < 0$, then $E_\alpha = X$ and is measurable since $\chi_{E_\alpha} = 1$ and 1 is measurable. Suppose that $\alpha \geqq 0$. If $\alpha = 0$, set $g = f$, while if $\alpha > 0$, let $g = (\alpha^{-1}f) - ((\alpha^{-1}f) \wedge 1)$. Since g is the difference of two functions in L^1, g is in L^1. In either case $g(x) > 0$ for $x \in E_\alpha$ and $g(x) = 0$ for $x \in X - E_\alpha$. For $n \in N$, put $h_n = 1 \wedge (ng)$. Then $h_n \in L^1$ and $h_n \uparrow \chi_{E_\alpha}$. Hence χ_{E_α} is measurable and so E_α is measurable.

PROPOSITION 14. *Let the constant function* 1 *be measurable with respect to* J, *and define a set function* μ *on the class* $\mathfrak{A}$ *of measurable sets by*

$$\mu(E) = J(\chi_E) \text{ when } E \text{ is integrable}$$

and

$$\mu(E) = \sup\{\mu(A) : A \subset E, A \text{ integrable}\}.$$

Then μ *is a measure on the* σ*-algebra* $\mathfrak{A}$.

Proof. We have $\mu(\varnothing) = J(0) = 0$. If A and B are integrable sets and $A \subset B$, the positivity of J implies that

$$0 \leqq \mu(A) \leqq \mu(B)$$

so that for all E in $\mathfrak{A}$,

$$\mu(E) = \sup\{\mu(A) : A \subset E, A \text{ integrable}\}.$$

Thus μ is montone for measurable sets. Now, if $(E_n)_{n=1}^\infty$ is a disjoint sequence in $\mathfrak{A}$,

$$E = \bigcup_{n=1}^\infty E_n$$

and A is integrable with $A \subset E$, we put $A_n = A \cap E_n$ so that each A_n is integrable and

$$\mu(A) = \sum_{n=1}^\infty \mu(A_n) \leqq \sum_{n=1}^\infty \mu(E_n),$$

by Proposition 4. Hence

$$\mu(E) = \sup_{A \subset E} \mu(A) \leqq \sum_{n=1}^\infty \mu(E_n).$$

The reverse inequality is immediate if $\mu(E) = +\infty$. If $\mu(E)$ is finite, then $\mu(E_n)$ is finite for each n and, given $\varepsilon > 0$, we can find integrable sets $A_n \subset E_n$ with

$$\mu(A_n) > \mu(E_n) - \frac{\varepsilon}{2^n}.$$

But then

$$\mu(E) \geqq \mu(\bigcup_{n=1}^{\infty} A_n) = \sum_{n=1}^{\infty} \mu(A_n) > \sum_{n=1}^{\infty} \mu(E_n) - \varepsilon,$$

and we must have

$$\mu(E) \geqq \sum_{n=1}^{\infty} \mu(E_n)$$

since $\varepsilon > 0$ is arbitrary. Therefore μ is countably additive on $\mathfrak{A}$. The proof is finished.

The following proposition states that the Daniell integral J on L^1 is equivalent with the integral with respect to the measure defined in Proposition 14; this result is due to M. H. Stone.

PROPOSITION 15. *Let the constant function* 1 *be measurable with respect to* J *and define the measure* μ *on the* σ*-algebra* $\mathfrak{A}$ *of measurable sets as in Proposition 14. Then a function* $f: X \to R_e^1$ *is in* L^1, *i.e.,* f *is integrable with respect to* J, *if and only if it is integrable with respect to this measure* μ. *Moreover*

$$J(f) = \int f d\mu \text{ for all } f \in L^1.$$

Proof. Consider a nonnegative function $f: X \to R_e^1$ in L^1 which is integrable with respect to J. For each pair (k, n) of positive integers, put

$$E_{k,n} = \left\{x \in X : f(x) > \frac{k}{2^n}\right\}.$$

Now, $E_{k,n} \in \mathfrak{A}$ and $\chi_{E_{k,n}} \in L^1$ (that is, $\mu(E_{k,n}) < \infty$) since

$$\chi_{E_{k,n}} = \chi_{E_{k,n}} \wedge \left(\frac{2^n}{k} f\right).$$

Let

$$h_n = \frac{1}{2^n} \sum_{k=1}^{2^{2n}} \chi_{E_{k,n}}$$

and observe that $(h_n)_{n=1}^{\infty}$ is a monotone sequence in L^1 which converges to f. Hence

$$J(f) = \lim_{n\to\infty} J(h_n).$$

But

$$J(h_n) = \frac{1}{2^n} \sum_{k=1}^{2^n} J(\chi_{E_{k,n}}) = \frac{1}{2^n} \sum_{k=1}^{2^{2n}} \mu(E_{k,n}) = \int h_n \, d\mu.$$

Since

$$\int f d\mu = \lim_{n\to\infty} \int h_n \, d\mu,$$

by the Monotone Convergence Theorem we have

$$J(f) = \int f d\mu$$

and f is integrable with respect to μ. Since an arbitrary f which is J-integrable is the difference of two nonnegative J-integrable functions, it follows that such an f must also be integrable with respect to μ and that

$$J(f) = \int f d\mu.$$

If f is a nonnegative function on X which is integrable with respect to μ, we construct $E_{k,n}$ and h_n as before. Since $\int f d\mu < \infty$, each $E_{k,n}$ has finite μ-measure and so $\chi_{E_{k,n}}$, and hence h_n belong to L_1. Since $h_n \uparrow f$ and

$$\lim_{n\to\infty} J(h_n) = \int f d\mu < \infty,$$

we have that $f \in L^1$ by Proposition 4. To see that each f which is integrable with respect to μ is also integrable with respect to J, we decompose f into the difference of two nonnegative functions, namely $f = f^+ - f^-$, where $f^+ = f \vee 0$ and $f^- = (-f) \vee 0$. The proof is finished.

The next question arising is that of uniqueness of the measure μ; the following proposition will show that under suitable conditions the measure μ is unique.

PROPOSITION 16. *Let L be a fixed linear lattice containing the constant function* 1 *and let $\mathfrak{B}$ be the smallest σ-algebra of subsets of X such that each function in L is measurable with respect to μ. Then for each Daniell integral J on L^1 there is a unique measure μ on $\mathfrak{B}$ such that*

$$J(f) = \int f d\mu \text{ for all } f \in L.$$

Proof. The existence of μ has been established in Proposition 15 and we only need to verify the uniqueness μ on $\mathfrak{B}$. Let $\mathfrak{A}$ be the σ-algebra of measurable sets given by Proposition 12. Proposition 13 asserts that each

f in L is measurable with respect to $\mathfrak{A}$ and so $\mathfrak{B} \subset \mathfrak{A}$. Since $1 \in L$, the functions χ_B are in L^1 for each B in $\mathfrak{A}$ and hence for each B in $\mathfrak{B}$. To prove the uniqueness it is enough to establish that for any such μ

$$\mu(B) = J(\chi_B) \text{ for all } B \in \mathfrak{B}.$$

If we let L' be the set of all functions on X which are measurable with respect to $\mathfrak{B}$ and integrable with respect to μ and put

$$J'(f) = \int f d\mu$$

for $f \in L'$, then Proposition 9 implies that

$$J'(f) = J(f)$$

for $f \in L^1 \cap L'$. But if $B \in \mathfrak{B}$, then $\chi_B \in L^1 \cap L'$, and so

$$\mu(B) = J'(\chi_B) = J(\chi_B).$$

Thus μ is indeed uniquely determined on $\mathfrak{B}$ by J and the proposition is proved.

Remark. In what follows we shall consider P. J. Daniell's method of integration for the case where the underlying set X is endowed with a certain topological structure; the discussion will culminate in a famous representation theorem due to F. Riesz.

A topological space X is said to be *Hausdorff* if for any two distinct points x and y there exist disjoint open sets G and H such that $x \in G$ and $y \in H$. A topological space X is called *locally compact* if for each point x in X there is an open set G containing x such that the closure of G, i.e., $\overline{G}$, is compact; thus a topological space is locally compact if and only if the collection of open sets with compact closures form a base for the topology of the space.

Let X be a locally compact Hausdorff space. It can be seen from Proposition 16 of Chapter 7 that X is then completely regular; a Hausdorff space is said to be completely regular if, for every point y in X and every closed set F not containing y, there is a real-valued continuous function f on X such that $0 \leqq f \leqq 1$, $f(y) = 0$ and $f = 1$ on F.

If f is a real-valued function on a locally compact Hausdorff space X, we define the *support* of f to be the closure of the set $\{x \in X : f(x) \neq 0\}$; f is said to have *compact support* if f vanishes outside some compact set. We shall write $C_c(X)$ for the class of all continuous real-valued functions on X with compact support. Clearly $C_c(X)$ is a normed linear space if we put $\|f\| = \sup\{|f(x)| : x \in X\}$ and we also use the fact that $C_c(X)$ is a linear lattice.

The class of *Baire sets* is defined as the smallest σ-algebra $\mathfrak{B}$ of subsets of X such that each function f in $C_c(X)$ is measurable with respect to $\mathfrak{B}$. Thus $\mathfrak{B}$ is the smallest σ-algebra which contains all sets of the form $\{x \in X : f(x) > \alpha\}, f \in C_c(X), \alpha \in R^1$. A measure μ is called a *Baire measure* on X if μ is defined on the σ-algebra $\mathfrak{B}$ of Baire subsets, and $\mu(K)$ is finite for each compact set in $\mathfrak{B}$.

PROPOSITION 17 (RIESZ). *Suppose that X is a locally compact Hausdorff space and I a positive linear functional on the space $C_c(X)$ of continuous functions $f : X \to R^1$ with compact support. Then there is a Baire measure μ on X such that*

$$I(f) = \int f d\mu \text{ for all } f \in C_c(X).$$

If X is compact and $C(X)$ is the set of all continuous functions $f : X \to R^1$, then there is a one-to-one correspondence between positive linear functionals J on $C(X)$ and finite Baire measures μ on X given by

$$J(f) = \int f d\mu.$$

Proof. We first verify that I must be a Daniell integral on $C_c(X)$. Let $f \in C_c(X)$, $(f_n)_{n=1}^{\infty}$ be an increasing sequence in $C_c(X)$ and

$$f(x) \leqq \lim_{n\to\infty} f_n(x)$$

for all $x \in X$. To show that

$$I(f) \leqq \lim_{n\to\infty} I(f_n),$$

it is enough to prove that

$$I(f) = \lim_{n\to\infty} I(g_n),$$

where $g_n = f \wedge f_n$, so that

$$f = \lim_{n\to\infty} g_n \leqq \lim_{n\to\infty} f_n.$$

But then, if we set $h_n = f - g_n$, we obtain a decreasing sequence of functions of $C_c(X)$ whose limit is zero. Let K be the support of h_1, then there is a function v in $C_c(X)$ which is nonnegative and satisfies $v(x) = 1$ for $x \in K$. For each $x \in K$, $\varepsilon > 0$, there is an n_x such that $h_{n_x}(x) < \frac{1}{2}\varepsilon$ and, since h_{n_x} is continuous, there is an open set G_x for which $x \in G_x$ and $h_{n_x}(t) < \varepsilon$ for $t \in G_x$. But K is compact and so there is a finite subcovering $G_{x_1}, \ldots, G_{x_s}$ of K. If $n_0 = \max\{n_{x_1}, \ldots, n_{x_s}\}$ we have $h_n(x) < \varepsilon$ for all x in K, $n \geqq n_0$. Therefore $0 \leqq h_n < \varepsilon v$ and so $0 \leqq I(h_n) < \varepsilon I(v)$. Since $\varepsilon > 0$ is arbitrary, we conclude that

$$\lim_{n\to\infty} I(h_n) = 0$$

and I is therefore a Daniell integral on $C_c(X)$.

We can now apply Proposition 15 to the extension J of I to $L^1 \supset C_c(X)$ to obtain a measure μ on the σ-algebra $\mathfrak{A}$ which contains the Baire sets and such that, for $f \in C_c(X)$,

$$I(f) = J(f) = \int f\,d\mu.$$

By considering the above function v which is in $C_c(X)$ and takes the value 1 on the compact set K, we obtain that

$$\mu(K) = J(\chi_K) \leqq J(v) = \int v\,d\mu < \infty,$$

so that the measure μ we have obtained is finite on compact sets.

When X is compact, $C_c(X)$ is the same as $C(X)$ and in this case the positive linear functionals on $C(X)$ correspond to finite Baire measures. Moreover, because of Proposition 16 there is uniqueness. The proof is finished.

We want to consider more general linear functionals on $C(X)$, namely, the set of all bounded linear functionals; it would be nice to express these as the difference of two positive linear functionals so that Proposition 17 can be used. It is indeed possible to do this as the next proposition will show. We note first that if L is a linear lattice of bounded real-valued functions on X, then L is a normed linear space with $\|f\| = \sup\{|f(x)| : x \in X\}$; a bounded linear functional F has a norm $\|F\| = \sup\{|F(f)| : \|f\| \leqq 1\}$.

Proposition 18. *Let L be a linear lattice of bounded functions mapping a set X into the real line R^1, and suppose that the constant function* 1 *belongs to L. Then for each bounded linear functional F on L there are two positive linear functionals F^+ and F^- such that $F = F^+ - F^-$ and $\|F\| = F^+(1) + F^-(1)$.*

Proof. For each nonnegative f in L, we define

$$F^+(f) = \sup\{F(\varphi) : 0 \leqq \varphi \leqq f\}.$$

Then $F^+(f) \geqq 0$, because $F(0) = 0$ and $F^+(f) \geqq F(f)$. Moreover, $F^+(cf) = cF^+(f)$ for $c \geqq 0$. Let f and g be two nonnegative functions in L. If $0 \leqq \varphi \leqq f$ and $0 \leqq \psi \leqq g$, then $0 \leqq \varphi + \psi \leqq f + g$ and so $F^+(f + g) \geqq F(\varphi) + F(\psi)$. Taking suprema over all such φ and ψ we obtain $F^+(f + g) \geqq F^+(f) + F^-(g)$. On the other hand, if $0 \leqq \psi \leqq f + g$, then $0 \leqq \psi \wedge f \leqq f$ and $0 \leqq \psi - (\psi \wedge f) \leqq g$ and so

$$F(\psi) = F(\psi \wedge f) + F(\psi - [\psi \wedge f]) \leqq F^+(f) + F^+(g).$$

Taking the suprema over all such ψ we get $F^+(f + g) \leqq F^+(f) + F^+(g)$ and so $F^+(f + g) = F^+(f) + F^+(g)$.

Let f be an arbitrary function in L and let p and q be two nonnegative constants such that $f + p$ and $f + q$ are nonnegative. Then $F^+(f + p + q) = F^+(f + p) + F^+(q) = F^+(f + q) + F^+(p)$. Hence $F^+(f + p) - F^+(p) = F^+(f + q) - F^+(q)$. Thus the value of $F^+(f + p) - F^+(p)$ is independent of the choice of p, and we define $F^+(f)$ to be this value. The functional F^+ is now defined on all of L and we have $F^+(f + g) = F^+(f) + F^+(g)$ and $F^+(cf) = cF^+(f)$ for $c \geqq 0$. Since $F^+(-f) + F^+(f) = F^+(0) = 0$, we have $F^+(-f) = -F^+(f)$ and F^+ is a linear functional on L.

Since $0 \leqq F^+(f)$ and $F(f) \leqq F^+(f)$ for $f \geqq 0$, both F^+ and the linear functional $F^- = F^+ - F$ are positive linear functionals; clearly $F = F^+ - F^-$.

We always have $\|F\| \leqq \|F^+\| + \|F^-\| = F^+(1) + F^-(1)$. To verify the inequality in the opposite direction, let h be any function in L such that $0 \leqq h \leqq 1$. Then $|2h - 1| \leqq 1$ and

$$\|F\| \geqq F(2h - 1) = 2F(h) - F(1).$$

Taking the supremum over all such h we get

$$\|F\| \geqq 2F^+(1) - F(1) = F^+(1) + F^-(1).$$

Hence $\|F\| = F^+(1) + F^-(1)$ and the proof is complete.

PROPOSITION 19 (RIESZ). *Let X be a compact Hausdorff space and $C(X)$ the set of all continuous functions $f: X \to R^1$. Then to each bounded linear functional F on $C(X)$ there corresponds a unique finite signed Baire measure ν on X such that*

$$F(f) = \int f d\nu$$

for each $C(X)$. Moreover, $\|F\| = |\nu|(X)$. In other words, the dual space of $C(X)$ is isometrically isomorphic to the space of all finite signed Baire measures on X with norm defined by $\|\nu\| = |\nu|(X)$.

Proof. If one starts with a finite signed Baire measure ν, then there is a Jordan decompostion $\nu = \nu^+ - \nu^-$ into the difference of two finite Baire measures. Evidently

$$F(f) = \int f d\nu^+ - \int f d\nu^-$$

then defines a bounded linear functional on $C(X)$ since each function f in $C(X)$ is bounded and measurable with respect to the class of Baire sets.

Conversely, given a bounded linear functional F on $C(X)$, this can be decomposed by Proposition 18 into the difference $F = F^+ - F^-$ of two positive linear functionals. Using Proposition 17 we find finite Baire measure μ_1 and μ_2 with

$$F^+(f) = \int f d\mu_1, \; F^-(f) = \int f d\mu_2 .$$

If we put $\nu = \mu_1 - \mu_2$, then ν is a finite Baire measure and

$$F(f) = \int f d\nu.$$

Since

$$|F(f)| \leqq \int |f| d|\nu| \leqq \|f\| \cdot |\nu|(X),$$

we see that $\|F\| \leqq |\nu|(X)$. On the other hand

$$|\nu|(X) \leqq \mu_1(X) + \mu_2(X) = F^+(1) + F^-(1) = \|F\|$$

and so we have $\|F\| = |\nu|(X)$.

Finally, to see that ν is uniquely determined by F, we assume that there are two signed measures ν_1 and ν_2 with

$$\int f d\nu_1 = \int f d\nu_2 \text{ for each } f \in C(X).$$

We decompose $\lambda = \nu_1 - \nu_2$ by use of the Jordan Decomposition Theorem to give $\lambda = \lambda^+ - \lambda^-$. Then $\int f d\lambda^+ = \int f d\lambda^-$ for all $f \in C(X)$. By Proposition 17 we must have $\lambda^+ = \lambda^-$. Hence $\lambda = 0$ and $\nu_1 = \nu_2$. The proof is finished.

Concluding Remarks

For a different approach to the Riesz Representation Theorem (i. e., Propositions 17 and 19), the reader might consult Chapters 2 and 6 of *Real and Complex Analysis* by Walter Rudin, McGraw-Hill, New York, 1966.

A definitive treatment of Daniell's method of integration is due to M. H. Stone (see M. H. Stone, "Notes on Integration," I - IV. *Proc. Nat. Acad. Sci. U.S.A.* **34** (1948); **35** (1949).) In Chapter 9 we offer a survey of the Stone-Daniell integral.

EXERCISES

1. Prove that the conditions (D), (D*), and (D**) listed in defining a Daniell integral are equivalent.
2. Consider the two examples of Daniell integrals given after the definition of a Daniell integral at the beginning of this chapter and supply the missing proofs.
3. The following exercise illustrates that the summation of an absolutely convergent series is an example of an integration process.

   Let X be the class of positive integers N, and let L be the class of functions $f : X \to R^1$ (i.e., real sequences) such that $f(n) = 0$ for all except a finite number of values of n. Define $I : L \to R^1$ by

   $$I(f) = \sum_{n=1}^{\infty} f(n).$$

   Is L a linear lattice and I a Daniell integral on L? If $f : X \to R_e^1$ is such that $f(n)$ is never $-\infty$ and $f(n) \geqq 0$ when $n > n_0$, where n_0 depends on f, then $f \in L^+$. Is L^+ composed exactly of all such functions? Show that $f : X \to R_e^1$ is integrable if and only if $f(n)$ is finite for each n and

   $$I(f) = \sum_{n=1}^{\infty} |f(n)| < \infty.$$

   Will in this case

   $$J(f) = \sum_{n=1}^{\infty} f(n)$$

   hold?
4. Take X and L as in Exercise 3, but define the integral I on L by

   $$I(f) = \sum_{n=1}^{\infty} \frac{f(n)}{n}.$$

   Find L^+ and L^1 and $J(f)$ when $f \in L^1$.
5. Define for any $f : X \to R_e^1$

   $$N(f) = \inf\{I(g) : g \in L^+, |f| \leqq g\}$$

   if there exists a $g \in L^+$ for which $|f| \leqq g$. Show that

   $$N(f) = J(|f|)$$

   for such an f. If no g of the required kind exists, define

   $$N(f) = +\infty.$$

Observe that $N(f) = I(f)$ when $f \in L^+$ and $f \geqq 0$.

Can the function N be used to construct a norm in the linear space formed from the functions f for which $N(f) < +\infty$?

6. Let T denote the class of all functions $f : X \to R_e^1$ such that $N(f) < +\infty$ (see Exercise 5). Prove the following three statements:
   (i) Let $(f_n)_{n=1}^{\infty}$ be a sequence in T such that $N(f_n - f_m) \to 0$ as $m \to \infty$ and $n \to \infty$. Then there is a function f in T such that $N(f_n - f) \to 0$.
   (ii) Suppose $f_n \in L^1$, $f \in T$ and $N(f_n - f) \to 0$. Then $f \in L^1$.
   (iii) If f and g belong to L^1, then $(f^2 + g^2)^{1/2}$ belongs to L^1.
7. Show that in a locally compact metric space the class of Baire sets coincides with the class of Borel sets (i.e., the smallest σ-algebra containing the topology of the space).
8. Let μ be a Baire measure on a locally compact space X and let H be the union of all open Baire sets G for which $\mu(G) = 0$. The complement $F = X - H$ is closed and called the *support* of μ. Show
   (i) if G is an open Baire set and $G \cap F$ is nonempty, then $\mu(G) > 0$;
   (ii) if K is a compact Baire set such that $K \cap F$ is empty, then $\mu(K) = 0$;
   (iii) if $f \in C_c(X)$ and $f \geqq 0$, $\int f d\mu = 0$ if and only if $f = 0$.
9. A *Radon measure* ψ on a locally compact space is defined as a linear functional on $C_c(X)$ which is continuous in the sense that, for each compact K, $\varepsilon > 0$, there is a $\delta > 0$ such that $|f(x)| < \delta$ for all x, with the support of f contained in K, implies that $|\psi(f)| < \varepsilon$.

   Verify that every positive linear functional is a Radon measure.

   For R^1 and the usual topology define

$$\psi(f) = \sum_{n=-\infty}^{\infty} (-1)^n f(n) \quad \text{for } f \in C_c(R^1).$$

   Show that ψ is a Radon measure, but that ψ does not correspond to any signed Baire measure.
10. Let L and F be as in Proposition 18. If G and H are two positive linear functionals on L such that $F = G - H$ and $G(1) + H(1) \leqq \|F\|$, show that then $G = F^+$ and $H = F^-$.
11. Let X be a locally compact Hausdorff space and $C_c(X)$ be the class of all continuous real-valued functions of compact support on X. Show that every positive linear functional I on $C_c(X)$ is a Daniell integral. [*Hints*: (a) Let $(f_n) \subset C_c(X)$ be a pointwise monotonically decreasing sequence and suppose that $f_n \to 0$ pointwise. For an arbitrary $\varepsilon > 0$, let $E_n = \{x \in X : f_n(x) \geqq \varepsilon\}$. Since f_n is continuous and vanishes outside a compact set of X, E_n, being a closed subset of the support of f_n, is compact for each $n \geqq 1$. Moreover, since $f_n(x) \to 0$ for each x,

$$\bigcap_{n \geqq 1} E_n = \varnothing .$$

Now the compactness of E_n implies that there is at least one m for which $E_m = \varnothing$. Hence

$$\sup_x |f_n(x)| \leqq \varepsilon$$

for all $n \geqq m$. This shows that $f_n \to 0$ in the metric of uniform convergence.

(b) Next we verify that I is uniformly bounded on a subspace $C_c(A)$ of $C_c(X)$, where A is a compact subset of X and $C_c(A)$ consists of those $f \in C_c(X)$ that vanish outside A. For this, let $g \in C_c^+(X) = \{f \in C_c(X) : f \geqq 0\}$ such that $g(x) \geqq 1$ for all $x \in A$. Then for any $f \in C_c(A)$,

$$|f(x)| \leqq \sup_{x \in X} |f(x)| g(x).$$

Hence

$$I(f) \leqq \sup_{x \in X} |f(x)| I(g)$$

and

$$\|I\| = \sup_{\|f\| \leqq 1} |I(f)| \leqq \sup_{x \in X} |f(x)| I(g) < \infty.$$

This shows that I is bounded on $C_c(A)$.
Finally, let $f_n \to 0$, where the sequence $(f_n)_{n=1}^{\infty}$ is monotonically decreasing and the convergence is pointwise. Then, by (a), pointwise convergence implies

$$\sup_x |f_n(x)| \to 0$$

as $n \to \infty$, where

$$\sup_x |f_n(x)|$$

is a monotonically decreasing sequence of real numbers. If $f_1 \in C_c(A)$, then $f_n \in C_c(A)$ for all $n \geqq 1$. Let $M > 0$ such that $\|I\| \leqq M$ on $C_c(A)$ by (b). Then

$$I(f_n) \leqq M \sup_x |f_n(x)| \to 0 \text{ as } n \to \infty.]$$

12. Let $(a_j)_{j=0}^{\infty}$ be a given sequence of real numbers. Show that there is an infinitely often differentiable function $f : [0, \infty) \to R^1$ whose support is contained in the interval $[0, 1]$ such that $f^{(j)}(0) = a_j$ for $j = 0, 1, 2, \ldots$.
[*Hints*: We first consider two lemmas.

LEMMA 1. *There is an infinitely often differentiable function* $g : [0, \infty) \to R^1$ *with* $g(0) = 1$, $0 \leqq g \leqq 1$, *and whose support is contained in the interval* $[0, a]$ *with* $a > 0$.

Proof. Let $h : R^1 \to R^1$ be defined by $h(x) = \exp(-1/x^2)$ when $x > 0$ and $h(x) = 0$ when $x \leqq 0$. Then define

$$g(x) = (h(a - x))/(h(a - x) + h(x - a/2)).$$

Clearly, g is infinitely often differentiable, $g(x) = 0$ for $x \geqq a$, $g(x) = 1$ for $x \leqq a/2$, and $0 \leqq g(x) \leqq 1$. This proves the lemma.

LEMMA 2. *For any* $n \geqq 1$ *and any* $\varepsilon > 0$, *there is an infinitely often differentiable function* $f : [0, \infty) \to R^1$ *whose support is contained in the interval* $[0, 1]$ *such that*

(*i*) $f^{(n)}(0) = 1$, $f^{(k)}(0) = 0$ *for* $k < n$;

(*ii*) $|f^{(k)}| < \varepsilon$ *for* $k < n$.

Proof. Take g as in Lemma 1 with support contained in $[0, a]$ and also a function v as in Lemma 1 with support contained in $[0, 1]$. Define

$$f(s) = v(s) \int_0^s \int_0^{t_n} \ldots \int_0^{t_3} \int_0^{t_2} g(t_1)\, dt_1 \ldots dt_n$$

and choose the number a small enough such that (*ii*) holds. Obviously (*i*) holds. This proves Lemma 2.

Returning now to the problem, we define

$$f = \sum_{j=0}^{\infty} c_j f_j,$$

where f_j and c_j are defined recursively: $f_j : [0, \infty) \to R^1$, f_j infinitely often differentiable and with support contained in $[0, 1]$, f_0 as in Lemma 2, c_0 arbitrary. Then by Lemma 2 we can choose f_j in the following way:

(1) $f_n^{(n)}(0) = 1$, $f_n^{(k)}(0) = 0$ for $k < n$;

(2) $c_n = a_n - (c_{n-1} f_{n-1}{}^{(n)}(0) + \ldots + c_0 f_0{}^{(n)}(0))$;

(3) $|c_n f_n^{(k)}| \leqq 1/2^n$ for $k < n$.

By (3), the series

$$\sum_{j=0}^{\infty} c_j f_j^{(k)}$$

converges uniformly for every k. On the other hand,

$$\sum_{j=0}^{\infty} c_j f_j(0) = 0.$$

Therefore, f is infinitely often differentiable and its support is contained in $[0, 1]$; moreover,

$$f^{(k)}(0) = \sum_{j=0}^{\infty} c_j f_j^{(k)}(0).$$

By (1),

$$\begin{aligned} f^{(k)}(0) &= \sum_{j=0}^{k} c_j f_j^{(k)}(0) \\ &= (c_0 f_0{}^{(k)}(0) + \ldots + c_{k-1} f_{k-1}{}^{(k)}(0)) + c_k f_k{}^{(k)}(0) \\ &= (a_k - c_k) + c_k = a_k .] \end{aligned}$$

13. Given a sequence of real numbers $(a_j)_{j=0}^{\infty}$, show that there is an infinitely often differentiable function $f : R^1 \to R^1$ with support contained in the interval $[-1, 1]$ such that $f^{(j)}(0) = a_j$ for $j = 0, 1, 2, \ldots$.
14. Show that there exist infinitely often differentiable functions $f : R^1 \to R^1$ that cannot be represented by power series.
[*Hint*: Take a sequence $(a_j)_{j=0}^{\infty}$ such that

$$\limsup_{n\to\infty}(|a_n|)^{1/n} = +\infty$$

and construct f as in the foregoing exercise.]

Chapter 9

The Stone-Daniell Integral (A Survey)

Section 1. Approximation of Continuous Functions

PROPOSITION 1 (K. WEIERSTRASS). *If f is a real-valued continuous function on the interval* $[0, 1]$, *then*

$$f(x) = \lim_{n\to\infty} \sum_{k=0}^{n} \binom{n}{k} f\left(\frac{k}{n}\right) x^k (1-x)^{n-k}$$

holds uniformly on $[0, 1]$.

Proof. We begin with some preliminary computations. For any real numbers p and q we have, by the binomial theorem,

$$\sum_{k=0}^{n} \binom{n}{k} p^k q^{n-k} = (p+q)^n \tag{1}$$

for $n \in N$. Differentiating with respect to p we obtain

$$\sum_{k=0}^{n} \binom{n}{k} k p^{k-1} q^{n-k} = n(p+q)^{n-1}$$

which implies

$$\sum_{k=0}^{n} \frac{k}{n} \binom{n}{k} p^k q^{n-k} = p(p+q)^{n-1} \tag{2}$$

for $n \in N$. Differentiating once more we have

$$\sum_{k=0}^{n} \frac{k^2}{n} \binom{n}{k} p^{k-1} q^{n-k} = p(n-1)(p+q)^{n-2} + (p+q)^{n-1}$$

and so

$$\sum_{k=0}^{n} \frac{k^2}{n} \binom{n}{k} p^k q^{n-k} = p^2\left(1 - \frac{1}{n}\right)(p+q)^{n-2} + \frac{p}{n}(p+q)^{n-1}. \tag{3}$$

Setting $p = x$ and $q = 1 - x$ with $x \in [0, 1]$, we see that (1), (2), and (3) yield

$$\sum_{k=0}^{n} \binom{n}{k} x^k (1 - x)^{n-k} = 1, \quad \sum_{k=0}^{n} \frac{k}{n} \binom{n}{k} x^k (1 - x)^{n-k} = x$$

and

$$\sum_{k=0}^{n} \frac{k^2}{n^2} \binom{n}{k} x^k (1 - x)^{n-k} = x^2 \left(1 - \frac{1}{n}\right) + \frac{x}{n}.$$

On expanding $((k/n) - x)^2$ we get from the foregoing three equations that

$$\sum_{k=0}^{n} \left(\frac{k}{n} - x\right)^2 \binom{n}{k} x^k (1 - x)^{n-k} = \frac{x(1 - x)}{n} \tag{4}$$

for $x \in [0, 1]$.

Observing that f is uniformly continuous on $[0, 1]$, we have: Given $\varepsilon > 0$ there exists $\delta > 0$ such that

$$|f(x) - f(y)| < \frac{\varepsilon}{2}$$

whenever $|x - y| < \delta$ with $x, y \in [0, 1]$. We now choose $n_0 \in N$ such that

$$\frac{1}{\sqrt[4]{n_0}} < \delta \tag{5}$$

and such that

$$\frac{1}{\sqrt[2]{n_0}} < \frac{\varepsilon}{4\|f\|}, \text{ where } \|f\| = \sup_{x \in [0,1]} |f(x)|. \tag{6}$$

(We assume, of course, that $\|f\| > 0$.) We fix $x \in [0, 1]$. Multiplying

$$\sum_{k=0}^{n} \binom{n}{k} x^k (1 - x)^{n-k} = 1$$

by $f(x)$ and subtracting

$$\sum_{k=0}^{n} f\left(\frac{k}{n}\right) \binom{n}{k} x^k (1 - x)^{n-k}$$

we obtain, for every $n \in N$,

$$\begin{aligned} f(x) - \sum_{k=0}^{n} f\left(\frac{k}{n}\right)\binom{n}{k}x^k(1-x)^{n-k} \\ = \sum_{k=0}^{n}\left(f(x) - f\left(\frac{k}{n}\right)\right)\binom{n}{k}x^k(1-x)^{n-k} = \Sigma' + \Sigma'', \end{aligned} \tag{7}$$

where $\sum'$ is the sum over those values of k such that

$$\left|\frac{k}{n} - x\right| < \frac{1}{\sqrt[4]{n}}, \tag{8}$$

while $\sum''$ is the sum over the other values of k. If k does not satisfy (8), that is, if

$$\left|\frac{k}{n} - x\right| \geqq \frac{1}{\sqrt[4]{n}},$$

then $(k - nx)^2 = n^2|(k/n) - x|^2 \geqq \sqrt[2]{n^3}$. Hence

$$\begin{aligned} |\Sigma''| &= \left|\Sigma''\left(f(x) - f\left(\frac{k}{n}\right)\right)\binom{n}{k}x^k(1-x)^{n-k}\right| \\ &\leqq \Sigma''\left(|f(x)| + \left|f\left(\frac{k}{n}\right)\right|\right)\binom{n}{k}x^k(1-x)^{n-k} \\ &\leqq 2\|f\|\ \Sigma''\binom{n}{k}x^k(1-x)^{n-k} \\ &\leqq \frac{2\|f\|}{\sqrt[2]{n^3}}\ \Sigma''\ (k - nx)^2\binom{n}{k}x^k(1-x)^{n-k} \\ &\leqq \frac{2\|f\|}{\sqrt[2]{n^3}} \sum_{k=0}^{n} (k - nx)^2\binom{n}{k}x^k(1-x)^{n-k}. \end{aligned}$$

Hence by (4)

$$|\Sigma''| \leqq \frac{2\|f\|}{\sqrt[2]{n^3}} nx(1-x) < \frac{2\|f\|}{\sqrt[2]{n}}$$

(observe that $x(1 - x) \leqq 1/4$ because $4x^2 - 4x + 1 = (2x - 1)^2 \geqq 0$). If $n \geqq n_0$, it follows from (6) that

$$\frac{1}{\sqrt[2]{n}} < \frac{\varepsilon}{4\|f\|}$$

and so

$$|\Sigma''| < \frac{\varepsilon}{2}.$$

Moreover, if $n \geqq n_0$ and k satisfies (8), then, by (5) and (8),

$$\left|\frac{k}{n} - x\right| < \delta \text{ and so } \left|f(x) - f\left(\frac{k}{n}\right)\right| < \frac{\varepsilon}{2}.$$

Thus

$$\begin{aligned}|\Sigma'| &= \left|\Sigma' \left(f(x) - f\left(\frac{k}{n}\right)\right)\binom{n}{k}x^k(1-x)^{n-k}\right| \\ &< \frac{\varepsilon}{2}\Sigma' \binom{n}{k}x^k(1-x)^{n-k}.\end{aligned}$$

But

$$\sum_{k=0}^{n} \binom{n}{k}x^k(1-x)^{n-k} = 1$$

and so

$$|\Sigma'| < \frac{\varepsilon}{2}.$$

From (7) we therefore see that

$$\begin{aligned}&\left|f(x) - \sum_{k=0}^{n} f\left(\frac{k}{n}\right)\binom{n}{k}x^k(1-x)^{n-k}\right| \\ &\qquad \leqq |\Sigma'| + |\Sigma''| < \frac{\varepsilon}{2} + \frac{\varepsilon}{2} = \varepsilon.\end{aligned}$$

Since x was any point in $[0, 1]$, and n any integer with $n \geqq n_0$, this shows that

$$\left|f(x) - \sum_{k=1}^{n} f\left(\frac{k}{n}\right)\binom{n}{k}x^k(1-x)^{n-k}\right| < \varepsilon$$

for all $x \in [0, 1]$ and $n \geqq n_0$. The proof is finished.

Remarks. The function

$$\sum_{k=0}^{n} f\left(\frac{k}{n}\right)\binom{n}{k}x^k(1-x)^{n-k}$$

is called the *Bernstein polynomial of f of degree n*. The foregoing proof is due to S. Bernstein.

Using Proposition 1 one may easily show that any continuous real-valued function on a bounded interval $[a, b]$ can be approximated uniformly by polynomials. Another proof of this fact may be based on the approximation of the special function $f(x) = |x|$. Assuming that $x \in [-1, 1]$ and putting $u = 1 - x^2$, we have

$$|x| = (1 - (1 - x^2))^{1/2} = (1 - u)^{1/2}.$$

We use now the fact that the binomial series

$$(1 - u)^{1/2} = 1 - \frac{1}{2}u - \frac{1}{2^2 \cdot 2!}u^2 - \frac{1 \cdot 3}{2^3 \cdot 3!}u^3 - \dots$$

is uniformly convergent in the closed interval $0 \leqq u \leqq 1$. Therefore, $(1 - u)^{1/2}$ admits uniform approximation in $0 \leqq u \leqq 1$ by polynomials in u, and $|x|$ in $-1 \leqq x \leqq 1$ by polynomials in x.

The graphical representation of $|x|$ is a polygonal line with one angle. We wish to show that any polygonal function admits uniform approximation by polynomials. First, this is true for $|x|$ on each interval $(-k, k)$, since $|x| = k \cdot |x/k|$ and $|x/k| \leqq 1$ on $(-k, k)$. Again, each function

$$L_c(x) = \begin{cases} 0 & \text{for } x \leqq c, \\ x - c & \text{for } x \geqq c, \end{cases}$$

may be approximated by polynomials on each interval $[a, b]$, since $L_c(x) = (1/2)((x - c) + |x - c|)$. Suppose now that g is any polygonal function on $[a, b]$ whose graph has angles at the basic points $a = a_0 < a_1 < \dots < a_n = b$. We shall see that g is a linear combination of the L_c. We put

$$g_0(x) = g(a) + C_0 L_{a_0}(x) + C_1 L_{a_1}(x) + \dots + C_{n-1} L_{a_{n-1}}(x),$$

and define the constants C_j by the equations

$$g_0(a_j) = g(a_j), \quad j = 0, 1, \dots, n.$$

The first of these equations is an identity, the second is $g(a) + C_0 L_{a_0}(a_1) = g_0(a_1)$ and defines C_0, the third defines C_1, and so on. The two polygonal functions g and g_0 coincide at all basic points and are therefore identical. This shows that any g admits approximation by polynomials.

Finally, if f is continuous on a compact interval $[a, b]$, it is uniformly continuous on it and hence can be approximated uniformly with an arbitrary small error by a polygonal function on it.

The foregoing method of proof is due to Lebesgue.

We shall now study the set $\mathfrak{C} = \mathfrak{C}(E; R^1)$ of continuous real-valued functions defined on a compact Hausdorff space E; the real-valued functions under consideration throughout this section are assumed always to be finite-valued. We shall moreover suppose that $\mathfrak{C}$ is endowed with the topology of uniform convergence; this topology is defined by the norm

$$\|f\| = \sup_{x \in E} |f(x)|;$$

with $\|f - g\|$ as distance $\mathfrak{C}$ is seen to be a complete metric space.

If H is a subset of $\mathfrak{C}$, we shall say that a continuous real-valued function f on E can be uniformly approximated by functions in H if f lies in the closure of H in the space $\mathfrak{C}$, i.e., if, for each $\varepsilon > 0$, there exists a function $g \in H$ such that $|f(x) - g(x)| \leqq \varepsilon$ for all $x \in E$. To say that every continuous real-valued function on E can be uniformly approximated by functions of H therefore means that H is dense in $\mathfrak{C}$.

On the set $\mathfrak{C}$, the relation $f \leqq g$ (which means that $f(x) \leqq g(x)$ for all $x \in E$) is an order relation, with respect to which $\mathfrak{C}$ is a lattice. Since $\||f| - |g|\| \leqq \|f - g\|$, we have that $f \to |f|$ is a uniformly continuous mapping of $\mathfrak{C}$ into itself and it follows that

$$(f, g) \to \sup(f, g) = \tfrac{1}{2}(f + g + |f - g|)$$

and

$$(f, g) \to \inf(f, g) = \tfrac{1}{2}(f + g - |f - g|)$$

are uniformly continuous on $\mathfrak{C} \times \mathfrak{C}$.

PROPOSITION 2. *Let E be a compact Hausdorff space and let H be a set of continuous real-valued functions defined on E. Let f be a continuous real-valued function on E such that for each $x \in E$ there exists a function $u_x \in H$ such that $u_x(x) > f(x)$ (resp. $u_x(x) < f(x)$). Then there exists a finite number of functions $u_{x_i} = f_i \in H$ $(1 \leqq i \leqq n)$ such that, if $v = \sup(f_1, \ldots, f_n)$ (resp. $w = \inf(f_1, \ldots, f_n)$), we have $v(x) > f(x)$ (resp. $w(x) < f(x)$) for all $x \in E$.*

Proof. For each $x \in E$, let G_x be the open set consisting of all $z \in E$ such that $u_x(z) > f(z)$ (resp. $u_x(z) < f(z)$). Since $x \in G_x$ by assumption, E is the union of the sets G_x as x runs through E. Since E is compact there exists a finite number of points x_i $(1 \leqq i \leqq n)$ such that the G_{x_i} cover E, and it is clear that the functions $f_i = u_{x_i}$ satisfy the conditions of the proposition. The proof is finished.

DEFINITION. *If H is a subset of $\mathfrak{C}$ we say that H is directed with respect to the relation $\leqq$ if for each f, $g \in H$ there exists an $h \in H$ such that $f \leqq h$ and $g \leqq h$; in an analogous manner we can define directedness with respect to the relation $\geqq$.*

PROPOSITION 3. *Let E be a compact Hausdorff space, and let H be a set of continuous real-valued functions on E which is directed with respect to the relation $\leqq$ (resp. $\geqq$). If $f = \sup(h : h \in H)$ (resp. $f = \inf(h : h \in H)$) is finite and continuous on E, then f can be uniformly approximated by functions belonging to H.*

Proof. Given any $\varepsilon > 0$, for each $x \in E$ there exists a function $u_x \in H$ such that $u_x(x) > f(x) - \varepsilon$. By Proposition 2 and the fact that H is directed with respect to the relation $\leqq$, there exists $g \in H$ such that $g(x) > f(x) - \varepsilon$ for all $x \in E$; on the other hand, we have $g(x) \leqq f(x)$ by definition, and thus the proposition is proved.

Application. The foregoing proposition, due to Dini, has the following important consequence: If an increasing (resp. decreasing) sequence of continuous real-valued functions on a compact Hausdorff space converges pointwise to a finite and continuous function on the space, then the sequence converges uniformly to this limit function. It is essential to assume compactness for the validity of the conclusion as the following example shows: Consider the decreasing sequence of functions $x/(n + x)$ for $x \geqq 0$.

PROPOSITION 4. *Let E be a compact Hausdorff space, and let H be a set of continuous real-valued functions on E such that, given any two functions $u \in H$, $v \in H$, the functions $\sup(u, v)$ and $\inf(u, v)$ are in H. Then a continuous real-valued function f on E can be uniformly approximated by functions belonging to H if and only if, for each real number $\varepsilon > 0$ and each pair x, y of points of E, there is a function $u_{x,y} \in H$ such that $|f(x) - u_{x,y}(x)| < \varepsilon$ and $|f(y) - u_{x,y}(y)| < \varepsilon$.*

Proof. It is easy to see that the condition is necessary; we shall show that it is also sufficient. For each $\varepsilon > 0$, we shall show that there is a function $g \in H$ such that, for all $z \in E$, $|f(z) - g(z)| < \varepsilon$. Let x be any point of E, and let H_x be the set of all functions $u \in H$ such that $u(x) < f(x) + \varepsilon$. By assumption, for each $y \in E$, the function $u_{x,y}$ belongs to H_x and we have $u_{x,y}(y) > f(y) - \varepsilon$. Hence, by Proposition 2, there is a finite number of functions of H_x, say, $h_1, \ldots, h_n$, such that $v_x(z) = \max(h_1(z), \ldots, h_n(z)) > f(z) - \varepsilon$ for all $z \in E$; on the other hand, we have $v_x(x) < f(x) + \varepsilon$ by the definition of H_x; finally, $v_x \in H$ by assumption. Proposition 2 therefore shows that there exists a finite number of functions v_{x_i} $(1 \leqq i \leqq m)$ such that $g(z) = \min(v_{x_1}(z), \ldots, v_{x_m}(z)) < f(z) + \varepsilon$ for all $z \in E$; but since we have $v_{x_i}(z) > f(z) - \varepsilon$ for all $z \in E$ and for every index i, we have also $g(z) > f(z) - \varepsilon$ for all $z \in E$. Since $g \in H$ by assumption, the proof is complete.

Remark. When the subset H satisfies the conditions of Proposition 4, it is a lattice with respect to the ordering $f \leqq g$. But a subset H of $\mathfrak{C}$ can be a lattice with respect to this ordering without it being necessarily the case that the supremum (resp. infimum) *in* H of two functions u, v of H is the same as their supremum (resp. infimum) *in* $\mathfrak{C}$; in other words, H in Proposition 4 is in fact a sublattice of $\mathfrak{C}$. An example is provided by the continuous convex mappings of a compact interval I of R^1 into R^1 which is not a sublattice of $\mathfrak{C}(I, R^1)$.

PROPOSITION 5. *Suppose that H is such that, whenever $u \in H$ and $v \in H$, we have* $\sup(u, v) \in H$ *and* $\inf(u, v) \in H$; *and is such that, given any two distinct points x, y of E and any two real numbers α, β, there is a function $g \in H$ such that $g(x) = \alpha$ and $g(y) = \beta$. Then every continuous real-valued function on E can be uniformly approximated by functions belonging to H.*

Remark. This proposition is a direct consequence of Proposition 4.

DEFINITION. *If X is any set, a set H of mappings of X into a set Y is said to separate the elements of a subset A of X if, given any two distinct elements x, y of A, there is a function $f \in H$ such that $f(x) \neq f(y)$.*

PROPOSITION 6 (M. H. STONE). *Let H be a subset of $\mathfrak{C}$ having the following properties*: (*1*) *H contains all constant functions*; (*2*) *H is a vector subspace of* $\mathfrak{C}$; (*3*) *if $h \in H$, then* $|h| = \max(h, -h) \in H$; (*4*) *H separates the points of E. Then every continuous real-valued function on E can be uniformly approximated by functions of H.*

Proof. It is enough to show that H satisfies the conditions of Proposition 5. By assumption, if $u \in H$ and $v \in H$, we have

$$\sup(u, v) = \tfrac{1}{2}(u + v + |u - v|) \in H$$

and

$$\inf(u, v) = \tfrac{1}{2}(u + v - |u - v|) \in H.$$

On the other hand, let x and y be any two distinct points of E, and let α, β be any two real numbers. By assumption, there is a function $h \in H$ such that $h(x) \neq h(y)$; say $h(x) = \gamma$ and $h(y) = \delta$. Since the constants belong to H, the function

$$g(z) = \alpha + (\beta - \alpha)\frac{h(z) - \gamma}{\delta - \gamma}$$

belongs to H and is such that $g(x) = \alpha$ and $g(y) = \beta$. The proof is finished.

Given a set H of real-valued functions defined on a set E, we say that a real-valued function of E is a polynomial with real coefficients in the functions of H, if it is of the form $x \to g(f_1(x), f_2(x), \ldots, f_n(x))$ where g is a polynomial in n indeterminates (n arbitrary) with real coefficients, and the f_i ($1 \leqq i \leqq n$) belong to H. Let H_0 denote the set of all polynomials in the functions of H.

In the following we take E to be a compact Hausdorff space and H to be a subset of $\mathfrak{C}$. Let $\overline{H}_o$ denote the closure of H_0 (with H_0 denoting the set of all polynomials in the functions of H as defined above) in $\mathfrak{C}$. It is clear that

$$H \subset H_0 \subset \overline{H}_0 \subset \mathfrak{C}.$$

The equation $\overline{H}_0 = \mathfrak{C}$ means that every continuous real-valued function on E can be approximated uniformly by polynomials (with real coefficients) in the functions of H. The following is M. H. Stone's generalization of the Weierstrass approximation theorem.

PROPOSITION 7. *Let E be a compact Hausdorff space and let H be a subset of $\mathfrak{C}$. Then $\overline{H}_0 = \mathfrak{C}$ if and only if H separates the points of E.*

Before we proceed to the proof of the proposition we shall consider two lemmas.

LEMMA 1. *For $\varepsilon > 0$ and $r > 0$ there is a polynomial P in the variable t with real coefficients such that (i) $P(0) = 0$ and (ii) $|P(t) - t^{1/2}| < \varepsilon$ and $0 \leqq t \leqq r$.*

LEMMA 2. *If $0 \leqq s < t$, then $t^{1/2} - s^{1/2} \leqq (t - s)^{1/2}$.*

The proof of Lemma 2 is fairly easy and we omit it. Using Proposition 1 we can readily prove Lemma 1; by Proposition 1 there exists a polynomial P_1 with $|P_1(t) - t^{1/2}| < \varepsilon/2$ on the interval in question and it is clear that $P(t) = P_1(t) - P_1(0)$ will have the desired properties. Another way of proving Lemma 1 runs as follows: We consider the sequence of polynomials

$$P_0(t) = 0,$$

$$P_{n+1}(t) = P_n(t) + \tfrac{1}{2}(t - (P_n(t))^2)$$

for $n = 0, 1, \ldots$ and show directly that in the interval $[0, 1]$ this sequence of polynomials is increasing and converges uniformly to $t^{1/2}$.

Indeed, for all $t \in [0, 1]$ we have

$$0 \leqq t^{1/2} - P_n(t) \leqq \frac{2t^{1/2}}{2 + nt^{1/2}} \tag{9}$$

and this in turn implies that $0 \leqq t^{1/2} - P_n(t) \leqq 2/n$. We verify (9) by induction on n. It is true for $n = 0$. If $n \geqq 0$ it follows from the inductive hypothesis (9) that $0 \leqq t^{1/2} - P_n(t) \leqq t^{1/2}$, hence $0 \leqq P_n(t) \leqq t^{1/2}$, and therefore from the definition of these polynomials we have

$$t^{1/2} - P_{n+1}(t) = (t^{1/2} - P_n(t))(1 - \tfrac{1}{2}(t^{1/2} + P_n(t))),$$

so that $t^{1/2} - P_{n+1}(t) \geqq 0$, and from (9)

$$\begin{aligned} t^{1/2} - P_{n+1}(t) &\leqq \frac{2t^{1/2}}{2 + nt^{1/2}}\left(1 - \frac{t^{1/2}}{2}\right) \\ &\leqq \frac{2t^{1/2}}{2 + nt^{1/2}}\left(1 - \frac{t^{1/2}}{2 + (n+1)t^{1/2}}\right) \\ &= \frac{2t^{1/2}}{2 + (n+1)t^{1/2}}. \end{aligned}$$

Having established Lemma 1 for the special case where the interval is [0, 1], it is easy to make the transition to any compact interval $[0, r]$.

After these preparations we turn to the proof of Proposition 7. Since H separates the points of E, so does $\overline{H}_0$; since H_0 is a vector space and contains all constants, the same can be said of $\overline{H}_0$. We shall show that $f \in \overline{H}_0$ implies $|f| \in \overline{H}_0$.

Indeed. If $f \in \overline{H}_0$, then $f^2 \in \overline{H}_0$ with $0 \leqq f^2 \leqq r_1$; this is so because f is bounded on E and the mapping $\lambda \to \lambda^2$ is uniformly continuous on a bounded interval of the λ-axis. For $\varepsilon > 0$ there exist $h_1, \ldots, h_n \in H$ and a polynomial $g(\lambda_1, \ldots, \lambda_n)$ with $p = g(h_1, \ldots, h_n)$ and $|p(x) - f^2(x)| < \varepsilon^2/8$ for all $x \in E$. In particular we have that $p(x) > -\varepsilon^2/8$ and as a consequence of this we obtain, for $q(x) = p(x) + \varepsilon^2/8$, on the one hand that $0 < q(x) < r_1 + \varepsilon^2/4 = r$ and on the other hand that $|q(x) - f^2(x)| < \varepsilon^2/4$ for all $x \in E$. By Lemma 2 we therefore may conclude that $|(q(x))^{1/2} - |f(x)|| < \varepsilon/2$. Lemma 1 provides us with a polynomial Q such that $|Q(q(x)) - (q(x))^{1/2}| < \varepsilon/2$ and consequently $|Q(q(x)) - |f(x)|| < \varepsilon$ for $x \in E$, i.e., $\|Q(q) - |f|\| < \varepsilon$ follows. Since $Q(q) \in H_0$, we may conclude from this that $|f| \in \overline{H}_0$.

It can be seen that $\overline{H}_0$ satisfies all conditions from (1) to (4) of Proposition 6. It therefore follows that $\mathfrak{C} = \overline{\overline{H}}_0 = \overline{H}_0$.

Finally, if H does not separate the points of E, the same may be claimed of H_0 and also of $\overline{H}_0$. However, since $\mathfrak{C}$ separates the points of E (this is so because a compact Hausdorff space is normal and there is, by Urysohn's Lemma, a function $f \in \mathfrak{C}$ with $f(x) = 1$ and $f(y) = 0$ whenever $x \neq y$) we have in this case $\overline{H}_0 \neq \mathfrak{C}$ and so the proposition is proved.

In the real q-dimensional Euclidean space R^q of points $\xi = (\xi_1, \ldots, \xi_q)$ let $\mathfrak{H}_0$ denote the system of all functions h which can be obtained from the projections $\pi_1(\xi) = \xi_i$, $i = 1, \ldots, q$, by the formation of finite linear combinations with real coefficients and by the formation of the absolute value.

$\mathfrak{H}_0$ is characterized as the smallest system of functions with the following properties:

(i) $\mathfrak{H}_o$ contains the projections $\pi_i(\xi) = \xi_i$ for $i = 1, \ldots, q$.
(ii) If a_1 and a_2 are real numbers and f_1 and f_2 are functions in $\mathfrak{H}_0$, then $a_1 f_1 + a_2 f_2$ also belongs to $\mathfrak{H}_0$.
(iii) If f is in $\mathfrak{H}_0$, then so is $|f|$.

If T denotes the subset of R^q defined by

$$|\xi_1| + \ldots + |\xi_q| = 1,$$

then the system $\mathfrak{H}_0'$ of functions $h' = h|T$ (here $h|T$ denotes the restriction of h to T), $h \in \mathfrak{H}_0$, satisfy all conditions of Proposition 6. Indeed, if c is any constant, then the function $c(|\xi_1| + \ldots + |\xi_q|) = c$ is in $\mathfrak{H}_0'$; moreover, the linearity of $\mathfrak{H}_0'$ as well as its closure relative to the formation of the absolute value is evident; finally, the functions $\pi_i(\xi) = \xi_i$ $(i = 1, \ldots, q)$ already form a set of functions which separate the points. We therefore have Proposition 8.

PROPOSITION 8. *Every continuous function on T can be uniformly approximated by functions of $\mathfrak{H}_0'$.*

The functions of $\mathfrak{H}_0$ are *positive homogeneous*, i.e., if $p \in \mathfrak{H}_0$ and $a \geqq 0$, then $p(a\xi) = ap(\xi)$, $\xi \in R^q$. A positive homogeneous function is already determined by its values on the set T introduced further above since $p(0) = 0$ and $p(\xi) = r(\xi)p(\xi/r(\xi))$ for $\xi \neq 0$, where $r(\xi) = |\xi_1| + \ldots + |\xi_q|$.

From Proposition 8 we obtain the following proposition.

PROPOSITION 9. *Every positive homogeneous function p on R^q can be approximated "relative uniformly in R^q", i.e., for $\varepsilon > 0$ there exists an $h \in \mathfrak{H}_0$ such that, for all $\xi \in R^q$,*

$$|p(\xi) - h(\xi)| < \varepsilon r(\xi).$$

In order to obtain similar approximation theorems for functions φ on R^q which are not positive homogeneous, we consider an extension of the function system $\mathfrak{H}_0$. Let $\mathfrak{H}_1$ denote the set of all functions h on R^q which are obtained upon repeated application of the following operations from the special functions $\pi_i(\xi) = \xi_i$ $(i = 1, \ldots, q)$:

1. $h_1, h_2 \to a_1 h_1 + a_2 h_2$ with real constants a_1, a_2;
2. $h \to |h|$;
3. $h \to \min(h, 1)$.

It is clear that $h(0) = 0$ for every function so obtained. We have the following approximation theorem.

PROPOSITION 10. *If φ is any continuous finite real-valued function on R^q with $\varphi(0) = 0$, then there exists a sequence $(h_n)_{n=1}^{\infty}$ of functions in $\mathfrak{H}_1$ which converges "monotonely away from zero" to φ. This means: For $\varphi(x) = 0$ we have $h_n(x) = 0$, $n = 1, 2, \ldots$; for $\varphi(x) \neq 0$, with $s = \operatorname{sgn} \varphi(x)$, we have $0 \leqq s \cdot h_1(x) \leqq s \cdot h_2(x) \leqq \ldots \leqq s \cdot h_n(x) \leqq \ldots$ with*

$$\lim_{n\to\infty} h_n(x) = \varphi(x).$$

The convergence is uniform in every bounded closed subset of R^q.

Proof. The proof will be done in steps.

Step 1: We observe that

$$\min(h, a) = a \min\left(\frac{h}{a}, 1\right)$$

belongs to $\mathfrak{H}_1$ if $h \in \mathfrak{H}_1$ and a is a positive constant.

Step 2: We next define several functions in $\mathfrak{H}_1$, where a, b denote positive constants:

$$\beta(\xi_1; b, a) = \min(\tfrac{1}{2}(\xi_1 + |\xi_1|), b + a) - \min(\tfrac{1}{2}(\xi_1 + |\xi_1|), b),$$

$$\tau_1(\xi_1) = \gamma(\xi_1; b, a) = \frac{1}{a}(\beta(\xi_1; b, a) - \beta(\xi_1; b + 2a, a)).$$

In addition we also have the function $\bar{\tau}_1(\xi_1) = \tau_1(-\xi_1)$. The functions τ_1 and $\bar{\tau}_1$ are "stump functions" for the case $q = 1$. The figure below illustrates the graph of $\gamma(\xi_1; b, a)$.

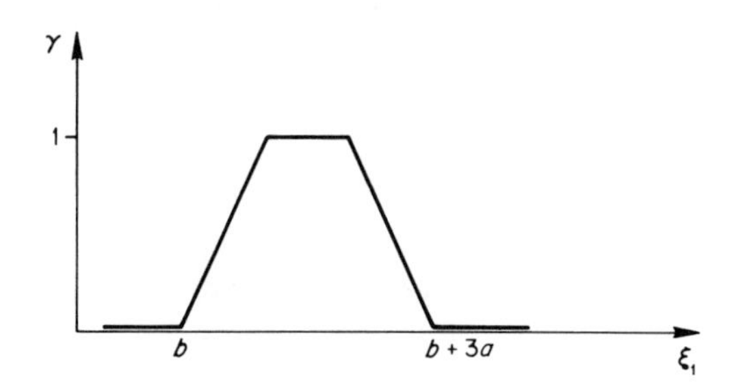

Step 3: For $q \geqq 2$ we construct the stump functions as follows:

$$\alpha(\xi_1, \ldots, \xi_q; b, a) = \min(\gamma(\xi_1; b, a), \ldots, \gamma(\xi_q; b, a)).$$

The figure below illustrates the level lines of $\alpha(\xi_1, \xi_2; b, a)$.

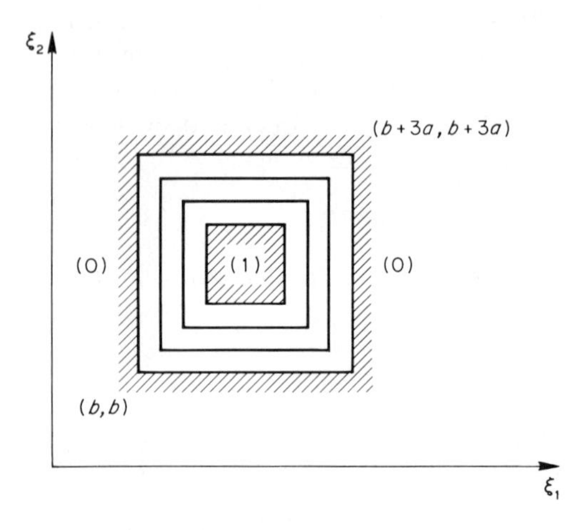

From these stump functions in turn we obtain the general "stump function" τ_q by a rotation

$$\xi_i \to \sum_k a_{ik} \xi_k$$

around the origin:

$$\tau(\xi) = \tau_q(\xi) = \alpha(\sum_k a_{1k} \xi_k, \ldots, \sum_k a_{qk} \xi_k; b, a)$$

with $\xi = (\xi_1, \ldots, \xi_q)$.

We have $0 \leqq \tau \leqq 1$; the open resp. closed nonempty cubes $\{\xi \in R^q : \tau(\xi) > 0\}$ resp. $\{\xi \in R^q : \tau(\xi) = 1\}$ we shall refer to as the *kernel* resp. *principal kernel of* τ.

Step 4: In the sequel we also need the function:

$$\omega(\xi_1, \xi_2) = \xi_1 + \xi_2 - \max(\min(\xi_1, \xi_2), 0) - \min(\max(\xi_1, \xi_2), 0);$$

if the signs of ξ_1 and ξ_2 coincide, then the value of this function is ξ_i with ξ_i being the one whose absolute value equals $\max(|\xi_1|, |\xi_2|)$ (and otherwise, i.e., if the signs of ξ_1 and ξ_2 are different, the value of the function is $\xi_1 + \xi_2$).

Step 5: Now to the approximation: The set

$$G = \{\xi \in R_q : \varphi(\xi) \neq 0\}$$

is open and does not contain the origin $\xi = 0$. Hence, for every point η of G and for every natural number n, there is a stump function τ and a constant $\zeta(\neq 0)$ with the following properties: (1) η belongs to the principal kernel K' of τ; (2) φ has everywhere in the kernel K of τ the same sign as ζ; (3) for the function $\varphi(\xi; \zeta; n) = \zeta\tau(\xi)$ we have

(a) $(\operatorname{sgn} \zeta)\varphi(\xi; \eta; n) \leqq (\operatorname{sgn} \zeta)\tau(\xi)$ for $\xi \in K$,

(b) $|\varphi(\xi; \eta; n) - \varphi(\xi)| < 1/n$ for $\xi \in K'$.

For fixed n we may represent G as the union of countably many K', e.g., $K'_{n1}, K'_{n2}, \ldots$, where every compact subset of G can be covered by a finite subfamily of these sets. We arrange these sets K'_{nm} as a simple sequence K'_1, $K'_2, \ldots$; let the corresponding $\varphi(\xi; \eta; n)$ be $\varphi_1(\xi)$, $\varphi_2(\xi)$, ... (the corresponding kernels are situated in G by (a)). In the sequence

$$h_1 = \varphi_1, \; h_2 = \omega(h_1, \varphi_2), \; h_3 = \omega(h_2, \varphi_3), \; \ldots$$

we have the desired approximating sequence.

Indeed: If B is a closed bounded subset of R^q, then φ is uniformly continuous on $B_0 = B \cap \{\xi \in R^q : \varphi(\xi) = 0\}$ in the sense that for every $\varepsilon > 0$ there exists an open set $U \supset B_0$ such that $|\varphi(\xi)| < \varepsilon$ for $\xi \in U$; by (a) we then also have that $|h_n(\xi)| < \varepsilon$ for all $\xi \in U$ and all $n \in N$. For the closed bounded set $B_1 = B - B \cap U$ we pick an index n_0 after having chosen an $n' > 1/\varepsilon$, such that among the $K'_1, \ldots, K'_{n_0}$ we already have all those finitely many $K'_{n'1}, K'_{n'2}, \ldots$ which have points in common with B_1 and cover B_1. Thus $|\varphi(\xi) - h_{n_0}(\xi)| < 1/n' < \varepsilon$ for $\xi \in B_1$ (by (b)) and so in general

$$|\varphi(\xi) - h_n(\xi)| < \varepsilon$$

holds for all $\xi \in B$ and $n \geqq n_0$. The claimed monotonicity property of the sequence $(h_n)_{n=1}^{\infty}$ is manifest by step 4 and (a). Thus everything is proved.

For arbitrary continuous functions φ on R^q we obtain an approximation theorem if in place of $\mathfrak{H}_1$ we consider the larger system $\mathfrak{H}_2$ which is obtained by using the projection functions $\pi_i(\xi) = \xi_i$ $(i = 1, \ldots, q)$ and the constant unit function 1 and then applying to these the operations of forming finite linear combinations with real coefficients and of forming the absolute value. In this case we can even construct a stump function which has the point 0 in its kernel so that the considerations in proving the foregoing proposition will be applicable to arbitrary continuous functions with the following result.

PROPOSITION 11. *If φ is a finite real-valued continuous function on R^q, then there exists a sequence of functions in $\mathfrak{H}_2$ which converges "monotonely away* *from zero" to φ; here the convergence is uniform on every compact subset of R^q.*

Remark. One may use Proposition 7 to show that if F is a closed subspace of a compact Hausdorff space E and f is a continuous real-valued function on f such that $|f(x)| \leqq a$ for all $x \in F$, then there is a continuous function g on E which coincides with f on F and is such that $|g(x)| \leqq a$ in E. We leave this as an exercise to the reader.

Section 2. Elementary and Norm Integrals.

Let $\mathcal{E}$ be a nonempty set of finite real-valued functions f defined on a common domain of definition A with A a fixed nonempty set. Let $\mathcal{E}$ have the following properties:

(1_c) $af \in \mathcal{E}$ whenever $f \in \mathcal{E}$ and a is a real number;
(2_c) $f_1 + f_2 \in \mathcal{E}$ whenever f_1, f_2 are in $\mathcal{E}$;
(3_c) $|f| \in \mathcal{E}$ whenever $f \in \mathcal{E}$.

It can be seen that $\mathcal{E}$ is a linear lattice under the natural order $\leqq$; the functions in $\mathcal{E}$ are called *elementary functions*.

Let E be a real-valued function defined on $\mathcal{E}$ with the following properties:

(4_c) $E(af) = aE(f)$ with a real and $f \in \mathcal{E}$;
(5_c) $E(f_1 + f_2) = E(f_1) + E(f_2)$ with f_1, f_2 in $\mathcal{E}$;
(6_c) $E(|f|) \geqq 0$ with $f \in \mathcal{E}$.

It is clear that E is a positive linear functional on $\mathcal{E}$; we call E a (*weak*) *elementary integral* and write $E|\mathcal{E}$.

Let $\mathfrak{G}$ denote the set of all extended real-valued functions g defined on A. For $g \in \mathfrak{G}$ we define the *norm of* g (associated with the elementary integral E) by

(a) $$\mathcal{N}(g) = \inf\{\sum_{n=1}^{\infty} E(|f_n|) : |g| \leqq \sum_{n=1}^{\infty} |f_n| \text{ with } f_n \in \mathcal{E}\},$$

provided that the set over which the infimum is formed is nonempty; in case there are no $f_1, f_2, \ldots$ in $\mathcal{E}$ such that

$$|g| \leqq \sum_{n=1}^{\infty} |f_n|,$$

we define

$$\mathcal{N}(g) = +\infty.$$

The following properties of $\mathcal{N}$ follow immediately from the foregoing definition: If $g, g_1, \ldots \in \mathfrak{G}$, then

(b) $0 \leqq \mathcal{N}(g) \leqq +\infty$;

(c) $\mathcal{N}(ag) = |a|\mathcal{N}(g)$ for finite real $a \neq 0$;

(d) $|g| \leqq \sum_{n=1}^{\infty} |g_n|$ implies $\mathcal{N}(g) \leqq \sum_{n=1}^{\infty} \mathcal{N}(g_n)$;

(e) $\mathcal{N}(|g|) = \mathcal{N}(g)$;

(f) $|g_1| \leqq |g_2|$ implies $\mathcal{N}(g_1) \leqq \mathcal{N}(g_2)$.

Indeed, (b), (e) and (f) are immediate consequences of the definition of $\mathcal{N}$. To see that (c) holds, we note that, for $a \neq 0$,

$$|ag| \leqq \sum_{n=1}^{\infty} |f_n| \text{ and } |g| \leqq \sum_{n=1}^{\infty} \frac{1}{|a|} |f_n|$$

are equivalent statements. To verify (d): Given $\varepsilon > 0$, there exist $(f_{n,k}) \subset \mathcal{E}$, $n = 1, 2, \ldots$ such that

$$|g_n| \leqq \sum_{k=1}^{\infty} |f_{n,k}| \text{ and } \mathcal{N}(g_n) > \sum_{k=1}^{\infty} E(|f_{n,k}|) - \varepsilon \cdot 2^{-n};$$

thus we have

$$|g| \leqq \sum_{n,k} |f_{n,k}| \text{ and } \mathcal{N}(g) \leqq \sum_{n,k} E(|f_{n,k}|) < \sum_{n=1}^{\infty} \mathcal{N}(g_n) + \varepsilon;$$

with $\varepsilon > 0$ the claim made in (d) follows.

Remark. The formal definition in (a) lends itself to the following informal presentation. Confining ourself to nonnegative functions, we may regard (a) as the condensed description of a measuring process. The set $\mathcal{E}$ provides us with a stock of measuring rods, the nonnegative elementary functions, by means of which the nonnegative functions in $\mathfrak{G}$ are to be gauged. Each measuring rod has a magnitude given by its elementary integral. The basic measuring process consists in choosing from stock such an infinite sequence of measuring rods $f_n = |f_n| \geqq 0$ that by addition they combine to surpass $g = |g| \geqq 0$ indicated in the inequality

$$g \leqq \sum_{n=1}^{\infty} f_n .$$

The real number

$$\lambda = \sum_{n=1}^{\infty} E(f_n)$$

obtained by adding together the magnitudes of the particular measuring rods thus employed is then accepted as an estimate, generally in excess, of the magnitude of g. Repetitions of this basic measuring process furnish successsively better estimates, convergent to the quantity $\mathscr{N}(g)$, when suitable precautions are taken. It is, of course, conceivable that the basic measuring process will fail to produce any real numbers as estimates of the magnitude of a particular function g, either because the inequality

$$g \leqq \sum_{n=1}^{\infty} f_n$$

cannot be realized, or because it implies the divergence of the series

$$\sum_{n=1}^{\infty} E(f_n).$$

Under these circumstances the formal definition requires that we take $\mathscr{N}(g) = +\infty$.

Let f be an elementary function. In order to have the equality $\mathscr{N}(f) = E(|f|)$ satisfied, we have to require of the elementary integral E the additional property that

$$(7_c) \qquad |f| \leqq \sum_{n=1}^{\infty} |f_n| \text{ implies } E(|f|) \leqq \sum_{n=1}^{\infty} E(|f_n|)$$

whenever $f, f_1, f_2, \ldots$ are in $\mathscr{E}$.

Incidentally, condition (7_c) holds if and only if the following is satisfied:

(g) $f \in \mathscr{E}$ implies $\mathscr{N}(f) = E(|f|)$.

Evidently $\mathscr{N}(f) \leqq E(|f|)$ holds for every $f \in \mathscr{E}$. On the one hand, if (7_c) is satisfied, then one also has that $E(|f|) \leqq \mathscr{N}(f)$ and thus condition (g) is fulfilled. If, on the other hand, (g) is satisfied, then we get on account of the infimum property of $\mathscr{N}(f)$ that (7_c) holds as well.

In the following we shall assume that the (weak) elementary integral E on $\mathscr{E}$ satisfies the conditions (1_c) to (7_c); we then refer to E simply as an *elementary integral* and drop the term "weak."

DEFINITIONS. *A function $g \in \mathfrak{G}$ is said to be a null function if $\mathscr{N}(g) = 0$. A subset T of A is said to be a null set if its characteristic function is a null function.*

As direct consequences of the properties (f) and (d) we see that any subset of a null set must be a null set and that the union of a countable number of null sets will again be a null set.

DEFINITION. *We shall say that a property holds almost everywhere on A if it holds for all $x \in A$ but a null set.*

PROPOSITION 1. *If $g \in \mathfrak{G}$ and $\mathcal{N}(g) < +\infty$, then g is finite almost everywhere.*

Proof. We shall show that the set $T = \{x \in A : g(x) = \pm\infty\}$ is a null set, i.e., the characteristic function of T, χ_T, is a null function. Let $T_M = \{x \in A : |g(x)| \geqq M\}$ for $0 < M < +\infty$. Since $T \subset T_M$, we have that $\mathcal{N}(\chi_T) \leqq \mathcal{N}(\chi_{T_M})$ and it will be enough to show that $\mathcal{N}(\chi_{T_M}) \to 0$ as $M \to +\infty$. By assumption there exist functions $f_1, f_2, \ldots$ in $\mathcal{E}$ such that

$$|g| \leqq |f_1| + |f_2| + \ldots \text{ and } \sum_n E(|f_n|) < +\infty.$$

Let

$$S_M = \{x \in A : M \leqq |f_1(x)| + |f_2(x)| + \ldots\}.$$

Since $T_M \subset S_M$ we see that

$$M\chi_{S_M} \leqq |f_1| + |f_2| + \ldots$$

and hence by (c) and (a)

$$M\mathcal{N}(\chi_{S_M}) = \mathcal{N}(M\chi_{S_M}) \leqq \sum_n E(|f_n|);$$

thus

$$\mathcal{N}(\chi_{S_M}) \to 0 \text{ as } M \to +\infty.$$

But

$$\mathcal{N}(\chi_{T_M}) \leqq \mathcal{N}(\chi_{S_M})$$

and so

$$\mathcal{N}(\chi_{T_M}) \to 0 \text{ as } M \to +\infty.$$

This finishes the proof.

PROPOSITION 2. *A function g in $\mathfrak{G}$ is a null function if and only if it is zero almost everywhere.*

Proof. Let $\mathcal{N}(g) = 0$. We consider the sets

$$T_j = \{x \in A : 2^{j-1} < |g(x)| \leqq 2^j\}$$

for $j = 0, \pm 1, \pm 2, \ldots$. Since

$$\{x \in A : g(x) \neq 0\} = \{x \in A : |g(x)| = +\infty\} \cup (\cup \{T_j : j = 0, \pm 1, \ldots\}),$$

and $2^{j-1}\chi_{T_j} < |g|$, we have that

$$2^{j-1}\mathscr{N}(\chi_{T_j}) \leqq \mathscr{N}(|g|) = \mathscr{N}(g) = 0.$$

Thus every T_j is a null set. By the foregoing proposition and by virtue of the fact that any countable union of null sets is again a null set, it follows that $\{x \in A : g(x) \neq 0\}$ is a null set.

Conversely, let $T = \{x \in A : g(x) \neq 0\}$ be a null set. For $g_m = m\chi_T$, with $m \in N$, we have $\mathscr{N}(g_m) \leqq m\mathscr{N}(\chi_T) = 0$, moreover $|g| \leqq |g_1| + |g_2| + \ldots$ and so

$$\mathscr{N}(g) \leqq \mathscr{N}(g_1) + \mathscr{N}(g_2) + \ldots = 0.$$

The proof is finished.

PROPOSITION 3. *If* $g_1 = g_2$ *almost everywhere, then* $\mathscr{N}(g_1) = \mathscr{N}(g_2)$.

Proof. By assumption $|g_1| \leqq |g_2| + g_3$ and $|g_2| \leqq |g_1| + g_3$, where g_3 denotes a nonnegative null function. Applying (d) to these inequalities gives the desired conclusion and the proof is complete.

On the basis of Proposition 3 we can extend our concept of a function: We shall speak of a function h on A whenever h is only defined almost everywhere on A. The set of all these functions we denote by $\mathfrak{H}$. Clearly, $\mathfrak{G} \subset \mathfrak{H}$. Regardless of how we extend h to all of A, the resulting function $g_h (\in \mathfrak{G})$ will give the same value for $\mathscr{N}(g_h)$ and we can therefore define in a unique manner $\mathscr{N}(h) = \mathscr{N}(g_h)$. At the same time we introduce in $\mathfrak{H}$ the equivalence relation:

(h) $h_1 - h_2$ if and only if g_{h_1} and g_{h_2} differ only on a null set.

Evidently, this defintion is independent of the choice of the g_{h_i} and moreover $h_1 = h_2$ implies that $\mathscr{N}(h_1) = \mathscr{N}(h_2)$.

Let $\mathfrak{L}$ be that part of $\mathfrak{H}$ which is characterized by the inequality $\mathscr{N}(h) < +\infty$. From Proposition 1 we know that the functions in $\mathfrak{L}$ are finite almost everywhere. Two functions in $\mathfrak{L}$ are equal if they are equal in the sense of the equivalence relation (h), i.e., if they are equal almost everywhere. It is clear that $\mathscr{E} \subset \mathfrak{L} \subset \mathfrak{H}$.

PROPOSITION 4. $\mathfrak{L}$ *is a normed linear space with norm* $\|h\| = \mathscr{N}(h)$.

Proof. If $-\infty < a < +\infty$ and $f_1, f_2 \in \mathfrak{L}$, then af_1 and $f_1 + f_2$ are defined almost everywhere and $\mathscr{N}(af_1) = |a|\mathscr{N}(f_1) < +\infty$, $\mathscr{N}(f_1 + f_2) \leqq \mathscr{N}(f_1) + \mathscr{N}(f_2) < +\infty$ (by (d)). We also note that $f_1, f_2 \in \mathfrak{L}$ implies that $|\mathscr{N}(f_1) - \mathscr{N}(f_2)| \leqq \mathscr{N}(f_1 - f_2)$; this is so because $\mathscr{N}(f_1) = \mathscr{N}((f_1$

$- f_2) + f_2) \leqq \mathcal{N}(f_1 - f_2) + \mathcal{N}(f_2)$ from which $\mathcal{N}(f_1) - \mathcal{N}(f_2) \leqq \mathcal{N}(f_1 - f_2)$ follows; by symmetry we get the other part of the inequality. The proof is finished.

DEFINITION. *A sequence $(f_n)_{n=1}^{\infty}$ in $\mathcal{I}$ is said to converge in norm to f in $\mathcal{I}$ if*

$$\lim_{n\to\infty} \mathcal{N}(f_n - f) = 0.$$

(We shall use the notation

$$f = \operatorname*{Lim}_n f_n \text{ or } f_n \underset{\mathcal{N}}{\rightarrow} f.)$$

PROPOSITION 5. *If $f_1, f_2, \ldots, f'_1, f'_2, \ldots$ and*

$$f = \operatorname*{Lim}_n f_n$$

belong to $\mathcal{I}$, then

(*1*) $f_n = f'_n$ *for* $n \in N$ *implies that* $\operatorname{Lim} f'_n = f$;
(*2*) *a real and* $\neq 0$ *implies that* $\operatorname{Lim}(af_n) = af$;
(*3*) $\operatorname{Lim} f'_n = f' \in \mathcal{I}$ *implies that* $\operatorname{Lim}(f_n + f'_n) = f + f'$;
(*4*) $\operatorname{Lim} |f_n| = |f|$.

Proof. Using $\mathcal{N}(f - f'_n) \leqq \mathcal{N}(f - f_n) + \mathcal{N}(f_n - f'_n) = \mathcal{N}(f - f_n)$ in (1), $\mathcal{N}(af_n - af) = |a|\mathcal{N}(f_n - f)$ in (2), $\mathcal{N}((f_n + f'_n) - (f + f')) \leqq \mathcal{N}(f_n - f) + \mathcal{N}(f'_n - f')$ in (3) and $\mathcal{N}(|f_n| - |f|) \leqq \mathcal{N}(|f_n - f|)$ in (4) the proposition is established.

The following two examples show that convergence in norm and convergence $\mathcal{N}$-almost everywhere must be distinguished. Let A be the unit interval $[0, 1]$ and $\mathcal{E}$ be the set of all step functions f on $[0, 1]$ and let $E(f)$ be the Riemann integral of f.

(I) Let $f_n(x) = n$ for $0 < x \leqq 1/n$ and $= 0$ elsewhere; then $f_n \to 0$, but not $\mathcal{N}(f_n) \to 0$ (because $\mathcal{N}(f_n) = 1$).
(II) Let $n = q_k 2^k + q_{k-1} 2^{k-1} + \ldots + q_0 2^0$ be the representation of the natural number n in binary form with $q_k = 1$; let $f_n(x) = 1$ for $n2^{-k} - 1 < x \leqq (n + 1)2^{-k} - 1$ and $= 0$ otherwise. Then we have that $\mathcal{N}(f_n) = 2^{-k} \to 0$, but not $f_n \to 0$.

PROPOSITION 6. *If*

$$f = \operatorname*{Lim}_n f_n,$$

then every subsequence (f_{n_i}) of (f_n) contains a subsequence (f_{m_j}) such that

$$f = \lim_j f_{m_j} \quad (\mathcal{N} - \text{a.e.}).$$

Proof. We select (f_{m_j}) from (f_{n_i}) in such a way that $\mathcal{N}(f_{m_{j+1}} - f_{m_j}) < 2^{-j}$ which is possible since $\mathcal{N}(f_n - f) \to 0$. We let $g = |f_{m_1}| + |f_{m_2} - f_{m_1}| + \ldots$ and obtain

$$\mathcal{N}(g) \leqq \mathcal{N}(f_{m_1}) + \mathcal{N}(f_{m_2} - f_{m_1}) + \ldots \leqq \mathcal{N}(f_{m_1}) + 2^{-1} + \ldots$$
$$= \mathcal{N}(f_{m_1}) + 1 < +\infty.$$

Since $\mathcal{N}(g) < +\infty$ it follows that g is finite $\mathcal{N}$-a.e. and so the function

$$\varphi = f_{m_1} + (f_{m_2} - f_{m_1}) + \ldots$$

is $\mathcal{N}$-a.e. absolutely and properly convergent, that is

$$\varphi = \lim_j f_{m_j} \quad (\mathcal{N} - \text{a.e.}).$$

From

$$\mathcal{N}(\varphi - f) \leqq \mathcal{N}(f_{m_j} - f) + \mathcal{N}(f_{m_{j+1}} - f_{m_j}) + \ldots$$
$$\leqq \mathcal{N}(f_{m_j} - f) + 2^{-(j-1)}$$

we conclude, on letting $j \to +\infty$, that $\mathcal{N}(\varphi - f) = 0$ and thus $\varphi = f$ ($\mathcal{N}$-a.e.). The proof is finished.

PROPOSITION 7. *The space* $\mathcal{G}$ *is complete.*

Proof. Let (f_n) be a Cauchy sequence. We pick a subsequence (f_{m_i}) from it such that $\mathcal{N}(f_{m_{i+1}} - f_{m_i}) < 2^{-i}$ and proceed as in the foregoing proof. In this manner we obtain in

$$\varphi = f_{m_i} + (f_{m_{i+1}} - f_{m_i}) + \ldots$$

a function (which is defined and finite $\mathcal{N}$-a.e. and is independent of i) of finite norm, namely, $\mathcal{N}(\varphi) \leqq \mathcal{N}(f_{m_1}) + 1$ and such that $\mathcal{N}(\varphi - f_{m_j}) \to 0$ as $j \to +\infty$. The proof is finished.

Let $f \in \mathcal{G}$. We define the functions

$$f^+ = \tfrac{1}{2}(|f| + f) \text{ and } f^- = \tfrac{1}{2}(|f| - f);$$

f^+ and f^- are nonnegative (almost everywhere) and are in $\mathcal{G}$. We may therefore define the function L on $\mathcal{G}$ given by

$$L(f) = \mathcal{N}(f^+) - \mathcal{N}(f^-) \text{ with } f \in \mathcal{G};$$

we observe that L is well-defined and finite.

PROPOSITION 8. *$L(f)$ is continuous in $\mathfrak{F}$ and $L(f) = E(f)$ whenever $f \in \mathscr{E}$.*

Proof. Since $|f^+ - g^+| \leqq |f - g|$ and $|f^- - g^-| \leqq |f - g|$ we have that $|L(f) - L(g)| \leqq |\mathscr{N}(f^+) - \mathscr{N}(g^+)| + |\mathscr{N}(f^-) - \mathscr{N}(g^-)| \leqq \mathscr{N}(f^+ - g^+) + \mathscr{N}(f^- - g^-) \leqq 2\mathscr{N}(f - g)$ and this shows the continuity of $L(f)$ in $\mathfrak{F}$. Using the fact that $f \in \mathscr{E}$ implies $\mathscr{N}(f) = E(|f|)$ (by (g)) and observing that $f^+ - f^- = f$ we see that $L(f) = \mathscr{N}(f^+) - \mathscr{N}(f^-) = E(f^+) - E(f^-) = E(f)$. Q.E.D.

By Proposition 8 we gained in the function L (on $\mathfrak{F}$) a continuous extension of the function E (on $\mathscr{E}$); we are now aiming for a maximal natural continuous extension of E to a positive linear functional. To this end we restrict ourselves to the consideration of the function space $\mathfrak{L}$ which is obtained by taking the closure of $\mathscr{E}$ in $\mathfrak{F}$. Thus $\mathfrak{L} \subset \mathfrak{F}$ and for an f in $\mathfrak{F}$ we have that $f \in \mathfrak{L}$ if and only if for every $\varepsilon > 0$ there exists a g in $\mathscr{E}$ such that $\mathscr{N}(f - g) < \varepsilon$. We may thus refer to the functions in $\mathfrak{L}$ as "*$\mathscr{N}$-almost elementary functions.*"

For every f in $\mathfrak{L}$ there exists a sequence $g_1, g_2, \ldots$ in $\mathscr{E}$ with

$$\operatorname{Lim}_k g_k = f.$$

This means that for $\varepsilon > 0$ there is an n_ε such that

$$\mathscr{N}(g_k - f) < \frac{\varepsilon}{2} \quad \text{for } k > n_\varepsilon,$$

and hence

$$|E(g_k) - E(g_j)| \leqq E(|g_k - g_j|) = \mathscr{N}(g_k - g_j) < \frac{\varepsilon}{2} + \frac{\varepsilon}{2} = \varepsilon$$

for (k and j) $> n_\varepsilon$. Thus

$$\lim\{E(g) : g \in \mathscr{E} \text{ and } \mathscr{N}(g - f) \to 0\}$$

exists for every f in $\mathfrak{L}$ and by Proposition 8 this limit is equal to $L(f)$.

DEFINITION. *The functions in $\mathfrak{L}$ are said to be $\mathscr{N}$-integrable and $L(f)$ is called the $\mathscr{N}$-integral (norm integral) of f, $f \in \mathfrak{L}$.*

PROPOSITION 9.

(*i*) *$\mathfrak{L}$ is a complete normed linear space with norm $\|f\| = \mathscr{N}(f)$ and is, moreover, a lattice under the natural ordering.*

(*ii*) *L is a continuous elementary integral in $\mathfrak{L}$.*

(*iii*) *If $f_k \in \mathfrak{L}$ and $f_k \geqq 0$ for $k \in N$, then the function f defined by*

$$f = \sum_k f_k$$

belongs to $\mathfrak{L}$ if and only if the series

$$\sum_k L(f_k)$$

is properly convergent (and in that case the last series equals $L(f)$).

Proof. We commence with part (i). Since $\mathfrak{L}$ is a subspace of $\mathfrak{I}$ we may apply Propositions 4 and 7; we note also that $|f| \in \mathfrak{L}$ follows from $f \in \mathfrak{L}$ (because with $f = \operatorname{Lim} g_k$ and $g_k \in \mathcal{E}$ we have that $|f| = \operatorname{Lim} |g_k|$ and $|g_k| \in \mathcal{E}$ on account of the inequality $\mathcal{N}(|g_k| - |f|) \leqq \mathcal{N}(g_k - f)$).

We next look at part (ii) of the proposition. Properties (1_c) to (3_c) follow immediately from part (i) of the proposition, properties (4_c) to (6_c) are obtained from the corresponding properties of E which remain preserved under taking of the limit, the continuity of L in $\mathfrak{L}$ is a consequence of Proposition 8 and, finally, property (7_c) follows from (d) and the fact that $\mathcal{N}(f) = L(|f|)$ for $f \in \mathfrak{L}$ (since $L(f) = \mathcal{N}(f^+) - \mathcal{N}(f^-)$ for $f \in \mathfrak{I}$ by definition).

To part (iii) of the proposition: If $f \in \mathfrak{L} \subset \mathfrak{I}$, then

$$\sum_1^m L(f_n) = L(\sum_1^m f_n) = \mathcal{N}(\sum_1^m f_n) \leqq \mathcal{N}(f) < +\infty,$$

so that

$$\sum_1^\infty L(f_n)$$

is a properly convergent series. Conversely, if

$$\sigma = \sum_1^\infty L(f_n) < +\infty,$$

then

$$\mathcal{N}(f) \leqq \sum_1^\infty \mathcal{N}(f_n) = \sigma$$

and so $f \in \mathfrak{I}$. This leads to

$$\mathcal{N}(f - \sum_1^m f_n) \leqq \sum_{m+1}^\infty \mathcal{N}(f_n) = \sum_{m+1}^\infty L(f_n) \to 0$$

for $m \to +\infty$; but $\mathfrak{L}$ is closed in $\mathfrak{I}$ and so $f \in \mathfrak{L}$. The continuity of L therefore implies that

$$L(f) = \lim_m L(\sum_1^m f_n) = \lim_m \sum_1^m L(f_n) = \sigma$$

and this finishes the proof.

Since $\mathfrak{L} \supset \mathcal{E}$ and $L|\mathcal{E} = E|\mathcal{E}$ we see that $L|\mathfrak{L}$ is an extension of $E|\mathcal{E}$. This extension (namely, $E|\mathcal{E} \to L|\mathfrak{L}$) applied to $L|\mathfrak{L}$ gives again $L|\mathfrak{L}$. We now prove a more general result.

PROPOSITION 10. *Let $E|\mathcal{E}$ and $E'|\mathcal{E}'$ be elementary integrals for functions defined on the same basic set A and suppose that the second is an extension of the first. Then we have the following relations among the corresponding norms $\mathcal{N}$, $\mathcal{N}'$, the domains of bounded norms $\mathfrak{I}$, $\mathfrak{I}'$, and the norm integrals L, L' and their domains $\mathfrak{L}$, $\mathfrak{L}'$: (1) $\mathcal{N}' \leqq \mathcal{N}$; (2) $\mathfrak{I}' \supset \mathfrak{I}$; (3) $\mathfrak{L}' \supset \mathfrak{L}$; (4) $L'|\mathfrak{L}'$ is an extension of $L|\mathfrak{L}$; (5) in particular, if $\mathcal{E} \subset \mathcal{E}' \subset \mathfrak{L}$, then we have $\mathcal{N}' = \mathcal{N}$, $\mathfrak{I}' = \mathfrak{I}$, $\mathfrak{L}' = \mathfrak{L}$, and $L' = L$.*

Proof. Concerning claims (1) and (2) we note that the set for which the infimum is formed in the determination of $\mathcal{N}'$ contains the set for which the infimum is taken in the determination of $\mathcal{N}$ and so $\mathcal{N}' \leqq \mathcal{N}$. From this in turn it follows that $\mathfrak{I}' \supset \mathfrak{I}$. Regarding claims (3) and (4) we observe: If $f \in \mathfrak{L}$, then there is a sequence of functions

$$(f_k) \subset \mathcal{E} \subset \mathcal{E}'$$

converging in norm $\mathcal{N}$; by (1) this sequence also converges in norm $\mathcal{N}'$ and so $f \in \mathfrak{L}'$. Here $L(f) = \lim E(f_k) = \lim E'(f_k) = L'(f)$. Finally, we consider claim (5). In view of (1) to (4) the most unfavorable special case arises when $E'|\mathcal{E}' = L|\mathfrak{L}$; we consider this case and show that the inclusions in (1) to (3) also hold in the opposite direction. Indeed, $\mathcal{N}'(g) \geqq \mathcal{N}(g)$ is trivial in case $\mathcal{N}'(g) = \infty$. If $\mathcal{N}'(g) < +\infty$, there exist, for given $\varepsilon > 0$, functions f_k in $\mathfrak{L}$ such that

$$\mathcal{N}'(g) > \sum_k L(|f_k|) - \varepsilon = \sum_k \mathcal{N}(f_k) - \varepsilon \text{ and } |g| \leqq \sum_k |f_k|.$$

Since $\mathcal{N}(f_k) < +\infty$, we have for suitable $f_{k,n}$ in $\mathcal{E}$ that

$$|f_k| \leqq \sum_n |f_{k,n}|$$

and

$$\mathcal{N}(f_k) > \sum_n E(|f_{k,n}|) - \varepsilon \cdot 2^{-k}$$

and so

$$|g| \leqq \sum_{k,n} |f_{k,n}|$$

and

$$\mathcal{N}'(g) > \sum_{k,n} E(|f_{k,n}|) - 2\varepsilon \geqq \mathcal{N}(g) - 2\varepsilon;$$

letting $\varepsilon \to 0$, we get $\mathcal{N}'(g) > \mathcal{N}(g)$. Thus $\mathfrak{I}' \subset \mathfrak{I}$ and so $\mathcal{N}' = \mathcal{N}$ and $\mathfrak{I}' = \mathfrak{I}$; since $\mathfrak{L}$ is closed in $\mathfrak{I}$, we get $\mathfrak{L}' \subset \mathfrak{L}$. By (4) we have that $L' = L$. The proof is finished.

If we pass from the formulation in terms of series limits in Proposition 9, part (iii), to the formulation in terms of sequential limits, we obtain the following *Monotone Convergence Theorem.*

PROPOSITION 11. *If $f_1 \leqq f_2 \leqq \ldots$ is a nondecreasing sequence of functions f_n in $\mathfrak{L}$ with $L(f_n) \leqq a < +\infty$ for $n \in N$, then*

$$f = \lim_n f_n \in \mathfrak{L}$$

and

$$L(f) = \lim_n L(f_n).$$

Indeed. Since $0 \leqq f - f_n = (f_{n+1} - f_n) + (f_{n+2} - f_{n+1}) + \ldots$, we obtain

$$\begin{aligned} 0 &\leqq \mathcal{N}(f - f_n) \leqq \mathcal{N}(f_{n+1} - f_n) + \mathcal{N}(f_{n+2} - f_{n+1}) + \ldots \\ &= L(f_{n+1} - f_n) + L(f_{n+2} - f_{n+1}) + \ldots \\ &= \lim_k L(f_k) - L(f_n) \to 0 \end{aligned}$$

for $n \to +\infty$. Thus $f \in \mathfrak{L}$. But $L|\mathfrak{L}$ is continuous (by Proposition 9, part (ii)) and the claimed equality follows.

A corollary of Proposition 11 is the following *Dominated Convergence Theorem.*

PROPOSITION 12. *If $g, f_1, f_2, \ldots$ belong to $\mathfrak{L}$ and if $f = \lim f_n \in \mathfrak{H}$ and, moreover, if $|f_n| \leqq |g|$ for $n \in N$, then $f \in \mathfrak{L}$ and*

$$L(f) = \lim_n L(f_n).$$

In place of a direct proof for Proposition 12, we shall prove three lemmas; it will be apparent that Lemma 3 is equivalent to Proposition 12.

LEMMA 1. *If $g \in \mathfrak{L}$, $g_n \in \mathfrak{L}$ with $g_n \leqq g$, $n \in N$, then*

$$s = \sup_n g_n \in \mathfrak{L}$$

and

$$L(s) \geqq \sup_n L(g_n).$$

Proof. Let $f_n = \max\{g_1, \ldots, g_n\} \leqq g$, then $f_n \in \mathfrak{L}$,

$$s = \lim_n f_n$$

and $L(f_n) \leqq L(g) < +\infty$ so that we can use Proposition 11 and obtain: $s \in \mathfrak{L}$ and

$$L(s) = \lim_n L(f_n) \geqq \lim_n \max\{L(g_1), \ldots, L(g_n)\} = \sup_n L(g_n).$$

The proof is finished.

LEMMA 2 (FATOU) . *If* $g \in \mathfrak{L}$, $g_n \in \mathfrak{L}$ *with* $g_n \leqq g$, $n \in N$, *and letting*

$$\lambda = \limsup_{n\to\infty} L(g_n) > -\infty,$$

then

$$\varphi = \limsup_{n\to\infty} g_n \in \mathfrak{L} \text{ and } L(\varphi) \geqq \lambda.$$

Proof. Let $\varphi = \lim \varphi_n$ with $\varphi_n = \sup\{g_n, g_{n+1}, \ldots\} \in \mathfrak{L}$ (by Lemma 1) and $\varphi_n \geqq \varphi_{n+1} \geqq g_{n+1}$. For $m > n$ we therefore have $L(\varphi_n) \geqq L(\varphi_m) \geqq L(g_m)$ so that $L(\varphi_n) \geqq \lambda > -\infty$. By Proposition 11 we therefore get that $\varphi \in \mathfrak{L}$ and $L(\varphi) = \lim L(\varphi_n) \geqq \lambda$. The proof is complete.

Remark. It is clear that similar lemmas hold for inf and lim inf.

LEMMA 3. *If* $g \in \mathfrak{L}$, $g_n \in \mathfrak{L}$ *with* $|g_n| \leqq g$, $n \in N$, *and if*

$$\lim_n g_n = \varphi$$

exists, then $\varphi \in \mathfrak{L}$ *and*

$$L(\varphi) = \lim_n L(g_n).$$

Proof. The assumption leads to $-g \leqq \pm g_n \leqq g$ (we shall keep both signs in the sequel). We get $L(-g) \leqq L(\pm g_n)$ and hence

$$\limsup_{n\to\infty} L(\pm g_n) > -\infty.$$

Using Lemma 2, we obtain

$$L(\pm\varphi) \geqq \limsup_{n\to\infty}(\pm L(g_n)),$$

or, upon separation of the cases,

$$\liminf_{n\to\infty} L(g_n) \geqq L(\varphi) \geqq \limsup_{n\to\infty} L(g_n),$$

which is only possible when we have equality. This finishes the proof.

Let Φ^q denote the family of all positive homogeneous continuous real-valued functions φ in the q variables $\xi_1, \ldots, \xi_q$; positive homogeneous means:

If $p > 0$, then $\varphi(p\xi_1, \ldots, p\xi_q) = p\varphi(\xi_1, \ldots, \xi_q)$ for arbitrary $\xi_1, \ldots, \xi_q$.

In Section 1 we studied the system $\mathfrak{H}_0$ of functions consisting of the functions $\pi_k(\xi) = \xi_k$ $(k = 1, \ldots, q)$ and all linear lattice combinations which can be formed from them. It is clear that Φ^q contains in particular the system $\mathfrak{H}_0$.

PROPOSITION 13. *If* $\varphi \in \Phi^q$ *and* $f_1, \ldots, f_q \in \mathfrak{L}$, *then* $\varphi(f_1, \ldots, f_q) \in \mathfrak{L}$.

Proof. By Proposition 8 of Section 1 every continuous function on $T = \{(\xi_1, \ldots, \xi_q) : |\xi_1| + \ldots + |\xi_q| = 1\}$ can be approximated uniformly by functions belonging to $\mathfrak{H}_0$: For $\varphi \in \Phi^q$ and $k > 0$ there is a ψ_k in $\mathfrak{H}_0$ such that $|\varphi - \psi_k| < 1/k$ on T. This means that

$$|\varphi - \psi_k| < (|\xi_1| + \ldots + |\xi_q|)/k.$$

If $f_1, \ldots, f_q$ are $\mathcal{N}$-integrable, then by part (i) of Proposition 9, $\psi_k(f_1, \ldots, f_q)$ is $\mathcal{N}$-integrable and since

$$\mathcal{N}(\varphi(f_1, \ldots, f_q) - \psi_k(f_1, \ldots, f_q)) \leqq (\mathcal{N}(f_1) + \ldots + \mathcal{N}(f_q))/k,$$

it follows, for $k \to \infty$, that $\varphi(f_1, \ldots, f_q)$ is $\mathcal{N}$-integrable. This finishes the proof.

Let X be a topological space. A family $\mathfrak{K}$ of real-valued functions on X is said to be *closed under pointwise limits* if $f \in \mathfrak{K}$ whenever f is a real-valued function on X and, for some sequence $(f_n) \subset \mathfrak{K}$,

$$f(x) = \lim_{n \to \infty} f_n(x)$$

for all $x \in X$.

Let $\mathfrak{R}_0$ be the set of all real-valued functions continuous on X. If α is an ordinal number such that $0 < \alpha < \Omega$, with Ω the smallest nondenumerable ordinal, define $\mathfrak{R}_\alpha$ to be the family of all functions f such that f is the pointwise limit of some sequence

$$(f_n) \subset \cup \{\mathfrak{R}_\beta : \beta \text{ is an ordinal number, } \beta < \alpha\}.$$

The functions in $\mathfrak{R}_\alpha$ are known as the *Baire functions of type* α. The family $\mathfrak{R}(X)$ of all *Baire functions on* X is defined to be

$$\mathfrak{R}(X) = \bigcup_{\alpha < \Omega} \mathfrak{R}_\alpha.$$

The sets $\mathfrak{R}_\alpha$ we shall call *Baire classes.*

We note that $\mathfrak{R}(X)$ may also be described as the intersection of all families $\mathfrak{K}$ of real-valued functions on X such that $\mathfrak{K}$ contains all real-valued continuous functions on X and $\mathfrak{K}$ is closed under pointwise limits. Observe that the set of all real-valued functions on X is such a class $\mathfrak{K}$.

PROPOSITION 14. *Let φ be a positive homogeneous Baire function of q real variables and suppose that $\varphi|T$ with $T = \{(\xi_1, \ldots, \xi_q) : |\xi_1| + \ldots + |\xi_q| = 1\}$ is a bounded Baire function, then $\varphi(f_1, \ldots, f_q)$ is $\mathcal{N}$-integrable whenever $f_1, \ldots,$ f_q are $\mathcal{N}$-integrable.*

Proof (by induction) . For the Baire class $\mathfrak{R}_0$ this proposition coincides with Proposition 13; suppose that the proposition is true for functions of every Baire class $< \alpha$ and let φ be a Baire function of type α with φ being positive homogeneous and $|\varphi| \leqq M$ on T so that $|\varphi| \leqq Mr$, where $r(x) = |\xi_1| + \ldots + |\xi_q|$ with $x = (\xi_1, \ldots, \xi_q)$. On T let $\varphi = \lim \varphi_n$, where φ_n are corresponding functions belonging to Baire classes $< \alpha$. We replace φ_n by

$$\psi_n(x) = \operatorname{med}\{-Mr(x), \varphi_n(x), Mr(x)\},$$

where $\operatorname{med}\{f, g, h\} = \min\{\max\{f, g\}, \max\{g, h\}, \max\{h, f\}\}$. The functions ψ_n also belong to Baire classes $< \alpha$ with $\lim \psi_n = \varphi$, and $|\psi_n(x)| \leqq Mr(x)$. By Proposition 13 $M(|f_1| + \ldots + |f_q|)$ is $\mathcal{N}$-integrable whenever $f_1, \ldots, f_q$ are $\mathcal{N}$-integrable and, by the inductive assumption, the $\psi_n(f_1, \ldots, f_q)$ are $\mathcal{N}$-integrable; thus Proposition 12 applies and we get that $\varphi(f_1, \ldots, f_q)$ is $\mathcal{N}$-integrable. The proof is finished.

In order to be able to free ourselves of the restriction that φ be positive homogeneous, as encountered in Proposition 14, we must impose in addition to the conditions (1_c) to (7_c) the following further requirement on $\mathcal{E}$:

(8_c) $\quad f \in \mathcal{E}$ implies that $\min\{f, 1\} \in \mathcal{E}$.

(Indeed, the somewhat weaker requirement, namely,

(8_c^*) $\quad f \in \mathcal{E}$ implies that $\min\{1, f\} \in \mathfrak{L}$

would suffice for the following.)

As a consequence of (8_c) we note the condition

$(8_c')$ $\quad f \in \mathfrak{L}$ implies that $\min\{1, f\} \in \mathfrak{L}$.

For, if $f = \lim f_n$ with $f_n \in \mathcal{E}$, then by (8_c) $\min\{1, f_n\} \in \mathcal{E}$. Thus $\operatorname{Lim} \min\{1, f_n\} = \min\{1, f\} \in \mathfrak{L}$; this is so since limit theorems similar to Proposition 5 hold for the max and the min of two functions and because $\mathfrak{L}$ is closed.

PROPOSITION 15. *Let the elementary integral E over $\mathcal{E}$ also satisfy the property (8_c). Then for every finite Baire function $\varphi(\xi_1, \ldots, \xi_q)$ in q real variables with $\varphi(0, \ldots, 0) = 0$ we have that the function $g = \varphi(f_1, \ldots, f_q)$ is $\mathcal{N}$-integrable* *whenever the functions* $f_1, \ldots, f_q$ *are* $\mathcal{N}$*-integrable and if, moreover, there is an* $\mathcal{N}$*-integrable function* h *such that* $|g| \leqq h$.

Proof. By Proposition 10 of Section 1 we may approximate every function of Baire class 0 with $\varphi(0) = 0$ by functions ψ_n belonging to the system $\mathfrak{H}_1$ in such a manner that in this approximation $|\psi_n| \leqq |\varphi|$ holds; thus $|\psi_n(f_1, \ldots, f_q)| \leqq h$ and we may apply Proposition 12 to see that $\varphi(f_1, \ldots, f_q)$ and $\psi_n(f_1, \ldots, f_q)$ are simultaneously $\mathcal{N}$-integrable.

We next suppose that the proposition is true for the functions belonging to Baire class $< \alpha$, where $\alpha > 0$, and we assume that φ is of Baire type α with $\varphi(0) = 0$ and $\varphi = \lim \varphi_n$, where the φ_n come from Baire classes $< \alpha$. We may assume here that $\varphi_n(0) = 0$ and, moreover, that $|\varphi_n(x)| \leqq nr(x)$ (otherwise we would take the functions

$$\varphi_n'(x) = \operatorname{med}\{-nr(x), \varphi_n(x), nr(x)\}$$

which are not of higher Baire classes). Since

$$|\varphi_n(f_1, \ldots, f_q)| \leqq n(|f_1| + \ldots + |f_q|),$$

the functions $\varphi_n(f_1, \ldots, f_q)$ are $\mathcal{N}$-integrable, but so are also

$$\psi_n = \operatorname{med}\{-h, \varphi_n(f_1, \ldots, f_q), h\}$$

with $|\psi_n| \leqq h$ and

$$\lim \psi_n = \varphi(f_1, \ldots, f_q);$$

here again Proposition 12 leads to the desired conclusion and the induction is complete. This finishes the proof.

We may drop the condition $\varphi(0) = 0$ in the foregoing proposition, if instead of (8_c) we impose the stronger requirement:

(9_c) The identically constant function 1 belongs to $\mathcal{E}$.

(In this connection observe: From $1 \in \mathcal{E}$ and $f \in \mathcal{E}$ follows $\min\{1, f\} = \frac{1}{2}(1 + f - |1 - f|) \in \mathcal{E}$.)

Proposition 16. *If the elementary integral* E *over* $\mathcal{E}$ *also satisfies the property* (9_c), *then Proposition 15 holds without the requirement* $\varphi(0) = 0$.

Proof. Indeed, referring to Proposition 11 of Section 1 it is clear that we may adopt the proof of Proposition 15 to the situation at hand.

In the set $\mathfrak{H}$ (introduced after Proposition 3) we consider the following two subsets

$$\mathfrak{M} = \{f \in \mathfrak{H} : (g \text{ and } h) \in \mathcal{E} \text{ implies } \operatorname{med}\{g, f, h\} \in \mathfrak{L}\}$$

and

$$\mathfrak{M}' = \{f \in \mathfrak{H} : (g \text{ and } h) \in \mathfrak{L} \text{ implies } \operatorname{med}\{g, f, h\} \in \mathfrak{L}\}.$$

Since $\mathscr{E} \subset \mathfrak{L}$ we have that $\mathfrak{M}' \subset \mathfrak{M}$.

PROPOSITION 17. $\mathfrak{M} = \mathfrak{M}'$.

Proof. If g and h are in $\mathfrak{L}$, then there exist sequences (g_n) and (h_n) in $\mathscr{E}$ with

$$\operatorname{Lim} g_n = \text{ and } \operatorname{Lim} h_n = h.$$

If $f \in \mathfrak{M}$, then $\operatorname{med}\{g_n, f, h_n\} \in \mathfrak{L}$; but $\operatorname{Lim} \operatorname{med}\{g_n, f, h_n\} = \operatorname{med}\{g, f, h\}$ (since limit theorems similar to Proposition 5 hold for the max and the min of two functions) and so $\operatorname{med}\{g, f, h\} \in \mathfrak{L}$ because $\mathfrak{L}$ is closed. This shows that the inclusion $\mathfrak{M} \subset \mathfrak{M}'$ is also valid and the proof is complete.

DEFINITION. *The functions in $\mathfrak{M}$ are called $\mathscr{N}$-measurable.*

PROPOSITION 18. *If two functions f_1 and f_2 differ only by a null function, then the $\mathscr{N}$-measurabilty of one implies the $\mathscr{N}$-measurability of the other.*

Proof. In the case at hand the functions

$$\Phi_1 = \operatorname{med}\{g, f_1, h\} \text{ and } \Phi_2 = \operatorname{med}\{g, f_2, h\}$$

differ only by a null function. This completes the proof.

We can see from the foregoing proposition that the concept of $\mathscr{N}$-measurability is compatible with the notion of equality introduced in $\mathfrak{H}$ by means of the relation (h).

Since $\mathfrak{L}$ is a lattice, the following proposition is immediate.

PROPOSITION 19. *Every $\mathscr{N}$-integrable function is $\mathscr{N}$-measurable.*

It is also easy to see the validity of the following result.

PROPOSITION 20. *If $f_1, f_2, \ldots \in \mathfrak{M}$ and if*

$$\lim_n f_n = f$$

in $\mathfrak{H}$ exists, then $f \in \mathfrak{M}$.

Proof. By assumption $\operatorname{med}\{g, f_n, h\} \in \mathfrak{L}$ for $g, h \in \mathfrak{L}$ and for $n \in N$. Since $\min\{g, h\} \leqq \operatorname{med}\{g, f_n, h\} \leqq \max\{g, h\}$ and since the two ends of this inequality are elements of $\mathfrak{L}$, Proposition 12 applies with the effect that

$$\lim_n \operatorname{med}\{g, f_n, h\} = \operatorname{med}\{g, f, h\} \in \mathfrak{L}.$$

This proves the proposition.

Proposition 21. *For* $f_1, f_2 \in \mathfrak{M}$ *and for all real* $\alpha \neq 0$ *the functions* αf_1, $\min\{f_1,f_2\}$, $\max\{f_1,f_2\}$, *and* $|f_1|$ *belong to* $\mathfrak{M}$; *moreover* $f_1 + f_2$ *belongs to* $\mathfrak{M}$ *provided that this sum is meaningful almost everywhere in* A.

Proof. The first claim follows from the equality $\alpha \operatorname{med}\{g/\alpha, f_1, h/\alpha\} = \operatorname{med}\{g, \alpha f_1, h\}$ with $g, h \in \mathfrak{L}$, the second claim results from the fact that $\mathfrak{L}$ is a lattice and that $\operatorname{med}\{g, \min\{f_1,f_2\}, h\} = \min\{\operatorname{med}\{g,f_1,h\}, \operatorname{med}\{g,f_2, h\}\}$ holds; the third claim is verified analogously and the fourth claim is a consequence of the preceding ones. The claim concerning the sum is established as follows: With g, $h \in \mathfrak{L}$, let $\varphi_{i,n} = \operatorname{med}\{-n(|g| + |h|), f_i, n(|g| + |h|)\} \in \mathfrak{L}$, $i = 1, 2$, $n \in N$, and let $\psi_n = \operatorname{med}\{g, \varphi_{1,n} + \varphi_{2,n}, h\} \in \mathfrak{L}$, so that $\lim_n \psi_n = \operatorname{med}\{g, f_1 + f_2, h\} \in \mathfrak{L}$; this is so because $\lim \varphi_{i,n}(x) = f_i(x)$ in case $|g(x)| + |h(x)| > 0$, and for the rest Proposition 12 again applies. This completes the proof.

Remark. $\mathfrak{M}$ is in fact a σ-complete lattice (i.e., every countable subset has a supremum and an infimum in $\mathfrak{M}$). Indeed, if for example

$$s = \sup\{f_1, f_2, \ldots\} = \lim_n \varphi_n$$

with $\varphi_n = \max\{f_1, \ldots, f_n\}$, then by Proposition 21 φ_n together with f_k are in $\mathfrak{M}$ and

$$\lim_n \varphi_n$$

converges on account of monotonicity, implying therefore that $s \in \mathfrak{M}$ by Proposition 20.

Proposition 22. *If* $f \geqq 0$, *then* $f \in \mathfrak{M}$ *if and only if* $\min\{f, g\} \in \mathfrak{L}$ *for every nonnegative function* g *in* $\mathfrak{L}$.

Proof. Let $f \in \mathfrak{M}$, $g \geqq 0$ and $g \in \mathfrak{L}$. Then $\operatorname{med}\{0, f, g\} = \min\{f, g\} \in \mathfrak{L}$. Suppose, conversely, the condition in the proposition is fulfilled. For arbitrary g, $h \in \mathfrak{L}$ and setting

$$\underline{m} = \min\{g, h\},\ \overline{m} = \max\{g, h\} \text{ and } m = \operatorname{med}\{g, f, h\}$$

we have

$$\begin{aligned} m &= \operatorname{med}\{\underline{m}, f, \overline{m}\} = \max\{\min\{\underline{m}, f\}, \min\{\overline{m}, f\}, \min\{\underline{m}, \overline{m}\}\} \\ &= \max\{\min\{\overline{m}, f\}, \underline{m}\}. \end{aligned}$$

But $\min\{\overline{m}, f\} = \min\{\max\{\overline{m}, 0\}, f\} + \min\{0, \overline{m}\}$ (as one may verify at once by considering the cases $\overline{m} \geqq 0$ and < 0) and $\max\{\overline{m}, 0\} \geqq 0$ and so $\min\{\overline{m}, f\} \in \mathfrak{L}$ which in turn implies that $m \in \mathfrak{L}$, completing the proof.

As an immediate consequence of the foregoing proposition we have the following.

PROPOSITION 23. *The constant function* 1 *is* $\mathcal{N}$*-measurable if and only if* $\min\{1, g\} \in \mathfrak{L}$ *for every nonnegative function* g *in* $\mathfrak{L}$.

Remark. The condition in the foregoing proposition is certainly fulfilled in case property (8_c) holds; this is so because (8_c) implies $(8'_c)$.

We are now going to give an example where $1 \notin \mathfrak{M}$: Let A be the closed unit interval $[0, 1]$, $\mathcal{E}$ the set of all functions of the form αx, where α is any finite real number, and $E(\alpha x) = \alpha/2$. The properties (1_c) to (7_c) are clearly satisfied. Since every limit of a convergent sequence of functions in $\mathcal{E}$ is again contained in $\mathcal{E}$, we see that $\mathfrak{L} = \mathcal{E}$; this shows that for example the function $\min\{1, 2x\}$ does not belong to $\mathfrak{L}$. By Proposition 23 therefore the identically constant function 1 cannot be $\mathcal{N}$-measurable.

For $0 \leqq x \leqq +\infty$ let $s_0(x) = \min\{1, x\}$ and for $n \in N$ let

$$s_n(x) = \begin{cases} s_{n-1}(x) \text{ for } 0 \leqq x \leqq \frac{n}{2}, \\ s_{n-1}(\frac{n}{2}) + (1/2^n)s_0(x - \frac{n}{2}) \text{ for } x \geqq \frac{n}{2}, \end{cases} \text{ and } s_n(-x) = -s_n(x).$$

We observe that

$$s(\xi) = \lim_n s_n(\xi)$$

defines a homeomorphism of the extended real numbers R_e^1 onto the interval $[-1, 1]$.

PROPOSITION 24. *If the functions* 1 *and* f *are* $\mathcal{N}$*-measurable, then so is* $s(f)$.

Proof. By definition $s_n(\xi)$ can be obtained from the functions 1 and ξ upon a finite number of linear lattice combinations (i.e., formation of linear combinations and formation of absolute values). By Proposition 21 we may therefore conclude that $s_n(f)$ is $\mathcal{N}$-measurable since 1 and f are. Invoking Proposition 20, we may thus conclude that

$$\lim_n s_n(f) = s(f)$$

is $\mathcal{N}$-measurable as well. The proof is finished.

Following Proposition 13 we defined the family of all Baire functions; to accommodate Baire functions which assume infinite values we permit that the operation of pointwise limits include improper convergence, i.e., infinite limits.

We shall use the mapping s whose definition precedes Proposition 24 in the following

DEFINITION. *By a Baire function* $\varphi(\lambda_1, \ldots, \lambda_q)$ *in the domain* $-\infty \leqq \lambda_i \leqq +\infty$, $i = 1, \ldots, q$, *we mean a function which by means of the transformation* $\mu_i = s(\lambda_i)$, $i = 1, \ldots, q$, *defined above becomes a Baire function* $\overline{\varphi}(\mu_1, \ldots, \mu_q)$ *on the cube* $W^q = \{(\mu_1, \ldots, \mu_q) : -1 \leqq \mu_i \leqq 1, i = 1, \ldots, q\}$; *thus* $\overline{\varphi}(s(\lambda_1), \ldots, s(\lambda_q)) = \varphi(\lambda_1, \ldots, \lambda_q)$.

PROPOSITION 25. *If* $1, f_1, \ldots, f_q$ *are* $\mathscr{N}$*-measurable and if* $\varphi(\lambda_1, \ldots, \lambda_q)$ *is a Baire function in the domain* $-\infty \leqq \lambda_i \leqq +\infty$, $i = 1, \ldots, q$, *then* $\varphi(f_1, \ldots, f_q)$ *is also* $\mathscr{N}$*-measurable.*

Proof. First of all, let $\overline{\varphi}$ be a Baire function of type 0, that is, let $\overline{\varphi}$ be a continuous function on W^q. By Proposition 6 of Section 1 we may approximate $\overline{\varphi}$ uniformly within arbitrarily small error by starting with the functions 1, $\mu_1, \ldots, \mu_q$ and forming real linear combinations a finite number of times and forming absolute values. Putting $\mu_i = s(f_i)$, we obtain (by Proposition 24) a sequence of functions $g_n \in \mathfrak{M}$ with $g = \lim g_n = \overline{\varphi}(s(f_1), \ldots, s(f_q)) \in \mathfrak{M}$ (on account of Proposition 20).

To complete the proof we must use transfinite induction; but there are no difficulties here on account of Proposition 20. We leave this part of the proof as an exercise to the reader.

As an application of the foregoing proposition we mention the following.

PROPOSITION 26. (*i*) *If* $1, t \in \mathfrak{M}$ *and* $t(x) \geqq 0$ *for* $x \in A$, *then we also have* $(t)^{1/2} \in \mathfrak{M}$.

(*ii*) *If* $1, f, g \in \mathfrak{M}$, $g \geqq 0$ *and* $\mathscr{N}$*-almost everywhere* $g > 0$, *then we have* $f/g \in \mathfrak{M}$.

Proof. $(t)^{1/2}$ is a Baire function and we can use Proposition 25; this establishes part (i). To verify part (ii), we set

$$f_n = \frac{nf}{ng + 1}, \quad n \in N.$$

By Proposition 25 we have that $f_n \in \mathfrak{M}$; since

$$\frac{f}{g} = \lim f_n \quad \mathscr{N}\text{-almost everywhere,}$$

we get by Proposition 20 that $f/g \in \mathfrak{M}$. This completes the proof.

PROPOSITION 27. *Each of the following conditions is necessary and sufficient* *for the $\mathcal{N}$-integrability of a function f:*

(a) $f \in \mathfrak{M}$ *and* $\mathcal{N}(f) < +\infty$;

(b) $f \in \mathfrak{M}$ *and there exist* $\mathcal{N}$*-integrable functions g and h with* $g \leqq f \leqq h$.

Proof. If f is $\mathcal{N}$-integrable, then by definition $\mathcal{N}(f) < +\infty$, and (b) is trivially satisfied with $g = h = f$. If, on the other hand, $\mathcal{N}(f) < +\infty$, then there exist $f_n \in \mathcal{E}$, $n \in N$, with

$$|f| \leqq \sum_n |f_n|$$

and

$$\sum_n E(|f_n|) < +\infty;$$

here

$$k = \sum_n |f_n| \in \mathfrak{L}$$

by part (iii) of Proposition 9, $g \leqq f \leqq h$ with $g = -k$ and $h = k$ in $\mathfrak{L}$. However, as soon as there are such g and h, we have on account of the $\mathcal{N}$-measurability of f immediately that $\mathfrak{L} \ni \operatorname{med}\{g, f, h\} = f$ and the proposition is proved.

Remark. The foregoing proposition may be expressed in abbreviated form by the equation

$$\mathfrak{L} = \mathfrak{M} \cap \mathfrak{I}.$$

PROPOSITION 28. (*a*) *If* $\mathcal{N}(f) < +\infty$, *then there is a* $g \in \mathfrak{L}$ *with* $|f| \leqq g$ *and* $\mathcal{N}(f) = L(g)$.

(*b*) *If* $1 \in \mathfrak{M}$ *and f is a characteristic function, then we may take in statement* (*a*) *for g a characteristic function as well.*

Proof. We first prove part (a). By assumption there exist elementary functions $f_{m,k}$, (m and k) $\in N$, with

$$|f| \leqq \sum_k |f_{m,k}| = g_m$$

and

$$\mathcal{N}(f) \leqq \sum_k E(|f_{m,k}|) \leqq \mathcal{N}(f) + 1/m,$$

where also

$$\mathcal{N}(f) \leqq \mathcal{N}(g_m) = L(g_m) \leqq \mathcal{N}(f) + 1/m;$$

thus $g_m \in \mathfrak{L}$. For

$$g = \lim_{m\to\infty} \min\{g_1, \ldots, g_m\}$$

we have by Proposition 11 $g \in \mathfrak{L}$ with $L(g) = \mathcal{N}(f)$ and $g \geqq |f|$.

We next prove part (b). Let f be a characteristic function with $\mathcal{N}(f) < +\infty$. The $g \in \mathfrak{L}$, which exists by part (a), we can replace by $g_1 = \min\{1, g\} \in \mathfrak{L}$ according to Proposition 23 and g_1 in turn we can replace by the $\mathcal{N}$-measurable $g_n = (g_1)^n$ for arbitrary $n = 2, 3, \ldots$ according to Proposition 25. We have $|f| \leqq g_n \leqq g_1$ and $g_n \in \mathfrak{L}$ by Proposition 27. Letting

$$h = \lim_n g_n,$$

we get $|f| \leqq h \in \mathfrak{L}$ by Proposition 11 and $\mathcal{N}(f) = L(h)$, where evidently h can only assume the values 0 and 1, that is, h is a characteristic function. The proof is now complete.

Section 3. Connections to Measure Theory

We commence with the function $\mathcal{N}(g)$, defined on the set $\mathfrak{G}$ of all real-valued functions on A. To each $Y \subset A$ we assign the characteristic function χ_Y and we set

$$m^*(Y) = \mathcal{N}(\chi_Y).$$

It is easy to see that m^* is an outer measure on $\mathcal{P}(A)$, the set of all subsets of A:

(1) $m^*(Y)$ is unambiguously defined for $Y \in \mathcal{P}(A)$ with $0 \leqq m^*(Y) \leqq +\infty$ and $m^*(\varnothing) = 0$, where $\varnothing$ is the empty set;

(2) for each sequence $Y_1, Y_2, \ldots$ in $\mathcal{P}(A)$ with

$$\bigcup_n Y_n \supset S \in \mathcal{P}(A)$$

we have

$$\sum_n m^*(Y_n) \geqq m^*(S).$$

Indeed, the unambiguity of $m^*(Y)$ is clear, the rest of (1) follows from property (b) of the norm $\mathcal{N}$ and the fact that $E(0) = 0$. The statement (2) follows from

$$\sum_n \chi_{Y_n} \geqq \chi_S$$

by property (d) of the norm $\mathscr{N}$. Properties (b) and (d) of the norm $\mathscr{N}$ were discussed at the beginning of Section 2.

We set

$$\mathfrak{m} = \{X : X \subset A \text{ and } \chi_X \in \mathfrak{M}\}$$

and call every X in $\mathfrak{m}$ *$\mathscr{N}$-measurable.* It is clear that

$$\mathfrak{k} = \{Y : Y \subset A \text{ and } \chi_Y \in \mathfrak{L}\}$$

is a subset of $\mathfrak{m}$ because $\mathfrak{L} \subset \mathfrak{M}$; the sets in $\mathfrak{k}$ we call *properly $\mathscr{N}$-measurable.*

We observe that $\mathfrak{k}$ is a *field of sets*, that is

(α) if $B, C \in \mathfrak{k}$, then $B \cup C \in \mathfrak{k}$;

(β) if $B, C \in \mathfrak{k}$ and $C \subset B$, then $B - C \in \mathfrak{k}$.

(Observe that a field of sets also satisfies

(γ) if $B, C \in \mathfrak{k}$, then $B \cap C \in \mathfrak{k}$;

(δ) if $B, C \in \mathfrak{k}$, then $B \triangle C \in \mathfrak{k}$.

To see this, we note the relations:

$$B \cup C = (B \cap C) \cup (B \triangle C),$$

$$B \triangle C = ((B \cup C) - B) \cup ((B \cup C) - C),$$

$$B \cap C = B - ((B \cup C) - C).)$$

For a fixed $X_0 \in \mathfrak{k}$, let

$$\mathfrak{k}_{X_0} = \{Z : \mathfrak{k} \ni Z \subset X_0\}.$$

It can be seen at once that $\mathfrak{k}_{X_0}$ (with $X_0 \in \mathfrak{k}$) is a σ-algebra of sets.

We now consider the function m on $\mathfrak{k}$ defined by

$$m(X) = L(\chi_X) \quad \text{for } X \in \mathfrak{k}.$$

If $Y \subset A$, we say that Y is *m-enclosable* if for each $\varepsilon > 0$ there exist two sets A_1, A_2 in $\mathfrak{k}$ with

$$A_1 \subset Y \subset A_2 \text{ and } m(A_2 - A_1) < \varepsilon.$$

The set function m on $\mathfrak{k}$ is said to be *complete* with respect $\mathscr{P}(A)$ if every m-enclosable set $Y \subset A$ belongs to $\mathfrak{k}$.

Proposition 1. *The function $m|\mathfrak{k}$, defined by*

$$m(X) = L(\chi_X) \quad \text{for } X \in \mathfrak{k},$$

is countably additive, nonnegative, and finite-valued (for brevity we say that m is a content) on the field of sets $\mathfrak{k}$. *In particular,* $m|\mathfrak{k}_{X_0}$ *(for fixed* $X_0 \in \mathfrak{k}$*) is a bounded measure on the σ-algebra* $\mathfrak{k}_{X_0} = \{Z : \mathfrak{k} \ni Z \subset X_0\}$. *Moreover,* $m|\mathfrak{k}$ *is a complete content with respect to the power set of A and* $m|\mathfrak{k}_{X_0}$ *is a complete measure with respect to the power set of* X_0 *(i.e., every m-enclosable set* $Y \subset X_0$ *belongs to* $\mathfrak{k}_{X_0}$*).*

Proof. Only the completeness needs closer scrutiny. If $Y \subset A$ is m-enclosable, i.e., if there exist for every $n \in N$ sets X'_n and X''_n in $\mathfrak{k}$ with

$$X'_n \subset Y \subset X''_n \text{ and } m(X''_n - X'_n) < \frac{1}{n},$$

then we obtain, taking

$$\underline{X} = \bigcup_n X'_n \text{ and } \overline{X} = \bigcap_n X''_n,$$

$\underline{X} \subset Y \subset \overline{X}$ and, using Proposition 12 in Section 2, we see that $\underline{X}, \overline{X} \in \mathfrak{k}$ and $m(\underline{X}) = m(\overline{X})$. Thus the functions $\chi_{\underline{X}}$ and $\chi_{\overline{X}}$ differ only by a null function and so $\chi_{\underline{X}}$ and χ_Y also differ only by a null function; hence $Y \in \mathfrak{k}$ and $m(Y) = m(\underline{X})$. The proof for the rest is trivial.

Remarks. One can easily see that the foregoing notion of completeness is equivalent with the usual notion of completeness according to which every subset of a measurable set of measure zero is again measurable and has measure zero.

In connection with the defining conditions (1) and (2) of outer measure met above, we observe: If (1) is satisfied, then (2) is equivalent with the simultaneous validity of the following two conditions:

(2′) m* is monotone in the same sense as the variable,

(2″) for any sequence $A_1, A_2, \ldots$ in $\mathcal{P}(A)$ we have $m^*(\cup A_n) \leqq \sum m^*(A_n)$.

However, (1) and (2″) together do not imply (2′). Let $B = C \cup D$ with C and D nonempty and disjoint. In the set of all subsets of B define $f(X) = 1$ for $\varnothing \neq X \subset C$ and $f(X) = 0$ otherwise. Then f satisfies (1) and (2″) but not (2′).

Example. Let $A = [0, 1]$ and let $\mathcal{E}$ be the set of all real functions f on A each of which vanishing at all but a finite number of points x; let $E(f) = \sum \{f(x) : x \in A\}$.

It is easy to see that properties (1_c) to (6_c) hold. $\mathcal{N}(g) = +\infty$, if g is nonzero at more than countably many points, otherwise $\mathcal{N}(g) = \sum \{|g(x)| : x \in A\}$, which may be considered as (a proper or improper) convergent series. Property (7_c) holds also; if $f, f_1, f_2, \ldots$ are nonnegative in $\mathcal{E}$ with

$$f \leqq \sum_n f_n,$$

then the summation of all inequalities

$$f(x) \leqq \sum_n f_n(x)$$

over the countably many points x, where $f(x)$ or an $f_n(x) \neq 0$, yields immediately

$$E(f) \leqq \sum_n E(f_n).$$

Only the identically vanishing function is a null function and only the empty set is a null set. $\mathfrak{I}$ consists of all functions f on A which $\neq 0$ on at most countably many points and for which moreover $\sum \{|f(x)| : x \in A\} < +\infty$ holds. Thus the function 1 is not in $\mathfrak{I}$. Convergence in norm in $\mathfrak{I}$ implies pointwise convergence. $\mathfrak{I} = \mathfrak{L}$ and $L(f) = \sum \{f(x) : x \in A\}$. Property (8_c) is fulfilled but property (9_c) is not. Here each function is $\mathscr{N}$-measurable in the sense of the definition following Proposition 17 of Section 2; the function 1 is $\mathscr{N}$-measurable, but not $\mathscr{N}$-integrable. All subsets of A are $\mathscr{N}$-measurable; in fact, $m(T) = +\infty$ if T is infinite and $m(T) = n$ if n is the number of elements of T. For later use we observe: There is no sequence $(f_n)_{n=1}^{\infty}$ of functions from $\mathfrak{L}$ for which

$$\{x : \sup_n |f_n(x)| = 0\}$$

is a null set because this would mean that at each point of A at least one f_n should not be zero; but this is impossible because the union of a countable family of countable sets is countable and the fact that the cardinality of A is c.

The connection between $\mathscr{N}$-measurable sets and the *additive decomposers of* m^*, i.e., those sets Z with

$$m^*(Y) = m^*(Y \cap Z) + m^*(Y - (Y \cap Z))$$

for all $Y \subset A$ with $m^*(Y) < +\infty$, is given by the following proposition.

PROPOSITION 2. *Every $\mathscr{N}$-measurable set is an additive decomposer of m^*; the converse holds if and only if* $1 \in \mathfrak{M}$.

Proof. Let X be $\mathscr{N}$-measurable. For an outer measure m^* we always have $m^*(Y) \leqq m^*(Y \cap X) + m^*(Y - (Y \cap X))$. In this inequality we have to prove equality under the assumption that $m^*(Y) = \mathscr{N}(\chi_Y) < +\infty$. By Proposition 28 of Section 2 there is a function $g \in \mathfrak{L}$ with $\chi_Y \leqq g$ and $m^*(Y) = L(g)$. Since $\chi_{Y \cap X} = \chi_X \chi_Y \leqq g\chi_X \leqq g$ and $g\chi_X \in \mathfrak{M}$

$$(g\chi_X = \lim_n \min\{n\chi_X, g\})$$

we have by Proposition 27 of Section 2 that $g\chi_X \in \mathfrak{L}$ and $m^*(Y \cap X) \leqq L(g\chi_X)$. Moreover, $\chi_{Y-(Y \cap X)} = \chi_Y(1 - \chi_X) \leqq g(1 - \chi_X) = g - g\chi_X \in \mathfrak{L}$ and so $m^*(Y - (Y \cap X)) \leqq L(g - g\chi_X)$. Addition of the foregoing inequalities yields $m^*(Y \cap X) + m^*(Y - (Y \cap X)) \leqq L(g\chi_X) + L(g - g\chi_X = L(g) = m^*(Y)$, proving the first part of the proposition.

Since the basic set A is always an additive decomposer of m^*, for the converse to be valid we must have that $1 = \chi_A \in \mathfrak{M}$. On the other hand, we assume that $1 \in \mathfrak{M}$ and that Z is an additive decomposer of m^* and we show that Z is $\mathscr{N}$-measurable by using the measurability criterion established in Proposition 22 of Section 2. To this end we let g be an arbitrary nonnegative $\mathscr{N}$-integrable function; in addition, we let $\varphi_n(x) = 0$ for $0 \leqq x < 1/n$ and $\varphi_n(x) = 1$ for $1/n \leqq x \leqq +\infty$ and we let $c_n = \varphi_n(g)$, $g_n = gc_n$, $n \in N$. By the definition of φ_n we have that $c_n \leqq ng$ and $g_n \leqq g$; thus c_n and g_n are $\mathscr{N}$-integrable by Propositions 25 and 27 of Section 2. Incidentally, c_n is the characteristic function of a set Y_n with $m^*(Y_n) = L(c_n)$. Since

$$\lim_n g_n = g,$$

we have

$$\min\{\chi_Z, g\} = \lim_n \min\{\chi_Z, g_n\} = \lim_n \min\{\chi_Z c_n, g\} \leqq g,$$

from which it follows that if $\chi_Z c_n$, $n \in N$, is in $\mathfrak{L}$, then so is also $\min\{\chi_Z, g\}$. It is therefore sufficient to show that $\chi_Z c_n \in \mathfrak{L}$. But $\chi_Z c_n = \chi_{Z \cap Y_n}$ and $m^*(Z \cap Y_n) \leqq m^*(Y_n) < +\infty$. Part (b) of Proposition 28 of Section 2 yields a characteristic function χ_{W_n} with

$$\chi_{Z \cap Y_n} \leqq \chi_{W_n} \in \mathfrak{L} \text{ and } \mathscr{N}(\chi_{Z \cap Y_n}) = L(\chi_{W_n}),$$

i.e., $Z \cap Y_n \subset W_n$ and $m^*(Z \cap Y_n) = m^*(W_n)$. Here we may assume that $W_n \subset Y_n$ and hence $m^*(W_n) < +\infty$ and $Z \cap Y_n = Z \cap W_n$; otherwise we replace W_n by $W_n \cap Y_n$. (Observe that $Z \cap Y_n \subset W_n \cap Y_n \subset W_n$ and hence also $m^*(Z \cap Y_n) = m^*(W_n \cap Y_n)$; furthermore, since $0 \leqq \chi_{W_n \cap Y_n} \leqq c_n \chi_{W_n} \leqq c_n \in \mathfrak{L}$ and $\chi_{W_n} \in \mathfrak{L}$, we have that $\chi_{W_n \cap Y_n} \in \mathfrak{L}$.) Now we obtain $m^*(W_n) = m^*(W_n \cap Z) + m^*(W_n - (W_n \cap Z)) \geqq m^*(W_n \cap Z) = m^*(Z \cap Y_n) = m^*(W_n)$, hence $m^*(W_n - (W_n \cap Z)) = 0$ and so $W_n - (W_n \cap Z) = W_n - (Z \cap Y_n)$ is a null set. Thus χ_{W_n} and $\chi_Z c_n$ differ only by a null function and so $\chi_Z c_n$ is $\mathscr{N}$-integrable as well. This finishes the proof.

Remark. In view of the foregoing proposition we have: If $1 \in \mathfrak{M}$, then the system of all $\mathscr{N}$-measurable sets is a σ-algebra.

PROPOSITION 3. *If* $1 \in \mathfrak{M}$, *then we have*:

(*a*) $\mathfrak{k} = \{Y : Y \in \mathfrak{m}$ *and* $m^*(Y) < +\infty\}$.

(*b*) $m^*(Y) = \min\{m(Z) : Y \subset Z \in \mathfrak{k}\}$ *provided that the set over which we take the min is not empty*; *otherwise* $m^*(Y) = +\infty$.

Proof. To part (a): If $m^*(Y) = \mathscr{N}(\chi_Y) < +\infty$, then Proposition 28 of Section 2 yields a χ_Z with $Y \subset Z \in \mathfrak{k}$ and $m^*(Y) = m(Z)$. If moreover $Y \in \mathfrak{m}$, then by Proposition 2 Y is an additive decomposer of m^*, hence $m^*(Z) = m^*(Y) + m^*(Z - Y)$, but $m^*(Z) = m(Z)$ and so $m^*(Z - Y) = 0$; χ_Z and χ_Y differ only by a null function and so $Y \in \mathfrak{k}$ if $Z \in \mathfrak{k}$. Hence $\{Y : Y \in \mathfrak{m}$ and $m^*(Y) < +\infty\} \subset \mathfrak{k}$. By Proposition 1 the inclusion in the other direction is evident.

To part (b): If $\{Y \subset Z \in \mathfrak{k}\} \neq \varnothing$, then clearly $m^*(Y) < +\infty$ and so by the proof of part (a) the claim in (b) also holds; if however the set in question is empty, then $m^*(Y)$ cannot be finite. This completes the proof.

Remark. By the foregoing proposition m^* is the outer measure which belongs to m. The set Z with $Y \subset Z \in \mathfrak{k}$ and $m^*(Y) = m(Z)$ introduced in the proof of part (a) is called the *equimeasurable hull of* Y.

PROPOSITION 4. *Let* $1 \in \mathfrak{M}$ *and* f *be finite and*

$$A(\alpha, \beta) = \{x \in A : \alpha \leqq f(x) < \beta\}.$$

If $A(\alpha, \beta)$ *is* $\mathscr{N}$*-measurable for all rationals* α, β, *then* f *is* $\mathscr{N}$*-measurable*; *if* f *is* $\mathscr{N}$*-measurable, then* $A(\alpha, \beta)$ *is* $\mathscr{N}$*-measurable for all real* α, β.

Proof. Let $\chi_{\alpha\beta}$ denote the characteristic function of the interval $[\alpha, \beta)$. If f is $\mathscr{N}$-measurable, then by Proposition 25 of Section 2 the function $\chi_{\alpha\beta}(f)$ is $\mathscr{N}$-measurable, i.e., the set $A(\alpha, \beta)$ is $\mathscr{N}$-measurable for arbitrary real numbers α and β. Conversely, if these sets are $\mathscr{N}$-measurable for all rationals α, β with $\alpha < \beta$, then we consider for every natural number p a monotone sequence $\ldots, \alpha_{-2}, \alpha_{-1}, \alpha_0, \alpha_1, \alpha_2, \ldots$ of rational numbers with $\alpha_n \to \pm\infty$ for $n \to \pm\infty$ and $\alpha_{n+1} - \alpha_n < 1/p$, and numbers μ_n with $\alpha_n \leqq \mu_n \leqq \alpha_{n+1}$ for $n = 0, \pm 1, \ldots$. Letting

$$\chi_n = \chi_{\alpha_n \alpha_{n+1}} \text{ and } f_p = \sum_{-\infty}^{+\infty} \mu_n \chi_n(f),$$

we have $|f_p(x) - f(x)| < 1/p$ for $x \in A$ and f_p is $\mathscr{N}$-measurable by Propositions 21 and 20 of Section 2; but

$$f(x) = \lim_p f_p(x)$$

for $x \in A$ and so f is $\mathscr{N}$-measurable also. This completes the proof.

Remarks. Under the general assumption that $1 \in \mathfrak{M}$ we have:

(1) If $f \in \mathfrak{L}$, then so also is $f\chi_{\alpha\beta}(f) \in \mathfrak{L}$. Indeed, $\chi_{\alpha\beta}(f)$ is measurable and so also is $f\chi_{\alpha\beta}(f)$ and since $-|f| \leqq f\chi_{\alpha\beta}(f) \leqq |f|$, then by Proposition 27 of Section 2 $f\chi_{\alpha\beta}(f) \in \mathfrak{L}$.

(2) If $f \in \mathfrak{L}$ and $\alpha < \beta$ but not $\alpha < 0 < \beta$, then $\chi_{\alpha\beta}(f) \in \mathfrak{L}$. Indeed, for example, let $0 < \alpha < \beta$; then $0 \leqq \chi_{\alpha\beta}(f) \leqq (1/\alpha)|f|$. The conclusion follows as before.

(3) If $f \in \mathfrak{L}$, then for every real $\alpha \neq 0$ the function $\varphi = \chi_{\{x \in A : f(x) = \alpha\}}$ belongs to $\mathfrak{L}$. This is so because φ is the limit of the sequence of measurable functions

$$\chi_{\{x \in A : \alpha_n \leqq f(x) < \beta_n\}}$$

with rational α_n, β_n and $\alpha_n \nearrow \alpha \swarrow \beta_n$ and is therefore measurable; moreover, φ is wedged between the integrable functions $-|f/\alpha|$ and $|f/\alpha|$.

PROPOSITION 5. *If* $1 \in \mathfrak{M}$ *and* $f \in \mathfrak{M}$, *then for every (proper or improper) real* r *the following sets are* $\mathscr{N}$*-measurable*: $\{x \in A : f(x) = r\}$, $\{x \in A : f(x) > r\}$, $\{x \in A : f(x) \neq r\}$, *and* $\{x \in A : f(x) \geqq r\}$.

Proof. For finite real r,

$$\chi_{\{x \in A : f(x) = r\}} = \lim\{\chi_{\{x \in A : r \leqq f(x) < r'\}} : r' \searrow r\}$$

is measurable by Proposition 4 above and by Proposition 20 of Section 2. Since $1 \in \mathfrak{M}$, we have by Proposition 22 of Section 2 that

$$\chi_{\{x \in A : f(x) \neq r\}} = 1 - \chi_{\{x \in A : f(x) = r\}} \in \mathfrak{M}.$$

Again by Proposition 22 of Section 2 we have $\max\{r, f\} \in \mathfrak{M}$, hence $\{x \in A : \max\{r, f\}(x) \neq r\}$, i.e., $\{x \in A : f(x) > r\}$, is measurable and with it is also

$$\chi_{\{x \in A : f(x) = +\infty\}} = \lim_{n \to \infty} \chi_{\{x \in A : f(x) > n\}}$$

and so forth. This concludes the proof.

DEFINITION. *Let* f *be a real-valued function on* A; f *is said to be continuous with respect to* $\mathscr{N}$*-decomposition if for each* $\varepsilon > 0$ *there exists a decomposition*

$$A = A_1 \cup A_2 \cup \dots$$

into countably many disjoint $\mathscr{N}$*-measurable sets* A_n *such that on each* A_n *the function* f *is bounded and its oscillation is less than* ε *(i.e.,* $\sup f(A_n) - \inf f(A_n) < \varepsilon$*).*

We can now give Proposition 4 another formulation as follows.

PROPOSITION 6. *If* $1 \in \mathfrak{M}$, *then for a finite function continuity with respect to* $\mathcal{N}$*-decomposition and* $\mathcal{N}$*-measurability are equivalent.*

Proof. If f is $\mathcal{N}$-measurable, then we have

$$A = \bigcup \left\{x \in A : \tfrac{m}{n} \leqq f(x) < \frac{m+1}{n} \text{ with } m = 0, \pm 1, \ldots \right\};$$

by Proposition 4 this is a decomposition of the required kind with $\varepsilon = 1/n$, $n \in N$.

If, conversely, $A = A_1 \cup A_2 \cup \ldots$ is a decomposition in the sense of the above definition with $\varepsilon = 1/n$, then A_n is measurable and so are χ_{A_n} and $\sup f(A_n)\chi_{A_n}$; by Proposition 20 of Section 2 we therefore see that

$$f_n = \sum_n \sup f(A_n)\chi_{A_n}$$

is measurable. Since $|f_n - f| \leqq 1/n$, we have

$$\lim_n f_n = f$$

and hence also $f \in \mathfrak{M}$. This completes the proof.

The following is a direct consequence of the foregoing proposition.

PROPOSITION 7. *If* $1 \in \mathfrak{M}$, f *is nonnegative, finite, and* $\mathcal{N}$*-integrable, then the statements*

$$m^*(\{x \in A : f(x) > 0\}) > 0 \text{ and } L(f) > 0$$

are equivalent.

Proof. If $m^*(\{x \in A : f(x) > 0\}) > 0$, then by Proposition 6

$$\begin{aligned} 0 &< m^*(\{x \in A : f(x) > 0\}) \\ &\leqq \sum_{n=1}^{\infty} m^*(A(n, n+1)) + \sum_{k+1}^{\infty} m^*\left(A\left(\frac{1}{k+1}, \frac{1}{k}\right)\right), \end{aligned}$$

so that at least one of the summands on the right must be positive, say for $A(\alpha, \beta)$. Taking into account item (2) in the Remarks following Proposition 4 we see that

$$0 < \alpha m^*(A(\alpha, \beta)) = \alpha \mathcal{N}(\chi_{\alpha,\beta}(f)) = L(\alpha \chi_{\alpha,\beta}(f)) \leqq L(f).$$

Conversely, from $L(f) > 0$ and $0 < \alpha = 1/n < n$ it follows that

$$m^*(\{x \in A : f(x) > 0\}) \geqq m^*(A(\alpha, n)) = L(\chi_{\alpha,n}(f))$$
$$= \alpha L(n\chi_{\alpha,n}(f)) \geqq \alpha L(f\chi_{\alpha,n}(f)) > 0$$

for sufficiently large n. This is so because for $n \to +\infty$ we have the bounded convergence $f\chi_{\alpha,n}(f) \to f$ and so $L(f\chi_{\alpha,n}(f)) \to L(f) > 0$. This proves the proposition.

PROPOSITION 8 (EGOROV). *If B is an $\mathscr{N}$-measurable subset of A with $m(B) < +\infty$ and $(f_n)_{n=1}^{\infty}$ a sequence of finite, $\mathscr{N}$-measurable functions on B which tend to the finite function f on B, then there is a decomposition*

$$B = B_0 \cup B_1 \cup B_2 \cup \ldots$$

into $\mathscr{N}$-measurable parts such that B_0 is an $\mathscr{N}$-null set and for $k \in N$ the sequence $(f_n)_{n=1}^{\infty}$ converges uniformly on B_k.

Proof. For $n \in N$, let

$$s_n(x) = \sup\{|f_{n'}(x) - f_{n''}(x)| : n', n'' \geqq n\}.$$

Then s_n is an $\mathscr{N}$-measurable function with

$$\lim_n s_n(x) = 0$$

for all $x \in B$. For $j \in N$ we form the $\mathscr{N}$-measurable sets

$$K_{n,j} = \{x : s_n(x) < j^{-1}\} \subset B.$$

For fixed j we have

$$K_{1,j} \subset K_{2,j} \subset \ldots \text{ and } \bigcup_k K_{k,j} = B.$$

Since m is σ-additive, we have

$$m(K_{k,j}) \to m(B) \quad \text{for } k \to +\infty.$$

For $j = 1$ we determine a k_1 with $m(B - K_{k_1,1}) \leqq 2^{-2}m(B)$, then a $k_2 > k_1$ with $m(K_{k_1,1} \cap (B - K_{k_2,2})) \leqq 2^{-3}m(B)$ and so forth. We let

$$B_1 = \bigcap_j K_{k_j,j}$$

and

$$B' = (B - K_{k_1,1}) \cup (K_{k_1,1} \cap (B - K_{k_2,2})) \cup \ldots;$$

we have the decomposition $B = B_1 \cup B'$, where

$$m(B') \leqq m(B)(2^{-2} + 2^{-3} + \ldots) = \tfrac{1}{2}m(B)$$

and on B_1 we have that s_n tends to zero uniformly.

We now repeat the process, described above for the set B, on the set B' and obtain a decomposition

$$B' = B_2 \cup B'' \text{ with } m(B'') \leqq \tfrac{1}{2}m(B') \leqq \tfrac{1}{4}m(B)$$

and with the sequence s_n tending uniformly to zero on B_2. Continuation of this process leads to a decomposition

$$B = (B_1 \cup B_2 \cup \dots) \cup B_0,$$

where

$$B_0 = B - (B_1 \cup B_2 \cup \dots)$$

and, since

$$m(B) \geqq m(B_1 \cup B_2 \cup \dots) = m(B_1) + m(B_2) + \dots \geqq m(B),$$

we see that B_0 is an $\mathcal{N}$-null set; moreover, on each individual B_j the s_n, and hence also the f_n, converge uniformly. The proof is finished.

Using the notation introduced in the proof of Proposition 4, we assign to each $\mathcal{N}$-measurable function f formally the "Lebesgue sums"

$$\Lambda_p(f) = \sum_n{}' \lambda_n \mathcal{N}(\chi_n(f)), \; p \in N,$$

where the α_n and χ_n satisfy the same conditions as in the proof of Proposition 4 and the λ_n are selected according to the following rule:

If $\alpha_n \geqq 1/p$ or $\alpha_{n+1} \leqq -1/p$, then let $\alpha_n \leqq \lambda_n \leqq \alpha_{n+1}$; otherwise let $\lambda_n = 0$.

The symbol $\sum'$ here denotes summation over all integers n with $\lambda_n \neq 0$.

PROPOSITION 9. *If* $1 \in \mathfrak{M}$ *and* f *is finite and* $\mathcal{N}$*-integrable, then the Lebesgue sums are absolutely and properly convergent and*

$$\lim_{p \to \infty} \Lambda_p(f) = L(f).$$

Proof. Let $f_p = \sum' \lambda_n \chi_n(f)$ and $\gamma_n = \min\{|\alpha_n|, |\alpha_{n+1}|\}$. Then for $\gamma_n \geqq 1/p$ (i.e., $\alpha_n \geqq 1/p$ or $\alpha_{n+1} \leqq -1/p$) we see that $|\lambda_n| \leqq \min\{|\alpha_n|, |\alpha_{n+1}|\} + 1/p \leqq 2\gamma_n$ and $|f| \geqq \gamma_n \chi_n(f)$ and hence $|f_p| \leqq 2 \sum' \gamma_n \chi_n(f) \leqq 2|f|$. Just as in the proof of Proposition 4 follows here too the $\mathcal{N}$-measurability of $\chi_n(f)$ and f_p; by part (a) of Proposition 27 in Section 2 now follows the $\mathcal{N}$-integrability of these functions. The foregoing inequality gives the bounded convergence of the series for f_p and so by Proposition 12 of Section 2 we have

$$L(f_p) = \sum' L(\lambda_n \chi_n(f)) = \Lambda_p(f);$$

we also have bounded convergence in $f_p \to f$ and so

$$L(f) = \lim_{p \to \infty} \Lambda_p(f).$$

This finishes the proof.

Remark. We call the limit of $\Lambda_p(f)$ the *Lebesgue integral of f over A with respect to m* and write $\int_A f\,dm$; the statement of the above proposition can therefore be written as $L(f) = \int_A f\,dm$.

If T is a topological space, we call a real-valued function on T a *K-function* if it is finite, continuous, and vanishes outside a compact subset of T. Every K-function is bounded and the set of all K-functions of a topological space forms a linear lattice.

PROPOSITION 10. *If T is a locally compact Hausdorff space, $\mathcal{E}$ the set of all K-functions on T, and E is a positive linear functional on $\mathcal{E}$, then $E|\mathcal{E}$ satisfies also the properties* (7_c) *and* (8_c).

Proof. Property (8_c) reads: if $f \in \mathcal{E}$, then $\min\{f, 1\} \in \mathcal{E}$; it is evident that this property is satisfied. We also recall property (7_c); it reads: if $f, f_1, f_2, \ldots$ are in $\mathcal{E}$ and

$$|f| \leqq \sum_{n=1}^{\infty} |f_n|,$$

then

$$E(|f|) \leqq \sum_{n=1}^{\infty} E(|f_n|).$$

To establish property (7_c) we proceed as follows: We take $f, f_1, f_2, \ldots$ in $\mathcal{E}$ with $|f| \leqq \sum |f_n|$ and show that there exists a $g \in \mathcal{E}$ with $g(x) \geqq 0$ for $x \in K = \overline{\{x : f(x) \neq 0\}}$ and the property that for each $\varepsilon > 0$ there is $\mu = \mu(\varepsilon)$ such that in general

$$|f| \leqq \sum_{n=1}^{\mu} |f_n| + \varepsilon g.$$

Indeed, if this is shown, then it follows that

$$E(|f|) \leqq \sum_{n=1}^{\mu} E(|f_n|) + \varepsilon E(g) \leqq \sum_{n=1}^{\infty} E(|f_n|) + \varepsilon E(g)$$

and from this we get (7_c) by letting $\varepsilon \to 0$.

Now to the construction of g. K is by assumption compact. To every $x \in K$ there corresponds a neighborhood U with compact closure $\overline{U}$. We can cover K already with finitely many of these U. The union V of these finitely many U is open and has a compact closure $\overline{V}$. $\overline{V}$ is normal and so there is a continuous function g on T with $0 \leqq g \leqq 1$, $\{x : g(x) = 0\} \supset T - V$ and $\{x : g(x) = 1\} \supset K$ by Urysohn's Lemma. This function g is a K-function.

Finally, we consider the sequence of K-functions

$$g_\mu = \max\{0, |f| - \sum_{n=1}^{\mu} |f_n|\}$$

which tends to zero nonincreasingly. Here $g_\mu(x) = 0$ for $x \in T - K$ and $\mu = 1, 2, \ldots$; moreover, we have uniform convergence by Dini's Theorem (see the Application following Proposition 3 in Section 1). Thus for $\varepsilon > 0$ there exists a $\mu' = \mu(\varepsilon)$ such that $0 \leqq g_\mu(x) < \varepsilon g(x)$ for $x \in T$ and $\mu > \mu'$. This completes the proof.

We have the following representation theorem.

PROPOSITION 11. *If T is a locally compact Hausdorff space and $E|\mathcal{E}$ is a positive linear functional on the set $\mathcal{E}$ of all K-functions of T, then there is an outer measure m^* on the family of all subsets of T such that, for each $f \in \mathcal{E}$, $E(f)$ is representable as Lebesgue integral of f with respect to m, in symbols*:

$$E(f) = \int_T f\,dm.$$

Proof. By Proposition 10 above and by Proposition 23 of Section 2 we have that $1 \in \mathfrak{M}$. Hence the various propositions concerning m^* established further above in this section apply; in particular we have that Proposition 9 is applicable since $f \in \mathcal{E}$ implies $f \in \mathfrak{L}$ and f is bounded. This finishes the proof.

On occasion, one requires of the elementary integrals under consideration not only that (8_c) or $1 \in \mathfrak{M}$ be satisfied, but also that the additional property, called *σ-finiteness*, be satisfied:

(10_c) There is a sequence $(g_n)_{n=1}^{\infty}$ of $\mathcal{N}$-integrable functions such that

$$\bigcap_n \{x : g_n(x) = 0\}$$

is an $\mathcal{N}$-null set.

Remark. Observe that (8_c) does not imply (10_c); in this connection see the example following Proposition 1. Simple examples show that (10_c) does not imply (8_c) either.

PROPOSITION 12. *Under the assumption that* $1 \in \mathfrak{M}$, *each of the following statements is equivalent with* (10_c):

(*a*) *There are (bounded) everywhere positive functions* $f_0 \in \mathfrak{L}$.

(*b*) *There is a sequence* $(f_n)_{n=1}^{\infty}$ *of nonnegative elementary functions for which*

$$\bigcap_n \{x : f_n(x) = 0\}$$

is an $\mathcal{N}$*-null set.*

(*c*) *Every* $\mathcal{N}$*-measurable set can be represented as a countable union of* $\mathcal{N}$*-measurable sets of finite measure.*

(*d*) *There is a sequence* $(g_n)_{n=1}^{\infty}$ *of functions in* $\mathcal{E}$ *with*

$$\sum_n |g_n| = +\infty$$

for all $x \in A$.

Proof. Concerning part (a): The statement is trivial if the basic set A itself is a null set. Otherwise we can leave out in the sequence $(g_n)_{n=1}^{\infty}$ of (10_c) all those functions for which $L(|g_n|) = 0$, since for such the set $\{x : g_n(x) = 0\}$ differs from A only by a null set. Thus, let $L(|g_n|) \neq 0$ for all n. Then, setting

$$f^* = \sum_n \frac{|g_n|}{2^n L(|g_n|)},$$

we have that $f^* \geqq 0$ and $f^* \in \mathfrak{L}$. Moreover,

$$\{x : f^*(x) = 0\} = \bigcap_n \{x : g_n(x) = 0\}$$

is a null set. If we therefore set

$$f_0(x) = \begin{cases} \min\{1, f^*(x)\} & \text{for } f^*(x) > 0 \text{ and} \\ 1 & \text{for } f^*(x) = 0, \end{cases}$$

then we have the desired function f_0 with $0 < f_0(x) \leqq 1$ for all $x \in A$. It is of course clear that the existence of such functions f_0 is sufficient for (10_c).

Concerning part (b): Observe that (b) is stronger than (10_c). On the other hand, (10_c) implies by (a) the existence of the function f_0 to which there corresponds a sequence $(\varphi_n)_{n=1}^{\infty}$ of elementary functions with

$$f_0 = \operatorname*{Lim}_n \varphi_n.$$

But then we also have

$$\operatorname*{Lim}_{n} |\varphi_n| = f_0$$

and in $f_n = \max\{|\varphi_1|, \ldots, |\varphi_n|\}$ we have the desired functions. Indeed, $0 \leqq f_n \in \mathcal{E}$ and $f_n(x) = 0$ for all n means that $\varphi_n(x) = 0$ for all n. But by Proposition 6 of Section 2 there is a subsequence $|\varphi_{n_1}|, |\varphi_{n_2}|, \ldots$ with

$$\lim_{j} |\varphi_{n_j}(x)| = f_0(x) > 0$$

for almost all x and so $f_n(x) = 0$ for all n can only occur at the points x of a null set.

Concerning part (c): Let (10_c) be satisfied. By Proposition 3 it follows that for a $Y \in \mathfrak{m}$ either $m^*(Y) = m(Y) < +\infty$ or $m^*(Y) = +\infty$. It is enough to pursue further the last case only. With χ_Y we also have that $\varphi_n = \min\{\chi_Y, nf_0\}$ is measurable and since $0 \leqq \varphi_n \leqq nf_0$ we have $\varphi_n \in \mathfrak{L}$. Letting

$$B_n = \{x : \varphi_n(x) = 1\},$$

we have that $\chi_{B_n} \in \mathfrak{L}$, i.e., B_n is measurable and has finite measure. Since φ_n converges monotonely to χ_Y, we have that

$$Y = \bigcup_n B_n$$

and we have (c).

Now, let (c) be satisfied. Since $1 \in \mathfrak{M}$, A is measurable; hence there exists a representation

$$A = \bigcup_n B_n$$

with $B_n \in \mathfrak{m}$ and $m(B_n) < +\infty$, where we may assume that the B_n are pairwise disjoint and $m(B_n) > 0$ (provided we do not have the trivial case $m(A) = 0$). We then put

$$f_0(x) = (2^n m(B_n))^{-1} \quad \text{for } x \in B_n,\ n \in N,$$

and we get that $f_0 > 0$, and by the Proposition 9, Section 2, we see that $L(f_0) \leqq 1$; thus (a) holds and so property (10_c) is satisfied.

Concerning part (d): If (d) holds, we put $f_n = |g_1| + \ldots + |g_n|$. Then $f_n \in \mathcal{E}$,

$$\sup_n f_n = \sum_k |g_k| \text{ and } \{x : \sup_n f_n(x) = 0\}$$

is the empty set, hence condition (b) holds. On the other hand, for the function f_0 in (a) we have, since $f_0 \in \mathfrak{L}$, the representation

$$\operatorname*{Lim}_k g_k = f_0$$

with $g_k \in \mathcal{E}$, or, if we envisage that a suitable subsequence has been chosen already, with

$$\lim_k g_k = f_0,$$

hence also

$$\lim_k |g_k| = |f_0| = f_0$$

$\mathcal{N}$-almost everywhere. Since $f_0 > 0$, we therefore have $|g_1| + |g_2| + \ldots = +\infty$ $\mathcal{N}$-almost everywhere with $|g_n| \in \mathcal{E}$. Since for every measurable set of finite measure and in particular for every null set Y there are sequences $(f_n)_{n=1}^{\infty}$ of elementary functions with $\sum |f_n| > 0$ for $x \in Y$ (because on account of $\mathcal{N}(\chi_Y) < +\infty$ there exists a sequence $(f_n)_{n=1}^{\infty}$ in $\mathcal{E}$ with $\chi_Y \leqq \sum |f_n|$), we obtain upon suitable choice of the f_n in $\mathcal{E}$ in the functions $h_n = |g_n| + |f_1| + \ldots + |f_n|$ a sequence in $\mathcal{E}$ with $\sum h_n(x) = +\infty$ for all x and so (d) is established. This completes the proof.

Remark. In the case of σ-finiteness the initially somewhat strange special stipulation in the definition of the norm, when there are no f_k in $\mathcal{E}$ with

$$|f| \leqq \sum_k |f_k|,$$

therefore no longer comes into play at all.

We now consider a generalization of the Dominated Convergence Theorem (see Proposition 12 of Section 2).

PROPOSITION 13. *Assume that* (10_c) *holds and that* $1 \in \mathfrak{M}$. *Suppose that for all* $n \in N$ *the functions* f_n, g_n *and* h_n *belong to* $\mathfrak{L}$ *and that* $f_n \leqq g_n \leqq h_n$ *holds* $\mathcal{N}$*-almost everywhere. Moreover, let*

$$\operatorname*{Lim}_n f_n = f \in \mathfrak{L},$$

$$\operatorname*{Lim}_n h_n = h \in \mathfrak{L}$$

and suppose that

$$\lim_n g_n = g$$

exists $\mathscr{N}$-almost everywhere. Then $g \in \mathfrak{L}$ and

$$L(g) = \lim_n L(g_n).$$

Proof. We have that $f \leqq g \leqq h$ holds $\mathscr{N}$-almost everywhere; we may assume without loss of generality that $f_n \leqq g \leqq h_n$ holds $\mathscr{N}$-almost everywhere as well. In order to realize this, we may simply replace f_n by $f - |f_n - f|$ and h_n by $h + |h_n - h|$.

Taking into account Proposition 6 of Section 2 and Propositions 8 and 12 (part (c)) from this section, we can decompose the basic set A into

$$A = A_0 \cup A_1 \cup A_2 \cup \ldots,$$

where A_0 is an $\mathscr{N}$-null set and for $r = 1, 2, \ldots$ every A_r is $\mathscr{N}$-measurable and of finite measure and such that the sequence $(g_n)_{n=1}^{\infty}$ converges uniformly on A_r. Letting

$$B_r = A_1 \cup \ldots \cup A_r \text{ and } C_r = A - B_r$$

and noting that g is $\mathscr{N}$-almost everywhere finite (because $f \leqq g \leqq h$), we obtain

$$\mathscr{N}(g - g_n) \leqq \mathscr{N}((g - g_n)\chi_{B_r}) + \mathscr{N}((g - g_n)\chi_{C_r}),$$

where

$$\begin{aligned}\mathscr{N}((g - g_n)\chi_{C_r}) = \mathscr{N}(|g - g_n|\chi_{C_r}) &\leqq \mathscr{N}((h_n - f_n)\chi_{C_r}) \\ &\leqq \mathscr{N}(f - f_n) + \mathscr{N}((f - h)\chi_{C_r}) + \mathscr{N}(h - h_n).\end{aligned}$$

On account of the proper convergence of the series

$$L((h - f)\chi_{A_1}) + L((h - f)\chi_{A_2}) + \ldots = L(h - f) = \mathscr{N}(h - f),$$

with the series remainders $\mathscr{N}((h - f)\chi_{C_r})$ we can, for every prescribed $\varepsilon > 0$, find an r' such that $\mathscr{N}((f - h)\chi_{C_{r'}}) < \varepsilon/4$. Since $g_n \to g$ uniformly on $B_{r'}$, we can find an n' such that $\mathscr{N}(|g - g_n|\chi_{B_{r'}}) < \varepsilon/4$ for $n > n'$; one may of course pick n' so large that $\mathscr{N}(f - f_n) < \varepsilon/4$ and $\mathscr{N}(h - h_n) < \varepsilon/4$ hold as well for $n > n'$. Thus $\mathscr{N}(g - g_n) < \varepsilon$ for $n > n'$ and so $g \in \mathfrak{L}$ and

$$L(g) = \lim_n L(g_n)$$

follow. The proof is finished.

PROPOSITION 14. *Let property* (10_c) *hold and* $1 \in \mathfrak{M}$. *Suppose also that an $\mathscr{N}$-measurable function Φ has the property that $\Phi f \in \mathfrak{L}$ whenever $f \in \mathfrak{L}$, where, moreover, $L(\Phi f) \geqq 0$ whenever $f \geqq 0$. Then $\Phi \geqq 0$ holds $\mathscr{N}$-almost everywhere.*

Proof. For $r < 0$ we have that $\{x : \Phi(x) < r\}$ is measurable; by part (c) of Proposition 12 this set is either an $\mathcal{N}$-null set or contains a measurable subset B of finite positive measure. If the latter were the case, we would have $\chi_B \in \mathfrak{L}$ and $L(\Phi\chi_B) \leqq rm(B) < 0$ in contradiction to the hypothesis of the proposition. Thus $\{x : \Phi(x) < r\}$ is a null set for all $r < 0$ and so the set $\{x : \Phi(x) < 0\}$ will also be a null set. The proof is finished.

PROPOSITION 15. *If φ is $\mathcal{N}$-measurable and bounded and if*

$$\mathcal{E} \ni g \geqq 0 \text{ implies } L(\varphi g) \geqq 0, \tag{1}$$

then $\varphi \geqq 0$ holds $\mathcal{N}$-almost everywhere.

Proof. To be able to apply Proposition 14, we show that

$$\mathfrak{L} \ni f \geqq 0 \text{ implies } L(\varphi f) \geqq 0 \tag{2}$$

holds whenever property (1) holds. Indeed, if $\mathfrak{L} \ni f \geqq 0$, we select $g_n \in \mathcal{E}$ with

$$\operatorname*{Lim}_n g_n = f.$$

Without loss of generality we may suppose that $g_n \geqq 0$ (or else we would use $\max\{g_n, 0\}$ in place of g_n in the following). Taking

$$M = \sup\{|\varphi(x)| : x \in A\} < +\infty,$$

we obtain that $|\varphi g_n| \leqq Mg_n \in \mathfrak{L}$ and hence by Proposition 13

$$\lim_n L(\varphi g_n) = L(\varphi g_n);$$

but $L(\varphi g_n) \geqq 0$ and so $L(\varphi f) \geqq 0$. This proves the proposition.

Section 4. The Function Spaces $\mathfrak{I}^p$ and $\mathfrak{L}^p$

For $p \geqq 1$ we define

$$\mathfrak{I}^P = \{f : f|f|^{p-1} \in \mathfrak{I}\} = \{f : \mathcal{N}(|f|^P) < +\infty\},$$

$$\mathfrak{L}^p = \{f : f|f|^{p-1} \in \mathfrak{L}\} \text{ and } \mathcal{N}_p(f) = (\mathcal{N}(|f|^P))^{1/p}.$$

It is clear that $\mathfrak{I}^1 = \mathfrak{I}$, $\mathfrak{L}^1 = \mathfrak{L}$, and $\mathcal{N}_1 = \mathcal{N}$.

PROPOSITION 1. *$\mathcal{N}_p$ satisfies the inequalities of Hölder and Minkowski*:

(*H*) *For $p > 1$, $p + q = pq$, $f \in \mathfrak{I}^p$, $g \in \mathfrak{I}^p$ we have*

$$\mathcal{N}(fg) \leqq \mathcal{N}_p(f)\mathcal{N}_q(g).$$

(*M*) *For* $p \geqq 1$, (f *and* g) $\in \mathfrak{I}^p$ *we have*

$$\mathcal{N}_p(f+g) \leqq \mathcal{N}_p(f) + \mathcal{N}_p(g).$$

Proof. First we consider (H). We may of course ignore the trivial case when $\alpha' = \mathcal{N}(|f|^p)$ or $\beta' = \mathcal{N}(|g|^q)$ is zero. For $\alpha, \beta > 0$ and taking

$$\gamma = \left(\left(\frac{p}{q}\right)^{1/p} + \left(\frac{q}{p}\right)^{1/q}\right)^{-1},$$

the function $\varphi(\xi) = \gamma(\alpha\xi^p + \beta\xi^{-q})$ over the interval $0 < \xi < +\infty$ has only at the point

$$\xi_0 = \left(\frac{\beta q}{\alpha p}\right)^{1/pq}$$

an absolute minimum by the rules of differential calculus and because $p + q = pq$; thus $\varphi(\xi_0) \leqq \varphi(\xi)$, or

$$\alpha^{1/p}\beta^{1/q} \leqq \gamma(\alpha\xi^p + \beta\xi^{-q}).$$

The last inequality clearly also holds for α or $\beta = 0$. We therefore obtain

$$|fg| \leqq \gamma(|f|^p\xi^p + |g|^q\xi^{-q})$$

from which we get that

$$\mathcal{N}(fg) \leqq \gamma(\alpha'\xi^p + \beta'\xi^{-q})$$

for all $\xi > 0$. For the special value

$$\xi = \xi^* = \left(\frac{\beta' q}{\alpha' p}\right)^{1/pq}$$

we get Hölder's Inequality.

Next we consider Minkowski's Inequality. The case $p = 1$ we have proved earlier (see property (d) of the norm $\mathcal{N}$, discussed at the beginning of Section 2). The case $p > 1$ we reduce to Hölder's Inequality. Since for reals x and y we have in general that $|x + y|^p \leqq 2^{p-1}(|x|^p + |y|^p)$ (because for positive x and y the quotient of both sides becomes extremal only for $x = y$), it follows that

$$\mathcal{N}(|f + g|^p) \leqq 2^{p-1}(\mathcal{N}(|f|^p) + \mathcal{N}(|g|^p));$$

this means that with f and g the function $f+g$ is also in $\mathfrak{Y}^p$. From $|f+g|^p \leqq (|f|+|g|)|f+g|^{p-1}$ we obtain by using Hölder's Inequality:

$$\begin{aligned}\mathscr{N}(|f+g|^p) &\leqq \mathscr{N}(|f||f+g|^{p-1}) + \mathscr{N}(|g||f+g|^{p-1})\\ &\leqq (\mathscr{N}_p(f) + \mathscr{N}_p(g))\mathscr{N}_q(|f+g|^{p-1})\end{aligned}$$

which yields Minkowski's Inequality on account of

$$\mathscr{N}_q(|f+g|^{p-1}) = (\mathscr{N}(|f+g|^p))^{1-1/p}.$$

This completes the proof.

We define in $\mathfrak{Y}^p$ equality by

$$f = g \text{ if and only if } \mathscr{N}_p(f-g) = 0.$$

Since $\mathscr{N}_p(h) = 0$ if and only if $\mathscr{N}(h) = 0$, we see that this definition of equality is identical with the one introduced in $\mathfrak{H}$. Minkowski's Inequality shows that $\mathfrak{Y}^p$ is a normed linear space under the norm $\mathscr{N}_p$. Moreover,

PROPOSITION 2. *$\mathfrak{Y}^p$ is complete.*

Proof. We need the inequality (due to S. Mazur)

$$2^{1-p}|x-y|^p \leqq |x\,|x|^{p-1} - y|\,y|^{p-1}| \leqq p|x-y|(|x|^{p-1} + |y|^{p-1}). \qquad (1)$$

To prove the first half of this inequality we put $x = \frac{1}{2} + r$, $y = -\frac{1}{2} + r$, which may be done on account of the homogeneity. Concerning the function $(\frac{1}{2}+r)|\frac{1}{2}+r|^{p-1} + (\frac{1}{2}-r)|\frac{1}{2}-r|^{p-1}$, it is easy to show that it is an even function and that it increases for $r > 0$ and hence assumes the minimum 2^{1-p} for $r = 0$. To prove the second half of the inequality, we again use the foregoing substitution. On account of the symmetry it will be enough to consider only the case where $r \geqq 0$. For $0 \leqq r \leqq 1/2$ the inequality in question reads:

$$(\tfrac{1}{2}+r)^p + (\tfrac{1}{2}-r)^p \leqq p((\tfrac{1}{2}+r)^{p-1} + (\tfrac{1}{2}-r)^{p-1});$$

but $r \leqq p - 1/2$ and so the inequality is seen to hold. For $r > 1/2$ we put

$$\frac{r-1/2}{r+1/2} = t;$$

hence $0 < t < 1$ and the inequality in question reads:

$$\frac{1+t}{1-t}\;\frac{1-t^{p-1}}{1+t^{p-1}} \leqq 2p-1.$$

The validity of this inequality is seen by noting that the function on the left-hand side is monotone in the interval $0 \leqq t \leqq 1$ and observing that the inequality is satisfied at the end points.

We next consider the mapping V which associates with every $f \in \mathcal{G}^p$ a function

$$V(f) = f|f|^{p-1}$$

from $\mathcal{G}$. V is a homeomorphism. Indeed, if $g = V(f) = |f|^p \operatorname{sign} f$, then $f = |g|^{1/p} \operatorname{sign} g = V^{-1}(g)$, the inverse of V, defined for all $g \in \mathcal{G}$. From the first half of inequality (1) follows for $g_1, g_2 \in \mathcal{G}$

$$|V^{-1}(g_1) - V^{-1}(g_2)|^p \leqq 2^{p-1}|g_1 - g_2|,$$

thus

$$\mathcal{N}_p(V^{-1}(g_1) - V^{-1}(g_2)) \leqq 2^{1/q}(\mathcal{N}(g_1 - g_2))^{1/p};$$

this shows the uniform continuity of V^{-1} in $\mathcal{G}^p$. From the second half of inequality (1) we get for $f_1, f_2 \in \mathcal{G}^p$:

$$|V(f_1) - V(f_2)| \leqq p|f_1 - f_2||f_1|^{p-1} + p|f_1 - f_2||f_2|^{p-1};$$

upon using Hölder's Inequality, we obtain

$$\begin{aligned}\mathcal{N}(V(f_1) - V(f_2)) &\leqq p\mathcal{N}_p(f_1 - f_2)(\mathcal{N}_q(|f_1|^{p-1}) + \mathcal{N}_q(|f_2|^{p-1})) \\ &= p\mathcal{N}_p(f_1 - f_2)(|\mathcal{N}_p(f_1)|^{p/q} + |\mathcal{N}_p(f_2)|^{p/q}).\end{aligned}$$

This shows that $V(f)$ is uniformly continuous on all bounded parts of $\mathcal{G}^p$.

Finally, we take a Cauchy sequence $(f_n)_{n=1}^{\infty}$ in $\mathcal{G}^p$ in the sense of the norm $\mathcal{N}_p$. By what we have shown above $(V(f_n))_{n=1}^{\infty}$ will be a Cauchy sequence in $\mathcal{G}$. But $\mathcal{G}$ is known to be complete by Proposition 7 of Section 2 and so

$$\operatorname*{Lim}_n V(f_n) = g$$

exists in $\mathcal{G}$. Since V^{-1} is continuous,

$$\lim_n f_n$$

exists in $\mathcal{G}^p$ and is equal to $V^{-1}(g)$. And so the completeness of $\mathcal{G}^p$ is established. Q.E.D.

Remark. From the foregoing proof we see that the spaces $\mathfrak{Y}^p$, $p \geqq 1$, are all homeomorphic to each other.

The results in the foregoing two propositions can be carried over to the spaces $\mathfrak{L}^p$ with the following improvements.

PROPOSITION 3. (H′) (HÖLDER'S INEQUALITY). *If $p > 1, p + q = pq, f \in \mathfrak{L}^p$, $g \in \mathfrak{L}^q$, then $fg \in \mathfrak{L}$ and*

$$|L(fg)| \leqq (L(|f|^p))^{1/p}(L(|g|^q))^{1/q}.$$

Here equality holds if and only if $f|f|^{p-1}$ and $g|g|^{q-1}$ are linearly dependent. In case $p = q = 2$, we have the inequality of Schwarz; in it equality holds only if f and g are proportional.

(M') (MINKOWSKI'S INEQUALITY). *If $p \geqq 1$, f and g belong to $\mathfrak{L}^p$, then $f + g \in \mathfrak{L}^p$ and*

$$(L(|f + g|^p))^{1/p} \leqq (L(|f|^p))^{1/p} + (L(|g|^p))^{1/p}.$$

Here equality holds in case $p = 1$ if and only if f and g have the same sign almost everywhere, in case $p > 1$ if and only if f and g are linearly dependent.

(L) $\mathfrak{L}^p$ is a (closed) linear subspace of $\mathfrak{Y}^p$.

Proof. We first consider (H′). The function

$$\varphi(x_1, x_2) = |x_1|^{1/p}|x_2|^{1/q}\operatorname{sign} x_1 x_2$$

is a positive homogeneous and continuous (hence Baire) function and so, by Proposition 14 of Section 2, $fg = \varphi(f', g')$ with $f' = f|f|^{p-1} \in \mathfrak{L}$ and $g' = g|g|^{q-1} \in \mathfrak{L}$ is also a member of $\mathfrak{L}$. From the inequality

$$|fg| \leqq \gamma(|f|^p \xi^{*p} + |g|^q \xi^{*-q})$$

derived in the proof of Proposition 1 it follows by L-integration that

$$\begin{aligned}|L(fg)| &\leqq L(|fg|) \leqq \gamma(L(|f|^p)\xi^{*p} + L(|g|^q)\xi^{*-q})\\ &= |L(|f|^p)|^{1/p}|L(|g|^q)|^{1/q}.\end{aligned}$$

We have equality between the end terms of these inequalities only if on the one hand $|L(fg)| = L(|fg|)$, i.e., $\pm L(fg) = L(|fg|)$, or $L(\pm fg - |fg|) = 0$, i.e., for a fixed sign $\pm fg - |fg|$ is a null function and, on the other hand, if the function $h = |fg| - \gamma(|f|^p \xi^{*p} + |g|^q \xi^{*-q})$ is a null function as well. In case the values of f and $g \neq 0$, then h will only be zero if

$$\xi^* = \left(\frac{|g|^q q}{|f|^p p}\right)^{1/pq}, \text{ or } |g|^q = \frac{p}{q}\xi^{*pq}|f|^p$$

according to the minimum considerations carried out in the proof of Proposition 1. It is evident that the same conclusion can be reached in case f or g is zero. Together with the above condition we obtain that for fixed sign almost everywhere $g|g|^{q-1} = \pm(p/q)\xi^{*pq}f|f|^{p-1}$ and so the necessity of the condition is verified. That the condition is also sufficient follows immediately from its derivation. In case $p = q = 2$ the last equation reads: $g|g| = \pm\xi^{*4}f|f|$; from it we get that $g = \pm\xi^{*2}f$.

Next we consider (M′). With the Baire function $\varphi(\lambda, \theta) = (|\lambda|^{1/p}\text{sign }\lambda + |\theta|^{1/p}\text{sign }\theta)||\lambda|^{1/p}\text{sign }\lambda + |\theta|^{1/p}\text{sign }\theta|^{p-1}$ and with $f' = f|f|^{p-1} \in \mathfrak{L}$, $g' = g|g|^{p-1} \in \mathfrak{L}$, that is, f' and $g' \in \mathfrak{L}^p$, we have $(f+g)|f+g|^{p-1} = \varphi(f', g')$ and so $f + g \in \mathfrak{L}^p$. The proof of the stated inequality proceeds as in Proposition 1. An inspection of the proof of Proposition 1 shows that equality can only occur if, on the one hand, almost everywhere $|f+g| = |f| + |g|$, which means that f and g have almost everywhere the same sign. In case $p = 1$ this is already sufficient. If, on the other hand, $p > 1$ and equality is to hold, we need in addition that

$$\mathscr{N}(|f||f+g|^{p-1}) = \mathscr{N}_p(|f|)\mathscr{N}_q(|f+g|^{p-1})$$

be satisfied. By (H′) we then get that $|f|^p$ and $|f+g|^{(p-1)q} = |f+g|^p$ are linearly dependent, and similarly for g that $|g|^p$ and $|f+g|^p$ are linearly dependent. As the trivial case that $f + g$ be a null function requires that f and g be null functions, we obtain linear dependence of $|f|^p$ and $|g|^p$, hence also linear dependence between $|f|$ and $|g|$; but the signs are equal and so there must be linear dependence between f and g themselves.

Finally, we consider (L). If $f \in \mathfrak{L}^p$, then αf and $|f|$ also belong to $\mathfrak{L}^p$ and so $\mathfrak{L}^p$ is a linear sublattice of $\mathfrak{I}^p$. The homeomorphism V studied in the proof of Proposition 2 maps $\mathfrak{L}^p$ onto $\mathfrak{L}$. Since $\mathfrak{L}$ is closed in $\mathfrak{I}$, so is $\mathfrak{L}^p$ in $\mathfrak{I}^p$. This ends the proof.

In the remainder of this section we assemble certain important properties of the space $\mathfrak{L}^2$. $\mathfrak{L}^2$ consists of all real-valued functions f, defined on some basic set A, such that $f|f| \in \mathfrak{L}$; L is the $\mathscr{N}$-integral which belongs to an elementary integral $E|\mathscr{E}$ over A. For $f \in \mathfrak{L}^2$, $L(f^2)$ is finite with the norm $\|f\| = (L(f^2))^{\frac{1}{2}}$ and the metric $\|f - g\|$ for any two elements f and g of $\mathfrak{L}^2$. For $f, g \in \mathfrak{L}^2$, the number

$$\langle f, g\rangle = L(fg)$$

is finite by Proposition 3, part (H′); it is called the *inner product* of f and g; the inner product is a bounded bilinear, symmetric function, that is, $\langle f,g\rangle = \langle g,f\rangle$, $\langle (f_1 + f_2), g\rangle = \langle f_1, g\rangle + \langle f_2, g\rangle$, $\langle \alpha f, g\rangle = \alpha\langle f, g\rangle$, $|\langle f, g\rangle| \leqq \|f\|\|g\|$ for arbitrary f, f_1, f_2, g belonging to $\mathfrak{L}^2$ and real α. From the last inequality follows the continuity of $\langle f, g\rangle$ in f and g. Moreover, by using $\|f\|^2 = \langle f,f\rangle$, we get the *parallelogram law*

$$\|f+g\|^2 + \|f-g\|^2 = 2(\|f\|^2 + \|g\|^2).$$

The functions f and g in $\mathfrak{L}^2$ are said to be *orthogonal* if $\langle f,g\rangle = 0$. It can be seen that $\langle f,f\rangle = 0$, or $\|f\| = 0$, if and only if $f = 0$ almost everywhere.

PROPOSITION 4. *Let* $E|\mathcal{E}$ *satisfy* (8_c) *(i.e.,* $f \in \mathcal{E}$ *implies* $\min\{f, 1\} \in \mathcal{E}$*), then, for* $f \in \mathfrak{L}^2$ *and* $n \in N$, *we have that* $f_n = \operatorname{med}\{-n, f, n\} \in \mathfrak{L}^2$ *and* $f_n \to f$ *in* $\mathfrak{L}^2$ *for* $n \to +\infty$.

Proof. Since $f|f| \in \mathfrak{L}$, so is $f_n|f_n| = \operatorname{med}\{-n^2, f|f|, n^2\}$ and thus $f_n \in \mathfrak{L}^2$. Moreover, $\varphi_n = (f_n - f)^2 \in \mathfrak{L}$ and $0 \leqq \varphi_{n+1} \leqq \varphi_n$ and the Monotone Convergence Theorem (see Proposition 11 of Section 2) is applicable. But

$$\lim_n \varphi_n(x)$$

for all x with finite $f(x)$ is zero $\mathcal{N}$-almost everywhere and hence is an $\mathcal{N}$-null function; it follows that

$$\|f_n - f\|^2 = L((f_n - f)^2) \to 0.$$

This completes the proof.

DEFINITION. *A subset* $\mathfrak{B}$ *of* $\mathfrak{L}^2$ *is called a linear manifold in* $\mathfrak{L}^2$ *if together with* f *and* g *all real linear combinations* $\alpha f + \beta g$ *also belong to* $\mathfrak{B}$.

PROPOSITION 5. *If the linear manifold* $\mathfrak{B}$ *is not dense in* $\mathfrak{L}^2$, *then there is an element* $h \neq 0$ *in* $\mathfrak{L}^2$ *which is orthogonal to all elements of* $\mathfrak{B}$.

Proof. We observe first that with $\mathfrak{B}$ also its closure $\overline{\mathfrak{B}}$ is a linear manifold. Indeed, if f_1 and f_2 belong to $\overline{\mathfrak{B}}$, then for any $\varepsilon > 0$ there are elements g_1 and g_2 in $\mathfrak{B}$ such that $\|f_i - g_i\| < \varepsilon$ for $i = 1, 2$; from this follows that $\|\alpha f_1 + \beta f_2 - (\alpha g_1 + \beta g_2)\| \leqq (|\alpha| + |\beta|)\varepsilon$ and for $\varepsilon \to 0$ we get $\alpha f_1 + \beta f_2 \in \overline{\mathfrak{B}}$. Since $\mathfrak{B}$ is not dense, there exists a $\varphi \in \mathfrak{L}^2 - \overline{\mathfrak{B}} \neq \varnothing$; here

$$d = \inf\{\|\varphi - f\| : f \in \mathfrak{B}\} > 0,$$

and there exists a minimal sequence $f_1, f_2, \ldots$ in $\mathfrak{B}$ with $\|\varphi - f_n\| \to d$. We show that the sequence $(f_n)_{n=1}^{\infty}$ is a Cauchy sequence and therefore will be convergent on account of the completeness of $\mathfrak{L}^2$. Indeed,

$$\begin{aligned}\left\|\frac{f_n - f_m}{2}\right\|^2 &= \left\|\frac{(f_n-\varphi) - (f_m - \varphi)}{2}\right\|^2 \\ &= \frac{1}{2}\|f_n - \varphi\|^2 + \frac{1}{2}\|f_m - \varphi\|^2 - \left\|\frac{f_n + f_m}{2} - \varphi\right\|^2 \\ &\leqq \frac{1}{2}(\|f_n - \varphi\|^2 + \|f_m - \varphi\|^2) - d^2,\end{aligned}$$

since $(f_n + f_m)/2 \in \mathfrak{B}$. The right-hand side of the last inequality tends to zero as $n, m \to \infty$. Thus there exists an $f^* \in \overline{\mathfrak{B}}$ with $f_n - \varphi \to f^* - \varphi$; it is clear that $\|f^* - \varphi\| = d$ holds. Now let f be an arbitrary element of $\mathfrak{B}$. For real γ we then have $f^* + \gamma f \in \overline{\mathfrak{B}}$, hence

$$\|\varphi - (f^* + \gamma f)\| \geqq \inf\{\|\varphi - g\| : g \in \overline{\mathfrak{B}}\} = d,$$

that is,

$$\begin{aligned} 0 &\leqq \|\varphi - (f^* + \gamma f)\|^2 - \|\varphi - f^*\|^2 \\ &= -2\gamma\langle(\varphi - f^*), f\rangle + \gamma^2\|f\|^2. \end{aligned}$$

Since this holds for every real γ, it is necessary that $\langle(\varphi - f^*), f\rangle = 0$. Taking $h = \varphi - f^*$, the proof is finished.

Definition. *A subset $\mathfrak{A}$ of $\mathfrak{L}^2$ is said to be a fundamental set in $\mathfrak{L}^2$ if the set of all linear combinations $\gamma_1 f_1 + \ldots + \gamma_k f_k$ of functions $f_1, \ldots, f_k$ in $\mathfrak{A}$ with real coefficients $\gamma_1, \ldots, \gamma_k$, $k \in N$, is dense in $\mathfrak{L}^2$.*

Remark. Some authors use the term "total set" in place of fundamental set.

Proposition 6. *A subset $\mathfrak{A}$ of $\mathfrak{L}^2$ is fundamental if and only if for any two functions f_1, f_2 in $\mathfrak{L}^2$ the following condition holds:*

$$\langle f_1, g\rangle = \langle f_2, g\rangle \textit{ for all } g \textit{ in } \mathfrak{A} \textit{ implies } f_1 = f_2. \tag{2}$$

Proof. If condition (2) is satisfied, then $\langle(f_1 - f_2), g\rangle = 0$ for all $g \in \mathfrak{A}$. But $\mathfrak{A}$ is fundamental in $\mathfrak{L}^2$ and the inner product is continuous; thus $\langle(f_1 - f_2), h\rangle = 0$ for all $h \in \mathfrak{L}^2$ and in particular for $h = f_1 - f_2$, implying that $f_1 - f_2 = 0$.

Now suppose that $\mathfrak{A}$ is not a fundamental set in $\mathfrak{L}^2$; then the linear manifold $\mathfrak{B}$ of all linear combinations of functions belonging to $\mathfrak{A}$ is not dense in $\mathfrak{L}^2$ and hence by Proposition 5 there exists an $h \neq 0$ with $\langle h, g\rangle = 0$ for all $g \in \mathfrak{B}$ and hence in particular for all $g \in \mathfrak{A}$. Thus $\langle 2h, g\rangle = 0 = \langle h, g\rangle$ for all $g \in \mathfrak{A}$, that is, condition (2) does not hold in general. This finishes the proof.

Remark. $\mathfrak{L}^2$ is itself a fundamental set and hence $\langle f_1, f\rangle = \langle f_2, f\rangle$ for all $f \in \mathfrak{L}^2$ implies $f_1 = f_2$.

Definition. *A real-valued function $M(f)$ is called a linear functional in $\mathfrak{L}^2$ if $M(\alpha f + \beta g) = \alpha M(f) + \beta M(g)$ holds for arbitrary real α, β, and f, g in $\mathfrak{L}^2$. M is said to be bounded if there exists a constant C_M such that $|M(f)| \leqq C_M\|f\|$ for all $f \in \mathfrak{L}^2$. A bounded linear functional is continuous and a continuous linear functional is bounded.*

PROPOSITION 7. *If $M(f)$ is a bounded linear functional in $\mathfrak{L}^2$, then there exists a unique element g^* in $\mathfrak{L}^2$ such that, for all $f \in \mathfrak{L}^2$, we have $M(f) = L(fg^*)$, where $\|g^*\| \leqq C_M$. If, moreover, M is positive, that is, $M(f) \geqq 0$ for $f \geqq 0$, then g^* is $\mathscr{N}$-almost everywhere nonnegative.*

Proof. The uniqueness of the representation is easy to see; by the Remark following Proposition 6 there can only be at most one such g^*. Next we show the existence of such a g^*.

Let $\mathfrak{B} = \{h : M(h) = 0\}$; since M is continuous and linear, $\mathfrak{B}$ is a closed linear manifold in $\mathfrak{L}^2$ and so is either identical with $\mathfrak{L}^2$, in which case we may set $g^* = 0$, or is a proper part of $\mathfrak{L}^2$, in which case by Proposition 5 there is an element $g(\neq 0)$ orthogonal to $\mathfrak{B}$, i.e., $\langle g, h\rangle = 0$ for all $h \in \mathfrak{B}$. For

$$g^* = \frac{M(g)}{\|g\|^2} g$$

we therefore have $M(h) = 0 = \langle h, g^*\rangle$ for all $h \in \mathfrak{B}$ and, moreover, $M(g) = \langle g, g^*\rangle$; for arbitrary $f \in \mathfrak{L}^2$ with

$$c = \frac{M(f)}{M(g)}$$

the function $f_1 = f - cg$ is in $\mathfrak{B}$ (since $M(f_1) = M(f) - cM(g) = 0$) and so $M(f) = M(f_1) + cM(g) = \langle f_1, g^*\rangle + c\langle g, g^*\rangle = \langle f, g^*\rangle$. For $f = g^*$ we get from the last inequality that

$$C_M \|g^*\| \geqq |M(g^*)| = |\langle g^*, g^*\rangle| = \|g^*\|^2,$$

that is, $\|g^*\| \leqq C_M$.

Finally, we assume that M is positive; if g^* is in $\mathfrak{L}^2$, then $|g^*|$ and $|g^*| - g^* = f^* \in \mathfrak{L}^2$ and $M(f^*) = L(f^*g^*) \geqq 0$. But $f^*g^* \leqq 0$ and so $L(f^*g^*) \leqq 0$ by the positivity of L; hence $0 = L(f^*g^*) = -L(-f^*g^*) = -\mathscr{N}(f^*g^*)$. This implies that $(|g^*| - g^*)g^*$ is an $\mathscr{N}$-null function and so g^* is nonnegative $\mathscr{N}$-almost everywhere. The proof is finished.

The real-valued function $B(f, g)$, f, $g \in \mathfrak{L}^2$, is called a *bilinear functional in* $\mathfrak{L}^2$ if $B(f, g)$ is a linear functional in f for each fixed g, and in g for each fixed f; $B(f, g)$ is said to be *bounded* if there exists a constant C_B such that $|B(f, g)| \leqq C_B \|f\| \|g\|$ for all f, $g \in \mathfrak{L}^2$. For example, $\langle f, g\rangle$ is a bounded bilinear functional. A mapping $T(f)$, $f \in \mathfrak{L}^2$, from $\mathfrak{L}^2$ into itself is called a *linear operator* if $T(\alpha f + \beta g) = \alpha T(f) + \beta T(g)$ for all real α, β and f, $g \in \mathfrak{L}^2$; T is said to be *bounded* if there is a constant C_T such that $\|T(f)\| \leqq C_T \|f\|$ for all $f \in \mathfrak{L}^2$.

PROPOSITION 8. *If $B(f,g)$ is a bounded bilinear functional in $\mathfrak{L}^2$, then there exists a bounded linear operator T mapping $\mathfrak{L}^2$ into itself such that $B(f,g) = L(fT(g))$ and $C_T \leqq C_B$.*

Proof. With fixed $g \in \mathfrak{L}^2$ we write $B(f,g) = M(f)$. Then $M(f)$ is a bounded linear functional, hence by Proposition 7 $M(f) = \langle f, g^* \rangle$ with some $g^* \in \mathfrak{L}^2$ and $\|g^*\| \leqq C_M = C_B\|g\|$. If we put $T(g) = g^*$ for $g \in \mathfrak{L}^2$, then T is uniquely defined by Proposition 7 and is a bounded operator by the inequality just proved with T mapping $\mathfrak{L}^2$ into $\mathfrak{L}^2$ and $C_T \leqq C_B$; here $B(f,g) = \langle f, T(g) \rangle$ for all f and g belonging to $\mathfrak{L}^2$. From the last equality it follows however that

$$\begin{aligned} \langle f, T(\alpha g_1 + \beta g_2) \rangle &= \alpha \langle f, T(g_1) \rangle + \beta \langle f, T(g_2) \rangle \\ &= \langle f, (\alpha T(g_1) + \beta T(g_2)) \rangle \end{aligned}$$

for fixed f and using the bilinearity of B; this being so for an arbitrary $f \in \mathfrak{L}^2$, we may conclude, using the Remark following Proposition 6, that

$$T(\alpha g_1 + \beta g_2) = \alpha T(g_1) + \beta T(g_2).$$

Thus T is seen to be linear and the proof is complete.

A bilinear functional $B(f,g)$, resp. a linear operator $T(f)$ in $\mathfrak{L}^2$ is said to be *positive* if for nonnegative f and g we have $B(f,g) \geqq 0$, resp. if $T(f)$ is $\mathcal{N}$-almost everywhere nonnegative.

PROPOSITION 9. *Let B be as in Proposition 8 and in addition assume that B is positive. Then the operator T introduced in Proposition 8 will also be positive.*

Proof. Indeed, for fixed nonnegative $g \in \mathfrak{L}^2$ we have that $B(f,g) = M(f)$ is positive; by Proposition 7 $g^* = T(g)$ is $\mathcal{N}$-almost everywhere nonnegative. But this is what we had to show.

Section 5. The Radon-Nikodym Theorem in Stone's Version

In this section we shall study the relations between different elementary integrals $E|\mathcal{E}, E'|\mathcal{E}, \ldots$ which are defined on the same set $\mathcal{E}$ of elementary functions f over a fixed basic set A and we consider the corresponding $\mathcal{N}$-integrals.

The set $\mathfrak{e}$ of all elementary integrals $E|\mathcal{E}$ over the same fixed domain $\mathcal{E}$ of elementary functions f over A forms a partially ordered commutative semigroup with respect to addition, admitting multiplication by positive real scalars: The functionals $E + E'$ and αE are defined by $(E + E')(f) = E(f) + E'(f)$ and $(\alpha E)(f) = \alpha E(f)$ for $f \in \mathcal{E}$; further, $E \leqq E'$ means $E(f) \leqq E'(f)$ for all nonnegative $f \in \mathcal{E}$. From $E = E' + E''$ follows E'

$\leqq E$ and $E'' \leqq E$; if $E' \leqq E$, then there is a unique E'', given by $E''(f) = E(f) - E'(f)$, with $f \in \mathcal{E}$, such that $E = E' + E''$.

The functionals and function spaces associated with the elementary integrals $E, E', \ldots, E_1, \ldots$ will be distinguished by the use of the corresponding primes and subscripts. Here we have in particular that $\mathcal{E} = \mathcal{E}'$ $= \ldots$ and $\mathfrak{G} = \mathfrak{G}' = \ldots$. With this convention in mind, we prove the following.

Proposition 1. *If $E \leqq E'$, then (i) $\mathcal{N}'(g) \leqq \mathcal{N}(g)$ for $g \in \mathfrak{G}$; (ii) $\mathfrak{H}' \supset \mathfrak{H}$; (iii) $\mathfrak{I}' \supset \mathfrak{I}$; (iv) $\mathfrak{L}' \supset \mathfrak{L}$; (v) $L'(f) \leqq L(f)$ for nonnegative f belonging to $\mathfrak{L}$; (vi) if the set $\{x : g_1(x) \neq g_2(x)\}$ is an $\mathcal{N}$-null set, then it is also an $\mathcal{N}'$-null set; (vii) $\mathfrak{M}' \supset \mathfrak{M}$.*

Proof. To (i) : By definition we can find for each value which serves for the formation of $\mathcal{N}(g)$ as infimum a nongreater one for the formation of $\mathcal{N}'(g)$.

To (ii): By part (i), $\mathcal{N}(g) = 0$ implies that $\mathcal{N}'(g) = 0$ and so every $\mathcal{N}$-null set will also be an $\mathcal{N}'$-null set; hence each member of $\mathfrak{H}$ is also a member of $\mathfrak{H}'$.

To (iii): By part (i), we have that $\mathcal{N}(f) < +\infty$ implies $\mathcal{N}'(f) < +\infty$, that is, $f \in \mathfrak{I}$ implies $f \in \mathfrak{I}'$.

To (iv): $\varphi \in \mathfrak{L}$ means that

$$\varphi = \mathcal{N} - \operatorname*{Lim}_n f_n$$

with $f_n \in \mathcal{E}$ by the definition of $\mathfrak{L}$; by part (i) we then also have that

$$\varphi = \mathcal{N}' - \operatorname*{Lim}_n f_n$$

and so $\varphi \in \mathfrak{L}'$.

To (v): The claim is once again a consequence of part (i) since for a nonnegative function the integral and the norm coincide.

To (vi): The claim is a consequence of part (ii).

To (vii): If $f \in \mathfrak{M}$, then $\operatorname{med}\{g, f, h\} \in \mathfrak{L} \subset \mathfrak{L}'$ for $g, h \in \mathcal{E}$ and thus $f \in \mathfrak{M}'$. The proof is finished.

Definition. *An elementary integral $E|\mathcal{E}$ satisfying condition (8_c) and the property*

(11_c) *There is a sequence of nonnegative functions f_k belonging to $\mathcal{E}$ with $E(f_1) + E(f_2) + \ldots = +\infty$ for all $x \in A$,*

we call a normal integral.

Observe that a normal integral is σ-finite in the sense of Section 3 and note that Propositions 12 and 13 of Section 3 are satisfied. Note that conditions (8_c) and (11_c) concern only the domain of definition of $\mathcal{E}$ and not E; this circumstance simplifies the following comparison studies.

PROPOSITION 2. *If $E|\mathcal{E}$ is a normal integral, $0 \leqq g \in \mathfrak{M}$ and $\sigma = \sup\{L(f) : f \in \mathfrak{L}$ and $0 \leqq f \leqq g\} < +\infty$, then $g \in \mathfrak{L}$ with $L(g) = \sigma$.*

Proof. Let $f_1, f_2, \ldots$ be a sequence satisfying (11_c) and let $g_n = \min\{f_1 + \ldots + f_n, g\}$. Then $(g_n)_{n=1}^{\infty}$ is a monotone nondecreasing sequence which converges to g with $L(g_n) \leqq \sigma$. By the Monotone Convergence Theorem (see Proposition 11 of Section 2) we get $g \in \mathfrak{L}$ and clearly $L(g) = \sigma$. This completes the proof.

In order to fix uniquely the product of two proper or improper real numbers, we agree to let the product be zero in case one of the factors is zero.

PROPOSITION 3. (RADON-NIKODYM). *If E and E' are normal integrals over $\mathcal{E}$ and $E' \leqq E$, then there exists one, and up to an $\mathcal{N}$-null set only one, real function Φ^* defined on A such that*

(*1*) $0 \leqq \Phi^* \leqq 1$ *and* $\Phi^* \in \mathfrak{M}$;
(*2*) $f \in \mathcal{E}$ *implies* $E'(f) = L(\Phi^* f)$;
(*3*) $f' \in \mathfrak{L}'$ *if and only if* $\Phi^* f' \in \mathfrak{L}$;
(*4*) $f' \in \mathfrak{L}'$ *implies* $L'(f') = L(\Phi^* f')$.

Proof. If $h, k \in \mathfrak{L}^2$, then $h, k \in \mathfrak{L} \subset \mathfrak{L}'$. Since $|L'(hk)| \leqq L'(|hk|) \leqq L(|hk|) \leqq (L(h^2))^{1/2}(L(k^2))^{1/2}$ (see Proposition 3 of Section 4), we have that $L'(hk)$ is a positive bounded bilinear functional in $\mathfrak{L}^2$ and so by Proposition 8 of Section 4 there is a positive bounded linear operator T mapping $\mathfrak{L}^2$ into itself such that $L'(hk) - L(hT(k))$. If g is a bounded function belonging to $\mathfrak{M}$, then gh and gk will also belong to $\mathfrak{L}^2$; indeed, for example, $gh|g||h|$ is $\mathcal{N}$-almost everywhere finite and, being a product of $\mathcal{N}$-measurable functions, is $\mathcal{N}$-measurable; moreover, $gh|g||h|$ is wedged in between $-rh^2$ and $+rh^2$ with $r = \sup\{[g^2(x)]^2 : x \in A\}$ and so must be $\mathcal{N}$-integrable by Proposition 27 of Section 2. The equations

$$L(hT(gk)) = L'(hgk) = L(hgT(k))$$

hold for all $h \in \mathfrak{L}^2$ and so, by Proposition 6 of Section 4, we have $T(gk) = gT(k)$ (here equality in the sense of $\mathcal{N}$-almost everywhere). To exploit the symmetry of g and k on the left-hand side of the last equation, we substitute for g a function $g_0 = f_0^{1/2} \in \mathfrak{L}^2$ according to part (a) of Proposition 12 of Section 3 and we substitute for k the functions $k_n = \text{med}\{-n, k, n\}$, $n \in N$, all of which are bounded for $k \in \mathfrak{L}^2$ and belongs to $\mathfrak{L}^2$. We obtain: $g_0 T(k_n) = T(g_0 k_n) = T(k_n g_0) = k_n T(g_0)$. Since $k_n \to k$ in $\mathfrak{L}^2$ by Proposition 4 of Section 4, we get by continuity considerations that $g_0 T(k) = kT(g_0)$ for $k \in \mathfrak{L}^2$. Setting $\Phi^* = T(g_0)/g_0$ we obtain an $\mathcal{N}$-measurable function; indeed, since $T(g_0) \in \mathfrak{L}^2$, we have $|T(g_0)|^2 \in \mathfrak{L} \subset \mathfrak{M}$ and, since $T(g_0) \geqq 0$ by Proposition 9 of Section 4, it follows by part

(i) of Proposition 26 of Section 2 that $T(g_0) \in \mathfrak{M}$ and in like manner that $g_0 \in \mathfrak{M}$, so that part (ii) of Proposition 26 of Section 2 is applicable. Thus we have $T(k) = \Phi^*k$. Now, if $f \in \mathfrak{L}$, then, letting $k = |f|^{1/2} \in \mathfrak{L}^2$ and $h = |f|^{1/2} \operatorname{sign} f$ in $\mathfrak{L}^2$, we get $f = hk$ and $\mathfrak{L} \ni hT(k) = h\Phi^*k = \Phi^*f$, that is,

$$(*) \qquad L'(f) = L'(hk) = L(hT(k)) = L(\Phi^*f).$$

Since $0 \leqq f \in \mathfrak{L}$, we see by part (v) of Proposition 1 that $0 \leqq L'(f) \leqq L(f)$, that is, $0 \leqq L(\Phi^*f) \leqq L(f)$ for all such f, and so, by Proposition 14 of Section 3, $0 \leqq \Phi^* \leqq 1$ $\mathscr{N}$-almost everywhere [because the right-hand side of the second last inequality may also be written as $0 \leqq L((1 - \Phi^*)f)$].

It is now easily seen that up to an $\mathscr{N}$-null set there can only be one such function Φ^*. Indeed, if Φ_1 and Φ_2 are $\mathscr{N}$-measurable functions with values in the interval $[0, 1]$ (so that $\Phi_1 f$ and $\Phi_2 f$ belong to $\mathfrak{L}$ whenever f belongs to $\mathfrak{L}$) and Φ_1 and Φ_2 enjoy the property that $L(\Phi_1 f) = L'(f) = L(\Phi_2 f)$ for all f in $\mathfrak{L}$, then it follows that $L((\Phi_1 - \Phi_2)f) = 0$ for all f in $\mathfrak{L}$ and so by Proposition 14 of Section 3 we have $\mathscr{N}$-almost everywhere both $\Phi_1 - \Phi_2 \leqq 0$ and $\geqq 0$, that is, $= 0$. Without restriction we can choose Φ^* such that $0 \leqq \Phi^*(x) \leqq 1$ for all $x \in A$. Thus parts (1) and (2) of the proposition are proven; for, in particular, the statement (*) further up in this proof holds for $f \in \mathscr{E}$.

Next, let $f' \in \mathfrak{L}'$. This means that for $\varepsilon > 0$ there are functions g, $f_1, f_2, \ldots$ belonging to $\mathscr{E}$ with

$$|f' - g| \leqq \sum_n |f_n|$$

and

$$\sum_n E'(|f_n|) \leqq \varepsilon.$$

Then

$$|L'(f') - L'(g)| \leqq L'(|f' - g|) \leqq \sum_n E'(|f_n|) \leqq \varepsilon$$

holds. But $|\Phi^*f' - \Phi^*g| \leqq \sum \Phi^*|f_n|$, that is,

$$\mathscr{N}(\Phi^*f' - \Phi^*g) \leqq \sum_n \mathscr{N}(\Phi^*|f_n|) = \sum_n L(\Phi^*|f_n|) = \sum_n L'(|f_n|)$$
$$= \sum_n E'(|f_n|) \leqq \varepsilon$$

and thus $\Phi^*f' \in \mathfrak{L}$ because $\Phi^*g \in \mathfrak{L}$. Furthermore it follows that $|L(\Phi^*f') - L'(g)| = |L(\Phi^*f' - \Phi^*g)| \leqq L(|\Phi^*f' - \Phi^*g|) = \mathscr{N}(\Phi^*f' - \Phi^*g) \leqq \varepsilon$, hence

$$|L(\Phi^* f') - L'(f')| \leqq |L(\Phi^* f') - L'(g)| + |L'(g) - L'(f')| \leqq 2\varepsilon,$$

and with $\varepsilon \to 0$ finally that $L(\Phi^* f') = L'(f')$. This proves claim (4) and the "only if" part in claim (3).

We now take up the "if" part in claim (3): (a) We show first that $P = \{x : \Phi^*(x) = 0\}$ is an $\mathscr{N}'$-null set. Indeed, since $\Phi^* \in \mathfrak{M}$, $\Phi^* \in \mathfrak{M}'$ by part (vii) of Proposition 1 and so, by Proposition 5 of Section 3 we obtain that $\chi_P \in \mathfrak{M}'$. Since by our general assumptions the basic set A is $\mathscr{N}'$-measurable (by (8_c) we have that $\chi_A = 1 \in \mathfrak{M}'$), we have by part (c) of Proposition 12 of Section 3 that

$$A = A_1 \cup A_2 \cup \ldots$$

where each A_k is measurable and of finite measure. The latter means that $\chi_{A_k} \in \mathfrak{L}'$. By Proposition 27 of Section 2 we therefore see that $\chi_P \chi_{A_k} \in \mathfrak{L}'$ and so by part (4), for all $k \in N$, $L'(\chi_P \chi_{A_k}) = L(\Phi^* \chi_P \chi_{A_k}) = 0$ because $\Phi^* \chi_P = 0$. From $P = (P \cap A_1) \cup (P \cap A_2) \cup \ldots$ it now follows that

$$L'(\chi_P) \leqq \sum_k L'(\chi_P \chi_{A_k}) = 0,$$

verifying the claim. (b) Since $\Phi^* f'$ is $\mathscr{N}$-almost everywhere finite, the same holds $\mathscr{N}'$-almost everywhere on account of (a); we therefore see that f' is $\mathscr{N}'$-almost everywhere finite.

Since $f' = |f'| - (|f'| - f')$, it will suffice to consider the case $f' \geqq 0$ only. Then in

$$h_n = \frac{n\Phi^* f'}{n\Phi^* + 1}, \qquad n \in N,$$

we have a sequence which converges nondecreasingly to an $\mathscr{N}'$-almost everywhere finite function h, where

$$h(x) = 0 \quad \text{for } x \in P,$$

$$h(x) = f'(x) \quad \text{for } x \in A - P$$

and $\Phi^* h = \Phi^* f'$. Since h_n as an $\mathscr{N}$-measurable function with $0 \leqq h_n \leqq n\Phi^* f' \in \mathfrak{L}$ also belongs to $\mathfrak{L}$, we have in addition that $h_n \in \mathfrak{L}'$; thus $L'(h_n) = L(\Phi^* h_n) \leqq L(\Phi^* f')$ because $h_n \leqq f'$. By the Monotone Convergence Theorem (see Proposition 11 of Section 2) we get that $h \in \mathfrak{L}'$. By (a) the functions h and f' differ only on an $\mathscr{N}'$-null set and so with h belonging to $\mathfrak{L}'$ so does f'. Thus everything is established and the proof is finished.

The foregoing proposition has the following converse.

PROPOSITION 4. *If the normal integral $E|\mathscr{E}$ and $\Phi \in \mathfrak{M}$ with $0 \leqq \Phi \leqq 1$ are given, then $E'|\mathscr{E}$, defined by*

$$E'(f) = L(\Phi f), \qquad f \in \mathcal{E},$$

is a normal integral with $E' \leqq E$ and the function Φ^* belonging to E' by virtue of Proposition 3 differs from Φ by an $\mathcal{N}$-null function only.

Proof. Properties (1_c) to (8_c) for $E'|\mathcal{E}$ follow immediately from the analogous properties of $L|\mathfrak{L}$; the normality of the integral E' is trivial. By (2) of Proposition 3 we have for Φ^* belonging to E':

$$L(\Phi^* f) = E'(f) = L(\Phi f) \text{ for } f \in \mathcal{E}.$$

By Proposition 14 of Section 3 we may conclude that $\Phi^* = \Phi$ ($\mathcal{N}$-almost everywhere). The proof is complete.

We take up some definitions:

Let $x \vee y$ and $x \wedge y$ denote $\sup\{x,y\}$ and $\inf\{x,y\}$, respectively. A lattice V is said to be *distributive* if and only if, for every $x, y, z \in V$,

$$x \wedge (y \vee z) = (x \wedge y) \vee (x \wedge z)$$

and

$$x \vee (y \wedge z) = (x \vee y) \wedge (x \vee z).$$

A lattice is said to be *σ-complete* if every countable subset of it has a supremum and an infimum and is called *conditionally σ-complete* if every countable subset of it which has an upper (lower) bound has a supremum (infimum).

From Propositions 3 and 4 follows the next proposition.

PROPOSITION 5. *The set $\mathfrak{e}^*$ of all normal integrals E over a fixed domain $\mathcal{E}$ is a conditionally σ-complete distributive lattice.*

Proof. The correspondence $E' \to \Phi^*$ of Proposition 3 is an isomorphism between the partially ordered sets $\mathfrak{e}^*_E = \{E' : E \geqq E' \in \mathfrak{e}^*\}$ and $\mathfrak{M}_1 = \{\Phi^* : \Phi^* \in \mathfrak{M} \text{ and } 0 \leqq \Phi^* \leqq 1\}$ identified modulo null functions, where the relations $E'_1 \leqq E'_2$ and $\Phi^*_1 \leqq \Phi^*_2$ correspond to each other. Since $\mathfrak{M}_1$ is σ-complete, so is $\mathfrak{e}^*_E$. If we do not restrict ourselves to subsets $\mathfrak{e}^*_E$ and consider $\mathfrak{e}^*$, only conditional σ-completeness remains. For example, $\mathfrak{e}^*\text{-}\sup\{E_1, E_2, \ldots\}$ exists if there is an $E \in \mathfrak{e}^*$ such that $E_i \leqq E$, $i \in N$; then one forms the function Φ^*_i corresponding to E_i by Proposition 3 and obtains $\mathfrak{e}^*\text{-}\sup\{E_1, E_2, \ldots\} = L(f \sup\{\Phi^*_1, \Phi^*_2, \ldots\})$. For given E_i, $i = 1, 2, 3$, one forms $E_4 = E_1 + E_2 + E_3$. Then to every E_i there is assigned a Φ^*_i with respect to E_4, $i = 1, 2, 3$; moreover we obtain the assignments

$$E_1 \wedge (E_2 \vee E_3) \to \Phi' = \min\{\Phi^*_1, \max\{\Phi^*_2, \Phi^*_3\}\},$$

$$(E_1 \wedge E_2) \vee (E_1 \wedge E_3) \to \Phi'' = \max\{\min\{\Phi_1^*, \Phi_2^*\}, \min\{\Phi_1^*, \Phi_3^*\}\}.$$

Since $\Phi' = \Phi''$ we have the validity of the distributive law in $\mathfrak{e}^*$. This completes the proof.

Let $E'|\mathcal{E}$ and $E''|\mathcal{E}$ be two elementary integrals over the same domain $\mathcal{E}$. It is convenient to introduce the following definition of a new order relation $\dashv$ between elementary integrals:

$E' \dashv E''$ *if and only if every* $\mathcal{N}''$*-null set is an* $\mathcal{N}'$*-null set.*

By Proposition 2 of Section 2 it is evident that the foregoing condition is equivalent with: Every $\mathcal{N}''$-null function is an $\mathcal{N}'$-null function. We observe that $E' \leqq E''$ implies $E' \dashv E''$ (by part (i) of Proposition 1) and that the relation $\dashv$ defined on $\mathfrak{e}^*$ is reflexive and transitive (but not necessarily antisymmetric).

Translating the relation $\dashv$ into the language of functions Φ^*, obtained by virtue of Proposition 3 from E, we have the following.

PROPOSITION 6. *If* E_1, E_2 *and* E *are normal integrals over* $\mathcal{E}$ *with* $E_i \leqq E$ *and if* Φ_i^* *are* Φ^* *functions corresponding to* E_i, $i = 1, 2$, *by virtue of Proposition 3, then the following two statements are equivalent*:

(*i*) $E_1 \dashv E_2$;

(*ii*) $Z = \{x : \Phi_2^*(x) = 0\} \cap \{x = \Phi_1^*(x) \neq 0\}$ *is an* $\mathcal{N}$*-null set.*

Proof. We have the representation $L_i(f) = L(\Phi_i^* f)$ for $f \in \mathfrak{L}_i$ and we put

$$K_i = \{x : \Phi_i^*(x) = 0\} \qquad (i = 1, 2).$$

Suppose that statement (ii) holds. If Z_2 is any $\mathcal{N}_2$-null set, then $L_2(\chi_{Z_2}) = 0$, or $L(\Phi_2^* \chi_{Z_2}) = 0$. By Proposition 2 of Section 2, $Z_2 \cap (A - K_2)$ is an $\mathcal{N}$-null set. Consider

$$Z_2 \cap K_2 = (Z_2 \cap K_2 \cap K_1) \cup (Z_2 \cap K_2) \cap (A - K_1).$$

The set $Z_2 \cap K_2 \cap K_1$, being a part of K_1, is an $\mathcal{N}_1$-null set; this can be seen from the proof of Proposition 3 (see claim (a) in paragraph 4 of the proof). The set $(Z_2 \cap K_2) \cap (A - K_1)$, being a part of Z, is an $\mathcal{N}_1$-null set also because Z is by assumption an $\mathcal{N}$-null set. Therefore Z_2 is an $\mathcal{N}_1$-null set and we have $E_1 \dashv E_2$.

Suppose, finally, that the set Z is not an $\mathcal{N}$-null set. Since Φ_1^* and Φ_2^* belong to $\mathfrak{M}$, we have that Z is $\mathcal{N}$-measurable (by Proposition 4 of Section 3); by part (c) of Proposition 12 of Section 3 there exists an $\mathcal{N}$-measurable subset Z_0 of Z with $0 < m(Z_0) < \infty$. Using part (a) of Proposition 12 of Section 3 we obtain

$$\chi_{Z_0} f_0 \in \mathfrak{L}_1 \text{ and } L_1(\chi_{Z_0} f_0) = L(\Phi_1^* \chi_{Z_0} f_0).$$

But $\{x : \Phi_1^* \chi_{Z_0} f_0(x) > 0\} = Z_0$ and hence by Proposition 2 of Section 2, $L_1(\chi_{Z_0} f_0) > 0$ and so $m_1^*(Z_0) > 0$, showing that Z_0 cannot be an $\mathscr{N}_1$-null set. Since Z_0 is a subset of Z and Z is an $\mathscr{N}_2$-null set, we conclude that the relation $E_1 \dashv E_2$ fails to be fulfilled. This ends the proof.

We now take up the Generalized Radon-Nikodym Theorem, that is a generalization of Propositions 3 and 4.

PROPOSITION 7. (*i*) *Let E' and E'' be normal integrals over $\mathscr{E}$ with $E' \dashv E''$. Then there exists up to $\mathscr{N}''$-null sets a unique nonnegative function $\psi \in \mathfrak{M}''$ with the following properties*:
(α) *$f' \in \mathfrak{L}'$ if and only if $\psi f' \in \mathfrak{L}''$*;
(β) *$f' \in \mathfrak{L}'$ implies $L'(f') = L''(\psi f')$; in particular*

$$E'(f) = L''(\psi f) \text{ for } f \in \mathscr{E}.$$

(*ii*) *If $E''|\mathscr{E}$ is a normal integral and ψ a nonnegative function belonging to $\mathfrak{M}''$ with the property that $\psi f \in \mathfrak{L}''$ for all $f \in \mathscr{E}$, then $E'(f) = L'(\psi f)$ defines a normal integral with $E' \dashv E''$.*

Proof. We consider (i) first and put $E = E' + E''$. We then have $E' \leqq E$ and $E'' \leqq E$; by Proposition 3 there are representations

$$E'(f) = L(\Phi' f) \text{ and } E''(f) = L(\Phi'' f)$$

for $f \in \mathscr{E}$, where $\Phi', \Phi'' \in \mathfrak{M} \subset \mathfrak{M}' \cap \mathfrak{M}''$ and $0 \leqq \Phi' \leqq 1$ and $0 \leqq \Phi'' \leqq 1$. Here $\Phi' + \Phi'' = 1$,

$$\{x : \Phi''(x) = 0\} \cap \{x : \Phi''(x) \neq 0\}$$

is an $\mathscr{N}$-null set and whence

$$\{x : \Phi''(x) = 0\}$$

is itself an $\mathscr{N}$-null set. Thus

$$\psi = \Phi'/\Phi''$$

is finite $\mathscr{N}$-almost everywhere, nonnegative, and $\psi \in \mathfrak{M} \subset \mathfrak{M}''$. By (3) of Proposition 3, $f' \in \mathfrak{L}'$ if and only if $\Phi' f' \in \mathfrak{L}$, that is, $\Phi''(\psi f') \in \mathfrak{L}$, which in turn is equivalent to $\psi f' \in \mathfrak{L}''$. Further, by (4) of Proposition 3, we have $L'(f') = L(\Phi' f') = L(\Phi''(\psi f')) = L''(\psi f')$.

Next, we consider (ii). Let E'' and ψ have the mentioned properties. We first observe that $E'(f) = L''(\psi f)$ defines a normal integral. Indeed, the properties (1_c) to (6_c) are clear on account of the corresponding properties of L'', (7_c) follows property (d) for $\mathcal{N}''$ (for properties (1_c) to (7_c) and property (d) consult the beginning of Section 2); the normality is trivial. We put $E = E' + E''$ and have the representations

$$E'(f) = L(\Phi' f) \text{ and } E''(f) = L(\Phi'' f)$$

for $f \in \mathcal{E}$. Moreover,

$$L(\Phi' f) = E'(f) = L''(\psi f) = L(\Phi''(\psi f))$$

for all f in $\mathcal{E}$. But this gives

$$L(\Phi' f) = L(\Phi''(\psi f))$$

for all f in $\mathfrak{L}$. Indeed: For such f let $(f_n)_{n=1}^{\infty}$ be a sequence from $\mathcal{E}$ with

$$\operatorname*{Lim}_n f_n = f$$

and (upon selection of a subsequence if necessary)

$$\lim_n f_n = f$$

$\mathcal{N}$-almost everywhere. Thus we also have

$$\lim_n \Phi''\psi f_n = \Phi''\psi f$$

and

$$\lim_n \Phi' f_n = \Phi' f$$

$\mathcal{N}$-almost everywhere. Since moreover

$$\begin{aligned} L(|\Phi''\psi f_m - \Phi''\psi f_n|) &= L(\Phi''\psi|f_m - f_n|) = L(\Phi'|f_m - f_n|) \\ &\leqq L(|f_m - f_n|), \end{aligned}$$

$$\operatorname*{Lim}_n \Phi''\psi f_n$$

and

$$\operatorname*{Lim}_n \Phi' f_n$$

exist and are equal to $\Phi''\psi f$ and $\Phi' f$, respectively. Here we have

$$L(\Phi''\psi f) = \lim_n L(\Phi''\psi f_n) = \lim_n L(\Phi' f_n) = L(\Phi' f);$$

but this is what we have claimed. Going on with the proof, we assume that $f_0 \in \mathfrak{L}$ is chosen according to (a) of Proposition 12 in Section 3; then by the above we have $\Phi''\psi f_0 \in \mathfrak{L}$, hence

$$\Phi''\psi = \frac{\Phi''\psi f_0}{f_0} \in \mathfrak{M}$$

(by Proposition 26 of Section 2) and so

$$\Omega = \Phi' - \Phi''\psi \in \mathfrak{M}$$

as well. From $L(\Omega f) = 0$ for all $f \in \mathfrak{L}$ we get by Proposition 14 of Section 3 that $\Omega = 0$, that is, $\Phi' = \Phi''\psi$ $\mathscr{N}$-almost everywhere. Thus

$$\{x : \Phi''(x) = 0\} \cap \{x : \Phi'(x) = 0\}$$

is an $\mathscr{N}$-null set and hence $E' \dashv E''$.

The essential uniqueness of ψ in (i) can be seen by the following argument: If ψ_1 is a function which satisfies (α) and (β), then

$$L'(f') = L(\Phi' f') = L''(\psi_1 f') = L''(\Phi''\psi_1 f')$$

for every $f' \in \mathfrak{L}'$ and a fortiori

$$L(\Phi' f) = L(\Phi''\psi_1 f) \text{ for } f \in \mathscr{E} \subset \mathfrak{L}'.$$

The last equation can be extended to all f in $\mathfrak{L}$ just as we did in the foregoing paragraph. Invoking Proposition 14 of Section 3, we see that $\Phi' - \Phi''\psi_1$ is an $\mathscr{N}$-null function and so

$$\psi_1 = \frac{\Phi'}{\Phi''} = \psi \quad (\mathscr{N}\text{-almost everywhere}).$$

This completes the proof.

The next proposition is similar to Proposition 5. In preparation for it, we give some definitions.

Let V be a lattice with $a, b \in V$ and $a \leqq b$. Then y is said to be a *relative complement of* x with respect to a and b if

$$x \wedge y = a \text{ and } x \vee y = b.$$

These equations can obviously only hold if x and y are between a and b. V

is said to be a *relatively complemented* lattice if for arbitrary $a, b \in V$ with $a \leqq b$, every element x between a and b has a relative complement y with respect to a and b.

By a *generalized Boolean* lattice we mean a distributive relatively complemented lattice.

Preceding Proposition 5 we have defined the notion of σ-completeness of a lattice and so it is clear what we mean by a σ-complete generalized Boolean lattice.

PROPOSITION 8. *The set* $\mathfrak{e}^*$ *of all normal integrals* $E|\mathcal{E}$ *over the fixed domain* $\mathcal{E}$ *becomes a generalized Boolean lattice under the order relation* $\dashv$ *and the equivalence definition*

$$E_1 = E_2 \text{ if and only if } (E_1 \dashv E_2 \text{ and } E_2 \dashv E_1);$$

each of its subsets

$$\{E' : \mathfrak{e}^* \ni E' \leqq E\},\ E \in \mathfrak{e}^*,$$

is a σ-complete generalized Boolean lattice.

Proof. We consider the subset

$$\{E' : \mathfrak{e}^* \ni E' \leqq E\}.$$

By Proposition 6 there exists in each equivalence class $[E']$ a representative E' with a corresponding function Φ' (by virtue of Proposition 7) which is at the same time a characteristic function. Under this association $E' \to \Phi'$ the relation $\dashv$ corresponds to the natural ordering of characteristic functions (modulo null function), that is, set inclusion in the system of all subsets of the basic set A modulo the σ-ideal of all null sets. Observing that the last system is a σ-complete generalized Boolean lattice completes the proof of the proposition.

We now consider the Lebesgue Decomposition Theorem.

PROPOSITION 9. *If E' and E'' are normal integrals over $\mathcal{E}$, then E' has a unique decomposition*

$$E' = E'_1 + E'_2$$

with $E'_1 \dashv E''$ *and* $\inf\{E'_2, E''\}(|f|) = 0$ *for all* $f \in \mathcal{E}$.

Proof. We put $E' + E'' = E$ and get by Proposition 3 the representations

$$E'(f) = L(\Phi' f) \text{ and } E''(f) = L(\Phi'' f)$$

and we set $\Phi_1'(x) = 0$ and $\Phi_2'(x) = \Phi'(x)$ for $\Phi''(x) = 0$ and put $\Phi_1'(x) = \Phi'(x)$ and $\Phi_2'(x) = 0$ for $\Phi''(x) \neq 0$. Then $\Phi_1' + \Phi_2' = \Phi'$,

$$\{x : \Phi''(x) = 0\} \cap \{x : \Phi_1'(x) \neq 0\}$$

is empty and $\inf\{\Phi_2, \Phi''\} = 0$; with $E_1'(f) = L(\Phi_i' f)$, $i = 1, 2$, we therefore have a representation of the desired kind. We still have to show that there is essentially only one such decomposition of Φ'. For a decomposition of E' of the desired kind there are in any case representations

$$E_i'(f) = L(\psi_i f)$$

and for the ψ_i we must have:

$$\Psi_1 + \Psi_2 = \Phi' \text{ and } \inf\{\Psi_2, \Phi''\} = 0$$

$\mathscr{N}$-almost everywhere and

$$\{x : \Phi''(x) = 0\} \cap \{x : \Psi_1(x) \neq 0\}$$

is an $\mathscr{N}$-null set. The last leads to $\Psi_1(x) = \Phi_1'(x)$ $\mathscr{N}$-almost everywhere for $\Phi''(x) = 0$ and the second last to $\Psi_2(x) = \Phi_2'(x)$ $\mathscr{N}$-almost everywhere for $\Phi''(x) \neq 0$, that is, $\Psi_1(x) = \Phi_1'(x)$ $\mathscr{N}$-almost everywhere for $\Phi''(x) \neq 0$. Thus $\Psi_1 = \Phi_1'$ and $\Psi_2 = \Phi_2'$ $\mathscr{N}$-almost everywhere. The proof is finished.

In the remainder of this section we shall derive from Proposition 7 the "change of variable" theorem for integrals as it occurs in the mapping of basic sets.

Let T be a single-valued mapping of the set A' into the set A, $T(a') \in A$ for $a' \in A'$, and moreover let $E|\mathscr{E}$, respectively $E'|\mathscr{E}'$, be normal integrals over the basic sets A, respectively A'. By $f'(a') = f(T(a'))$, $a' \in A'$, to every real function f (on A) there is uniquely associated a real function f' (on A') as pre-image of f, which we for brevity denote by $f(T)$. If we assume that

$$(\#) \qquad f(T) \in \mathfrak{L}' \text{ for } f \in \mathscr{E}$$

then we obtain an additional normal integral $E_1|\mathscr{E}$ over $\mathscr{E}$ by setting $E_1(f) = L'(f(T))$. Indeed, the properties (1_c), (2_c), (3_c), (8_c), and (11_c) are evident as properties of $\mathscr{E}$, properties (4_c) to (7_c) are consequences of corresponding properties of L'. If now on A

$$(\#\#) \qquad E_1 \dashv E,$$

then by Proposition 7 there exists a nonnegative function Ψ belonging to $\mathfrak{M}$ with

$$L_1(g) = L(g \cdot \Psi)$$

for $g \in \mathfrak{L}_1$ (if and only if $g \cdot \Psi \in \mathfrak{L}$). In the particular case, where $g \in \mathcal{E}$, this yields the transformation formula

$$(\#\#\#) \qquad L'(g(T)) = L(g \cdot \Psi).$$

However, this formula has significance beyond the mentioned special case for we have:

$$g \in \mathfrak{L}_1 \text{ implies } g(T) \in \mathfrak{L}' \text{ and } L_1(g) = L'(g(T)).$$

Indeed, from $g \in \mathfrak{L}_1$ follows the existence of $g_n \in \mathcal{E}$ with $\mathcal{N}_1(g - g_n) < 1/n$, $n \in N$, hence with

$$|g - g_n| \leqq \sum_k g_{n,k},$$

$0 \leqq g_{n,k} \in \mathcal{E}$ for $k \in N$, and

$$\sum_k E_1(g_{n,k}) < 1/n.$$

But then we also have

$$|g(T) - g_n(T)| \leqq \sum_k g_{n,k}(T)$$

and, using (#)

$$\mathcal{N}'(g(T) - g_n(T)) \leqq \sum_k \mathcal{N}'(g_{n,k}(T)) = \sum_k L'(g_{n,k}(T))$$
$$= \sum_k E_1(g_{n,k}) < \frac{1}{n},$$

and since $g_n(T) \in \mathfrak{L}'$ it follows that $g(T) \in \mathfrak{L}'$. Thus

$$L_1(g) = \lim_n E_1(g_n) = \lim_n L'(g_n(T)) = L'(g(T)).$$

We summarize the foregoing in the following proposition.

PROPOSITION 10. *Under the assumption (#) and (##) we have formula (###) provided that $g \cdot \Psi \in \mathfrak{L}$ holds, that is, provided that the right-hand side of (###) is meaningful.*

Remark. The above considerations show that $\mathcal{N}(f) = 0$ implies $\mathcal{N}_1(f) = 0$ which in turn implies $\mathcal{N}'(f(T)) = 0$; (##) therefore has as consequence that the pre-image $f(T)$ of every $\mathcal{N}$-null function f is an $\mathcal{N}'$-null function.

To obtain the validity of the transformation formula (# # #) also for the case where the left-hand side is assumed to be meaningful, we make the additional assumption that g (defined on A) is measurable: $g \in \mathfrak{M}$, which then implies that $g\Psi \in \mathfrak{M}$. In order to be able to use Proposition 2, we assume that

$$\sup\{L(f) : f \in \mathfrak{L} \text{ and } 0 \leqq f \leqq g\Psi\} = +\infty$$

If we set

$$f'(x) = \frac{f(x)}{\Psi(x)} \quad \text{for } \Psi(x) < 0$$

and otherwise $f'(x) = 0$, then $0 \leqq f' \leqq g$ and $f = f'\Psi \in \mathfrak{L}$, hence, by Proposition 10, $L'(f'(T)) = L(f)$; but $f'(T) \leqq g(T)$ and so $L(f) \leqq L'(g(T))$; thus $+\infty \leqq L'(g(T))$ in contradiction with $g(T) \in \mathfrak{L}'$. Therefore the supremum in question is finite and by Proposition 2 we have $g\Psi \in \mathfrak{L}$. Now we may use Proposition 10 and get formula (# # #).

The case where g changes sign is reduced to the case already discussed by decomposing g into its positive and negative parts. Thus we have the following.

Proposition 11. *If T is a single-valued mapping of the set A' into the set A and if, moreover, $E'|\mathcal{E}'$ and $E|\mathcal{E}$ are normal integrals over A' and A, respectively, as basic sets such that relations (#) and (# #) hold, then there is a nonnegative function $\Psi \in \mathfrak{M}$ such that the transformation formula*

$$L'(g(T)) = L(g \cdot \Psi)$$

holds for every function $g \in \mathfrak{M}$ provided that one side of the equation is meaningful. (In that case the other side is meaningful also and both sides are equal).

Section 6. Fubini's Theorem in Stone's Version

For $i = 1, 2$, let $\mathcal{E}_i$ be a linear lattice of elementary functions f_i with domain of definition X_i and let $E_i|\mathcal{E}_i$ be an elementary integral, i.e., we assume that both elementary integrals satisfy conditions (1_c) to (7_c) introduced at the beginning of Section 2. For brevity we shall write $E_i f_i$ in place of $E_i(f_i)$ provided that this does not lead to misunderstanding and we shall use similar notation for other functionals. We shall consider functions f whose domain of definition X_3 is the Cartesian product set $X_3 = X_1 \times X_2$ and in the following we shall make use of the simplifying idea that one may view f with $x_1 \in X_1$ held fixed as a function on X_2 (in other words, we identify the partial function f on $\{x_1\} \times X_2$ with a function on X_2). Let $\mathcal{E}_1 * \mathcal{E}_2$ denote the set of all functions f on X_3 for which on the one hand f with fixed but arbitrary $x_1 \in X_1$ as a function on X_2 belongs to $\mathcal{E}_2$ so that the function $E_2 f$ with X_1 as domain of definition exists and for which on the other hand $E_2 f \in \mathcal{E}_1$ so that the number $E_1 E_2 f = E_1(E_2 f)$ is defined.

We note that the set $\mathcal{E}_1 * \mathcal{E}_2$ does not satisfy condition (3_c) in general. For example, if $E_i|\mathcal{E}_i$ is the Riemann integral on the interval $[0, 1]$ for $i = 1, 2$ and

$$f(x_1, x_2) = \begin{cases} 0 & \text{for } x_1 = 0 \\ (1/x_1)\operatorname{sign}((1/2) - x_2) & \text{for } 0 < x_1 \leqq 1, \end{cases}$$

then $f \in \mathcal{E}_1 * \mathcal{E}_2$ (with $E_1 E_2 f = 0$) but not $|f| \in \mathcal{E}_1 * \mathcal{E}_2$.

On account of the apparent shortcomings of $\mathcal{E}_1 * \mathcal{E}_2$ we shall make the following restrictive assumptions:

(1) $X_3 = X_1 \times X_2$;

(2) for $i = 1, 2, 3$, on X_i there is given a linear lattice of elementary functions $\mathcal{E}_i$ and with it is given an elementary integral $E_i|\mathcal{E}_i$ such that

$$(*) \qquad \mathcal{E}_3 \subset \mathcal{E}_1 * \mathcal{E}_2 \text{ and } E_3|\mathcal{E}_3 = E_1 E_2|\mathcal{E}_3 \text{ holds.}$$

We stipulate that to $E_i|\mathcal{E}_i$ belong the norm $\mathcal{N}_i$, the function systems $\mathfrak{G}_i$, $\mathfrak{H}_i$ and $\mathfrak{I}_i$, and, finally, the $\mathcal{N}_i$-integral L_i with the corresponding function space $\mathfrak{L}_i$, for $i = 1, 2, 3$.

By $\mathfrak{L}_1 * \mathfrak{L}_2$ we denote the set of all functions f with domain of definition $X_3 = X_1 \times X_2$ satisfying the following requirements:

(i) for f there is an $\mathcal{N}_1$-null set $X_{1,0} \subset X_1$ such that for every fixed $x_1 \in X_1 - X_{1,0}$ the function f as function on X_2 is equal to a function in $\mathfrak{L}_2$;

(ii) The real-valued function $L_2 f$ with domain of definition $X_1 - X_{1,0}$ equals a function in $\mathfrak{L}_1$. (Equality in points (i) and (ii) is understood in the sense of equality almost everywhere).

Proposition 1. *If* (*) *holds, then, for arbitrary* $f \in \mathfrak{G}_3$;

$$\mathcal{N}_3 f \geqq \mathcal{N}_1 \mathcal{N}_2 f.$$

Proof. We need only to consider the case where $\mathcal{N}_3 f < +\infty$; in the case where $\mathcal{N}_3 f = +\infty$ there is nothing to prove.

If $\mathcal{N}_3 f < +\infty$ and $\varepsilon > 0$ is prescribed, we can select $f_n \in \mathcal{E}_3$, $n \in N$, in such a manner that

$$|f| \leqq \sum_n |f_n|$$

and

$$\sum_n E_1 E_2 |f_n| = \sum_n E_3 |f_n| < \mathscr{N}_3 f + \varepsilon$$

hold. On the other hand, using properties (d) and (g) of the norm discussed early in Section 2 we see that

$$\mathscr{N}_2 f \leqq \sum_n \mathscr{N}_2 f_n = \sum_n E_2 |f_n|$$

and so

$$\mathscr{N}_1 \mathscr{N}_2 f \leqq \sum_n \mathscr{N}_1 \mathscr{N}_2 |f_n| = \sum_n E_1 E_2 |f_n| < \mathscr{N}_3 f + \varepsilon$$

follows; letting $\varepsilon \to 0$, we get the assertion and the proof is finished.

Remark. From the foregoing it follows that $\mathscr{N}_1 \mathscr{N}_2 f$ in $\mathfrak{H}_3$ and $L_1 L_2 f$ in $\mathfrak{L}_1 * \mathfrak{L}_2$ are uniquely defined. Indeed, $f = f'$ in the sense of equality in $\mathfrak{H}_3$ means that

$$\mathscr{N}_3 \chi = 0$$

for

$$\chi = \chi_{\{(x_1,x_2): f(x_1,x_2) \neq f'(x_1,x_2)\}}.$$

However, by the foregoing result this means that the number

$$\mathscr{N}_1 \mathscr{N}_2 \chi = 0,$$

or the function

$$\mathscr{N}_2 \chi = 0$$

on $X_1 - Y_1$, where Y_1 is an $\mathscr{N}_1$-null set such that $\{x_2 : f(x_1, x_2) \neq f'(x_1, x_2)\}$ for $x_1 \in X - Y_1$, is an $\mathscr{N}_2$-null set. We therefore have that $\mathscr{N}_2 f = \mathscr{N}_2 f'$ for $x_1 \in X_1 - Y_1$; this in turn means that $\mathscr{N}_1 \mathscr{N}_2 f = \mathscr{N}_1 \mathscr{N}_2 f'$. If, moreover, $f, f' \in \mathfrak{L}_1 * \mathfrak{L}_2$ (with corresponding $\mathscr{N}_1$-null sets $X_{1,0}$ and $X'_{1,0}$) then we obtain that

$$L_2 f(x_1) = L_2 f'(x_1)$$

for $x_1 \in X_1 - (Y_1 \cup X_{1,0} \cup X'_{1,0})$ and so $L_1 L_2 f = L_1 L_2 f'$. But this is what we wanted to show.

PROPOSITION 2 (FUBINI-STONE). *If the elementary integrals $E_i | \mathscr{E}_i, i = 1, 2, 3,$*

satisfy the conditions (*), *then the corresponding norm integrals* $L_i|\mathfrak{L}_i$ *satisfy the related conditions*

$$\mathfrak{L}_3 \subset \mathfrak{L}_1 * \mathfrak{L}_2 \text{ and } L_3 f = L_1 L_2 f$$

for $f \in \mathfrak{L}_3$.

Proof. Let $f \in \mathfrak{L}_3$. Then there is an $f_n \in \mathcal{E}_3$, $n \in N$, such that $\mathcal{N}_3(f - f_n) < 2^{-n}$. Since

$$g = \sum_n \mathcal{N}_2(f - f_n) \in \mathfrak{G}_1,$$

we get by Proposition 1 that

$$\mathcal{N}_1 g \leqq \sum_n \mathcal{N}_1 \mathcal{N}_2(f - f_n) \leqq \sum_n \mathcal{N}_3(f - f_n) < \sum_n 2^{-n} = 1;$$

thus $g \in \mathfrak{I}_1$, i.e., $0 \leqq g(x_1) < +\infty$ for $x_1 \in X_1 - X_{1,0}$, where $X_{1,0}$ denotes an $\mathcal{N}_1$-null set. For fixed $x_1 \in X_1 - X_{1,0}$ we therefore have that

$$\lim_n \mathcal{N}_2(f - f_n) = 0$$

and so f as function on X_2 belongs to $\mathfrak{L}_2$; for f_n with x_1 fixed belongs to $\mathcal{E}_2$. We consider $h = L_2 f$ as function belonging to $\mathfrak{H}_1$ (the values in $X_{1,0}$ do not matter) and we obtain

$$|h - E_2 f_n| = |L_2(f - f_n)| \leqq \mathcal{N}_2(f - f_n)$$

so that

$$\mathcal{N}_1(h - E_2 f_n) \leqq \mathcal{N}_1 \mathcal{N}_2(f - f_n) \leqq \mathcal{N}_3(f - f_n) < 2^{-n};$$

thus, since $E_2 f_n \in \mathcal{E}_1$, by the definition of $\mathfrak{L}_1$ we have that $h \in \mathfrak{L}_1$. Finally, we have

$$L_1 L_2 f = L_1 h = \lim_n E_1 E_2 f_n = \lim_n E_3 f_n = L_3 f.$$

The proof is finished.

There is a partial converse to the foregoing proposition.

PROPOSITION 3. *Let the elementary integrals* $E_i|\mathcal{E}_i$, $i = 1, 2, 3$, *satisfy the conditions* (*). *If the function f with domain of definition* X_3 *is* $\mathcal{N}_3$*-measurable and if for* $n \in N$ *there are functions* f_n *in* $\mathfrak{I}_3$, *i.e., with* $\mathcal{N}_3 f_n < +\infty$ *and with*

$$|f| \leqq \sum_n |f_n|,$$

then $|f| \in \mathfrak{L}_1 * \mathfrak{L}_2$ *already implies* $f \in \mathfrak{L}_3$.

Proof. By part (a) of Proposition 28 in Section 2 we may without restriction assume that $f_n \in \mathfrak{L}_3$. Since $|f_n| \in \mathfrak{M}_3$ and

$$s_n = |f_1| + \ldots + |f_n| \in \mathfrak{L}_3,$$

we have by Proposition 22 of Section 2 that $g_n = \min\{|f|, s_n\} \in \mathfrak{L}_3$. Since $0 \leqq g_n \leqq |f|$, we have that $\mathcal{N}_2 g_n \leqq \mathcal{N}_2 f$ and $\mathcal{N}_1 \mathcal{N}_2 g_n \leqq \mathcal{N}_1 \mathcal{N}_2 f$. From $|f| \in \mathfrak{L}_1 * \mathfrak{L}_2$ follows the existence of an $\mathcal{N}_1$-null set $X_{1,0}$ such that $L_2|f| = \mathcal{N}_2 f$ for fixed $x_1 \in X_1 - X_{1,0}$ and in addition that $\mathcal{N}_2 f \in \mathfrak{L}_1$. Thus $\mathcal{N}_1 \mathcal{N}_2 f = L_1 \mathcal{N}_2 f < +\infty$. On the other hand, $\mathcal{N}_3 f \geqq L_3 g_n \geqq L_3 g_{n-1}$ holds. From this we may conclude that

$$\mathcal{N}_3 f = \lim_n L_3 g_n$$

in case

$$\lim_n L_3 g_n = +\infty;$$

if, on the other hand, this limit is finite, the Monotone Convergence Theorem (see Proposition 11 of Section 2) can be used (as g_n tends monotonely to $|f|$) to establish the same equality. Applying Proposition 2, we get $L_3 g_n = \mathcal{N}_1 \mathcal{N}_2 g_n \leqq \mathcal{N}_1 \mathcal{N}_2 f$ which leads to $\mathcal{N}_3 f \leqq \mathcal{N}_1 \mathcal{N}_2 f < +\infty$ upon letting $n \to \infty$. Finally we obtain

$$f \in \mathfrak{I}_3 \cap \mathfrak{M}_3 = \mathfrak{L}_3$$

(by Proposition 27 of Section 2) and the proof is complete.

EXERCISES

1. Let f be a continuous real-valued function on $[0, 1]$. Show that there is a monotone decreasing sequence $(P_n)_{n=1}^{\infty}$ of polynomials which converges uniformly to f on $[0, 1]$.
   [*Hints*: Let $(e_n)_{n=1}^{\infty}$ be any sequence of positive real numbers such that

   $$\sum_{n=1}^{\infty} e_n = 1.$$

   By Proposition 1 of Section 1 we can find polynomials $P_1, P_2, \ldots$ such that for all $x \in [0, 1]$

   $$|f(x) + (1 - \tfrac{1}{2}e_1) - P_1(x)| < \tfrac{1}{2}e_1$$

   $$|f(x) + (1 - e_1 - \tfrac{1}{2}e_2) - P_2(x)| < \tfrac{1}{2}e_2$$

   . . . .

   $$|f(x) + (1 - e_1 - \cdots - e_{n-1} - \tfrac{1}{2}e_n) - P_n(x)| < \tfrac{1}{2}e_n.$$

Letting $f_0(x) = f(x) + 1$ and

$$f_n(x) = f(x) + (1 - \sum_{k=1}^{n} e_k) \qquad \text{for } n \geqq 1,$$

we have

$$f_0(x) \geqq P_1(x) \geqq f_1(x) \geqq P_2(x) \geqq \cdots \geqq f_n(x)P_{n+1}(x) \geqq \cdots.$$

But $f_n \to f$ uniformly. Observe that we may replace "decreasing" by "increasing" and the conclusion will of course still be valid.]

2. Let f be a continuous real-valued function on the interval $[-\pi/2, \pi/2]$. Is it possible to approximate f uniformly on this interval by linear combinations of $(\sin x)^k$ with $k = 0, 1, 2, \ldots$? Consider also the same question for $(\cos x)^k$.
3. Approximation of continuous real-valued functions defined on a product of compact Hausdorff spaces: Let $(X_t)_{t \in T}$ be a family of compact Hausdorff spaces and let

$$X = \bigtimes_{t \in T} X_t.$$

Show that every continuous real-valued function on X can be uniformly approximated by sums of a finite number of functions of the form

$$(x_t) \to \bigtimes_{s \in S} u_s(x_s),$$

where S is an (arbitrary) finite subset of T and u_s is a continuous real-valued function on X_s for each $s \in S$.

[*Hints*: Consider the set H of "functions of one variable" $(x_t) \to u_s(x_s)$ (any $s \in T$) which are continuous on X. This set separates the points of X, for if $x = (x_t)$ and $y = (y_t)$ are two distinct points of X, there exists $s \in T$ such that $x_s \neq y_s$ and there exists a continuous real-valued function h_s on X_s such that $h_s(x_s) \neq h_s(y_s)$. The function

$$x \to h_s(pr_s x)$$

then belongs to H and takes distinct values at x and y. Since every polynomial in the functions H is of the form stated in the claim, the result follows from Proposition 7 of Section 1.]

4. Stone-Weierstrass approximation theorem for complex-valued functions: Let X be a compact Hausdorff space. Let H be a family of complex-valued continuous functions on X which separates the points of X, contains the constant function 1 and contains, with every f, its

complex conjugate $\bar{f}(x) = \overline{f(x)}$. Then every continuous function g of X into the complex number field C can be approximated uniformly by polynomials in the functions $f \in H$ and their conjugates $\bar{f}$ with coefficients in C.

[*Hints*: Let g: $X \to C$ be continuous. Show that the real-valued functions $\operatorname{Re} g$ and $\operatorname{Im} g$ can be approximated by polynomials in functions belonging to H. For $f \in H$, we have $\operatorname{Re} f = (f + \bar{f})/2$ and $\operatorname{Im} f = (f - \bar{f})/2i$, and so $\operatorname{Re} f$ and $\operatorname{Im} f$ are polynomials in the functions from H. The family of functions H_0, consisting of all $\operatorname{Re} f$ and $\operatorname{Im} f$ for $f \in H$, is a separating family on X, for if $x, y \in X$ and $f(x) \neq f(y)$, then either $\operatorname{Re} f(x) \neq \operatorname{Re} f(y)$ or $\operatorname{Im} f(x) \neq \operatorname{Im} f(y)$. Also H_0 contains the function 1. It can therefore be seen that there are polynomials P and Q in functions from H_0 which approximate arbitrarily closely in the sense of uniform approximation $\operatorname{Re} g$ and $\operatorname{Im} g$, respectively, and so $P + iQ$ will give us the desired uniform approximation for g itself.]

5. Show that any complex-valued continuous function f: $R^1 \to C$ with period 2π can be approximated uniformly by a sequence of trigonometric polynomials of the form

$$\sum_k c_k e^{ikx}$$

with $k = 0, \pm 1, \pm 2, \ldots, \pm n$.

6. Prove the statement in the Remark at the end of Section 1.
7. Show that the number of odd binomial coefficients in any finite expansion is a power of 2.

[*Hints*: Recall Pascal's "triangle" for binomial coefficients:

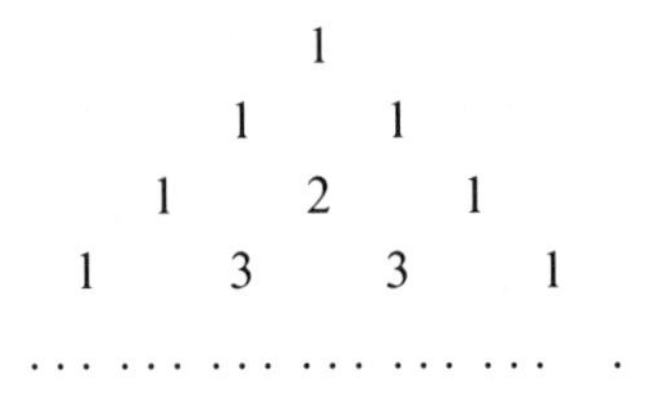

We choose to denote odd numbers by a minus sign and even numbers by a plus sign. Since the sum of two odd numbers is an even number, the sum of an odd and an even number is an odd number, etc., we can manipulate by use of the simple rules "minus times minus gives plus," "minus times plus gives minus," etc. Each row in Pascal's "triangle" begins and ends with the number 1 (i.e., a minus sign) and the other terms in a row are calculated from terms in the preceding row by use of the equation

$$\binom{n}{k} + \binom{n}{k+1} = \binom{n+1}{k+1}$$

for $n > k$. In our minus-plus scheme no such involved calculations are needed. Using the scheme proposed, we get:

$$\begin{array}{c}
- \\
- \quad - \\
- \quad + \quad - \\
- \quad - \quad - \quad - \\
- \quad + \quad + \quad + \quad - \\
- \quad - \quad + \quad + \quad - \quad - \\
- \quad + \quad - \quad + \quad - \quad + \quad - \\
- \quad - \quad - \quad - \quad - \quad - \quad - \quad - \\
- \quad + \quad + \quad + \quad + \quad + \quad + \quad + \quad - \\
- \quad - \quad + \quad + \quad + \quad + \quad + \quad + \quad - \quad - \\
- \quad + \quad - \quad + \quad + \quad + \quad + \quad + \quad - \quad + \quad - \\
\cdots\cdots\cdots\cdots\cdots\cdots\cdots\cdots\cdots\cdots\cdots\cdots\cdots\cdots\cdots\cdots\cdots\cdots\cdots\cdots
\end{array}$$

It is clear by inspection that the claim made holds for the "triangle of first order":

$$\begin{array}{c}
- \\
- \quad - \,.
\end{array}$$

Observe next that the "triangle of second order," i.e.,

$$\begin{array}{c}
- \\
- \quad - \\
- \quad + \quad - \\
- \quad - \quad - \quad - \,,
\end{array}$$

is essentially made up of three "triangles of first order" with the two lower ones situated symmetrically (which is of course as it should be on account of the equality

$$\binom{n}{k} = \binom{n}{n-k}).$$

Next, observe that the "triangle of third order," i.e.,

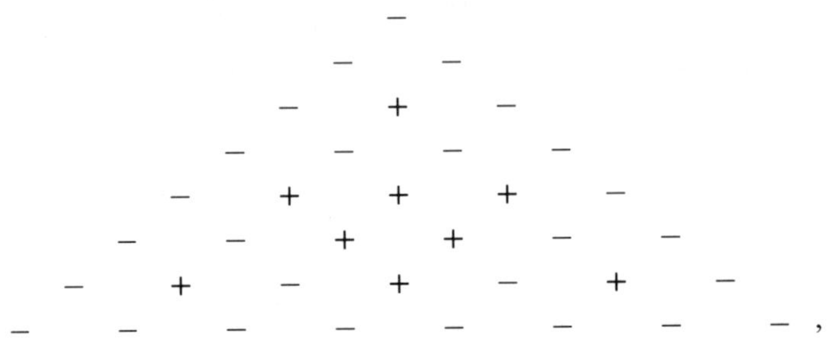

is essentially made up of three "triangles of second order" with the two lower ones situated symmetrically, etc. A moment's reflection shows us that we have indeed an induction going and that therefore the claim is pretty clear.]

8. Let μ be a measure on an algebra $\mathfrak{A}$ of subsets of an arbitrary set A and let $\mathscr{E}$ be the family consisting of those functions which are finite linear combinations of characteristic functions of sets in $\mathfrak{A}$ with finite measure, and let E be integration with respect to μ. Show that $E|\mathscr{E}$ is a (weak) elementary integral.
9. Consider the following generalization of the foregoing example: Let $\mathfrak{A}$ be a field of subsets of an arbitrary set A (for the definition of the notion "field" see the beginning of Section 3) and let μ be a finite-valued, nonnegative, finitely additive set function on $\mathfrak{A}$. We define an elementary function f on A as follows: Let $X_1, \ldots, X_n$ be a finite number of pairwise disjoint sets in $\mathfrak{A}$ and let $y_1, \ldots, y_n$ be finite real numbers; we put $f(x) = y_j$ for $x \in X_j$, $j = 1, \ldots, n$ and we put $f(x) = 0$ for $x \notin X_1 \cup \ldots \cup X_n$ and we set $E(f) = y_1 \mu(X_1) + \ldots + y_n \mu(X_n)$.
   (i) Show that E is a (weak) elementary integral on the set $\mathscr{E}$ of elementary functions f.
   (ii) Show that the norm $\mathscr{N}(f)$ for an f in $\mathscr{E}$ is the infimum of all sums

$$\sum_k a_k \mu(A_k),$$

   where a_k is a nonnegative real number and the A_k denote members of $\mathfrak{A}$, $k \in N$, with

$$|f| \leqq \sum_k a_k \chi_{A_k}.$$

   (iii) Verify that $E|\mathscr{E}$ is an elementary integral (i.e., it satisfies condition (7_c) of Section 2) if and only if μ is σ-additive.

(iv) Prove that a subset B of A is an $\mathscr{N}$-null set if and only if B can be covered by a countable number of sets B_k in $\mathfrak{A}$ such that

$$\sum_k \mu(B_k)$$

is arbitrarily small.

10. A real linear space X is said to be a vector lattice if X is a lattice with respect to the partial order relation $\leqq$ and satisfies the following conditions: For $x, y \in X$,
    (a) if $x \leqq y$, then $x + z \leqq y + z$ for every $z \in X$,
    (b) if $x \leqq y$, then $\lambda x \leqq \lambda y$ for every real number $\lambda > 0$.

    Show that the set $\mathscr{E}$ of function defined at the beginning of Section 2 of this chapter is a vector lattice under the natural order $\leqq$.
11. Let $C_0[0, \infty)$ (resp. $C_0(-\infty, \infty)$) be the complete metric space under the supremum distance of all real-valued continuous functions x on $[0, \infty)$ (resp. $-\infty, \infty)$) satisfying

$$\lim_{t\to\infty} x(t) = 0$$

$$(\text{resp. } \lim_{t\to\infty} x(t) = \lim_{t\to-\infty} x(t) = 0).$$

    Show: For any fixed $\alpha > 0$, the set of all functions of the form $e^{-\alpha t} p(t)$ (resp. $e^{-\alpha t^2} p(t)$), where $p(t)$ is a polynomial, is dense in $C_0[0, \infty)$ (resp. $C_0(-\infty, \infty)$).
12. For $i = 1, 2$, let $E_i|\mathscr{E}_i$ be (weak) elementary integrals of the type described in Exercise 8. Does the set of finite real linear combinations of characteristic functions of measurable rectangles form a system $\mathscr{E}_3$ of the desired kind in the sense of Section 6?

Chapter 10

Normed Linear Spaces

Let X be a linear space over the field K of real or complex numbers. A *norm* on X, denoted by $\| \ \|$, is a real-valued function on X with the following properties: For all $x, y \in X$ and $\alpha \in K$,

(i) $\|x\| \geqq 0$ for all $x \in X$,
(ii) $x \neq 0$ implies $\|x\| \neq 0$,
(iii) $\|\alpha x\| = |\alpha| \|x\|$,
(iv) $\|x + y\| \leqq \|x\| + \|y\|$.

The linear space X, together with a norm on X, is called a *normed linear space*. When the scalars over X are the reals, we shall refer to X as a *real* normed linear space.

A normed linear space becomes a metric space if we define the distance $\rho(x,y)$ as $\|x - y\|$, and we call a normed linear space a *Banach space* if it is complete in this metric.

It follows from properties (iii) and (iv) of the norm that

$$|\|x\| - \|y\|| \leqq \|x \pm y\| \leqq \|x\| + \|y\|;$$

it is therefore clear that $\|x\|$ is a continuous function of x.

Euclidean n-dimensional space is a real Banach space with the norm of a point $x = (x_1, x_2, \ldots, x_n)$ taken to be its ordinary length

$$(\sum_{j=1}^{n} x_j^2)^{1/2}.$$

Using the sequential version of Minkowski's Inequality it is easy to see that it remains a Banach space if the norm of x is changed to

$$\|x\|_p = (\sum_{j=1}^{n} |x_j|^p)^{1/p}$$

for any fixed $p \geqq 1$. The L^p spaces of integration theory with norm

$$\|f\|_p = (\int |f|^p)^{1/p}$$

and the set of all bounded continuous real-valued functions on a topological space S with norm

$$\|f\| = \sup\{|f(x)| : x \in S\}$$

provide further examples of Banach spaces.

PROPOSITION 1. *Every incomplete normed linear space has a completion.*

Proof. Let $\tilde{X}$ be the completion of the metric space (X, ρ) (see Proposition 30 of Chapter 7). We have to show vector addition and multiplication by a scalar as well as the norm of X can be extended to $\tilde{X}$ in such a way that $\tilde{X}$ becomes a normed linear space.

Let $(x_n)_{n=1}^{\infty}$ and $(y_n)_{n=1}^{\infty}$ be Cauchy sequences in X which define the elements $\tilde{x} \in \tilde{X}$ and $\tilde{y} \in \tilde{X}$, respectively. Since

$$\|\alpha x_n - \alpha x_m\| = |\alpha| \|x_n - x_m\|,$$

$$\begin{aligned} \|(x_n + y_n) - (x_m + y_m)\| &= \|(x_n - x_m) + (y_n - y_m)\| \\ &\leqq \|x_n - x_m\| + \|y_n - y_m\|, \end{aligned}$$

we may conclude that $(\alpha x_n)_{n=1}^{\infty}$ and $(x_n + y_n)_{n=1}^{\infty}$ are Cauchy sequences also. We denote by $\alpha\tilde{x}$ and $\tilde{x} + \tilde{y}$ the elements of $\tilde{X}$ so defined. It is easily seen that we have introduced in $\tilde{X}$ a well-defined vector addition and scalar multiplication and that $\tilde{X}$ has become a normed linear space. From $\rho(x, y) = \|x - y\|$ we get that $\rho(\alpha x, \alpha y) = |\alpha| \rho(x, y)$ and $\rho(x + z, y + z) = \rho(x, y)$ for all x, y, $z \in X$ and $\alpha \in K$. By passage to the limit we see that these relations remain preserved for all $\alpha \in K$ and $\tilde{x}$, $\tilde{y}$, $\tilde{z} \in \tilde{X}$. But then the function

$$\|\tilde{x}\| = d(\tilde{x}, \tilde{0}),$$

with $\tilde{x} \in \tilde{X}$, will be a norm on $\tilde{X}$. The triangle inequality for example can be verified as follows:

$$\begin{aligned} \|\tilde{x} + \tilde{y}\| &= d(\tilde{x} + \tilde{y}, \tilde{0}) \leqq d(\tilde{x} + \tilde{y}, \tilde{y}) + d(\tilde{y}, \tilde{0}) \\ &= d(\tilde{x}, \tilde{0}) + d(\tilde{y}, \tilde{0}) = \|\tilde{x}\| + \|\tilde{y}\|. \end{aligned}$$

Thus $\tilde{X}$ is a normed linear space, where

$$d(\tilde{x}, \tilde{y}) = d(\tilde{x} - \tilde{y}, \tilde{0}) = \|\tilde{x} - \tilde{y}\|$$

for all $\tilde{x}$, $\tilde{y} \in \tilde{X}$. Moreover, by construction $\tilde{X}$ is a complete metric space, i.e., $\tilde{X}$ is a Banach space. The proof is finished.

Let A and B be linear spaces over K. A function T from A into B is said to be a *linear operator* if

(i) $T(x + y) = T(x) + T(y)$ and

(ii) $T(\alpha x) = \alpha T(x)$ for all $x, y \in A$ and $\alpha \in K$.

If A and B are normed linear spaces, a linear operator from A into B is called *bounded* if there exists a nonnegative real number C such that

(iii) $\|T(x)\| \leqq C\|x\|$ for all $x \in A$.

In this case we define the *norm of T* to be the infimum of the set of all C's which satisfy (iii) and we denote by $\|T\|$ the norm of the linear operator T. It is easy to verify that

$$\|T\| = \sup\left\{\frac{\|T(x)\|}{\|x\|} : x \in A, x \neq 0\right\}$$
$$= \sup\{\|T(x)\| : x \in A, \|x\| = 1\}$$
$$= \sup\{\|T(x)\| : x \in A, \|x\| \leqq 1\}$$

and

$$\|T(x)\| \leqq \|T\| \; \|x\|$$

for all $x \in A$.

PROPOSITION 2. *If T is a linear operator mapping a normed linear space A into a normed linear space B, then the following conditions are equivalent:*

(*a*) *T is continuous.*

(*b*) *T is continuous at one point.*

(*c*) *T is bounded.*

Proof. If T is continuous at x_0, then there is a positive constant M such that $\|T(x - x_0)\| = \|T(x) - T(x_0)\| \leqq 1$ when $\|x - x_0\| \leqq M$. Thus $\|T(z)\| \leqq 1$ whenever $\|z\| \leqq M$, and, for any nonzero y,

$$\|T(y)\| = (\|y\|/M)\|T(y(M/\|y\|))\| \leqq \|y\|/M,$$

which is condition (c) with $C = 1/M$. But then

$$\|T(x) - T(x_1)\| = \|T(x - x_1)\| \leqq C\|x - x_1\| < \varepsilon$$

whenever $\|x - x_1\| < \varepsilon/C$, and T is continuous at every point x_1. This completes the proof.

PROPOSITION 3. *Let A and B be normed linear spaces over K and let $\mathcal{B}(A, B)$ denote the set of all bounded linear operators from A into B. Then, with pointwise linear operations and the operator norm, $\mathcal{B}(A, B)$ is a normed linear space. Moreover, if B is a Banach space, so is $\mathcal{B}(A, B)$.*

Proof. If $T, S \in \mathcal{B}(A, B)$, then $\alpha T \in \mathcal{B}(A, B)$ and $T + S \in \mathcal{B}(A, B)$ and we have

$$\|\alpha T\| = |\alpha| \|T\|, \qquad \|T + S\| \leqq \|T\| + \|S\|;$$

the verification of this is based on the relations

$$\frac{\|\alpha T(x)\|}{\|x\|} = |\alpha| \frac{\|T(x)\|}{\|x\|},$$

$$\frac{\|(T + S)(x)\|}{\|x\|} \leqq \frac{\|T(x)\|}{\|x\|} + \frac{\|S(x)\|}{\|x\|} \leqq \|T\| + \|S\|.$$

We now assume that B is complete and we let $(T_n)_{n=1}^{\infty}$ be a Cauchy sequence in $\mathcal{B}(A, B)$. Then for every $\varepsilon > 0$, there exists an $n(\varepsilon)$ such that

$$\|T_n - T_m\| < \varepsilon$$

for all n, $m > n(\varepsilon)$. Thus, for every $x \in A$,

$$(*) \qquad \|T_n(x) - T_m(x)\| = \|(T_n - T_m)(x)\| < \varepsilon \|x\|$$

for all n, $m > n(\varepsilon)$. Therefore $(T_n(x))_{n=1}^{\infty}$ is seen to be a Cauchy sequence in B; since B is complete, this sequence has a limit in B. We put

$$T(x) = \lim_{n \to \infty} T_n(x).$$

It is clear that the operator T so defined is linear. We shall show that T belongs to $\mathcal{B}(A, B)$ and is the limit of the sequence $(T_n)_{n=1}^{\infty}$.

For $n \to \infty$, we obtain from (*)

$$\|T(x) - T_m(x)\| \leqq \varepsilon \|x\|$$

for all $m > n(\varepsilon)$. Hence $T - T_m \in \mathcal{B}(A, B)$ and

$$(**) \qquad \|T - T_m\| \leqq \varepsilon$$

for $m > n(\varepsilon)$. But then

$$T = T_m + (T - T_m) \in \mathcal{B}(A, B).$$

From (**) we moreover see that $\|T - T_m\| \to 0$ for $m \to \infty$ and the proof is finished.

Proposition 4 (Open Mapping Theorem). *If A and B are Banach spaces and T is a bounded linear operator which maps A onto B, then T is an open mapping, i.e., the image under T of an open set in A is open in B.*

Proof. Let G be an open set in A and let $y \in T(G)$ so that $y = T(x)$ for some $x \in G$. Since G is open, there exists a sphere $S(x, \delta) = \{z \in A : \|z - x\| \leqq \delta\} \subset G$, whence $T(S(x, \delta)) \subset T(G)$. Provided we can show that there is a sphere $S(y) \subset T(S(x, \delta))$ we will have $S(y) \subset T(G)$, so that $T(G)$ will be open. To ensure the provision it is enough to prove that there exists a sphere $S(\theta) \subset T(S_0)$, where $S_0 = \{x \in A : \|x\| < 1\}$ and $S(\theta)$ is a sphere centered at the origin θ in B. For, having proved this, $T(S(x, \delta) - x)$ will contain a sphere about the origin θ, whence $T(S(x, \delta))$ will contain a sphere about y.

We now proceed to prove that there exists $S(\theta) \subset T(S_0)$. Let

$$S_k = S(\theta, 2^{-k}) = \{x : \|x\| < 2^{-k}\},$$

$k = 0, 1, 2, \ldots$. Now if $x \in A$, we write $x = k(x/k)$, where $k = [2\|x\|] + 1$; the square bracket denoting the integer part of $2\|x\|$. Then $x/k \in S_1$ and so $x \in kS_1$. Hence

$$A = \bigcup_{k=1}^{\infty} kS_1.$$

Since T is onto, $T(A) = B$, we get

$$B = \bigcup_{k=1}^{\infty} kT(S_1).$$

But B is complete and so of second category, whence there exists k such that $kT(S_1)$ is not nowhere dense. Thus $\overline{T(S_1)}$ contains some sphere $S(b, r)$, say. We shall now verify that

$$(\#) \qquad S(\theta, r) \subset \overline{T(S_0)}.$$

We have

$$S(\theta, r) \subset \overline{T(S_1)} - b \subset \overline{T(S_1)} - \overline{T(S_1)} \subset 2\overline{T(S_1)} = \overline{T(S_0)}$$

which proves (#). In the above chain of inclusions we have used the facts that $b \in \overline{T(S_1)}$, $\overline{T(S_1)}$ is convex [i.e., if $x, y \in \overline{T(S_1)}$, $\lambda + \mu = 1$, $\lambda \geqq 0$, $\mu \geqq 0$ imply $\lambda x + \mu y \in \overline{T(S_1)}$] and that $-y \in \overline{T(S_1)}$ whenever $y \in \overline{T(S_1)}$.

From (#) it now follows immediately that

$$(\#\#) \qquad S(\theta, r2^{-n}) \subset \overline{T(S_n)}.$$

Finally, we show that $S(\theta, r/2) \subset T(S_0)$. Take $y \in S(\theta, r/2)$. Then $y \in \overline{T(S_1)}$ by (##). Hence $\|y - y_1\| < r/4$ for some $y_1 \in T(S_1)$. Also, $y - y_1 \in S(\theta, r/4) \subset \overline{T(S_2)}$ and so $\|y - y_1 - y_2\| < r/8$ for some $y_2 \in T(S_2)$. Continuing, we find

$$(\#\#\#) \qquad \|y - \sum_{k=1}^{n} y_k\| < r/2^{n+1},$$

where $y_k \in T(S_k)$, whence $y_k = T(x_k)$ for some $x_k \in S_k$. From (# # #) we get

$$y = \sum T(x_k),$$

and since $\|x_k\| < 2^{-k}$, $\sum \|x_k\| < 1$, we see that $\sum x_k = x$, say (for the latter see Exercise 1 of Chapter 5). Also $\|x\| \leqq \sum \|x_k\| < 1$, $x \in S_0$, and by the continuity of T,

$$T(x) = \sum T(x_k) = y.$$

Thus $y \in S(\theta, r/2)$ implies $y = T(x)$ for some $x \in S_0$, i.e., $y \in T(S_0)$, which proves that $S(\theta, r/2) \subset T(S_0)$. From our earlier remarks this inclusion is enough to allow us to deduce the proposition and the proof is therefore finished.

The following two propositions are direct corollaries of the Open Mapping Theorem.

PROPOSITION 5. *If A and B are Banach spaces and T is a one-to-one continuous linear operator from A onto B, then T^{-1} is continuous.*

Proof. If U is open in A (A = range of T^{-1}), then $(T^{-1})^{-1}(U) = T(U)$ is open in B (B = domain of definition of T^{-1}).

PROPOSITION 6. *If E is a linear space over K and $\| \ \|$ and $\| \ \|'$ are two Banach space norms for E, then the metric topologies induced on E by $\| \ \|$ and $\| \ \|'$ are identical (i.e., there are two positive real numbers a and b such that $a\|x\| \leqq \|x\|' \leqq b\|x\|$ for all $x \in E$) if and only if there is a positive constant α such that*

$$\alpha\|x\| \geqq \|x\|'$$

for all $x \in E$.

Proof. Consider the identity mapping on E as a linear operator from the Banach space $(E, \| \ \|)$ onto the Banach space $(E, \| \ \|')$. The details are left to the reader.

Let A and B be normed linear spaces and let T: $A \to B$ be a linear operator. Then the subset $\{(x, T(x)) : x \in A\}$ of $A \times B$ is called the *graph of T*. We may turn $A \times B$ into a normed linear space with

$$\|(x, y)\| = (\|x\| + \|y\|),$$

defining $(x,y) + (x',y') = (x + x', y + y')$, $\alpha(x,y) = (\alpha x, \alpha y)$. It is clear that $(x_n, y_n) \to (x,y)$ if and only if $x_n \to x$ and $y_n \to y$. Hence if A, B are Banach spaces, then $A \times B$ is also a Banach space.

PROPOSITION 7 (CLOSED GRAPH THEOREM). *Let A and B be Banach spaces and let T: $A \to B$ be a linear operator. Then T is continuous if and only if its graph is closed.*

Proof. Let T be continuous and let $G(T)$ be the graph of T. We show that $G(T)$ is closed (actually the linearity of T is superfluous in this part of the proposition). Let $(x,y) \in \overline{G(T)}$. Then there exists $x_n \in A$ such that $x_n \to x$ and $T(x_n) \to y$. But $T(x_n) \to T(x)$ and so $y = T(x)$, whence $(x,y) = (x, T(x)) \in G(T)$.

Conversely, suppose that $G(T)$ is closed. Then $G(T)$ is a Banach space, being a closed subspace of the Banach space $A \times B$. Let P_1 and P_2 be the projections of $G(T)$ into A and B respectively, i.e.,

$$P_1(x, T(x)) = x \text{ and } P_2(x, T(x)) = T(x)$$

for all $x \in A$. We have

$$\|P_1(x, T(x))\| = \|x\| \leqq \|(x, T(x))\|$$

and

$$\|P_2(x, T(x))\| = \|T(x)\| \leqq \|(x, T(x))\|$$

and therefore P_1 and P_2 are continuous linear operators. Since T is single-valued and the domain of definition of T is A, P_1 is one-to-one and onto A. By Proposition 5 we see that P_1^{-1} is continuous. Clearly $T = P_2 \circ P_1^{-1}$, and so T is continuous. This completes the proof.

Remark. It is easy to see that the graph of the linear operator T: $A \to B$ is closed if and only if T has the property that whenever $(x_n)_{n=1}^{\infty}$ is a sequence in A which converges in A to some point x and $(T(x_n))_{n=1}^{\infty}$ converges in B to a point y, then $T(x) = y$. A linear operator T: $A \to B$ whose graph is closed is called a *closed operator*.

PROPOSITION 8. *Let B be a Banach space and let S be a nonempty set. Let Λ denote the set of all functions λ from S into B such that*

$$\sup_{s \in S} \|\lambda(s)\| < \infty$$

and let $\|\lambda\|$ denote this supremum. Then Λ, with pointwise linear operations and the above norm, is a Banach space.

The proof of the foregoing proposition is virtually the same as that given for Proposition 24 of Chapter 7.

PROPOSITION 9 (UNIFORM BOUNDEDNESS PRINCIPLE). *Let A and B be Banach spaces and let $\{T_s : s \in S\}$ be a nonempty family of bounded linear operators from A into B such that*

$$\sup_{s \in S} \|T_s(x)\| < \infty$$

for every $x \in A$. Then

$$\sup_{s \in S} \|T_s\| < \infty.$$

Proof. Let Λ be as in Proposition 8. Define a mapping $V: A \to \Lambda$ by

$$V(x)(s) = T_s(x)$$

for $x \in A$, $s \in S$. Since the family $\{T_s : s \in S\}$ is assumed to be pointwise bounded, we see that $V(x) \in \Lambda$ for each $x \in A$. It is clear that V is linear. We verify that the graph of V is closed.

Suppose that $x_n \to x$ in A and that $V(x_n) \to \lambda$ in Λ. For each $s \in S$, we have

$$\begin{aligned} \|\lambda(s) - V(x)(s)\| &\leqq \|\lambda(s) - V(x_n)(s)\| + \|V(x_n)(s) - V(x)(s)\| \\ &\leqq \|\lambda - V(x_n)\| + \|T_s(x_n) - T_s(x)\|. \end{aligned}$$

Since T_s is continuous, the last expression has limit 0 as $n \to \infty$. Thus $\lambda(s) = V(x)(s)$ for all $s \in S$, and so $\lambda = V(x)$. This shows that the graph of V is closed.

Since V has a closed graph, V is continuous by Proposition 7. Thus

$$\|V\| = \sup\{\|V(x)\| : \|x\| \leqq 1\} < \infty.$$

Since

$$\begin{aligned} \|T_s\| &= \sup\{\|T_s(x)\| : \|x\| \leqq 1\} \\ &= \sup\{\|V(x)(s)\| : \|x\| \leqq 1\} \\ &= \sup\{\|V(x)\| : \|x\| \leqq 1\} \\ &= \|V\| \end{aligned}$$

for every $s \in S$, we conclude that

$$\sup_{s \in S} \|T_s\| \leqq \|V\| < \infty.$$

This finishes the proof.

Remark. An alternative way of proving Proposition 9 is to make use of Exercise 3 of Chapter 7. We leave the details for the Exercise section at the end of this chapter.

PROPOSITION 10 (BANACH-STEINHAUS THEOREM). *Let A and B be Banach spaces and let $(T_n)_{n=1}^{\infty}$ be a pointwise convergent sequence of bounded linear operators from A into B. Then the mapping $T: A \to B$ defined by*

$$T(x) = \lim_{n\to\infty} T_n(x)$$

is a bounded linear operator.

Proof. T is clearly linear. We also see that, for each $x \in A$,

$$\sup\{\|T_n(x)\| : n \in N\} < \infty.$$

By Proposition 9 there exists a positive constant M such that $\|T_n\| \leqq M$ for all $n \in N$. Thus

$$\|T(x)\| = \lim_{n\to\infty} \|T_n(x)\| \leqq M\|x\|$$

for $x \in A$ and so T is bounded and $\|T\| \leqq M$. The proof is finished.

Let E be a linear space over K where K denotes either the field of real numbers or the field of complex numbers. A *linear functional* on E is a linear operator from E into K, where K is looked upon as a one-dimensional linear space over itself. If E is a normed linear space and the absolute value is used as a norm on K, we denote by E^* the space of all bounded linear functionals on E, that is, $E^* = \mathscr{B}(E, K)$. The space E^* is called the *conjugate space of E*; the conjugate space E^{**} of the space E^* is called the *second conjugate space* of E and so forth. By Proposition 3, the conjugate spaces E^*, E^{**}, ... are seen to be Banach spaces.

With E a normed linear space, we define the so-called *natural mapping of E into E^{**}* as follows: For $x \in E$, we define $\hat{x}$ on E^* by $\hat{x}(f) = f(x)$. It is easy to see that each $\hat{x}$ is a linear functional on E^* and that the mapping $x \to \hat{x}$ is a linear operator. Also

$$\begin{aligned}\sup\{|\hat{x}(f)| : f \in E^*, \|f\| \leqq 1\} &= \sup\{|f(x)| : f \in E^*, \|f\| \leqq 1\}\\ &\leqq \sup\{\|f\|\|x\| : f \in E^*, \|f\| \leqq 1\}\\ &\leqq \|x\|\end{aligned}$$

for each $x \in E$. Thus the mapping $x \to \hat{x}$ is a bounded linear operator from E into E^{**} of norm $\leqq 1$.

PROPOSITION 11 (HAHN-BANACH EXTENSION THEOREM). *Suppose that E is a real linear space and M is a linear subspace of E. Let p be a real-valued* *function on E with the following properties*:

(*i*) $p(x + y) \leqq p(x) + p(y)$,

(*ii*) $p(\alpha x) = \alpha p(x)$

for all $x, y \in E$ *and all positive real numbers* α. *If* f *is a linear functional defined on* M *such that*

$$f(x) \leqq p(x)$$

for all $x \in M$, *then there exists a linear functional* F *which is an extension of* f *to all of* E *such that*

$$F(x) \leqq p(x)$$

for all $x \in E$.

Proof. Let $\mathcal{P}$ be the set of all real-valued functions g such that the domain of definition of g is a linear subspace of E, g is linear, g extends f, in symbols, $f \subset g$ [observe that a function is a set of ordered pairs], and $g(x) \leqq p(x)$ for all x in the domain of definition of g. $\mathcal{P}$ is nonempty, since $f \in \mathcal{P}$. Partially order $\mathcal{P}$ by having $g \subset h$ mean that h is a linear extension of g. Let $\mathfrak{C}$ be any chain in $\mathcal{P}$ and let $g = \cup\, \mathfrak{C}$. Then $g \in \mathcal{P}$. Using Zorn's Lemma (see Chapter 7) we see that $\mathcal{P}$ contains a maximal member, say F. It remains to prove that the domain of definition of F, in symbols, dom F, is all of E.

Assume that there exists a $y \in E$ which is not in dom F. Define

$$H = \{x + \alpha y : x \in \text{dom } F, \alpha \in R^1\}.$$

Then H is a linear subspace of E and dom F is a proper subset of H. Let c be a fixed, but arbitrary, real number and define g on H by

$$g(x + \alpha y) = F(x) + \alpha c.$$

Then g is well-defined; if $x_1 + \alpha_1 y = x_2 + \alpha_2 y$, where $x_1, x_2 \in$ dom F and $\alpha_1, \alpha_2 \in R^1$, then $(\alpha_1 - \alpha_2)y = x_2 - x_1 \in$ dom F and so $\alpha_1 = \alpha_2$ and $x_1 = x_2$. Evidently g is a linear functional and $F \subsetneq g$. If we select c in such a way that $g(x) \leqq p(x)$ for all $x \in H$, then g will belong to $\mathcal{P}$, which of course contradicts the maximality of F. We will now show that c can indeed be so selected.

The requirement is that

$$F(x) + \alpha c = g(x + \alpha y) \leqq p(x + \alpha y)$$

for all $x \in$ dom F, $\alpha \in R^1$. Since F is linear and p is positive homogeneous, we see that the above requirement amounts to

$$F\left(\frac{x}{\alpha}\right) + c \leqq p\left(\frac{x}{\alpha} + y\right) \text{ for } x \in \operatorname{dom} F \text{ and } \alpha > 0,$$

and

$$F\left(\frac{x}{\alpha}\right) + c \geqq -p\left(-\frac{x}{\alpha} - y\right) \text{ for } x \in \operatorname{dom} F \text{ and } \alpha < 0.$$

Thus it is sufficient to have that

$$F(w) - p(w - y) \leqq c \leqq -F(v) + p(v + y)$$

for all $w, v \in \operatorname{dom} F$. But we do have

$$F(w) + F(v) = F(w + v) \leqq p(w + v) \leqq p(w - y) + p(v + y)$$

for all $w, v \in \operatorname{dom} F$. We let

$$a = \sup\{F(w) - p(w - y) : w \in \operatorname{dom} F\}$$

and

$$b = \inf\{-F(v) + p(v + y) : v \in \operatorname{dom} F\}.$$

Evidently, $a \leqq b$ holds. Upon taking c to be any real number with $a \leqq c \leqq b$, the construction is finished and the proof is complete.

The following proposition is a direct consequence of the Hahn-Banach Theorem.

PROPOSITION 12. *Let E and M be as in Proposition 11. If $f \in M^*$, then there exists $F \in E^*$ such that F extends f and $\|F\| = \|f\|$.*

Proof. Define p on E by

$$p(x) = \|f\|\|x\|.$$

Then p is subadditive and positive homogeneous on E and we have

$$f(x) \leqq |f(x)| \leqq p(x)$$

for all $x \in M$. By Proposition 11, there exists a linear functional F on E which extends f and $F(x) \leqq p(x)$ for $x \in E$. Moreover, it is clear that $F \in E^*$ and $\|F\| \leqq \|f\|$. To show that $\|F\| = \|f\|$, we observe that

$$\begin{aligned} \|F\| &= \sup\{|F(x)| : x \in E, \|x\| \leqq 1\} \\ &\geqq \sup\{|F(x)| : x \in M, \|x\| \leqq 1\} \\ &= \sup\{|f(x)| : x \in M, \|x\| \leqq 1\} = \|f\| \end{aligned}$$

and so the proof is finished.

Next, we consider the complex version of the Hahn-Banach theorem.

PROPOSITION 13. *Suppose that E is a linear space over the complex number field. Let M be a linear subspace of E and let p be a real-valued function on E with the following properties*:

(*i*) $p(x + y) \leqq p(x) + p(y)$

(*ii*) $p(\alpha x) = |\alpha| p(x)$

for all $x, y \in E$ and all complex numbers α. If f is a linear functional on M such that

$$|f(x)| \leqq p(x)$$

for all $x \in M$, then there exists a linear functional F which is an extension of f to all of E such that

$$|F(x)| \leqq p(x)$$

for all $x \in E$.

Proof. The idea of the proof is to reduce the complex case to the real case and apply Proposition 11. We do this now. Write

$$f(x) = \operatorname{Re} f(x) + i \operatorname{Im} f(x)$$

for $x \in M$, where $\operatorname{Re} f(x)$ and $\operatorname{Im} f(x)$ denote the real and imaginary parts of $f(x)$, respectively. Since $f(ix) = if(x)$, we see that

$$\operatorname{Re} f(ix) + i \operatorname{Im} f(ix) = -\operatorname{Im} f(x) + i \operatorname{Re} f(x).$$

Thus $\operatorname{Im} f(x) = -\operatorname{Re} f(ix)$ and

$$(*) \qquad f(x) = \operatorname{Re} f(x) - i \operatorname{Re} f(ix).$$

Let E_r be E considered as a linear space over the reals. As sets, $E = E_r$. Now $\operatorname{Re} f$ is a linear functional on M considered as a subspace of E_r and

$$|\operatorname{Re} f(x)| \leqq |f(x)| \leqq p(x)$$

for all $x \in M$. Hence, by what we have already proved, there exists a linear extension G of $\operatorname{Re} f$ to all of E_r such that

$$|G(x)| \leqq p(x)$$

for $x \in E_r$. In view of equation (*), define F on E by

$$F(x) = G(x) - i\,G(ix)$$

for $x \in E$. A simple calculation shows that F is a linear extension of f to all of E. Given $x \in E$, write $F(x)$ in polar form

$$F(x) = |f(x)|e^{i\theta}.$$

Then

$$|F(x)| = F(e^{-i\theta}x) = \text{Re } F(e^{-i\theta}x) = G(e^{-i\theta}x) \leqq p(e^{-i\theta}x) = p(x).$$

Thus the proposition is proved.

In analogy to Proposition 12 we have the following proposition.

PROPOSITION 14. *Let E be a complex normed linear space and let M be a linear subspace of E. If $f \in M^*$, then there exists $F \in M^*$ such that F extends f and $\|F\| = \|f\|$.*

PROPOSITION 15. *Let M be a linear subspace of a normed linear space E. Given $z \in E$ and*

$$\text{dist}(z, M) = \inf_{m \in M} \|z - m\| = d > 0,$$

then there exists $F \in E^*$ such that $F(M) = \{0\}$, $F(z) = d$, and $\|F\| = 1$. In particular, if $M = \{0\}$, then $F(z) = \|z\|$.

Proof. Let

$$S = \{m + \alpha z : m \in M, \alpha \text{ being any scalar}\}.$$

Then S is a linear subspace of E. Define f on S by

$$f(m + \alpha z) = \alpha d.$$

It can be seen that f is a well-defined linear functional on S such that $f(S) = \{0\}$ and $f(z) = d$. Also

$$\begin{aligned} \|f\| &= \sup\left\{ \frac{|\alpha d|}{\|m + \alpha z\|} : m + \alpha z \in S, \|m + \alpha z\| \neq 0 \right\} \\ &= \sup\left\{ \frac{d}{\|-y + z\|} : y \in M \right\} \\ &= \frac{d}{d} = 1. \end{aligned}$$

Using the Hahn-Banach Theorem (either Proposition 11 or Proposition 13, depending on whether the underlying scalar field is the real or the complex number field) we obtain the required functional $F \in E^*$.

Remarks. Let π denote the natural mapping $x \to \hat{x}$ of E into E^{**}. As we have already seen, π is a bounded linear operator from E into E^{**} and

$\|\pi\| \leqq 1$. Let x be any nonzero element of E. Using Proposition 15 we see that there is an element $F \in E^*$ such that $\|F\| = 1$ and $F(x) = \|x\|$. Thus

$$\|x\| = F(x) \leqq \sup\{|f(x)| : f \in E^*, \|f\| = 1\} = \|\pi(x)\| \leqq \|x\|$$

and so $\|\pi(x)\| = \|x\|$. Moreover $\|\pi(0)\| = 0 = \|0\|$. This means that π is a norm-preserving linear operator from E into E^{**}. Moreover, if $x \neq y$ in E, then

$$\|\pi(x) - \pi(y)\| = \|\pi(x - y)\| = \|x - y\| \neq 0$$

and so π is seen to be a one-to-one map. We see therefore that a normed linear space E is a subspace of E^{**}; if the mapping π is onto E^{**}, the space E is said to be *reflexive*. Since E^{**} is complete and π is an isometry, i.e., π preserves distance, every reflexive normed linear space is a Banach space. For example, the L^p spaces with $1 < p < \infty$ are reflexive. Finally, using the concept of the natural mapping π of E into E^{**} we can easily see that every normed linear space can be embedded in a complete normed linear space, namely, in its second conjugate space.

Let K denote the field of real or complex numbers. A linear space E over K is called a *pre-Hilbert space* if there is defined a function taking $E \times E$ into K, denoted by $\langle\ ,\ \rangle$ and called *inner product*, which satisfies the postulates

(1) $\langle x, x\rangle \geqq 0$; $\langle x, x\rangle = 0$ only if $x = 0$;
(2) $\langle y, x\rangle = \overline{\langle x, y\rangle}$;
(3) $\langle \alpha x, y\rangle = \alpha\langle x, y\rangle$;
(4) $\langle x_1 + x_2, y\rangle = \langle x_1, y\rangle + \langle x_2, y\rangle$

for all x, x_1, x_2, y in E and all $\alpha \in K$; here $\overline{\langle x, y\rangle}$ denotes the complex conjugate of $\langle x, y\rangle$.

It is easy to see that the inner product satisfies the additional conditions:

$$\langle x, \alpha y\rangle = \bar{\alpha}\langle x, y\rangle$$

and

$$\langle x, y_1 + y_2\rangle = \langle x, y_1\rangle + \langle x, y_2\rangle.$$

PROPOSITION 16 (INEQUALITY OF CAUCHY-SCHWARZ). *In any pre-Hilbert space we have*

$$|\langle x, y\rangle|^2 \leqq \langle x, x\rangle\langle y, y\rangle.$$

Proof. For $y = 0$, the inequality is obvious. If $y \neq 0$, then

$$\langle x,x\rangle - \bar{\alpha}\langle x,y\rangle - \alpha\overline{\langle x,y\rangle} + \alpha\bar{\alpha}\langle y,y\rangle = \langle x - \alpha y, x - \alpha y\rangle \geqq 0$$

implies for $\alpha = \langle x,y\rangle/\langle y,y\rangle$ the inequality in question and the proof is finished.

In a pre-Hilbert space one can introduce a norm by setting

$$\|x\| = \sqrt{\langle x,y\rangle}\,.$$

Indeed, $\|x\| \geqq 0$ and $\|x\| = 0$ if and only if $x = 0$. Moreover

$$\|\alpha x\| = |\alpha|\|x\|.$$

To show that $\|x + y\| \leqq \|x\| + \|y\|$, we note that (using the inequality of Cauchy-Schwarz):

$$\begin{aligned}\|x + y\|^2 &= \langle x + y, x + y\rangle = \langle x,x\rangle + \langle x,y\rangle + \langle y,x\rangle + \langle y,y\rangle \\ &\leqq \langle x,x\rangle + 2\sqrt{\langle x,x\rangle\langle y,y\rangle} + \langle y,y\rangle = (\|x\| + \|y\|)^2.\end{aligned}$$

PROPOSITION 17. *In a pre-Hilbert space the following two identities hold*:
(*i*) $\|x + y\|^2 + \|x - y\|^2 = 2(\|x\|^2 + \|y\|^2)$;
(*ii*) $\langle x,y\rangle = \frac{1}{4}\{\|x + y\|^2 - \|x - y\|^2 + i\|x + iy\|^2 - i\|x - iy\|^2\}$.
Identity (*i*) *is called the "parallelogram law" and identity* (*ii*) *is called the "polarization identity."*

Proof. To show (i):

$$\begin{aligned}\|x + y\|^2 &= \langle x + y, x + y\rangle = \langle x,x\rangle + \langle x,y\rangle + \langle y,x\rangle + \langle y,y\rangle \\ &= \|x\|^2 + \|y\|^2 + \langle x,y\rangle + \langle y,x\rangle;\end{aligned}$$

replacing y by $-y$, we get

$$\|x - y\|^2 = \|x\|^2 + \|y\|^2 - \langle x,y\rangle - \langle y,x\rangle.$$

To show (ii): Since

$$\|x + y\|^2 = \|x\|^2 + \|y\|^2 + \langle x,y\rangle + \langle y,x\rangle,$$

we get on replacing y by $-y$, iy, and $-iy$, respectively,

$$\|x - y\|^2 = \|x\|^2 + \|y\|^2 - \langle x,y\rangle - \langle y,x\rangle;$$

$$\|x + iy\|^2 = \|x\|^2 + \|y\|^2 - i\langle x,y\rangle + i\langle y,x\rangle;$$

$$\|x - iy\|^2 = \|x\|^2 + \|y\|^2 + i\langle x,y\rangle - i\langle y,x\rangle.$$

This completes the proof.

PROPOSITION 18 (V. NEUMANN-JORDAN). *Let E be a normed linear space. If the norm on E satisfies the "parallelogram law," i.e., if for all $x, y \in E$ we have*

$$\|x + y\|^2 + \|x - y\|^2 = 2(\|x\|^2 + \|y\|^2),$$

then there is a unique inner product on $E \times E$ into K such that

$$\|x\| = \sqrt{\langle x, x\rangle}$$

for all $x \in E$.

Proof. Take

$$\langle x, y\rangle = \tfrac{1}{4}\{\|x + y\|^2 - \|x - y\|^2 + i\|x + iy\|^2 - i\|x - iy\|^2\}.$$

Then clearly

$$\langle x, x\rangle = \|x\|^2$$

for every $x \in E$. Moreover, it is easy to see that

$$\langle ix, y\rangle = i\langle x, y\rangle$$

and

$$\langle x, iy\rangle = -i\langle x, y\rangle.$$

Using the "parallelogram law," for any complex number α, we have

$$\begin{aligned}\langle \alpha x, y\rangle &+ \langle x, \alpha y\rangle \\ &= \tfrac{1}{8}\Gamma\{\|x + y\|^2 - \|x - y\|^2 + i\|x + iy\|^2 - i\|x - iy\|^2\} \\ &= (\alpha + \overline{\alpha})\langle x, y\rangle,\end{aligned}$$

where

$$\Gamma = |\alpha + 1|^2 - |\alpha - 1|^2;$$

note that

$$|\alpha + 1|^2 - |\alpha - 1|^2 = 2(\alpha + \overline{\alpha})$$

for any complex number α. Putting $\alpha = i\lambda$ for an arbitrary real λ, we obtain

$$\langle i\lambda x, y\rangle + \langle x, i\lambda y\rangle = 0$$

and hence $\langle \lambda x, y\rangle = \langle x, \lambda y\rangle$ for every real number λ. Putting $\alpha = \lambda$ for an arbitrary real number λ, we obtain therefore

$$2\langle \lambda x, y\rangle = \langle \lambda x, y\rangle + \langle x, \lambda y\rangle = 2\lambda\langle x, y\rangle$$

and hence $\langle \lambda x, y\rangle = \lambda\langle x, y\rangle$ for any real number λ. We conclude that

$$\langle \alpha x, y\rangle = \alpha\langle x, y\rangle$$

for every complex number α. Using the "parallelogram law" we conclude further that

$$\langle x, z\rangle + \langle y, z\rangle = \tfrac{1}{2}\langle x + y, 2z\rangle = \langle x + y, z\rangle.$$

Thus $\langle x, y\rangle$, for $x, y \in E$, is an inner product such that

$$\langle x, x\rangle = \|x\|^2$$

for every $x \in E$. The uniqueness of such an inner product is evident from the polarization identity and the proof is complete.

A complete pre-Hilbert space is called a *Hilbert space*.

It is clear that every Hilbert space is also a Banach space. In view of what we have just proved, we can see that a Banach space is a Hilbert space if and only if its Banach space norm satisfies the "parallelogram law."

In an L^2 space we can introduce in a natural way an inner product by defining

$$\langle f, g\rangle = \sqrt{\int fg}$$

(if f and g are complex-valued, then we set

$$\langle f, g\rangle = \sqrt{\int f\bar{g}}).$$

In the second half of Section 4 of Chapter 9 we already had occasion to consider some of the Hilbert space properties of L^2.

Let x and y be vectors of a pre-Hilbert space $\mathfrak{H}$. We say that x is *orthogonal* to y and write $x \perp y$ if $\langle x, y\rangle = 0$; if $x \perp y$, then $y \perp x$, and $x \perp x$ if and only if $x = 0$. Let $\mathfrak{M}$ be a subset of a pre-Hilbert space $\mathfrak{H}$. We denote by $\mathfrak{M}^\perp$ the set of all vectors x in $\mathfrak{H}$ orthogonal to every vector of $\mathfrak{M}$. $\mathfrak{M}^\perp$ is sometimes also denoted by $\mathfrak{H} \ominus \mathfrak{M}$. $\mathfrak{M}^\perp$ is clearly a closed subspace.

PROPOSITION 19. *Let $\mathfrak{M}$ be a closed linear subspace of a Hilbert space $\mathfrak{H}$. Then $\mathfrak{M}^\perp$ is also a closed linear subspace of $\mathfrak{H}$, and $\mathfrak{M}^\perp$ is called the orthogonal complement of $\mathfrak{M}$. Any vector $x \in \mathfrak{H}$ can be decomposed uniquely in the form*

$$x = m + n, \text{ where } m \in \mathfrak{M} \text{ and } n \in \mathfrak{M}^\perp. \tag{1}$$

The element m in (1) *is called the orthogonal projection of x onto $\mathfrak{M}$ and will be denoted by $P_{\mathfrak{M}}(x)$; $P_{\mathfrak{M}}$ is called the projection operator onto $\mathfrak{M}$. We have thus, because $\mathfrak{M} \subseteqq (\mathfrak{M}^\perp)^\perp$,*

$$x = P_{\mathfrak{M}}(x) + P_{\mathfrak{M}^\perp}(x), \text{ that is, } I = P_{\mathfrak{M}} + P_{\mathfrak{M}^\perp}, \tag{1'}$$

with I the identity operator.

Proof. The linearity of $\mathfrak{M}^\perp$ is a consequence of the linearity in x of the inner product $\langle x,y \rangle$. $\mathfrak{M}^\perp$ is closed because the inner product is a continuous mapping. The uniqueness of the decomposition (1) is clear since a vector orthogonal to itself is the zero vector.

To show the possibility of the decomposition (1) we may assume that $\mathfrak{M} \neq \mathfrak{H}$ and $x \notin \mathfrak{M}$, for, if $x \in \mathfrak{M}$, we have the trivial decomposition with $m = x$ and $n = 0$. Thus, since $\mathfrak{M}$ is closed and $x \notin \mathfrak{M}$, we have

$$d = \sup_{m \in \mathfrak{M}} \|x - m\| > 0.$$

Let $(m_k)_{k=1}^\infty \subseteq \mathfrak{M}$ be a minimizing sequence, i.e.,

$$\lim_{k \to \infty} \|x - m_k\| = d.$$

Then $(m_k)_{k=1}^\infty$ is a Cauchy sequence. Indeed, by the parallelogram law we obtain

$$\begin{aligned}
\|m_k - m_j\|^2 &= \|(x - m_j) - (x - m_k)\|^2 \\
&= 2(\|x - m_j\|^2 + \|x - m_k\|^2) - \|2x - m_j - m_k\|^2 \\
&= 2(\|x - m_j\|^2 + \|x - m_k\|^2) \\
&\quad - 4\|x - (m_j + m_k)/2\|^2 \\
&\leqq 2(\|x - m_j\|^2 + \|x - m_k\|^2) - 4d^2 \\
&\quad \left(\text{since } \frac{m_j + m_k}{2} \in \mathfrak{M}\right).
\end{aligned}$$

By the completeness of the Hilbert space $\mathfrak{H}$, there exists an element $m \in \mathfrak{H}$ such that

$$\lim_{j\to\infty} m_j = m.$$

We have $m \in \mathfrak{M}$ since $\mathfrak{M}$ is closed. Also, by the continuity of the norm, we have $\|x - m\| = d$.

Write $x = m + (x - m)$. Putting $n = x - m$, we have to show that $n \in \mathfrak{M}^\perp$. For any $m' \in \mathfrak{M}$ and any real number α, we have $(m + \alpha m') \in \mathfrak{M}$ and so

$$\begin{aligned} d^2 &\leqq \|x - m - \alpha m'\|^2 = \langle n - \alpha m', n - m' \rangle \\ &= \|n\|^2 - \alpha\langle n, m'\rangle - \alpha\langle m', n\rangle + \alpha^2 \|m'\|^2 \end{aligned}$$

Since $\|n\| = d$, this gives

$$0 \leqq -2\alpha \operatorname{Re} \langle n, m\rangle + \alpha^2 \|m'\|^2$$

for every real α. Hence $\operatorname{Re} \langle n, m'\rangle = 0$ for every $m' \in \mathfrak{M}$. Replacing m' by im', we obtain $\operatorname{Im} \langle n, m'\rangle = 0$ and so $\langle n, m'\rangle = 0$ for every $m' \in \mathfrak{M}$. This completes the proof.

Remark. The foregoing proposition implies the corollary: If $\mathfrak{M}$ is a closed linear subspace of $\mathfrak{H}$, then $\mathfrak{M} = (\mathfrak{M}^\perp)^\perp = \mathfrak{M}^{\perp\perp}$.

PROPOSITION 20. *The projection operator $P = P_{\mathfrak{M}}$ is a bounded linear operator such that*

$$P = P^2 \text{ (idempotent)} \tag{2}$$

$$\langle Px, y\rangle = \langle x, Py\rangle \text{ (symmetric)}. \tag{3}$$

Conversely, a bounded linear operator P on a Hilbert space $\mathfrak{H}$ into itself satisfying (2) and (3) is a projection operator onto $\mathfrak{M}$ = range of P.

Proof. (2) is clear from the definition of the orthogonal projection. We have, by (1′) and $P_{\mathfrak{M}}(x) \perp P_{\mathfrak{M}^\perp}(y)$,

$$\begin{aligned} \langle P_{\mathfrak{M}}(x), y\rangle &= \langle P_{\mathfrak{M}}(x), P_{\mathfrak{M}^\perp}(y) + P_{\mathfrak{M}}(y)\rangle = \langle P_{\mathfrak{M}}(x), P_{\mathfrak{M}}(y)\rangle \\ &= \langle P_{\mathfrak{M}}(x) + P_{\mathfrak{M}^\perp}(x), P_{\mathfrak{M}}(y)\rangle = \langle x, P_{\mathfrak{M}}(y)\rangle. \end{aligned}$$

Next let $y = x + z$, $x \in \mathfrak{M}$, $z \in \mathfrak{M}^\perp$ and $w = u + v$, $u \in \mathfrak{M}$, $v \in \mathfrak{M}^\perp$, then $y = w = (x + u) + (z + v)$ with $x + u \in \mathfrak{M}$, $z + v \in \mathfrak{M}^\perp$ and so, by the uniqueness of the decomposition (1),

$$P_{\mathfrak{M}}(y + w) = P_{\mathfrak{M}}(y) + P_{\mathfrak{M}}(w);$$

similarly we obtain $P_{\mathfrak{M}}(\alpha y) = \alpha P_{\mathfrak{M}}(y)$. The boundedness of the operator $P_{\mathfrak{M}}$ is proved by

$$\begin{aligned}\|x\| &= \|P_{\mathfrak{M}}(x) + P_{\mathfrak{M}^\perp}(x)\|^2 \\ &= \langle P_{\mathfrak{M}}(x) + P_{\mathfrak{M}^\perp}(x), P_{\mathfrak{M}}(x) + P_{\mathfrak{M}^\perp}(x)\rangle \\ &= \|P_{\mathfrak{M}}(x)\|^2 + \|P_{\mathfrak{M}^\perp}(x)\|^2 \geqq \|P_{\mathfrak{M}}(x)\|^2.\end{aligned}$$

Hence, in particular, we have

$$\|P_{\mathfrak{M}}\| \leqq 1. \tag{4}$$

The converse part of the proposition is proved as follows. The set $\mathfrak{M}$ = range of P is a linear subspace, since P is a linear operator. The statement $x \in \mathfrak{M}$ is equivalent to the existence of a certain $y \in \mathfrak{H}$ such that $x = P(y)$ and this in turn is equivalent by (2) to $x = P(y) = P^2(y) = P(x)$. Therefore $x \in \mathfrak{M}$ is equivalent to $x = P(x)$. $\mathfrak{M}$ is a closed subspace; for $x_n \in \mathfrak{M}$,

$$\lim_{n\to\infty} x_n = y$$

imply by the continuity of P and $x_n = P(x_n)$:

$$\lim_{n\to\infty} x_n = \lim_{n\to\infty} P(x_n) = P(y) \text{ so that } y = P(y).$$

We have to verify that $P = P_{\mathfrak{M}}$. If $x \in \mathfrak{M}$, we have $P(x) = x = P_{\mathfrak{M}}(x)$; and if $y \in \mathfrak{M}^\perp$, we have $P_{\mathfrak{M}}(y) = 0$. Moreover, in the latter case, $\langle P(y), P(y)\rangle = \langle y, P^2(y)\rangle = \langle y, P(y)\rangle = 0$ and so $P(y) = 0$. Thus we obtain, for any $y \in \mathfrak{H}$, $P(y) = P(P_{\mathfrak{M}}(y) + P_{\mathfrak{M}^\perp}(y)) = PP_{\mathfrak{M}}(y) + PP_{\mathfrak{M}^\perp}(y) = P_{\mathfrak{M}}(y) + 0$ that is $P(y) = P_{\mathfrak{M}}(y)$. This completes the proof.

PROPOSITION 21. *A bounded linear operator P on a Hilbert space $\mathfrak{H}$ into itself is a projection operator if and only if P satisfies $P = P^2$ and $\|P\| \leqq 1$.*

Proof. We need to prove the "if" part only. Set $\mathfrak{M}$ = range of P and $\mathfrak{N} = \{y : P(y) = 0\}$. As in the proof of Proposition 20, $\mathfrak{M}$ is a closed linear subspace and $x \in \mathfrak{M}$ is equivalent to $x = P(x)$. $\mathfrak{N}$ is also a closed linear subspace by force of the continuity of P. In the decomposition $x = P(x) + (I - P)(x)$, we have $P(x) \in \mathfrak{N}$ and $(I - P)(x) \in \mathfrak{N}$. The latter assertion is clear from $P(I - P) = P - P^2 = 0$.

We thus have to prove that $\mathfrak{N} = \mathfrak{M}^\perp$. For every $x \in \mathfrak{H}$, $y = P(x) - x \in \mathfrak{N}$ by $P^2 = P$. Hence, if, in particular, $x \in \mathfrak{N}^\perp$, then $P(x) = x + y$ with $\langle x, y\rangle = 0$. It follows then that $\|x\|^2 \geqq \|Px\|^2 = \|x\|^2 + \|y\|^2$ so that $y = 0$. We have proved therefore that $x \in \mathfrak{N}^\perp$ implies $x = P(x)$, that is, $\mathfrak{N}^\perp \subseteqq \mathfrak{M}$ = range of P. Let, conversely, $z \in \mathfrak{M}$ = range of P, so that $z = P(z)$. Then we have the orthogonal decomposition $z = y + x$, $y \in \mathfrak{N}$, $x \in \mathfrak{N}^\perp$, and so $z = P(z) = P(y) + P(x) = P(x) - x$, the last equality being already proved.

This shows that $\mathfrak{M}$ = range of $P \subseteqq \mathfrak{N}^{\perp}$. We have thus obtained $\mathfrak{M} = \mathfrak{N}^{\perp}$, and so, by $\mathfrak{N} = (\mathfrak{N}^{\perp})^{\perp}$, $\mathfrak{N} = \mathfrak{M}^{\perp}$. The proof is finished.

PROPOSITION 22 (RIESZ REPRESENTATION THEOREM). *Let $\mathfrak{H}$ be a Hilbert space and f a bounded linear functional on $\mathfrak{H}$. Then there exists a uniquely determined vector y_f of $\mathfrak{H}$ such that*

$$f(x) = \langle x, y_f \rangle \text{ for all } x \in \mathfrak{H}, \text{ and } \|f\| = \|y_f\|.$$

Conversely, any vector $y \in \mathfrak{H}$ defines a bounded linear functional f_y on $\mathfrak{H}$ by

$$f_y(x) = \langle x, y \rangle \text{ for all } x \in \mathfrak{H}, \text{ and } \|f_y\| = \|y\|.$$

Proof. The uniqueness of y_f is clear since $\langle x, z \rangle = 0$ for all $x \in \mathfrak{H}$ implies $z = 0$. To prove its existence, consider the null space $\mathfrak{N} = \mathfrak{N}_f = \{x \in X : f(x) = 0\}$ of f. Since f is continuous and linear, $\mathfrak{N}$ is a closed linear subspace. The proposition is trivial in the case when $\mathfrak{N} = \mathfrak{H}$; we take in this case, $y_f = 0$. Suppose $\mathfrak{N} \neq \mathfrak{H}$. Then there exists a $y_0 \neq 0$ which belongs to $\mathfrak{N}^{\perp}$; this is so by Proposition 13. Define

$$y_g = \frac{\overline{f(y_0)}}{\|y_0\|^2} y_0$$

We shall show that this y_g meets the condition of the proposition. First, if $x \in \mathfrak{N}$, then $f(x) = \langle x, y_f \rangle$ since both sides vanish. Next if x is of the form $x = \alpha y_0$, then we have

$$\langle x, y_f \rangle = \langle \alpha y_0, y_f \rangle = \left\langle \alpha y_0, \frac{\overline{f(y_0)}}{\|y_0\|^2} y_0 \right\rangle = \alpha f(y_0) = f(\alpha y_0) = f(x).$$

Since $f(x)$ and $\langle x, y_f \rangle$ are both linear in x, the equality $f(x) = \langle x, y_f \rangle$, $x \in \mathfrak{H}$, is proved if we have shown that $\mathfrak{H}$ is spanned by $\mathfrak{N}$ and y_0. To show the last assertion, we write, recalling that $f(y_f) \neq 0$,

$$x = \left(x - \frac{f(x)}{f(y_f)} y_f \right) + \frac{f(x)}{f(y_f)} y_f.$$

The first summand is an element of $\mathfrak{N}$, since

$$f\left(x - \frac{f(x)}{f(y_f)} y_f \right) = f(x) - \frac{f(x)}{f(y_f)} f(y_f) = 0.$$

We have therefore proved the representation $f(x) = \langle x, y_f \rangle$.

Therefore, we have

$$\|f\| = \sup_{\|x\| \leqq 1} |f(x)| = \sup_{\|x\| \leqq 1} |\langle x, y_f \rangle| \leqq \sup_{\|x\| \leqq 1} \|x\| \cdot \|y_f\| = \|y_f\|,$$

and also

$$\|f\| = \sup_{\|x\| \leqq 1} |f(x)| \geqq |f(y_f)/\|y_f\|| = \left\langle \frac{y_f}{\|y_f\|}, y_f \right\rangle = \|y_f\|.$$

Hence we have verified that $\|f\| = \|y_f\|$.

Finally, the converse part of the proposition is clear from

$$|f_y(x)| = |\langle x, y\rangle| \leqq \|x\|\|y\|.$$

The proof is finished.

A set S of vectors in a pre-Hilbert space $\mathfrak{H}$ is called an *orthogonal set*, if $x \perp y$ for each pair of distinct vectors, $x, y \in S$. If, in addition, $\|x\| = 1$ for each $x \in S$, then the set S is called an *orthonormal set*. An orthonormal set S of a Hilbert space $\mathfrak{H}$ is called a *complete orthonormal set* or an *orthogonal base* of $\mathfrak{H}$, if no orthonormal set of $\mathfrak{H}$ contains S as a proper subset.

PROPOSITION 23. *A Hilbert space $\mathfrak{H}$ (having a nonzero vector) has at least one complete orthonormal system. Moreover, if S is any orthonormal set in $\mathfrak{H}$, there is a complete orthonormal system containing S as a subset.*

Proof. Let S be an orthonormal set in $\mathfrak{H}$. Such a set certainly exists; for example, if $x \neq 0$, the set consisting only of $x/\|x\|$ is orthonormal. We consider the totality $\mathscr{S}$ of orthonormal sets which contain S as a subset. $\mathscr{S}$ is partially ordered by writing $S_1 < S_2$ for the inclusion relation $S_1 \subset S_2$. Let $\mathscr{S}'$ be a linearly ordered subsystem of $\mathscr{S}$, then

$$\bigcup_{S' \in \mathscr{S}'} S'$$

is an orthonormal set and an upper bound of $\mathscr{S}'$. Thus, by Zorn's Lemma there exists a maximal element S_0 of $\mathscr{S}$. This orthonormal set S_0 contains S and by maximality it must be a complete orthonormal system. The proof is complete.

PROPOSITION 24. *Let $S = \{x_\alpha : \alpha \in \mathfrak{A}\}$ be a complete orthonormal system of a Hilbert space $\mathfrak{H}$. For any $f \in \mathfrak{H}$, we define its Fourier coefficients (with respect to S):*

$$f_\alpha = \langle f, x_\alpha \rangle. \tag{5}$$

Then we have Parseval's Equality:

$$\|f\|^2 = \sum_{\alpha \in \mathfrak{A}} |f_\alpha|^2. \tag{6}$$

Proof. First we prove *Bessel's Inequality*:

$$\sum_{\alpha \in \mathfrak{A}} |f_\alpha|^2 \leqq \|f\|^2. \tag{6'}$$

Let $\alpha_1, \alpha_2, \ldots, \alpha_n$ be any finite set of α's. For any finite set of complex numbers $c_{\alpha_1}, c_{\alpha_2}, \ldots, c_{\alpha_n}$, we have, by the orthonormality of $\{x_\alpha : \alpha \in \mathfrak{A}\}$,

$$\begin{aligned} \|f - \sum_{j=1}^{n} c_{\alpha_j} x_{\alpha_j}\|^2 &= \langle f - \sum_{j=1}^{n} c_{\alpha_j} x_{\alpha_j}, f - \sum_{j=1}^{n} c_{\alpha_j} x_{\alpha_j} \rangle \\ &= \|f\|^2 - \sum_{j=1}^{n} c_{\alpha_j} \bar{f}_{\alpha_j} - \sum_{j=1}^{n} \bar{c}_{\alpha_j} f_{\alpha_j} + \sum_{j=1}^{n} |c_{\alpha_j}|^2 \\ &= \|f\|^2 - \sum_{j=1}^{n} |f_{\alpha_j}|^2 + \sum_{j=1}^{n} |f_{\alpha_j} - c_{\alpha_j}|^2. \end{aligned}$$

Hence the minimum of

$$\|f - \sum_{j=1}^{n} c_{\alpha_j} x_{\alpha_j}\|^2,$$

for fixed $\alpha_1, \alpha_2, \ldots, \alpha_n$, is attained when $c_{\alpha_j} = f_{\alpha_j}$ $(j = 1, 2, \ldots, n)$. We have thus

$$\begin{aligned} &\|f - \sum_{j=1}^{n} f_{\alpha_j} x_{\alpha_j}\|^2 = \|f\|^2 - \sum_{j=1}^{n} |f_{\alpha_j}|^2 \\ &\text{and hence } \sum_{j=1}^{n} |f_{\alpha_j}|^2 \leqq \|f\|^2. \end{aligned} \tag{7}$$

By the arbitrariness of $\alpha_1, \alpha_2, \ldots, \alpha_n$, we see that Bessel's Inequality (6′) is true, and $f_\alpha \neq 0$ for at most a countable number of α's, say $\alpha_1, \alpha_2, \ldots$. We then show that

$$f = \lim_{n \to \infty} \sum_{j=1}^{n} f_{\alpha_j} x_{\alpha_j}.$$

First, the sequence

$$\left(\sum_{j=1}^{n} f_{\alpha_j} x_{\alpha_j}\right)_{n=1}^{\infty}$$

is a Cauchy sequence, since, by the orthonormality of $(x_\alpha)_{\alpha \in \mathfrak{A}}$.

$$\|\sum_{j=k}^{n} f_{\alpha_j} x_{\alpha_j}\|^2 = \sum_{j=k}^{n} |f_{\alpha_j}|^2$$

which tends, by (7) proved above, to 0 as $k \to \infty$. We set

$$\lim_{n \to \infty} \sum_{j=1}^{n} f_{\alpha_j} x_{\alpha_j} = f',$$

and prove that $(f - f')$ is orthogonal to every vector of S. By the continuity of the inner product, we have

$$\langle f - f', x_{\alpha_j} \rangle = \lim_{n\to\infty} \langle f - \sum_{k=1}^{n} f_{\alpha_k} x_{\alpha_k}, x_{\alpha_k} \rangle = f_{\alpha_j} - f_{\alpha_j} = 0,$$

and, when $\alpha \neq \alpha_j$ $(j = 1, 2, \ldots)$,

$$\langle f - f', x_\alpha \rangle = \lim_{n\to\infty} \langle f - \sum_{k=1}^{n} f_{\alpha_k} x_{\alpha_k}, x_\alpha \rangle = 0 - 0 = 0.$$

Thus, by the completeness of the orthonormal system $S = \{x_\alpha : \alpha \in \mathfrak{A}\}$, we must have $(f - f') = 0$. Hence by (7) and the continuity of the norm, we have

$$\begin{aligned} 0 &= \lim_{n\to\infty} \|f - \sum_{j=1}^{n} f_{\alpha_j} x_{\alpha_j}\|^2 = \|f\|^2 - \lim_{n\to\infty} \sum_{j=1}^{n} |f_{\alpha_j}|^2 \\ &= \|f\|^2 - \sum_{\alpha\in\mathfrak{A}} |f_\alpha|^2 . \end{aligned}$$

The proof is complete.

Remarks. The set $\{(1/\sqrt{2\pi})e^{int} : n = 0, \pm 1, \pm 2, \ldots\}$ is a complete orthonormal system in $L^2([0, 2\pi])$. Bessel's Inequality implies that any vector f of a Hilbert space can have at most a countable number of nonzero Fourier coefficients with respect to a given orthonormal system. Indeed let S be any orthonormal system. Then the number of elements $s \in S$ such that $|\langle f, s\rangle| \geqq 1/n$ cannot exceed $n^2 \|f\|^2$ by Bessel's Inequality. Since $|\langle f, s\rangle| \geqq 1/n$ for some n if $\langle f, s\rangle \neq 0$, the set $s \in S$ such that $\langle f, s\rangle \neq 0$ is a countable union of finite sets.

The proof of the following proposition is left as an exercise.

PROPOSITION 25. *For any orthonormal system (x_α) in a Hilbert space $\mathfrak{H}$ the following statements are equivalent*:

(*i*) *(x_α) is complete in $\mathfrak{H}$*;

(*ii*) *every vector $f \in \mathfrak{H}$ can be represented in the form*

$$f = \sum_\alpha f_\alpha x_\alpha,$$

where

$$f_\alpha = \langle f, x_\alpha \rangle;$$

(*iii*) *for every vector $f \in \mathfrak{H}$ we have*

$$\|f\|^2 = \sum_\alpha |f_\alpha|^2;$$

(*iv*) *for every two vectors* f, $g \in \mathfrak{H}$ *we have*

$$\langle f, g\rangle = \sum_{\alpha} f_\alpha g_\alpha,$$

where

$$f_\alpha = \langle f, x_\alpha\rangle$$

and

$$g_\alpha = \langle g, x_\alpha\rangle.$$

PROPOSITION 26. *Every complete orthonormal system in a Hilbert space has the same cardinality (called dimension of the space).*

Proof. In a space of finite dimension a complete orthonormal system forms a basis of the linear space so that the number of system vectors equals the dimension of the space. Hence we can restrict attention in what follows here to Hilbert spaces $\mathfrak{H}$ which are not finite dimensional.

Let (x_α) and (x'_β) be two complete orthonormal systems in $\mathfrak{H}$ and let a and b denote their cardinality, respectively. For fixed α the inner product $\langle x_\alpha, x'_\beta\rangle$ has to be different from zero for at least one β because (x'_β) is a complete orthonormal system. On the other hand, there can be at most a countable number of β's such that $\langle x_\alpha, x'_\beta\rangle \neq 0$ with α fixed. We associate with each x_α those x'_β for which $\langle x_\alpha, x'_\beta\rangle$ is not zero. This correspondence from (x_α) onto (x'_β) is at least single-valued and at most countably multivalued. Hence $a \geqq b$. By the symmetry of the situation $a \leqq b$, and hence $a = b$. Note: The correspondence $(x_\alpha) \to (x'_\beta)$ is onto for if it was not, then there would exist a (necessarily nonzero) vector in the system which would be orthogonal to the complete system (x_α); but this is impossible.

PROPOSITION 27. *Two Hilbert spaces are isometric if and only if they have the same dimension.*

Proof. The condition is of course necessary. To see that the condition is sufficient, suppose that $\mathfrak{H}'$ and $\mathfrak{H}''$ have the same dimension. Let (x'_α) and (x''_β) be complete orthonormal systems in $\mathfrak{H}'$ and $\mathfrak{H}''$, respectively. Since both systems have the same cardinality, we can choose for both systems the same indexing set. To each vector

$$f' = \sum_{\alpha} f'_\alpha x'_\alpha$$

from $\mathfrak{H}'$ we associate the vector

$$f'' = \sum_{\alpha} f'_{\alpha} x''_{\alpha}$$

of $\mathfrak{H}''$, where we used the same Fourier coefficients. This correspondence of $\mathfrak{H}'$ onto $\mathfrak{H}''$ is isometric.

PROPOSITION 28. *There are Hilbert spaces of any dimension.*

Proof. Let S be a nonempty set of cardinal a. Consider the Cartesian product

$$\bigtimes_{s \in S} C_s$$

where C_s is a copy of the complex plane. In other words,

$$\bigtimes_{s \in S} C_s$$

is the set of all functions x on S such that $x(s) \in C_s$. We now define a subset of

$$\bigtimes_{s \in S} C_s$$

by requiring that

(1) the set of s for which $x(s) \neq 0$ is at most countable,
(2) the series

$$\sum_{s \in S} |x(s)|^2$$

converges.

We denote by l_a^2 this subset. Under the usual pointwise linear operations

$$(x + y)(s) = x(s) + y(s)$$

$$(\lambda x)(s) = \lambda(x(s))$$

l_a^2 forms a linear space.

Let

$$\langle x, y \rangle = \sum_{s \in S} x(s)\overline{y(s)}$$

(its convergence follows from $|x(s)y(s)| \leqq (1/2)(|x(s)|^2 + |y(s)|^2))$, so that l_a^2 is a Hilbert space in which the functions e_{s_0} (as s_0 runs through S)

$$e_{s_0}(s) = \begin{cases} 1 \text{ when } s = s_0 \\ 0 \text{ when } s \neq s_0 \end{cases}$$

form a complete orthonormal system, so that l_a^2 has in fact dimension a. This completes the construction. Q.E.D.

Let $\mathfrak{H}$ be a Hilbert space over the field K of real or complex numbers. A nonempty set M of vectors in $\mathfrak{H}$ is termed a *linear manifold* in $\mathfrak{H}$ if, for every pair of vectors $f, g \in M$ and every pair of scalars $\alpha, \beta \in K$, the vector $\alpha f + \beta g$ also belongs to M. In other words, a linear manifold M in $\mathfrak{H}$ is a linear subspace of $\mathfrak{H}$ which is not necessarily closed.

By the *opening* of two linear manifolds in a Hilbert space $\mathfrak{H}$ we mean the norm of the difference of the projection operators which map $\mathfrak{H}$ onto the closure of these linear manifolds.

Denoting the opening of the linear manifolds M_1 and M_2 by $Q(M_1, M_2)$, we therefore have

$$Q(M_1, M_2) = \|P_1 - P_2\|,$$

where P_1 and P_2 are the projection operators which map onto the subspaces $\overline{M}_1$ and $\overline{M}_2$, respectively. By definition

$$Q(M_1, M_2) = Q(\overline{M}_1, \overline{M}_2) = Q(\mathfrak{H} \ominus M_1, \mathfrak{H} \ominus M_2).$$

Let I denote the identity operator. For any $h \in \mathfrak{H}$ we have (omitting brackets around the variable)

$$(P_2 - P_1)h = P_2(I - P_1)h - (I - P_2)P_1 h.$$

Since the vectors $P_2(I - P_1)h$ and $(I - P_2)P_1 h$ are orthogonal, we see that

$$\|(P_2 - P_1)h\|^2 = \|P_2(I - P_1)h\|^2 + \|(I - P_2)P_1 h\|^2$$
$$\leqslant \|(I - P_1)h\|^2 + \|P_1 h\|^2 = \|h\|^2$$

holds. The inequality (1) shows that

$$0 \leqslant Q(M_1, M_2) \leqslant 1.$$

The opening of two linear manifolds is actually equal to 1 if one of these manifolds contains a nonzero vector which is orthogonal to the other manifold.

PROPOSITION 29. *If the opening of two linear manifolds M_1 and M_2 is less than 1, then*

$$\dim M_1 = \dim M_2.$$

(*By the dimension of a linear manifold M in $\mathfrak{H}$ we mean the dimension of its closure $\overline{M}$ in $\mathfrak{H}$.*)

Proof. By the remark made above, it suffices to show that the inequality $\dim M_2 > \dim M_1$ implies the existence of a nonzero vector in $\overline{M}_2$ which is orthogonal to $\overline{M}_1$. To see that this is actually the case we project $\overline{M}_1$ onto $\overline{M}_2$. We obtain the subspace $G = P_2\overline{M}_1$ whose dimension, of course, does not exceed the dimension of $\overline{M}_1$ and therefore is less than the dimension of $\overline{M}_2$. Hence there is in $\overline{M}_2 \ominus G$ a nonzero vector, which means that in $\overline{M}_2$ there is a nonzero vector that is orthogonal to G. This vector will also be orthogonal to $\overline{M}_1$ because $\overline{M}_1 \ominus G$ is orthogonal to $\overline{M}_2$. This proves the proposition.

The next object for consideration is the formula

$$Q(M_1, M_2) = \max\{\sup_{f\in\overline{M}_2, \|f\|=1} \|(I - P_1)f\|, \sup_{g\in\overline{M}_1, \|g\|=1} \|(I - P_2)g\|\}. \tag{8}$$

We note first of all that the quantity

$$\|(I - P_1)f\| = \operatorname{dist}[f, \overline{M}_1]$$

represents the distance between the element f and $\overline{M}_1$:

$$\operatorname{dist}[f, \overline{M}_1] = \inf_{g\in M_1} \|f - g\|$$

and therefore we can express (8) in the form

$$Q(M_1, M_2) = \max\{\sup_{f\in\overline{M}_2, \|f\|=1} \operatorname{dist}[f, \overline{M}_1], \sup_{g\in\overline{M}_1, \|g\|=1} \operatorname{dist}[g, \overline{M}_2]\}.$$

We now turn to establishing equation (8).

$$\begin{aligned} Q(M_1, M_2) &= \sup_{h\in\mathfrak{H}} \frac{\|(P_2 - P_1)h\|}{\|h\|} \\ &= \sup_{h\in\mathfrak{H}} \frac{\sqrt{\|P_2(I - P_1)h\|^2 + \|(I - P_2)P_1h\|^2}}{\|h\|}. \end{aligned} \tag{9}$$

Therefore,

$$\begin{aligned} Q(M_1, M_2) &\geqslant \sup_{h\in\overline{M}_1} \frac{\sqrt{\|P_2(I - P_1)h\|^2 + \|(I - P_2)P_1h\|^2}}{\|h\|} \\ &= \sup_{h\in\overline{M}_1} \frac{\|(I - P_2)h\|}{\|h\|} = r_2. \end{aligned}$$

In the same manner we obtain

$$Q(M_1, M_2) \geqslant \sup_{h \in \overline{M}_1} \frac{\|(I - P_1)h\|}{\|h\|} = r_1.$$

Consequently

$$Q(M_1, M_2) \geqslant \max\{r_1, r_2\}.$$

We show next that the inequality sign in the last relation can be inverted. By the definition of the number r_2 we have

$$\|(I - P_2)P_1\|^2 \leqq r_2^2 \|P_1 h\|^2. \tag{10}$$

On the other hand

$$\begin{aligned} \|P_2(I - P_1)h\|^2 &= \langle P_2(I - P_1)h, P_2(I - P_1)h\rangle \\ &= \langle P_2(I - P_1)h, (I - P_1)h\rangle \\ &= \langle P_2(I - P_1)h, (I - P_1)^2 h\rangle \\ &= \langle (I - P_1)P_2(I - P_1)h, (I - P_1)h\rangle \\ &\leqq \|(I - P_1)P_2(I - P_1)h\| \cdot \|(I - P_1)h\| \end{aligned}$$

and hence by the definition of r_1

$$\|P_2(I - P_1)h\|^2 \leqq r_1^2 \|P_2(I - P_1)h\| \cdot \|(I - P_1)h\|.$$

Therefore

$$\|P_2(I - P_1)h\| \leqq r_1 \|(I - P_1)h\|. \tag{11}$$

Using (10) and (11) we get

$$\begin{aligned} \|(I - P_2)P_1 h\|^2 + \|P_2(I - P_1)h\|^2 &\leqq r_2^2 \|P_1 h\|^2 + r_1^2 \|(I - P_1)h\|^2 \\ &\leqq \max\{r_1^2, r_2^2\} \, [\|P_1 h\|^2 + \|(I - P_1)h\|^2] \\ &= \|h\|^2 \max\{r_1^2, r_2^2\} \end{aligned}$$

so that by (9):

$$Q(M_1, M_2) \leqq \max\{r_1, r_2\}$$

and we get what we set out to verify.

In the following we assume that $\mathfrak{H}$ is a Hilbert space over the complex number field.

We say that Ω is a *bilinear functional* on a Hilbert space $\mathfrak{H}$ if to every pair of elements $f, g \in \mathfrak{H}$ corresponds a complex number $\Omega(f,g)$ such that

(a) $\Omega(\alpha_1 f_1 + \alpha_2 f_2, g) = \alpha_1 \Omega(f_1, g) + \alpha_2 \Omega(f_2, g)$,

(b) $\Omega(f, \beta_1 g_1 + \beta_2 g_2) = \bar{\beta}_1 \Omega(f, g_1) + \bar{\beta}_2 \Omega(f, g_2)$,

(c) $\sup\{|\Omega(f,g)| : \|f\| \leqq 1, \|g\| \leqq 1\} < \infty$

holds.

An example of the bilinear functional is the inner product $\langle f,g \rangle$.

The number

$$\sup\{|\Omega(f,g)| : \|f\| \leqq 1, \quad \|g\| \leqq 1\}$$

is called the *norm of the bilinear functional* and is denoted by $\|\Omega\|$.

As a consequence of the Riesz Representation Theorem (see Proposition 22) we prove the following.

PROPOSITION 30. *Every bilinear functional Ω on a Hilbert space $\mathfrak{H}$ is of the form*

$$\Omega(f,g) = \langle A(f), g \rangle,$$

where A is a bounded linear operator mapping $\mathfrak{H}$ into itself; here A is determined uniquely by Ω and $\|A\| = \|\Omega\|$.

Proof. First we verify the uniqueness of the operator A. Suppose that for arbitrary $f, g \in \mathfrak{H}$ we had

$$\Omega(f,g) = \langle A'(f), g \rangle,$$

$$\Omega(f,g) = \langle A''(f), g \rangle,$$

then we would obtain for arbitrary $f, g \in \mathfrak{H}$ that

$$\langle A'(f) - A''(f), g \rangle = 0.$$

From this in turn it would follow that

$$A'(f) - A''(f) = 0,$$

i.e., $A' = A''$.

Now to the representation of Ω. We consider a fixed f. Then $\overline{\Omega(f,g)}$ is a linear functional defined on all of $\mathfrak{H}$ in the variable g. By Proposition 22 there exists an element h uniquely determined by f such that

$$\overline{\Omega(f,g)} = \langle g, h \rangle$$

or

$$\Omega(f, g) = \langle h, g \rangle$$

for arbitrary $g \in \mathfrak{H}$. To each $f \in \mathfrak{H}$ there corresponds an element h. Hence $h = A(f)$ and

$$\Omega(f, g) = \langle A(f), g \rangle.$$

Since

$$\Omega(\alpha_1 f_1 + \alpha_2 f_2, g) = \alpha_1 \Omega(f_1, g) + \alpha_2 \Omega(f_2, g),$$

we have for arbitrary $g \in \mathfrak{H}$,

$$\langle A(\alpha_1 f_1 + \alpha_2 f_2) - \alpha_1 A(f_1) - \alpha_2 A(f_2), g \rangle = 0.$$

Since $g \in \mathfrak{H}$ is arbitrary, we conclude that

$$A(\alpha_1 f_1 + \alpha_2 f_2) = \alpha_1 A(f_1) + \alpha_2 A(f_2)$$

and the linearity of A is established.

The domain of definition of the operator A is the entire space $\mathfrak{H}$. Since

$$|\langle A(f), g \rangle| \leqq \|A(f)\| \|g\|,$$

we have

$$\|\Omega\| = \sup \frac{|\Omega(f, g)|}{\|f\| \cdot \|g\|} = \sup \frac{|\langle A(f), g \rangle|}{\|f\| \cdot \|g\|} \leqq \sup \frac{\|A(f)\|}{\|f\|}.$$

On the other hand,

$$\|\Omega\| = \sup \frac{|\langle A(f), g \rangle|}{\|f\| \cdot \|g\|} \geqq \frac{\langle A(f), A(f) \rangle}{\|f\| \cdot \|A(f)\|} = \sup \frac{\|A(f)\|}{\|f\|}.$$

These relations show that the operator A is bounded and that the equation $\|A\| = \|\Omega\|$ holds. The proof is finished.

Let $A: \mathfrak{H} \to \mathfrak{H}$ be a bounded linear operator. The expression

$$\langle f, A(g) \rangle$$

is clearly a bilinear functional of f and g with the norm $\|A\|$. In view of the foregoing proposition one may define uniquely a bounded linear operator $A^* : \mathfrak{H} \to \mathfrak{H}$ with norm $\|A^*\| = \|A\|$ such that

$$\langle f, A(g) \rangle = \langle A^*(f), g \rangle$$

for all $f, g \in \mathfrak{H}$ holds.

Hence to every bounded linear operator A defined on all of $\mathfrak{H}$ there corresponds an analogous operator A^* with the same norm and the property that for all $f, g \in \mathfrak{H}$

$$\langle f, A(g)\rangle = \langle A^*(f), g\rangle$$

holds. This operator A^* is said to be the *adjoint operator* of A.

In the foregoing definition of the adjoint of a bounded linear operator A defined everywhere in $\mathfrak{H}$, we started from the fact that each element $g \in \mathfrak{H}$ determines uniquely an element g^* such that, for all $f \in \mathfrak{H}$, the equation

$$\langle A(f), g\rangle = \langle f, g^*\rangle$$

is satisfied. If T is a linear mapping whose domain of definition $\mathfrak{D}(T)$ and whose range $\mathcal{R}(T)$ are linear manifolds in $\mathfrak{H}$, we can again consider the inner product

$$\langle T(f), g\rangle,$$

where f runs through $\mathfrak{D}(T)$. But one may no longer assert that for every element g the expression

$$\langle T(f), g\rangle,$$

as a function of the vector $f \in \mathfrak{D}(T)$, is representable in the form

$$\langle f, g^*\rangle.$$

However, in general, there exist some pairs g and g^* for which the equation

$$\langle T(f), g\rangle = \langle f, g^*\rangle$$

holds for all $f \in \mathfrak{D}(T)$. Indeed, this equation holds at least for $g = g^* = 0$.

Only on the basis of the existence of pairs g, g^* for which the equation

$$\langle T(f), g\rangle = \langle f, g^*\rangle$$

holds for all $f \in \mathfrak{D}(T)$ one may not as yet define an operator T^* which is adjoint to T. It is also necessary that the element g^* be determined uniquely by the element g. This last requirement is satisfied if and only if $\mathfrak{D}(T)$ is dense in $\mathfrak{H}$. Indeed, if $\mathfrak{D}(T)$ is not dense in $\mathfrak{H}$, then there is a nonzero element h which is orthogonal to $\mathfrak{D}(T)$; then

$$\langle T(f), g\rangle = \langle f, g^*\rangle$$

for all $f \in \mathfrak{D}(T)$ implies that

$$\langle T(f), g \rangle = \langle f, g^* + h \rangle$$

for all $f \in \mathfrak{D}(T)$. On the other hand, if $\mathfrak{D}(T)$ is dense in $\mathfrak{H}$ and if, for each $f \in \mathfrak{D}(T)$,

$$\langle T(f), g \rangle = \langle f, g_1^* \rangle,$$

$$\langle T(f), g \rangle = \langle f, g_2^* \rangle,$$

then, for each $f \in \mathfrak{D}(T)$,

$$\langle f, g_1^* - g_2^* \rangle = 0.$$

This implies that $g_1^* = g_2^*$.

Thus, if $\mathfrak{D}(T)$ is dense in $\mathfrak{H}$, then the operator T has an adjoint operator T^*. The domain of definition of T^*, i.e., $\mathfrak{D}(T^*)$, is defined as follows: $g \in \mathfrak{D}(T^*)$ if and only if there exists a vector g^* such that

$$\langle T(f), g \rangle = \langle f, g^* \rangle$$

is satisfied for $f \in \mathfrak{D}(T)$. For each such pair g and g^*,

$$T^*(g) = g^*.$$

It is clear that this operator T^* is also a linear map; in fact, if it is defined for g_1 and for g_2 we have

$$\begin{aligned}\langle T(f), \alpha_1 g_1 + \alpha_2 g_2 \rangle &= \bar{\alpha}_1 \langle T(f), g_1 \rangle + \bar{\alpha}_2 \langle T(f), g_2 \rangle \\ &= \bar{\alpha}_1 \langle f, T^*(g_1) \rangle + \bar{\alpha}_2 \langle f, T^*(g_2) \rangle \\ &= \langle f, \alpha_1 T^*(g_1) + \alpha_2 T^*(g_2) \rangle\end{aligned}$$

for all $f \in \mathfrak{D}(T)$; hence $\alpha_1 g_1 + \alpha_2 g_2$ also belong to $\mathfrak{D}(T^*)$ and we have

$$T^*(\alpha_1 g_1 + \alpha_2 g_2) = \alpha_1 T^*(g_1) + \alpha_2 T^*(g_2).$$

Another property of T^* is its closedness: If $(g_n)_{n=1}^{\infty}$ is a sequence of elements in $\mathfrak{D}(T^*)$ such that $g_n \to g$ and $T^*(g_n) \to h$, then $g \in \mathfrak{D}(T^*)$ and we have that $T^*(g) = h$.

This is a direct consequence of the fact that the inner product is a continuous function of its factors (by the Cauchy-Schwarz Inequality); in fact,

$$\langle T(f), g \rangle = \lim_{n\to\infty} \langle T(f), g_n \rangle = \lim_{n\to\infty} \langle f, T^*(g_n) \rangle = \langle f, h \rangle.$$

Addendum. The Möbius Transform

We will now consider linear operators in a Hilbert space $\mathfrak{H}$ over the field of complex numbers whose domain of definition is no longer assumed to be the entire space but only a linear manifold in $\mathfrak{H}$ and whose range is in $\mathfrak{H}$ as well.

If S and T are such operators and λ is a complex number, we define λS and $S + T$ by

$$\mathfrak{D}(\lambda S) = \mathfrak{D}(S), \quad (\lambda S)(f) = \lambda S(f), \quad f \in \mathfrak{D}(S);$$

$$\mathfrak{D}(S + T) = \mathfrak{D}(S) \cap \mathfrak{D}(T),$$

$$(S + T)(f) = S(f) + T(f),$$

$$f \in \mathfrak{D}(S) \cap \mathfrak{D}(T).$$

To define the product ST and the inverse S^{-1} we proceed as follows:

$$(ST)(f) = S(T(f))$$

provided that $T(f)$ and $S(T(f))$ are defined.

If $S(f) = S(g)$ only for $f = g$, then S^{-1} is defined thus:

$$\mathfrak{D}(S^{-1}) = \mathscr{R}(S) \text{ and } S^{-1}(S(f)) = f.$$

For the existence of S^{-1} it is enough to require that $S(f) = 0$ only for $f = 0$.

Since S and T are linear maps, then so are the various new maps just defined.

We say that S is *closed* if

$$(f_n)_{n=1}^{\infty} \subset \mathfrak{D}(S), \ f_n \to f \text{ and } S(f_n) \to g$$

imply

$$f \in \mathfrak{D}(S) \text{ and } S(f) = g.$$

We say that T is an *extension* of S if $\mathfrak{D}(T) \supset \mathfrak{D}(S)$ and $S(f) = T(f)$ for every $f \in \mathfrak{D}(S)$.

A linear operator S in a Hilbert space $\mathfrak{H}$ is called a *Hermitian* operator

$$\langle S(f), g \rangle = \langle f, S(g) \rangle$$

for all f, $g \in \mathfrak{D}(S)$ and is called a *symmetric* operator if it is a Hermitian operator and $\mathfrak{D}(S)$ is dense in $\mathfrak{H}$; if S is a symmetric operator and $S = S^*$, then S is called a *self-adjoint* operator. S is called an *isometric* operator if

$$\langle S(f), S(g) \rangle = \langle f, g \rangle$$

for all $f, g \in \mathfrak{D}(S)$; an isometric operator S is called a *unitary* operator if $\mathfrak{D}(S) = \mathcal{R}(S) = \mathfrak{H}$, where $\mathcal{R}(S)$ denotes the range of S.

We say that the linear operator S is *bounded* (in $\mathfrak{D}(S)$) if

$$\sup\{\|S(f)\| : f \in \mathfrak{D}(S), \|f\| \leqq 1\} < \infty.$$

The left member of this inequality is called the *norm* of S (in $\mathfrak{D}(S)$) and is denoted by $\|S\|$. It is clear that $\|S\|$ can also be defined as

$$\|S\| = \sup\{\|S(f)\| : f \in \mathfrak{D}(S), \|f\| = 1\}$$

$$= \sup\left\{\frac{\|S(f)\|}{\|f\|} : f \in \mathfrak{D}(S)\right\}.$$

The following claims are easily checked as well: A bounded linear operator is continuous. If a linear operator is continuous at one point, then it is bounded. A bounded linear operator S can be extended uniquely to $\overline{\mathfrak{D}(S)}$ with preservation of the bound.

A complex number λ is called a *point of regular type* for the linear operator T if there is a number $k = k(\lambda) > 0$ such that

$$\|(T - \lambda I)f\| \geqq k\|f\|$$

for all $f \in \mathfrak{D}(T)$. If λ is a point of regular type for T and if $\mathcal{R}(T - \lambda I) = \mathfrak{H}$, we call λ a *regular point* for T.

Remark. All nonreal complex numbers are points of regular type for a Hermitian operator and regular points for a self-adjoint operator; if all nonreal complex numbers are regular points for a Hermitian operator, then this operator will be self-adjoint. We leave the verification of this Remark to the reader.

We shall now consider the Möbius transformation of complex analysis, namely,

$$w = (az + b)(cz + d)^{-1} \qquad ad - bc \neq 0,$$

and look at the corresponding Möbius transforms of linear operators.

DEFINITION. *Suppose that A is a linear operator in a Hilbert space $\mathfrak{H}$ over the complex number field and let I denote the identity operator in $\mathfrak{H}$. Assume that for some complex numbers c and d the operator $cA = dI$ is one-one* (*that is, $h \in \mathfrak{D}(A)$ and $cAh + dh = 0$ imply $h = 0$*). *For any complex numbers a and b such that $ad - bc \neq 0$ we define a linear operator B on $\mathcal{R}(cA + dI)$ by*

$$Bf = aAg + bg, \tag{1}$$

where

$$f = cAg + dg \qquad (g \in \mathfrak{D}(A)); \tag{2}$$

that is,

$$B = (aA + bI)(cA + dI)^{-1}, \tag{3}$$

and we shall call the operator B a Möbius transform of the operator A.

Remark. Without loss of generality we can assume that $ad - bc = 1$; we shall do this.

PROPOSITION 1. *If the operator B is a Möbius transform of the operator A as expressed in formula (3), then the operator A is given by the formula*

$$A = (dB - bI)(-cB + aI)^{-1}.$$

Proof. Solving the system of equations

$$cAg + dg = f$$

$$aAg + bg = Bf$$

for Ag and g, we obtain

$$Ag = \frac{\begin{vmatrix} f & d \\ Bf & b \end{vmatrix}}{\begin{vmatrix} c & d \\ a & b \end{vmatrix}} = dBf - bf$$

and

$$g = \frac{\begin{vmatrix} c & f \\ a & Bf \end{vmatrix}}{\begin{vmatrix} c & d \\ a & b \end{vmatrix}} = -cBf + af$$

for $g \in \mathfrak{D}(A)$ and $f \in \mathfrak{D}(B)$. Moreover, the operator $-cB + aI$ is one-one, because if $g = 0$, then $0 = dBf - bf$ and $0 = g = -cBf + af$; hence $0 = (ad - bc)f = f$ and the proof is complete.

PROPOSITION 2. *If the operators A and B are Möbius transforms of each other, then the closedness of one implies the closedness of the other.*

Proof. It will be enough to show that if A is a closed operator, so is B. Suppose then that the operator A is closed. If the sequence $(f_n)_{n=1}^{\infty}$ with $f_n \in \mathfrak{D}(B)$ for $n = 1, 2, \ldots$ and the sequence $(Bf_n)_{n=1}^{\infty}$ converge to f_0 and h_0, respectively, then the sequences, for $n = 1, 2, \ldots,$

$$g_n = -cBf_n + af_n$$

$$Ag_n = dBf_n - bf_n$$

converge to g_0 and q_0, respectively, and $g_n \in \mathfrak{D}(A)$. Since the operator A is closed, $g_0 \in \mathfrak{D}(A)$ and $q_0 = Ag_0$. By (2) and (1):

$$f_n = cAg_n + dg_n$$

$$Bf_n = aAg_n + bg_n.$$

Passing to the limit as $n \to \infty$ we get

$$f_0 = cq_0 + dg_0 = cAg_0 + dg_0$$

$$h_0 = aq_0 + bg_0 = aAg_0 + bg_0$$

and hence $f_0 \in \mathfrak{D}(B)$ and $h_0 = Bf_0$. But this means that the operator B is closed and the proof is finished.

COROLLARY. *If A is a closed linear operator and λ is a point of regular type for A, then $\mathcal{R}(A - \lambda I)$ is a subspace.*

Proof. Since λ is a point of regular type for A, there exists a number $k = k(\lambda) > 0$ such that the inequality

$$\|(A - \lambda I)f\| \geqq k\|f\|$$

holds for all $f \in \mathfrak{D}(A)$. This implies that the operator $(A - \lambda I)^{-1}$, acting on $\mathcal{R}(A - \lambda I)$, is bounded. By the foregoing proposition the operator $(A - \lambda I)^{-1}$ is closed. We, therefore, have that the operator $(A - \lambda I)^{-1}$ is defined on a closed linear manifold.

PROPOSITION 3. *If the numbers a, b, c, and d are real and the operator A is Hermitian (resp. self-adjoint), then the Möbius transform B of A will be Hermitian (resp. self-adjoint) also.*

Proof. Let A be Hermitian. An easy calculation shows that

$$\langle aAg_1 + bg_1, cAg_2 + dg_2\rangle = \langle cAg_1 + dg_1, aAg_2 + bg_2\rangle$$

holds for any $g_1, g_2 \in \mathfrak{D}(A)$. Using formulas (2) and (1) we obtain:

$$\langle Bf_1, f_2\rangle = \langle f_1, Bf_2\rangle$$

for $f_1, f_2 \in \mathfrak{D}(B)$. This shows that B is Hermitian as well.

To verify the second part of the proposition, we assume that A is self-adjoint and show that for any nonreal number λ the set $\mathcal{R}(B - \lambda I)$ coincides with the entire space $\mathfrak{H}$; this however means that the Hermitian operator B is in fact self-adjoint.

By formulas (2) and (1) the set $\mathcal{R}(B - \lambda I)$ consists of elements of the form:

$$aAg + bg - \lambda(cAg + dg) = (a - \lambda c)Ag + (b - \lambda d)g$$

with $g \in \mathfrak{D}(A)$. Since a, b, c, and d are real numbers, then for nonreal λ:

$$a - \lambda c \neq 0 \text{ and } \mu = -(b - \lambda d)(a - \lambda c)^{-1}$$

are nonreal as well, and therefore

$$\mathcal{R}(B - \lambda I) = \mathcal{R}(A - \mu I) = \mathfrak{H}.$$

This proves the proposition.

Proposition 4. *If $c = \bar{a}$, $d = \bar{b}$ and the operator A is Hermitian (resp. self-adjoint), then the Möbius transform B of A is isometric (resp. unitary).*

Proof. If A is Hermitian, a simple computation shows that for any $g_1, g_2 \in \mathfrak{D}(A)$:

$$\langle aAg_1 + bg_1, aAg_2 + bg_2 \rangle = \langle cAg_1 + dg_1, cAg_2 + dg_2 \rangle.$$

This in turn means that $\langle Bf_1, Bf_2 \rangle = \langle f_1, f_2 \rangle$ for $f_1, f_2 \in \mathfrak{D}(B)$; therefore the operator B is isometric.

Next, let the operator A be self-adjoint. Then the domain of definition of B,

$$\mathfrak{D}(B) = \mathcal{R}(\bar{a}A = \bar{b}I),$$

and the range of B,

$$\mathcal{R}(B) = \mathcal{R}(aA + bI),$$

coincide with the entire space $\mathfrak{H}$ and the isometric operator B is unitary. This proves the proposition.

Analogously one shows the converse statement.

Proposition 5. *If $c = \bar{a}$, $d = \bar{b}$ and the operator B is isometric (resp. unitary), then the operator A is Hermitian (resp. self-adjoint).*

From the definition of Möbius transform we get the next proposition directly.

PROPOSITION 6. *Suppose that B is a Möbius transform of A:*

$$B = (aA + bI)(cA + dI)^{-1}.$$

If $\hat{A}$ is an extension of the operator A and if the operator $c\hat{A} + dI$ is one-one (that is, $h \in \mathfrak{D}(\hat{A})$ and $c\hat{A}h + dh = 0$ implies $f = 0$), then the Möbius transform of the operator A:

$$\hat{B} = (a\hat{A} + bI)(c\hat{A} + dI)^{-1}$$

is an extension of the operator B.

PROPOSITION 7. *Let T be a Möbius transform of a symmetric operator S:*

$$T = (aS + bI)(cS + dI)^{-1}$$

where a, b, c, d are real numbers and $ad - bc \neq 0$. Then given any self-adjoint extension $\hat{T}$ of T there is a self-adjoint extension $\hat{S}$ of S given by

$$\hat{S} = (d\hat{T} - bI)(-c\hat{T} + aI)^{-1}.$$

Proof. By Propositions 1, 5, and 6 it will suffice to show that $c\hat{T}f - af = 0$ implies $f = 0$. Thus, let $c\hat{T}f - af = 0$; then $\langle c\hat{T}f - af, h\rangle = 0$ for $h \in \mathfrak{H}$. In particular we have $\langle c\hat{T}f - af, w\rangle = 0$, where $w \in \mathfrak{D}(T)$. Therefore

$$\langle f, c\hat{T}w - aw\rangle = \langle f, cTw - aw\rangle = 0$$

which means that $\langle f, g\rangle = 0$ for $g \in \mathfrak{D}(S)$. But $\mathfrak{D}(S)$ is dense in $\mathfrak{H}$ as S is a symmetric operator. It follows that $f = 0$ and the proposition is proved.

Remark. Setting $a = d = 1$, $b = -i$, and $c = i$, the Möbius transform becomes the *Cayley transform* which is of considerable importance in the study of the spectral theory of linear operators in a Hilbert space.

EXERCISES

1. Using Exercise 3 of Chapter 7, prove the following version of Proposition 9: If F is an arbitrary family of continuous linear operators from a Banach space A into a normed linear space B such that for each $x \in A$ there exists a number M_x satisfying $\|T(x)\| \leqq M_x$ for all $T \in F$, then there is a real number M such that $\|T\| \leqq M$ holds for all $T \in F$.

[*Hints*: For each T the function f defined by $f(x) = \|T(x)\|$ is a real-valued continuous function on A. Since the family of these functions is bounded at each x in A and A is complete, there is by Exercise 3 of Chapter 7 an open set U in A on which these functions are uniformly bounded. Thus there is a constant M_1 such that $\|T(x)\| \leqq M_1$ for all $x \in U$. Let $y \in U$. Since U is open, there is a sphere S of radius δ centered at y and contained in U. If $\|z\| \leqq \delta$, then $T(z) = T(y - z) - T(y)$ with $y + z$ in $S \subset U$. Hence

$$\|T(z)\| \leqq \|T(y + z)\| + \|T(y)\| \leqq M_1 + M_y.$$

Consequently,

$$\|T\| \leqq \frac{M_1 + M_y}{\delta}$$

for all T in F.]

2. An interesting application of the Uniform Boundedness Principle is to prove that there are continuous functions of period 2π whose Fourier series diverge at a given point.

   To this end, let A be the Banach space of such functions with $\|x\| = \max|x(t)|$. The n-th partial sum of the Fourier series of $x = x(t)$ at $t = 0$ may be written in the form

$$f_n(x) = \frac{1}{2\pi} \int_0^{2\pi} x(t) D_n(t)\, dt,$$

where

$$D_n(t) = \frac{\sin(n + \frac{1}{2})t}{\sin \frac{1}{2}t}.$$

Now, for each $x \in A$,

$$|f_n(x)| \leqq \frac{1}{2\pi} \int_0^{2\pi} |D_n(t)|\, dt \cdot \|x\|.$$

Clearly, f_n is linear and it is bounded with

$$\|f_n\| \leqq \frac{1}{2\pi} \int_0^{2\pi} |D_n(t)|\, dt.$$

Let y be the function which is $+1$ for those t for which $D_n(t) \geqq 0$ and which is -1 for those t for which $D_n(t) < 0$. For every $\varepsilon > 0$, y may be modified to a continuous x of norm 1 so that

$$\left| f_n(x) - \frac{1}{2\pi} \int_0^{2\pi} \left| D_n(t) \right| dt \right| = \frac{1}{2\pi} \left| \int_0^{2\pi} (x(t) - y(t)) D_n(t)\, dt \right| < \varepsilon.$$

It follows that

$$\|f_n\| = \frac{1}{2\pi} \int_0^{2\pi} |D_n(t)|\, dt.$$

But the sequence (in n)

$$\int_0^{2\pi} \left| \frac{\sin(n + \frac{1}{2})t}{\sin \frac{1}{2}t} \right| dt$$

is unbounded; since $\sin v \leqq v$, $0 \leqq v \leqq \pi$, this follows if we show that the sequence (in n)

$$\int_0^{\pi} \left| \frac{\sin(2n+1)v}{v} \right| dv$$

is unbounded. Indeed, with $a_n = (k+1)\pi/(2n+1)$ and $b_n = k\pi/(2n+1)$, we get

$$\begin{aligned}
\int_0^{\pi} \left| \frac{\sin(2n+1)v}{v} \right| dv &= \sum_{k=0}^{2n} \int_{a_n}^{b_n} \left| \frac{\sin(2n+1)v}{v} \right| dv \\
&\geqq \sum_{k=0}^{2n} \frac{2n+1}{(k+1)\pi} \int_{a_n}^{b_n} |\sin(2n+1)v|\, dv \\
&= \sum_{k=0}^{2n} \frac{1}{(k+1)\pi} \int_{k\pi}^{(k+1)\pi} |\sin s|\, ds \\
&= \frac{2}{\pi} \sum_{k=0}^{2n} \frac{1}{k+1}.
\end{aligned}$$

Thus the sequence $(\|f_n\|)_{n=1}^{\infty}$ is unbounded and so there is an $x \in A$ whose Fourier series diverges at $t = 0$.

3. Let B be the Banach space of all continuous real-valued functions f on the interval $[0, 1]$ with $\|f\| = \max\{|f(x)| : x \in [0, 1]\}$ and let A denote the linear subspace of B consisting of those functions f which have continuous derivatives f' on $[0, 1]$. Prove that the operator $T: A \to B$ defined by $T(f) = f'$ for every $f \in A$ is linear and has a closed graph in $A \times B$. Also show that T is not continuous (hence A is not complete).
4. Let $C([0, 1])$ be the space B defined in the foregoing exercise. Show: If F is a continuous linear functional on $C([0, 1])$, there exists a function g of bounded variation on $[0, 1]$ such that

$$F(f) = \int_0^1 f\,dg, \qquad f \in C([0, 1])$$

and

$$\|F\| = V_0^1(g),$$

where $V_0^1(g)$ denotes the total variation of g on $[0, 1]$.
[*Hints*: We consider $C([0, 1])$ as a subspace of the space $M([0, 1])$ of all bounded real-valued functions on $[0, 1]$ with the supremum norm and apply the Hahn-Banach theorem to extend F to all of $M([0, 1])$, preserving the norm $\|F\|$. We may now apply F to the functions

$$h_x(t) = \begin{cases} 1 & \text{for } 0 \leqq t < x, \\ 0 & \text{for } x \leqq t \leqq 1, \end{cases}$$
$$h_0(t) = 0 \text{ for } 0 \leqq t \leqq 1,$$

and define

$$g(x) = F(h_x) \quad \text{for } 0 \leqq x \leqq 1.$$

It is easy to verify that $V_0^1(g) = \|F\|$.

Now, if $f \in C([0, 1])$, we define a sequence of functions $f_n \in M([0, 1])$ such that $f_n \to f$ in the space $M([0, 1])$ and the values $F(f_n)$ are approximating sums for the Riemann-Stieltjes integral $\int_0^1 f\,dg$. Namely,

$$f_n(t) = \sum_{k=1}^{n} f\left(\frac{k}{n}\right)(h_{k/n}(t) - h_{(k-1)/n}(t)).$$

It is evident that f_n converges uniformly to f. Now

$$F(f_n) = \sum_{k=1}^{n} f\left(\frac{k}{n}\right)(F(h_{k/n}) - F(h_{(k-1)/n}))$$
$$= \sum_{k=1}^{n} f\left(\frac{k}{n}\right)\left(g\left(\frac{k}{n}\right) - g\left(\frac{k-1}{n}\right)\right),$$

so that $F(f_n)$ converges to $\int_0^1 f dg$. On the other hand, $F(f_n)$ converges uniformly to $F(f)$ by the continuity of F. Hence $F(f) = \int_0^1 f dg$. Observe that this proof can be easily modified for the case $C([a, b])$ with $b - a$ finite.]

5. Let E be a normed linear space (real or complex). A sequence $(x_k)_{k=1}^{\infty}$ in E is said to be *complete* if $L(x_k) = 0$, $n = 1, 2, \ldots$, $L \in E^*$, implies $L = 0$. A sequence in E is said to be *closed* if every element of E can be approximated in norm arbitrarily closely by finite linear combinations of the elements of the sequence. Prove: A sequence $(x_k)_{k=1}^{\infty}$ of elements in E is closed if and only if it is complete.
6. Let $E = C([a, b])$ with $b - a$ finite and

$$\|h\| = \max\{|h(x)| : x \in [a, b]\}.$$

By the Weierstrass approximation theorem (Chapter 9, Section 1) the powers $1, x, x^2, \ldots$, are a closed sequence in the space. For a given $g \in C([a, b])$ the linear functional

$$L(h) = \int_a^b h(x)g(x)\,dx$$

is in E^*. Hence $\int_a^b x^n g(x)\,dx = 0$, $n = 0, 1, 2, \ldots$, implies $g = 0$.

7. Let E be a normed linear space and let $(x_k)_{k=1}^{\infty}$ be a closed sequence in E. Then a second sequence $(y_k)_{k=1}^{\infty}$ is closed in E if and only if each x_k can be approximated arbitrarily closely by linear combinations of the y_k, i.e., the property of closure is transitive.
8. (i) Let E be a normed linear space and suppose that the sequence $(x_n)_{n=1}^{\infty}$ is closed in E. If $(y_n)_{n=1}^{\infty}$ is a sequence such that for some number c, $0 \leqq c < 1$, and for all finite sequences of constants $a_1, a_2, \ldots, a_n$, we have

$$(*) \qquad \left\|\sum_{k=1}^{n} a_k(x_k - y_k)\right\| \leqq c\left\|\sum_{k=1}^{n} a_k x_k\right\|,$$

show that $(y_n)_{n=1}^{\infty}$ is also in E.

(ii) Prove: If (*) holds with $0 < c < \frac{1}{2}$, then $(y_n)_{n=1}^{\infty}$ is a closed sequence if and only if $(x_n)_{n=1}^{\infty}$ is a closed sequence.

9. Let $\mathfrak{H}$ be a Hilbert space and let $(x_k)_{k=1}^{\infty}$ be a complete orthonormal sequence. Let $(y_k)_{k=1}^{\infty}$ be any sequence such that

$$(\#) \qquad \left\|\sum_{k=1}^{n} a_k(x_k - y_k)\right\|^2 \leqq c \sum_{k=1}^{n} |a_k|^2,$$

with $0 \leqq c \leqq 1$, for every sequence $(a_k)_{k=1}^{\infty}$ of complex numbers. Prove that $(y_k)_{k=1}^{\infty}$ is also complete. In particular, show that $(\#)$ holds if

$$\sum_{k=1}^{n} \|x_k - y_k\|^2 = c < 1.$$

10. Let $\mathfrak{H}$ be a Hilbert space and $(x_n)_{n=1}^{\infty}$ be a complete orthonormal system. Let $(y_n)_{n=1}^{\infty}$ be a second orthonormal system. If

$$\sum_{n=1}^{\infty} \|x_n - y_n\|^2 < \infty,$$

show that $(y_n)_{n=1}^{\infty}$ is also complete.
[*Hints*: Observe, if for some $m \geqq 0$

$$\sum_{k=m+1}^{\infty} \|x_k - y_k\|^2 < 1,$$

then the system $x_1, x_2, \ldots, x_m, y_{m+1}, y_{m+2}, \ldots$ is complete in $\mathfrak{H}$. Moreover, if for some m,

$$\sum_{k=m+1}^{\infty} \|x_k - y_k\|^2 < 1,$$

and if we set

$$z_k = x_k - \sum_{j=m+1}^{\infty} \langle x_k, y_j \rangle y_j$$

with $k = 1, 2, \ldots, m$, then the system $z_1, \ldots, z_m, y_{m+1}, y_{m+2}, \ldots$ is complete in $\mathfrak{H}$.

To prove the assertion in the exercise, select m so large that

$$\sum_{k=m+1}^{\infty} \|x_k - y_k\|^2 < 1.$$

Let S denote the orthogonal complement of $\{y_{m+1}, y_{m+2}, \ldots\}$, i.e., the set of all elements of $\mathfrak{H}$ orthogonal to these elements. For $j \geqq m+1$, and for $1 \leqq k \leqq m$,

$$\langle z_k, y_j \rangle = \langle x_k, y_j \rangle - \sum_{p=m+1}^{\infty} \langle x_k, y_p \rangle \langle y_p, y_j \rangle = \langle x_k, y_j \rangle - \langle x_k, y_j \rangle$$
$$= 0.$$

Hence $z_1, \ldots, z_m \in S$. Since $z_1, \ldots, z_m, y_{m+1}, \ldots$ is complete, S cannot contain elements other than linear combinations of $z_1, \ldots, z_m$. Thus, S is a finite dimensional space of dimension $\leqq m$. Note further that the elements $y_1, \ldots, y_m$ are in S. Since they are orthonormal, they are linearly independent and hence span S. The elements $z_1, \ldots, z_m$ are therefore linear combinations of $y_1, \ldots, y_m$. If, therefore $\langle w, y_k \rangle = 0$, $k = 1, 2, \ldots,$ it follows that w is orthogonal to $z_1, \ldots, z_m, y_{m+1}, \ldots,$ and hence $w = 0$.]

11. The theorems of Ch. Müntz:
    (i) Let $\{x^p\}$ be a given infinite set of distinct powers with $p > -\frac{1}{2}$. In order that the system be closed in $L^2(0, 1)$, it is necessary and sufficient that the exponents $\{p\}$ contain a sequence (p_i) such that either

$$\lim_{i\to\infty} p_i = -\tfrac{1}{2}, \qquad \sum_{i=1}^{\infty} (p_i + \tfrac{1}{2}) = \infty,$$

or

$$\lim_{i\to\infty} p_i = p, \qquad -\tfrac{1}{2} < p < \infty,$$

or

$$\lim_{i\to\infty} p_i = \infty, \qquad p_i \neq 0, \qquad \sum_{i=1}^{\infty} \frac{1}{p_i} = \infty.$$

    (ii) Let (p_i) be a sequence of distinct nonnegative numbers. In order that the sequence $x^{p_1}, x^{p_2}, \ldots$ be closed in $C([0, 1])$, it is sufficient that
    (a) One of the p_i's is 0 and that (p_i) contains a sequence (p_{i_j}) for which

$$\lim_{j\to\infty} p_{i_j} = \infty$$

and

$$\sum_{\substack{j=1 \\ p_{i_j \neq 0}}}^{\infty} \frac{1}{p_{i_j}} = \infty,$$

or that
    (b) One of the p_i's is 0 and that (p_i) contains a sequence (p_{i_j}) for which

$$\lim_{j\to\infty} p_{i_j} = p,$$

$0 < p < \infty$.

(For details of proof consult Chapter 1 of N. I. Achieser, *Theory of Approximation*, Ungar, New York, 1956.)

12. Show: On a finite dimensional linear space E all norms are equivalent, i.e., if $\| \ \|$ and $\| \ \|'$ are two norms on E, then there exist positive reals a and b such that $a\|x\| < \|x\|' < b\|x\|$ for all $x \in E$. Finite dimensional normed linear spaces are thus seen to be complete.
13. Prove: The unit sphere $S = \{x \in X : \|x\| \leqq 1\}$ of a Banach space X is compact if and only if X is of finite dimension.
[*Hints*: The "if" part is a consequence of the Bolzano-Weierstrass theorem asserting that a bounded closed set in R^n is compact.
The "only if" part: Since S is compact, there is a finite sequence of points a_i $(i = 1, \ldots, n)$ such that S is contained in the union of open spheres of center a_i and radius $\frac{1}{2}$. Let V be the finite dimensional subspace generated by the a_i's. Show by contradiction that $V = X$. Suppose there is an $x \in X$ such that $x \notin V$. Now V is finite dimensional and therefore closed. Hence
$$\operatorname{dist}(x, V) = \inf\{\|x - y\| : y \in V\} = c > 0;$$
by the definition of $\operatorname{dist}(x, V)$ there exists in V a point y such that $c \leqq \|x - y\| < 3c/2$. Let $z = (x - y)/\|x - y\|$; we have $\|z\| = 1$, hence there is an index i such that $\|z - a_i\| \leqq \frac{1}{2}$. We write
$$x = y + \|x - y\|z = y + \|x - y\|a_i + \|x - y\|(z - a_i)$$
and observe that $y + \|x - y\|a_i \in V$. By the definition of $\operatorname{dist}(x, V)$ we have therefore $\|x - y\| \cdot \|z - a_i\| \geqq c$, hence $\|x - y\| \geqq 2c$, which contradicts the choice of y since $c \neq 0$.]
14. Let X and Y be Banach spaces and let $T : X \to Y$ be a linear operator such that $g \circ T \in X^*$ for every $g \in Y^*$. Show that T is continuous.
15. Let X and Y be normed linear spaces and let T be a bounded linear operator from X into Y. For $g \in Y^*$, define $T^*(g) = g \circ T$. Show that
   (i) T^* is a bounded linear operator from Y^* into X^*;
   (ii) T^* is one-to-one if and only if $T(X)$ is dense in Y;
   (iii) T is one-to-one if and only if $T^*(Y^*)$ separates points of X.
16. Let X be a normed linear space and M be a closed linear subspace of X. Consider the quotient space
$$X/M = \{x + M : x \in X\},$$
where linear operations are defined by
$$(x + M) + (y + M) = (x + y) + M$$

and

$$\alpha(x + M) = (\alpha x) + M.$$

Define the *quotient norm* on X/M by

$$\|x + M\| = \inf\{\|x + m\| : m \in M\}$$

and show that this defines a norm on X/M. Verify also:

(i) if X is a Banach space, then so is X/M.

(ii) if M and X/M are both complete, then so is X.

17. Let X be a normed linear space and Y be a Banach space over the same field of scalars K. Let T be a linear map with $\overline{\mathcal{D}(T)} = X$ and $\mathcal{R}(T) \subset Y$ (here $\mathcal{D}(T)$ denotes the domain of definition of T and $\mathcal{R}(T)$ denotes the range of T) such that there is a constant C with

$$\|T(x)\| \leqq C\|x\| \text{ for all } x \in \mathcal{D}(T).$$

Show that T can be extended uniquely to $\tilde{T}$ such that

$$\|\tilde{T}(x)\| \leqq C\|x\| \text{ for all } x \in X.$$

18. Show: All nonreal complex numbers are points of regular type for a Hermitian operator and regular points for a self-adjoint operator; if all nonreal complex numbers are regular points for a Hermitian operator, then this operator will be self-adjoint.

19. If $1 \leqq p < \infty$, the space l_p is the set of all numerical sequences (real or complex) $x = (x_1, x_2, \ldots)$ satisfying

$$\sum_{j=1}^{\infty} |x_j|^p < \infty \text{ with norm } \|x\| = (\sum_{j=1}^{\infty} |x_j|^p)^{1/p}.$$

Show that the l_p spaces (under the usual coordinate-wise linear operations) are Banach spaces and that l_2 is a Hilbert space with inner product

$$\langle x, y \rangle = \sum_{j=1}^{\infty} x_j \overline{y_j}.$$

Observe also: If $1 \leqq p_1 < p_2 < \infty$, then $l_{p_1} \subset l_{p_2}$ (*while* $L^{p_1} \supset L^{p_2}$). To see that this inclusion is in fact proper, let $p = p_2/(1 + \varepsilon)$; then $p_1 < p < p_2$ and the sequence

$$x_j = (j)^{-1/p} \qquad (j = 1, 2, \ldots)$$

is an element of l_{p_2} but not of l_{p_1}.
By l_∞ we denote the space of numerical sequences endowed with the norm $\|x\| = \sup\{|x_j| : j \in N\}$; l_∞ is also a Banach space. The spaces l_p with $1 \leqq p < \infty$ are separable, while the space l_∞ is not.

20. (Riesz-Fischer): If the series

$$\left|\frac{a_0}{2}\right| + \sum_{k=1}^{\infty} (|a_k|^2 + |b_k|^2)$$

converges, then there exists a function $f \in L^2([0, 2\pi])$ with Fourier coefficients a_k, b_k $(k = 0, 1, 2, \ldots)$, i.e., the equations

$$a_k = \frac{1}{\pi} \int_0^{2\pi} f(t) \cos kt\, dt$$

$$b_k = \frac{1}{\pi} \int_0^{2\pi} f(t) \sin kt\, dt$$

$(k = 0, 1, 2, \ldots)$ hold. This function f is determined uniquely up to measure 0.

21. Prove Proposition 25.
22. Show that Hilbert spaces are reflexive.
23. Verify: If E^* is separable, then so is E (the converse is false; note that l_∞ is the conjugate space of l_1).
24. Show: The points of regular type for an operator A from an open set in the complex plane.
[*Hint*: If λ_0 is a point of regular type for the operator A, then for $|\lambda - \lambda_0| < k_{\lambda_0}$ we have

$$\|(A - \lambda I)f\| \geqq \|(A - \lambda_0 I)f\| - |\lambda - \lambda_0| \cdot \|f\| \geqq k_\lambda \|f\|,$$

where $f \in \mathfrak{D}(A)$ and $k_\lambda = k_{\lambda_0} - |\lambda - \lambda_0|$.]

25. Let D denote a connected region of the complex plane consisting of points of regular type for the operator A. Then the orthogonal complement $\mathfrak{N}_\lambda$ of $\mathcal{R}(A - \lambda I)$ in $\mathfrak{H}$ will have the same dimension for all $\lambda \in D$.
[*Hints*: We shall show that for each point $\lambda_0 \in D$ we can find a neighborhood W such that the dimension of $\mathfrak{N}_\lambda$ and the dimension of $\mathfrak{N}_{\lambda_0}$ will be equal for all $\lambda \in W$; from this the claim will follow by the Heine-Borel theorem. Let W be a neighborhood of the point λ_0 with radius $(1/3)k_{\lambda_0}$. Then for $\lambda \in W$ we get

$$\|(A - \lambda I)f\| \geqq \|(A - \lambda_0 I)f\| - |\lambda - \lambda_0| \cdot \|f\| > \left(\frac{2}{3}\right)k_{\lambda_0}\|f\|$$

for all $f \in \mathfrak{D}(A)$ and

$$\|(A - \lambda I)f - (A - \lambda_0 I)f\| = |\lambda - \lambda_0| \cdot \|f\| < \frac{1}{2}\|(A - \lambda I)f\|,$$

$$\|(A - \lambda I)f - (A - \lambda_0 I)f\| < \frac{1}{3}\|(A - \lambda_0 I)f\|.$$

This shows that the opening of the subspaces $\overline{\mathcal{R}(A - \lambda_0 I)}$ and $\overline{\mathcal{R}(A - \lambda I)}$ is not larger than 1/2 for all $\lambda \in W$ by formula (8). Hence by Proposition 29 the dimension of $\mathfrak{N}_\lambda$ $(\lambda \in W)$ equals the dimension of the subspace $\mathfrak{N}_{\lambda_0}$.]

26. Let $(X, \mathfrak{A}, \mu)$ be a measure space and let $\mathfrak{A}_0$ be the set of all $A \in \mathfrak{A}$ such that $\mu(A) < \infty$. Define $d(A_1, A_2) = \mu((A_1 \cup A_2) - (A_1 \cap A_2))$ and show that $(\mathfrak{A}_0, d)$ gives rise to a metric if we identify two sets A_1 and A_2 of $\mathfrak{A}_0$ when $\mu((A_1 \cup A_2) - (A_1 \cap A_2)) = 0$. Show that $(\mathfrak{A}_0, d)$ is complete. (Compare this with Exercise 2 of Chapter 1.)
[*Hint*: Observe that

$$d(A_1, A_2) = \int_X |\chi_{A_1} - \chi_{A_2}|\, d\mu.]$$

27. Let $(X, \mathfrak{B}, \mu)$ be an arbitrary measure space and $\mathcal{G}$ be a collection of μ-integrable functions. We say that $\mathcal{G}$ is a *uniformly absolutely continuous* collection, if, for any $\varepsilon > 0$, there is a $\delta > 0$ such that

$$\mu(E) < \delta \quad \text{implies} \quad \int_E |g|\, d\mu < \varepsilon \quad \text{for all } g \in \mathcal{G}.$$

We say that $\mathcal{G}$ is *equicontinuous at* $\varnothing$ if, for any $\varepsilon > 0$ and any decreasing sequence $(E_j)_{j=1}^{\infty}$ of measurable sets with

$$\bigcap_{j=1}^{\infty} E_j = \varnothing,$$

there is a j_0 such that

$$\int_{E_j} |g|\, d\mu < \varepsilon \quad \text{for all } g \in \mathcal{G} \quad \text{and all } j \geqq j_0.$$

Show:

(i) If the collection $\mathcal{G}$ of integrable functions is equicontinuous at $\varnothing$, then it is uniformly absolutely continuous.

(ii) Let $(g_n)_{n=1}^{\infty}$ be a sequence of measurable functions, and for some p, $1 \leqq p < \infty$, assume that $|g_n|^p$ is integrable for all n. Then $(g_n)_{n=1}^{\infty}$ converges in L^p-norm if and only if it converges in measure and $(|g_n|^p)_{n=1}^{\infty}$ is equicontinuous at $\varnothing$.

(iii) Let $(g_n)_{n=1}^{\infty}$ be a sequence of measurable functions that converges in measure. Let h be a measurable function such that $|h|^p$ is integrable for some p, $1 \leqq p < \infty$, and assume that $|g_n| \leqq h$ (μ-a.e.) for all n. Then $(g_n)_{n=1}^{\infty}$ converges in L^p-norm.

28. Let $(X, \mathfrak{B}, \mu)$ be an arbitrary finite measure space and $\mathcal{G}$ be a collection of μ-integrable functions. We say that the collection $\mathcal{G}$ is *uniformly integrable* if, for any $\varepsilon > 0$, there exists a constant $k(\varepsilon)$ such that, for all $g \in \mathcal{G}$,

$$\int_V |g|\, d\mu < \varepsilon,$$

where $V = \{x \in X : |g(x)| > k(\varepsilon)\}$.
Show:

(i) The collection $\mathcal{G}$ of integrable functions is uniformly absolutely continuous if and only if it is equicontinuous at $\varnothing$.

(ii) Let $\mathcal{G}$ be a collection of integrable functions, and assume that there is an integrable function h such that $|g| \leqq h$ (μ-a.e.) for all $g \in \mathcal{G}$. Then $\mathcal{G}$ is uniformly integrable.

(iii) A collection $\mathcal{G}$ of integrable functions is uniformly integrable if and only if it is uniformly absolutely continuous and

$$\sup_{\mathcal{G}} \int_X |g|\, d\mu < \infty.$$

(iv) Let $(g_n)_{n=1}^{\infty}$ be a sequence of measurable functions, and, for some p, $1 \leqq p < \infty$, assume that $|g_n|^p$ is integrable for all n. Then $(g_n)_{n=1}^{\infty}$ converges in L^p-norm if and only if it converges in measure and $(|g_n|^p)_{n=1}^{\infty}$ is uniformly integrable.

29. Read the first chapter of the book: *Almost Periodic Functions* by C. Corduneanu, John Wiley & Sons, New York, 1968.

Index